AF903822

ADVANCES IN PROTEIN CHEMISTRY AND STRUCTURAL BIOLOGY

Volume 87

Structural and Mechanistic Enzymology: Bringing Together Experiments and Computing

ADVANCES IN PROTEIN CHEMISTRY AND STRUCTURAL BIOLOGY

Structural and Mechanistic Enzymology: Bringing Together Experiments and Computing

EDITED BY

Christo Christov
Department of Biomedical Sciences
School of Life Sciences
Northumbria University
Newcastle upon Tyne
NE2 8ST, UK

Tatyana Karabencheva-Christova
Department of Biomedical Sciences
School of Life Sciences
Northumbria University
Newcastle upon Tyne
NE2 8ST, UK

ELSEVIER

AMSTERDAM • BOSTON • HEIDELBERG • LONDON
NEW YORK • OXFORD • PARIS • SAN DIEGO
SAN FRANCISCO • SINGAPORE • SYDNEY • TOKYO
Academic Press is an imprint of Elsevier

Academic Press is an imprint of Elsevier
525 B Street, Suite 1900, San Diego, CA 92101–4495, USA
225 Wyman Street, Waltham, MA 02451, USA
32, Jamestown Road, London NW1 7BY, UK
Linacre House, Jordan Hill, Oxford OX2 8DP, UK
Radarweg 29, PO Box 211, 1000 AE Amsterdam, The Netherlands

First edition 2012

Copyright © 2012 Elsevier Inc. All rights reserved.

No part of this publication may be reproduced, stored in a retrieval system or transmitted in any form or by any means electronic, mechanical, photocopying, recording or otherwise without the prior written permission of the publisher.

Permissions may be sought directly from Elsevier's Science & Technology Rights Department in Oxford,
UK: phone: (+44) (0) 1865 843830; fax: (+44) (0) 1865 853333;
email: permissions@elsevier.com.
Alternatively you can submit your request online by visiting the Elsevier web site at http://elsevier.com/locate/permissions, and selecting, *Obtaining permission to use Elsevier material.*

Notice
No responsibility is assumed by the publisher for any injury and/or damage to persons or property as a matter of products liability, negligence or otherwise, or from any use or operation of any methods, products, instructions or ideas contained in the material herein. Because of rapid advances in the medical sciences, in particular, independent verification of diagnoses and drug dosages should be made.

ISBN: 978-0-12-398312-1
ISSN: 1876-1623

For information on all Academic Press publications
visit our website at store.elsevier.com

Working together to grow
libraries in developing countries

www.elsevier.com | www.bookaid.org | www.sabre.org

ELSEVIER BOOK AID International Sabre Foundation

Contents

Structural and Computational Enzymology: Bringing Experiments and Computations Together

Tatyana Karabencheva and Christo Christov

Structural and Functional Analysis of Proteins by High-Speed Atomic Force Microscopy

Arivazhagan Rajendran, Masayuki Endo and Hiroshi Sugiyama

Reaction Intermediates Discovered in Crystal Structures of Enzymes

Irene T. Weber, Johnson Agniswamy, Guoxing Fu, Chen-Hsiang Shen and Robert W. Harrison

Structural, Functional, and Mutagenesis Studies of UDP-Glycosyltransferases

Vatsala Malik and Gary W. Black

Vipoxin and Its Components: Structure–Function Relationship

S. D. Petrova, V. N. Atanasov and K. Balashev

Experimental and Theoretical Studies of Enzyme-Catalyzed Hydrogen-Transfer Reactions

Zhen Wang, Daniel Roston and Amnon Kohen

Protein Flexibility and Enzymatic Catalysis

M. Kokkinidis, N. M. Glykos and V. E. Fadouloglou

Molecular Modeling of Inhibitors of Human DNA Methyltransferase with a Crystal Structure: Discovery of a Novel DNMT1 Inhibitor

Jakyung Yoo, Joo Hee Kim, Keith D. Robertson and José L. Medina-Franco

In Silico Strategies Toward Enzyme Function and Dynamics

Sílvia G. Estácio

Dynamic Simulations as a Complement to Experimental Studies of Enzyme Mechanisms

Andrew Christofferon, Lifeng Zhao and Qiang Pei

The Increasing Role of QM/MM in Drug Discovery

Alessio Lodola and Marco De Vivo

Allostery and Binding Cooperativity of the Catalytic Subunit of Protein Kinase A by NMR Spectroscopy and Molecular Dynamics Simulations

Larry R. Masterson, Alessandro Cembran, Lei Shi and Gianluigi Veglia

Using a Combination of Computational and Experimental Techniques to Understand the Molecular Basis for Protein Allostery

WANTING JIAO AND EMILY J. PARKER

STRUCTURAL AND COMPUTATIONAL ENZYMOLOGY: BRINGING EXPERIMENTS AND COMPUTATIONS TOGETHER

By TATYANA KARABENCHEVA AND CHRISTO CHRISTOV

Department of Biomedical Sciences, School of Life Sciences, Northumbria University at Newcastle, Newcastle upon Tyne, United Kingdom

Abstract

In this introductory chapter, we present how the experimental and computational strategies in enzyme research are developed and how they complement each other to provide better insights for understanding enzyme structures and mechanisms.

Enzymes, biological catalysts are of great interest for biosciences, medicine, pharmacy, and biotechnology (Berg et al., 2002). Historically, enzymology started with determining the roles of enzymes in physiological processes and passed along with identifying individual enzymes, their purification, characterization, and exploration of the enzyme kinetics. The development of crystallography, NMR, and spectroscopic methods such us UV absorption, fluorescence, circular dichroism, electron paramagnetic resonance, magnetic circular dichroism provided in-depth insight into understanding enzyme structure–function relationships such as substrate and cofactor binding, reaction mechanisms, inhibition, alosteric regulation, kinetic phenomena, and mutation effects (Neidig and Solomon, 2005; Boehr et al., 2006; Carey, 2006; Derike Smiley and Hammes, 2006; Karabencheva and Christov, 2010). The aforementioned methods have provided invaluable contribution into the analyzing protein structure, active site geometry and elucidated important electronic structure features of proteins (especially in the case of metalloenzymes). Further application of instrumental methods contributed for revealing conformational flexibility and time-dependent behavior of proteins (dynamics) and their influence on the enzymatic functions.

ADVANCES IN PROTEIN CHEMISTRY AND STRUCTURAL BIOLOGY, Vol. 87
DOI: 10.1016/B978-0-12-398312-1.00001-9

Copyright 2012, Elsevier Inc.
All rights reserved.

In parallel to the accumulation of important experimental results, a new strategy and philosophy for analyzing chemical and boichemical processes emerged and became routinely applied nowadays—exploring and understanding the nature of molecular structures, interactions, reactivity, and dynamics using computers (*in silico*). Enzyme structure and mechanisms became a central target for the computational methods and were intensively explored with a focus on different sides of enzyme structures, mechanisms, inhibition, and regulation (Mulholland, 2001; Garcia-Viloca et al., 2004; Karplus and Kuriyan, 2005; Gao et al., 2006; Warshel et al., 2006). The exponential growth in computer power and the building of massive parallel supercomputers powerfully increased the applicability of the computational methods and the biological significance of their results. The recent development of graphical processor units is likely to provide the next level of increasing computer power and will help for getting even more thorough insights into biological processes.

Two main areas of computational methods have been developed: those based on the quantum mechanics (QM) (Friesner, 2005) and those based on the classical mechanics, called molecular mechanics (MM) (Adcock and McCammon, 2006). The QM-based computational methods explicitly analyze the electronic structure (structure, energies, orbitals, vibrations, etc.) and thus are used for exploring the reaction mechanisms of enzymes. They are very computationally demanding and could be applied practically for up to several hundreds of atoms. The MM-based methods do not treat the electronic behavior of the molecules explicitly. These methods are computationally very effective and require a set of parameters known as *force fields*. Therefore, MM methods are very useful for studying conformational changes and determining the energies of large molecular systems such as solvated or membrane-located enzymes, without going into electronic structural details. QM and MM methods can synergize into hybrid schemes such as combined quantum mechanical and molecular mechanical methods (QM/MM) (Warshel and Levitt, 1976), which have been successfully applied to investigate a large number of enzyme reactions, correctly describing changes in the electronic structure (chemically relevant groups from the enzyme active site and the substrate) with QM method while calculating the structure, energy, and interactions of the rest of the enzyme molecule at MM level (Friesner and Guallar, 2005; Senn and Thiel, 2009).

Both strategies for investigation in the structural and mechanistic enzymology developed to some extent independently; however, over the past years, a trend emerged for strengthening their integration. This combination not only brings together computational and experimental approaches on the same enzymatic issues but also provides complementary insights into the investigated properties and has a powerful synergy effect.

The present thematic volume focuses on recent success in structural and mechanistic enzymology with main emphasis on explaining the enzyme structure–function relationships by both the experimental and computational approaches. The selected outstanding contributions demonstrate how the application of a variety of experimental techniques and modeling methods helps for understanding of enzyme dynamics, mechanism, inhibition, and can be translated into drug design.

References

Adcock, S. A., McCammon, J. A. (2006). Molecular dynamics: survey of methods for simulating the activity of proteins. *Chem. Rev.* **106**, 1589–1615.

Berg, J. M., Tymoczko, J. L., Stryer, L. (2002). Biochemistry. W.H. Freeman and Co., New York.

Boehr, D., Dyson, H. J., Wright, P. E. (2006). An NMR perspective on enzyme dynamics. *Chem. Rev.* **106**, 3055–3079.

Carey, P. R. (2006). Spectroscopic characterization of distortion in enzyme complexes. *Chem. Rev.* **106**, 3043–3054.

Derike Smiley, R., Hammes, G. G. (2006). Single molecule studies of enzyme mechanisms. *Chem. Rev.* **106**, 3080–3094.

Friesner, R. A. (2005). Ab initio quantum chemistry: methodology and applications. *Proc. Natl. Acad. Sci. USA* **102**, 6648–6653.

Friesner, R. A., Guallar, V. (2005). Ab initio quantum chemical and mixed quantum mechanics/molecular mechanics (QM/MM) methods for studying enzymatic catalysis. *Annu. Rev. Phys. Chem.* **56**, 389–427.

Gao, J. L., Ma, S. H., Major, D. T., Nam, K., Pu, J. Z., Truhlar, D. G. (2006). Mechanisms and free energies of enzymatic reactions. *Chem. Rev.* **106**, 3188–3209.

Garcia-Viloca, M., Gao, J., Karplus, M., Truhlar, D. G. (2004). How enzymes work: analysis by modern rate theory and computer simulations. *Science* **303**, 186–195.

Karabencheva, T., Christov, C. (2004). Mechanisms of protein circular dichroism: Insights from computational modelling. *Adv. Protein Chem. Struct. Biol.* **80**, 85–115.

Karplus, M., Kuriyan, J. (2005). Molecular dynamics and protein function. *Proc. Natl. Acad. Sci. USA* **102**, 6679–6685.

Mulholland, A. J. (2001). The QM/MM approach to enzymatic reactions. In: Theoretical Biochemistry—Processes and Properties of Biological Systems, Eriksson, L. A. (Ed.), pp. 597–653. Elsevier, Amsterdam.

Neidig, M. L., Solomon, E. I. (2005). Structure-function correlations in oxygen activating non-heme iron enzymes. *Chem. Commun.* **47**, 5843–5863.

Senn, H. M., Thiel, W. (2009). QM/MM methods for biomolecular systems. *Angew. Chem. Int. Ed.* **48**, 1198–1229.
Warshel, A., Levitt, M. (1976). Theoretical studies of enzymic reactions—dielectric, electrostatic and steric stabilization of carbonium-ion in reaction of lysozyme. *J. Mol. Biol.* **103**, 227–249.
Warshel, A., Sharma, P. K., Kato, M., Xiang, Y., Liu, H. B., Olsson, M. H. M. (2006). Electrostatic basis for enzyme catalysis. *Chem. Rev.* **106**, 3210–3235.

STRUCTURAL AND FUNCTIONAL ANALYSIS OF PROTEINS BY HIGH-SPEED ATOMIC FORCE MICROSCOPY

By ARIVAZHAGAN RAJENDRAN,[*] MASAYUKI ENDO,[†,‡] AND HIROSHI SUGIYAMA[*,†,‡]

[*]Department of Chemistry, Graduate School of Science, Kyoto University, Kitashirakawa-oiwakecho, Sakyo-ku, Kyoto, Japan
[†]Institute for Integrated Cell-Material Sciences (iCeMS), Kyoto University, Yoshida-ushinomiyacho, Sakyo-ku, Kyoto, Japan
[‡]CREST, Japan Science and Technology Corporation (JST), Sanbancho, Chiyoda-ku, Tokyo, Japan

Abbreviations

ADP adenosine diphosphate
Amp antimicrobial peptide
ATP adenosine triphosphate

DOI: 10.1016/B978-0-12-398312-1.00002-0

Copyright 2012, Elsevier Inc. All rights reserved.

BER	base-excision repair
bR	bacteriorhodopsin
CaM	calcium modulated protein (calmodulin)
CBD	cellulose-binding domain
CD	catalytic domain
DPTAP	1,2-dipalmitoyl-3-trimethylammonium-propane
dsDNA	double-stranded DNA
EC	extracellular side
FtsZ	filamenting temperature-sensitive mutant Z
GTP	guanosine triphosphate
hOGG1	8-oxoguanine glycosylase
HS-AFM	high-speed atomic force microscopy
K_d	dissociation constant
PDG	T4 pyrimidine dimer glycosylase
PM	purple membrane
REase	restriction endonuclease
Tr	*Trichoderma reesei*

Abstract

Proteins are dynamic in nature, work at the single-molecule level, and facilitate several biological functions. The structure of a protein is closely associated with its function; thus, a large number of structural analyses of proteins were performed using techniques such as X-ray crystallography and NMR. Although these methods provide structural information, they often fail because of difficulties in crystallizing the proteins that are complexed with other biomolecules. Moreover, these techniques do not allow the observation of structural changes in the active form of the molecule. Single-molecule fluorescence techniques have been used for the direct observation of protein functions; however, they only reveal the dynamics of individual fluorescent spots, rather than the structural changes that occur over the entire protein. The recent development of high-speed atomic force microscopy (HS-AFM) overcame this problem and allowed the observation of the structural dynamics of proteins and other biomacromolecules directly and in real time. In this chapter, we describe the HS-AFM analysis of the dynamic molecular processes in photoactivated bacteriorhodopsin, membrane-mediated protein–protein interactions, ATP-induced conformational changes in purinergic receptors, the two-dimensional crystal structure of streptavidin, the nature of FtsZ polymers, the role of ClpX in the regulation of FtsZ polymer dynamics, the function of restriction enzymes, the action of motor proteins, the movement of *Tr*Cel7A on crystalline cellulose substrates, and the antimicrobial peptide activity on individual bacterial cells.

I. Introduction

Proteins are biochemical compounds consisting of 20 types of amino acids linked via peptide bonds and are typically folded into a globular or fibrous form (Schulz and Schirmer, 1979). They are dynamic in nature, work at the single-molecule level, and facilitate various biological functions. The structure of a protein greatly influences its function. Therefore, a large number of structural analyses of proteins were performed using techniques such as X-ray crystallography (Drenth, 1999) and NMR (Roberts and Lian, 2011). Although the combination of conventional biochemical techniques with structural studies involving atomic resolution has been useful for analyzing the function of proteins, the thorough characterization of their function cannot be achieved because of difficulties in crystallizing proteins that are complexed with biomacromolecules, such as DNA. Furthermore, dynamics is one of the essential features of biological systems. For instance, the biological function of a protein is closely associated with its ability to undergo structural changes when the molecule is in action. It is not possible to monitor such changes using these static and ensemble-averaging techniques. Single-particle tracking experiments of labeled proteins were performed using optical microscopy to understand the function of proteins in real time (Park et al., 2007; Joo et al., 2008). However, this reveals the dynamic behavior of individual fluorescent spots, and not that of the whole protein molecule. Furthermore, it provides information only on the labeled biomolecules, and the molecular environment is not observed simultaneously.

Atomic force microscopy (AFM) is a versatile technique that is being used increasingly in the study of biomacromolecules at the unimolecular level (Binnig et al., 1986; Butt et al., 1990; Rajendran et al., 2011b). The advantage of AFM is that the analysis can be performed in physiological conditions. Nevertheless, such investigations have been restricted by innate slow scan speed, as the scanning rates of commercially available AFMs usually range from several seconds to minutes per frame, whereas biological reactions take place on a subsecond time scale. Over the past decade, various efforts have been dedicated to overcome this longstanding problem (Viani et al., 1999, 2000; Hansma et al., 2006; Picco et al., 2007) and an extraordinary improvement was achieved by Ando et al. (2001, 2008a,b). Those authors developed a high-speed atomic force microscopy (HS-AFM) that has a miniaturized cantilever and scanning stage, which reduces the mechanical response time of the feedback system and prevents the onset of resonant motion during high-speed imaging. Its maximum imaging rate of 33 frames/s enables direct imaging of the structure and dynamics of proteins and other biomolecular processes in real time. In addition, the range of operation of HS-AFM is suitable for the analyses of various biomolecules, ranging from a single DNA fragment or DNA nanostructures (Rajendran et al., 2011a) to a whole cell (Fantner et al., 2010), and from sizes of a few nanometers to several micrometers (Fig. 1). However, other techniques work either at the nanometer (for instance, X-ray crystallography) or the micrometer scale (light microscopy).

This chapter aims to summarize recent studies that used HS-AFM to study the dynamics of protein structures and their functions. Briefly, we explain the dynamic molecular process in photoactivated bacteriorhodopsin (bR), membrane-mediated protein–protein interactions, ATP-induced conformational changes in purinergic receptors, the two-dimensional crystal structure of streptavidin, the nature of FtsZ polymers, the role of ClpX in the regulation of FtsZ polymer dynamics, the molecular mechanism of restriction enzymes (such as *Apa*I and *Eco*RII), the action of motor proteins (such as myosin V and F_1-ATPase), and the movement of cellobiohydrolase I (Cel7A) *Trichoderma reesei* (*Tr*) on a crystalline cellulose substrate. Single-molecule analysis using HS-AFM is not limited to biomolecules such as proteins, as it can also be performed at the single-cell level, which is demonstrated by the recent report on the kinetics of antimicrobial peptide activity on individual bacterial cells. Furthermore, an introductory note on DNA origami as a novel substrate for protein analysis is provided.

II. Photoactivated Bacteriorhodopsin

bR is a 26-kDa protein used by archaea, the most notable one being Halobacteria (Haupts et al., 1999). The protein can occupy up to ~50% of the surface area of the archaeal cell. It acts as a light-driven proton pump, that is, it captures light energy and uses it to transfer protons vectorially across the membrane from the cytoplasmic side to the outside of the cell against a concentration gradient. The resulting proton gradient is

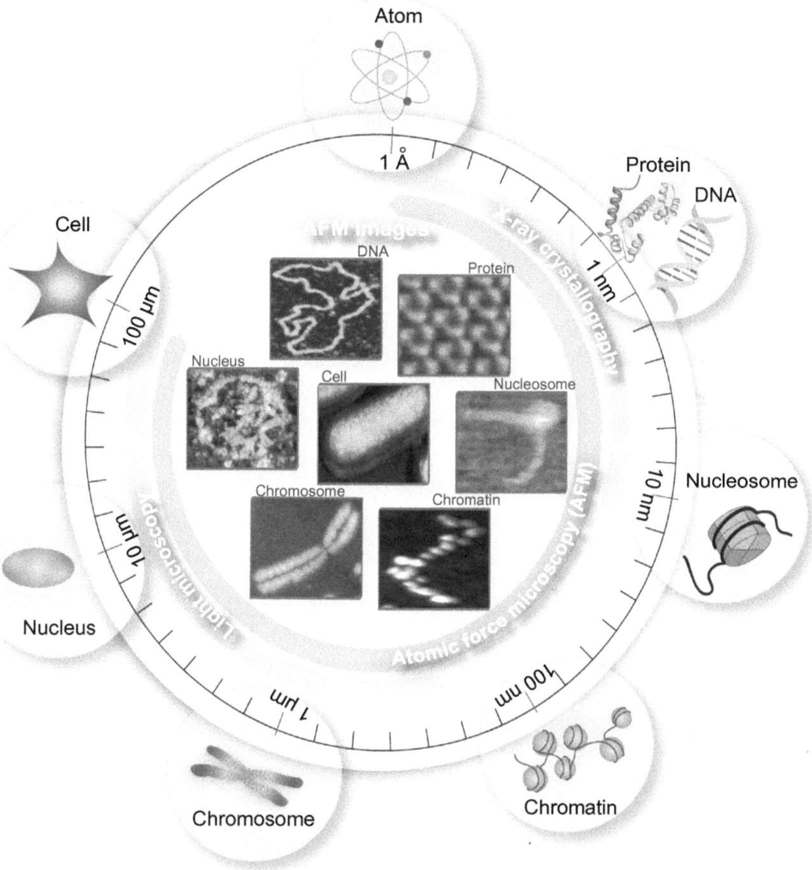

FIG. 1. Graphical representation of the range of operation of AFM. It is suitable for the analysis of various biomolecules, ranging from a single DNA fragment (nm scale) to a whole cell (several µm) with atomic resolution. Other techniques work either at the nanometer (for instance, X-ray crystallography) or micrometer (light microscopy) scale. AFM images adapted from DNA: Yuri (2011), protein: Shibata et al. (2011), nucleosome: Miyagi et al. (2011), chromatin: Karymov et al. (2001), chromosome: Tatsuo et al. (2008), nucleus: Bucchianico and Poma (2010), and cell: Fantner et al. (2010). (For color version of this figure, the reader is referred to the online version of this chapter.)

subsequently converted into chemical energy. bR is an integral membrane protein usually found in two-dimensional crystalline patches known as the purple membrane (PM). The repeating element of the hexagonal lattice comprises three identical protein chains, each rotated by 120° relative to others. Each chain in bR is composed of seven transmembrane α-helices termed A–G and contains one molecule of retinal buried deep within the protein, which is the typical structure of retinylidene proteins. It is the retinal molecule that changes its conformation when absorbing green light (absorption maximum of 568 nm), resulting in photoisomerization from the all-*trans* to the 13-*cis* conformations of retinal, with subsequent primary proton pumping action, which triggers a cascade of conformational changes in the surrounding protein. It is covalently bound to Lys216 in the chromophore via a protonated Schiff base. After photoisomerization of the retinal molecule, Asp85 becomes an acceptor of the donor proton from the retinal molecule/Schiff base. This releases a proton from a holding site into the extracellular (EC) side of the membrane. Reprotonation of the retinal molecule by Asp96 restores its original isomerized form. This results in a second proton being released to the EC side. Asp85 releases its proton into the holding site, where a new cycle may begin.

The crystal structures of the ground (PDB: 1C3W) and light-activated (PDB: 1DZE) states of bR are shown in Fig. 2. During the photocycle, a

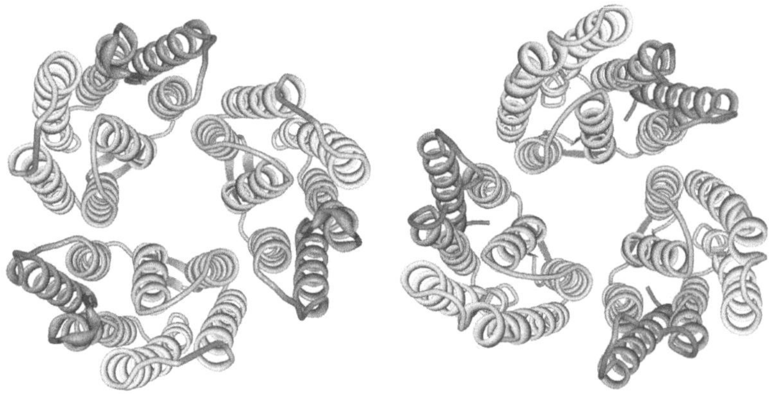

FIG. 2. The X-ray crystal structures of bR in the ground (left, PDB: 1C3W) and photoactivated excited (right, PDB: 1DZE) states. (For color version of this figure, the reader is referred to the online version of this chapter.)

serious of spectral intermediates occur in the order of J, K, L, M, N, and O. M is the only intermediate that contains a deprotonated Schiff base. The photo-induced conformational changes in bR have been studied using various methods (Subramaniam et al., 1993) and the common understanding regarding the structure of bR during the photocycle is that the proton channel on the cytoplasmic surface is opened by the tilting of helix F away from the protein center. This alteration is followed by rearrangement of the interhelix E–F loop, resulting in large-scale conformational changes in the M and N intermediate states. Nevertheless, clear information regarding these conformational changes had not been derived, and direct and real time observations of the dynamic nature of bR were required.

AFM has been used as a powerful tool to visualize nanometer-scale objects in various environments, including physiological conditions, which is suitable for the *in singulo* analysis of several biomolecular processes in real time. Taking advantage of this, Shibata et al. used their recently developed and speed-enhanced HS-AFM for the dynamic analysis of the structural changes in bR that are triggered by light as an external stimulus (Shibata et al., 2010, 2011). The photocycle of wild-type bR at neutral pH is very fast and proceeds within $\sim$10 ms; such fast dynamics is difficult to track using the HS-AFM that is currently available. To slow down the photocycle, those authors used the D96N bR mutant, which has a longer photocycle of $\sim$10 s at neutral pH while retaining its proton pumping ability. As shown in Fig. 3, the individual protein molecules can be seen clearly in successive AFM images of the D96N bR mutant at the cytoplasmic surface, captured at a rate of 1 frame/s. Under the illumination of 532 nm green light, a part of each bR molecule is displaced anticlockwise ($7.4 \pm 2.2°$) and outward from the trimer center (15 and 28 s in Fig. 3A). As a result, the three nearest neighboring bR monomers, each of which belongs to a different adjacent trimer, are brought into contact with each other. The mass center analysis of dark-illumination cycles and of the activated bR depicts the displacement of the center of mass after exposure to green light (Fig. 3B). The average displacement was 0.69 ± 0.15 nm, which is consistent with the results of previous EPR studies (Thorgeirsson et al., 1997; Xiao et al., 2000). The protrusion observed in the AFM topographs reflects the interhelical loop structures located on the cytoplasmic surface. Compared with the atomic model of the α-helical cytoplasmic ends in the unphotolysed state (A–G in Fig. 3C and D), the

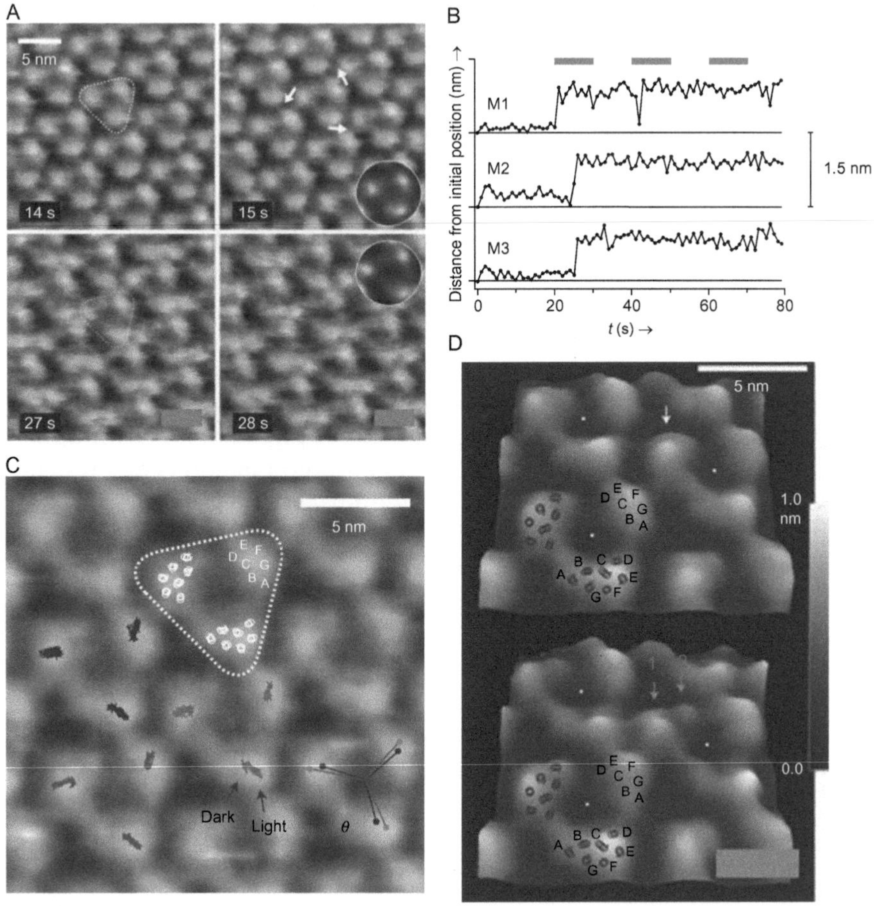

FIG. 3. (A) HS-AFM images of the D96N bR mutant at the cytoplasmic surface. White triangles indicate the bR trimer and white arrows indicate the direction of the conformational change in bR (15 s). The insets in the images at 15 and 28 s are averaged images of a bR trimer captured in the dark and under green light illumination, respectively. (B) Displacement of centers of mass for three D96N bR molecules measured as a function of time at pH 8. (C) Traces (red, blue, and black marks) of the mass center positions under dark-illumination cycles are superimposed on the image of D96N in the dark. (D) Surface maps of the magnified images in the dark (top) and under illumination (bottom). White dots denote the position of each trimer center and the white arrow indicates a monomer in the dark. Upon illumination, the topography of the monomer splits into major and minor protrusions, as indicated by green arrows 1 and 2, respectively. The green bars indicate application of 532 nm green light. Images adapted from Shibata et al. (2010, 2011). (For interpretation of the references to color in this figure legend, the reader is referred to the online version of this chapter.)

protruding parts of bR around helices E and F displace outward from the trimer center under green light. Thus, the protrusion movement visualized here is assigned to the displacement of the E–F loop. Consequently, it is divided into two sections (major and minor protrusions, as indicated by green arrows 1 and 2 in Fig. 3D, respectively), and the minor protrusion likely corresponds to the position around helices A and B. Moreover, a cysteine accessibility study using D96N indicated the displacement of the E–F loop in the M state, which confirms that the HS-AFM images captured the light-driven ground-to-M-state transition.

After it is formed via exposure to green light, M_{410} can be driven back to the ground state (bR_{570}) by exposure to blue light. The spectroscopically detected reversal did not indicate the reversal of the entire protein structure. In particular, it is not clear whether the protein structure is tightly coupled to the conformation of retinal. Such an issue is also addressed by imaging D96N bR directly under alternate illumination with green (532 nm) and blue (408 nm) light using HS-AFM. The cytoplasmic surface under the alternate illumination with green and blue light is shown in Fig. 4A. It is unambiguous that the green light induced changes

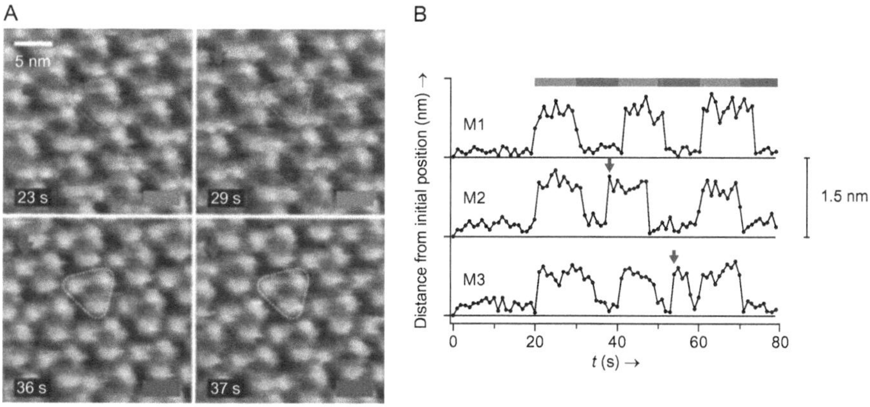

FIG. 4. HS-AFM images of the D96N bR mutant at the cytoplasmic surface illuminated by green light, followed by blue light (A). Displacement of centers of mass for three D96N bR molecules measured at pH 8 (B). The green and blue bars indicate the application of 532 nm green and 408 nm blue light, respectively. bR activation by blue light is indicated by blue arrows. Figure adapted from Shibata et al. (2011). (For interpretation of the references to color in this figure legend, the reader is referred to the online version of this chapter.)

are recovered after exposure to the blue light. This is further confirmed by the center of mass analysis, as shown in Fig. 4B. In addition, the decay analysis of the activated state confirmed that the conformation of D96N bR in the activated state observed using HS-AFM corresponds solely to the M_{410} intermediate, which absorbs blue light efficiently.

III. Membrane-Mediated Protein–Protein Interaction

As explained before, PM consists of a large amount of crystalline bR, which works together with ATPase for proton pumping and adenosine triphosphate (ATP) synthesis. bR was studied extensively using conventional AFM (Müller et al., 1995) and, more recently, using HS-AFM (Shibata et al., 2010, 2011); in contrast, the dynamics of ATPase in the PM remained unclear. For the first time, HS-AFM was used to image the edge regions of the native PM from *Halobacterium salinarum*, and the moving unlabeled membrane proteins were illustrated on ATP synthase c-rings (Casuso et al., 2010). The ring-shaped molecules with a diameter of 65 Å in the edge region of PM were identified as ATP synthase c-rings. This was further confirmed by mass spectrometry. As shown in Fig. 5, the c-rings assemble into dimers that exhibit a dynamic behavior. In some cases, the dimers are in close contact, whereas in other cases they are elongated. The measured center-to-center distances of tight and elongated dimers are ~80 and ~120 Å, respectively. The elongated dimers dissociate into monomers and reassemble into the initial dimer state. Furthermore, the HS-AFM data provided the experimental basis for the hydrophobic mismatch that contributes to the interaction between membrane proteins.

IV. Nucleotide-Dependent Conformational Changes

ATP is an incredible source of intracellular energy. It delivers energy and controls several biological processes and reactions. In addition, it can act as an intercellular signal by binding membrane receptors. In addition to ATP, other nucleotides, such as ADP, play key roles in numerous functions of proteins in biological systems. This part of the book chapter describes two examples of nucleotide-dependent protein conformational changes and dynamics.

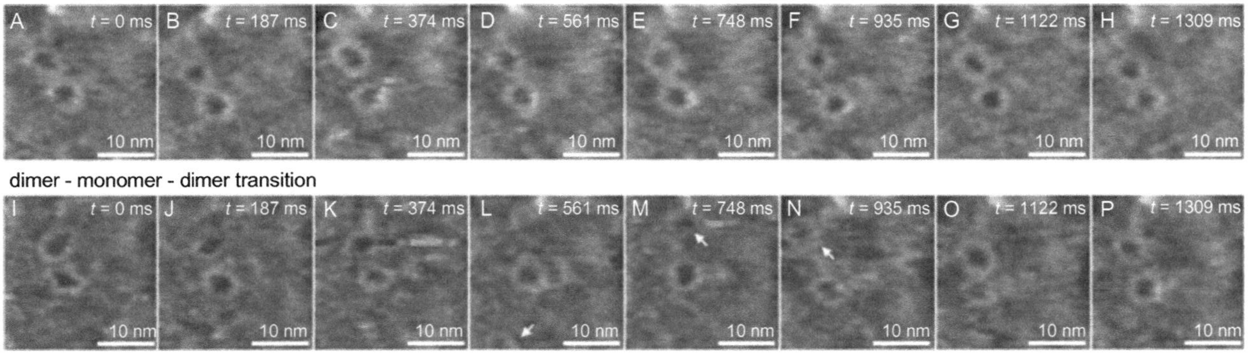

FIG. 5. HS-AFM images of the diffusion movement of the ATP synthase c-rings in native purple membrane. Two c-rings in a dimer over a period of 1309 ms are given in the top row images. Dimer–monomer–dimer transitions are given in the bottom row images. Arrows in (L), (M), and (N) indicate the dissociating and associating monomers. Scan rate, 187 ms/frame. Figure adapted from Casuso et al. (2010, 2011).

A. *ATP-Induced Conformational Changes in $P2X_4R$ Protein*

Purinergic receptors that bind nucleotides, including ATP, are known as P2 receptors. These EC nucleotide receptors are divided into two types: (i) ligand-gated cationic channels (P2X) and (ii) G protein-coupled (P2Y) receptors (Chen et al., 1995). P2X receptors (P2XRs) form a family of at least seven subtypes ($P2X_{1-7}$) comprising 379 ($P2X_6$) to 595 ($P2X_7$) amino acids (Khakh and Alan North, 2006). One functional P2XR channel is composed of three subunits (Fig. 6A). The subunits all share a common topology, and each subunit has a large EC loop containing the ATP-binding site, two transmembrane domains, and intracellular amino and carboxyl terminal tails (Khakh and Alan North, 2006). A six-stranded antiparallel β-pleated sheet structure is believed to exist in the EC domain. This region coordinates ATP binding and the allosteric coupling of the conformational changes in the ATP-binding domain, with corresponding changes at the transmembrane channel gate through a linker region. In addition, P2XRs have different permeability states (Cockcroft and Gomperts, 1979); in the presence of calcium, this protein trimer only opens a small cation-permeable channel pore, whereas in the absence of calcium, it opens the pore wider, which allows the transit of larger molecules, such as ethidium bromide. Though several studies were carried out on this protein, the receptor structure and the conformational changes it experiences after ATP binding remain to be clarified.

The direct observation of the ATP-induced conformational changes in a single $P2X_4$ receptor was reported recently with the help of HS-AFM in combination with single-particle averaging (Shinozaki et al., 2009). In that study, the rat $P2X_4R$ protein was expressed in human 1321N1 astrocytoma cells and used after purification. The protein was immobilized on a mica surface pretreated with a high concentration of KCl and time-lapse imaging was performed at a rate of 2 frames/s. The results are summarized in Fig. 6B. Before the addition of ATP, the protein exhibited a circular structure (−2.5 to 0.0 s), whereas after the addition of ATP an immediate structural change was observed; the circular structure was converted into a trimeric form within 0.5 s. Furthermore, the distance between individual lobes gradually increased and a pore-dilation-like structure was observed within 5 s. Washing off the ATP led to the reversal of the pore-dilation-like structure to the circular structure, which indicated the reversibility of the system. In addition to the analysis performed on a bare mica surface,

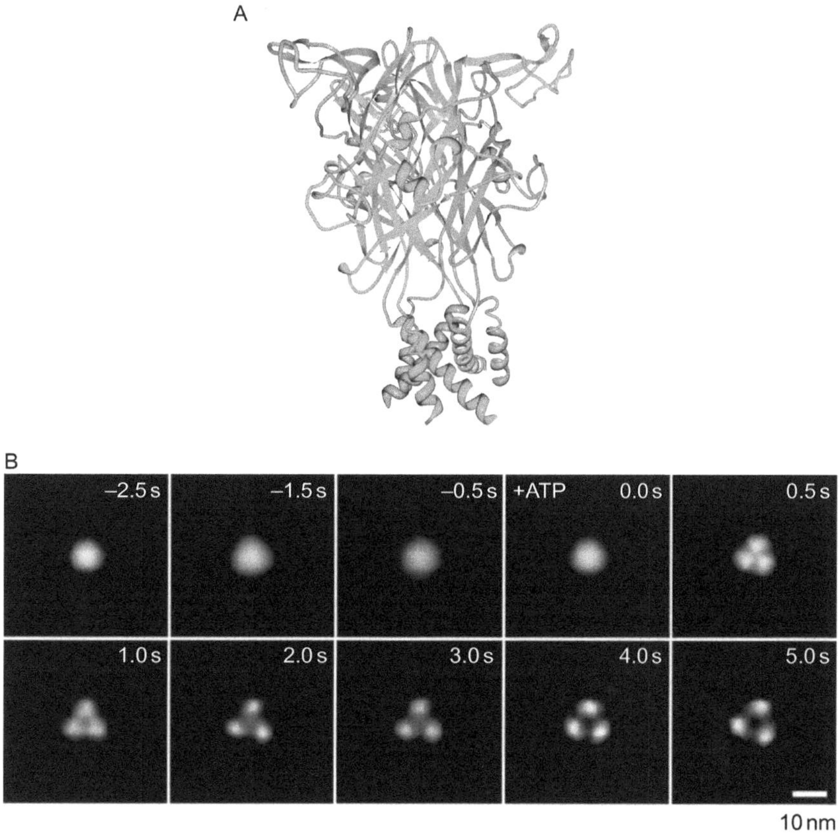

FIG. 6. (A) Crystal structure of the trimeric form of the $P2X_4$ receptor (PDB: 3I5D) in the closed state. (B) Time-lapse images of ATP-induced structural changes in $P2X_4R$. Before ATP addition, the protein had a circular shape (−2.5 to 0.0 s). After ATP addition, the protein structure changed to a trimer structure within 0.5 s. The changes continued up to 5 s and the molecule adopted a pore-dilation-like conformation. Ten $P2X_4R$ particles were averaged for each frame. HS-AFM images were adapted from Shinozaki et al. (2009). (For color version of this figure, the reader is referred to the online version of this chapter.)

the analysis was carried out by reconstituting the protein in a lipid bilayer. This study provided similar results to those obtained on the bare mica surface, indicating that the protein is inserted in an upward orientation in the membrane and that the β-pleated sheet structure lies on the surface of the membrane and responds to ATP. Moreover, a dye-uptake analysis in the absence and presence of calcium evidenced that the trimeric structure obtained at 0.5 s and the pore-dilation-like structure obtained at 5 s correspond to the cation and dye-permeable states, respectively.

B. ATP/ADP-Dependent Conformational Changes in Chaperonin GroEL

GroEL belongs to the chaperonin family of molecular chaperones. It is found in a large variety of bacteria. To obtain the functionally active 3D structure, newly synthesized or denatured proteins must undergo a series of folding processes (Hightower, 1991; Hartl and Hayer-Hartl, 2002). In prokaryotes and eukaryotes, this is achieved with the help of molecular chaperones. The *Escherichia coli* chaperonin GroEL–GroES is one of the well-studied chaperone systems (Grallert and Buchner, 2001). GroEL is composed of 14 identical 57 kDa subunits that form two homoheptameric rings (each with a large central cavity) that are stacked back to back, whereas GroES is a single homoheptameric ring (Fig. 7A) (Chaudhry et al., 2003). The chaperonin reaction begins with the binding of the unfolded substrate protein and GroES to one side of GroEL in the presence of nucleotides. This results in a dramatic conformational change of the GroEL protein, from a closed to an open state. Once the chaperonin complex GroEL/unfolded protein/GroES is formed, the unfolded protein is refolded into the functional conformation inside the large central cavity of GroEL. Subsequently, the complex disintegrates via the dissociation of GroES, which results in the release of the trapped protein. In turn, this leads to a conformational switching of GroEL back to its closed state.

HS-AFM analysis was performed recently to gain a clear idea of the chaperonin action and to shed light on the fast kinetics involved in this process (Yokokawa et al., 2006a). GroEL was immobilized on an unmodified mica surface and imaged in liquid conditions. The analysis revealed that the protein is packed and aligned in an end-up orientation, forming a layered structure. The central channel of the GroEL ring is seen clearly in the image (Fig. 7B). Subsequently, to study nucleotide-induced conformational changes, an imaging study was performed using a buffer that contained ADP. In this condition, some taller features were observed on the GroEL layer, with a height difference of $\sim$1 nm. This indicated the ADP-induced conformational change in GroEL from the closed to the open state. The reverse process of open-to-closed conformation was also observed. Furthermore, these conformational changes were observed for

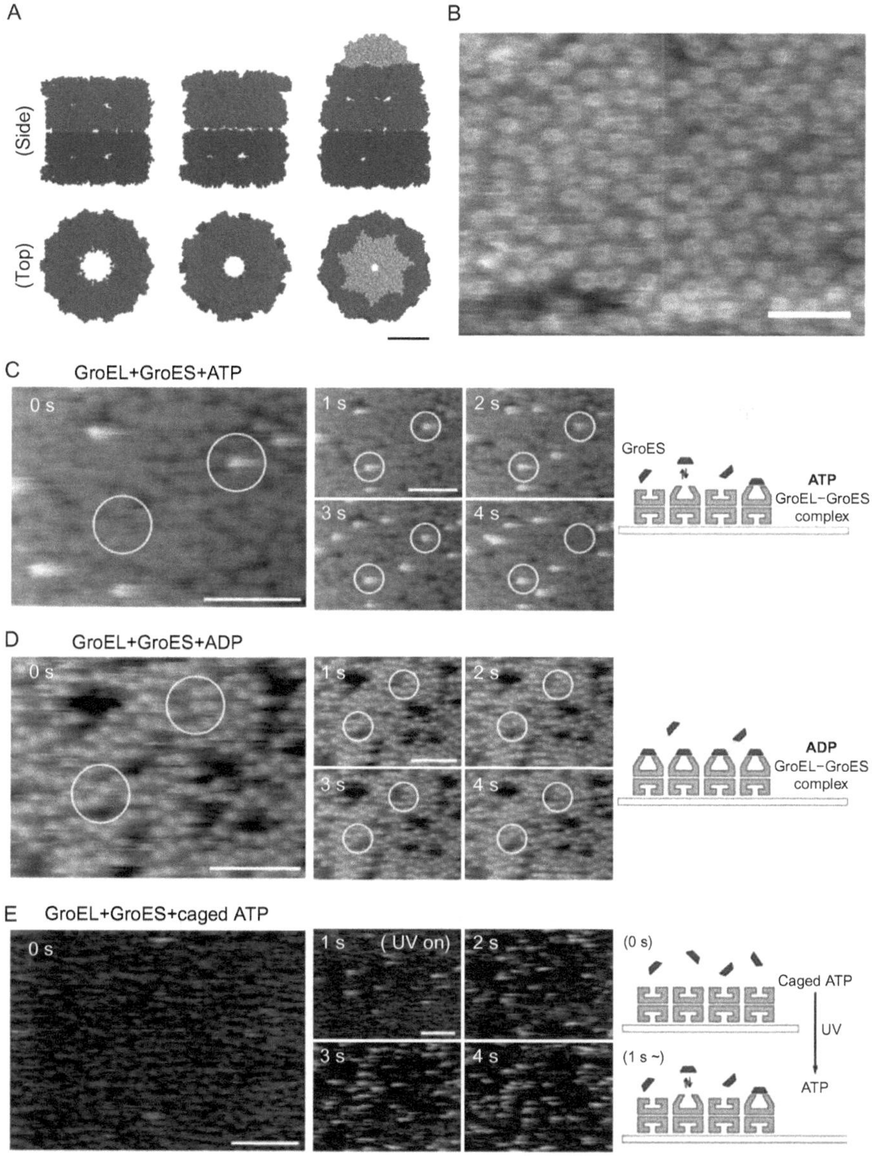

FIG. 7. (A) Crystal structure of GroEL (left, PDB: 1GRL), cryo-EM map of the GroEL–ATP complex (middle, PDB: 2C7E), and crystal structure of the GroEL–GroES–ADP complex (PDB: 1AON). Red, green, and yellow represent the lower GroEL ring,

other nucleotides, such as ATP and nonhydrolyzable ATP analogs, indicating that ATP hydrolysis is not necessary for such conformational changes. After characterizing these nucleotide-induced conformational changes, the dynamics of the GroEL–GroES interactions in the presence of nucleotides was analyzed. In the absence of nucleotides, the addition of GroES to the imaging solution did not cause any apparent change in the surface features of the GroEL layer, which is consistent with the results described in previous reports (Chandrasekhar et al., 1986). In contrast, in the presence of ATP and GroES, features that were much taller than those of the open conformation of GroEL were observed (Fig. 7C, 0 s). The analysis of the height profiles yielded two different peaks near 3.9 ± 0.6 and 1.3 ± 0.5 nm, which correspond to the GroEL/ATP/GroES complex and the GroEL open state, respectively. A series of time-lapse images indicated the repeated association and dissociation of GroES (Fig. 7C, 0–4 s). Immediately after the dissociation of GroES, GroEL was switched back to the closed conformation (90%).

These studies were also performed in the presence of ADP, which led to the observation of extremely stable complexes of GroEL/ADP/GroES (Fig. 7D). The dissociation of GroES was hardly observed in this condition and the lifetime of this complex was greater than 1 min, which agrees well with results reported previously (Motojima and Yoshida, 2003). Caged ATP was used to observe the ATP-induced GroES binding directly. The uncaging of the caged ATP was performed with UV-light illumination during the AFM scanning. After irradiation, GroES quickly began to bind GroEL reversibly (Fig. 7E). Finally, the kinetics of the binding and unbinding events in the presence of various nucleotides was studied in detail.

---

upper GroEL ring, and GroES, respectively. Scale bar, 5 nm. (B) HS-AFM image of GroEL immobilized on a mica surface. Scale bar, 50 nm. (C) Time-lapse sequence HS-AFM images of GroEL immobilized on a mica surface acquired in a buffer containing ATP and GroES. (D) Images acquired in the presence of ADP and GroES. (E) Images recorded in the presence of caged ATP and GroES. The schematic explanation of the binding event in each case is also given. Scan rate, 1 frame/s; scale bar, 100 nm; Z-scale, 10 nm. Images adapted from Yokokawa et al. (2006a). (For interpretation of the references to color in this figure legend, the reader is referred to the online version of this chapter.)

V. Two-Dimensional Crystal Structure of Streptavidin

The structural analysis of proteins was further extended to streptavidin crystals. Streptavidin is a 60-kDa protein that was purified from the well-known bacterium *Streptomyces avidinii*. It is a homotetrameric protein with dihedral *D*2 symmetry (Fig. 8A; PDB: 1STP). It has four sites with an

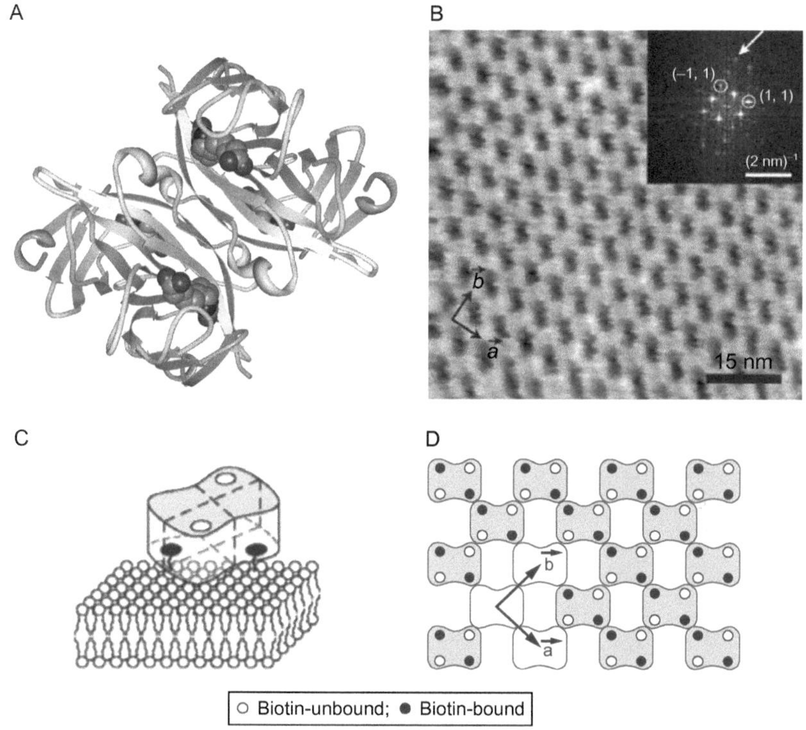

Fig. 8. (A) Crystal structure of tetrameric streptavidin (ribbon) with bound biotin molecules (spheres, PDB: 1STP) (B) HS-AFM image of a streptavidin crystal in its *C*222 symmetry. The calculated diffraction pattern of the AFM image is shown in the inset. The arrow represents the weakest measurable diffraction spot at (−1, 3) (1.9 nm resolution). (C) Schematic representation of a tetrameric streptavidin on a biotinylated lipid bilayer. (D) Drawing of streptavidin arrays in a *C*222 crystal. Unit lattice vectors are indicated: the *a*-axis includes rows of contiguous biotin-bound subunits, whereas the *b*-axis includes rows of contiguous biotin-unbound subunits. Images adapted from Ando et al. (2008a) and Yamamoto et al. (2008). (For color version of this figure, the reader is referred to the online version of this chapter.)

extraordinarily high affinity for the nutrient termed biotin (also known as vitamin B7). The dissociation constant (K_d) of the binding of biotin to streptavidin is in the order of $\sim 10^{-14}$ M, which is one of the strongest noncovalent interactions known in nature. Because of this strong binding, this system is the focus of a number of studies in the field of bionanotechnology and molecular biology aimed at determining the nature of the particular intermolecular interactions that give rise to the tight binding.

A. *Anisotropic Diffusion of Point Defects*

The behavior of lattice defects both in two- and three-dimensional (2D and 3D) crystals has been a subject of great interest in many fields (Queisser and Haller, 1998; Wang et al., 1999). These defect analyses were also applied to protein crystallography, as these defects may limit the resolution of protein structures obtained using diffraction methods (Chernov, 2003). Furthermore, the analysis of the dynamics of individual point defects may offer deep insights not only into protein–protein interactions in crystals but also into the mechanism of the crystal growth. The crystal structure of streptavidin bound to biotin was first solved by Hendrickson et al. (1989); since then, numerous analyses of the crystal structure of the streptavidin–biotin complex have been performed, which shed light on the origins of the remarkable affinity of these molecules. The complex is easily crystallized in 2D forms on biotinylated lipid layers, which is considered an ideal model system for investigating 2D crystal growth on lipid layers (Blankenburg et al., 1989) and protein crystallization in general. Though each streptavidin subunit specifically binds to one biotin, on the biotinylated lipid layers two of the four biotin binding sites face the lipid layers and are occupied by biotin, whereas the other two are exposed toward the aqueous environment and, hence, are biotin free. Previous studies on the ordered 2D arrays of streptavidin (Blankenburg et al., 1989; Ku et al., 1993; Scheuring et al., 1999; Lou et al., 2007) revealed that it self-assembles in three distinct crystalline arrangements, namely *P*1, *P*2, and *C*222, depending on the crystallization conditions. To date, several attempts have been made to visualize directly the morphological changes in protein crystals using successive AFM imaging (Durbin and Carlson, 1992; Yau et al., 2000; Malkin and McPherson, 2002). However, because of the slow image acquisition speed, it is difficult to follow the dynamic processes in 2D crystals, as they occur within a few seconds or even faster.

In this regard, the recently developed HS-AFM was successfully used to address the issues related to the defect diffusion in the orthorhombic $C222$ crystal of streptavidin formed on biotin-containing supported lipid bilayers (Yamamoto et al., 2008). The 2D migration of defects was successfully traced at an imaging rate of 2 frames/s. The HS-AFM image of the streptavidin 2D crystal is shown in Fig. 8B. The periodic structure exhibits $C222$ symmetry with lattice constants of $a = 5.9 \pm 0.1$ nm, $b = 5.9 \pm 0.3$ nm, and $\gamma = 92 \pm 3°$, which is consistent with the values reported previously and estimated at the air/water interface (Yatcilla et al., 1998). The calculated diffraction pattern indicated a lateral resolution better than 2 nm (Fig. 8B, inset); because of this resolution, the orientation of the rectangular streptavidin molecules was clearly identified. Moreover, the arrangement of biotin-bound and biotin-unbound subunits in the crystal axis coordinates can be determined in HS-AFM images. In the $C222$ crystal (Fig. 8C and D), the intermolecular contacts between biotin-bound subunits are aligned along one crystal axis (a-axis), whereas the contacts between biotin-unbound subunits are aligned along the other axis (b-axis).

Next, monovacancy defects were produced in the streptavidin 2D crystal by increasing the tapping force of the oscillating tip onto the sample, and the diffusion of these point defects in the crystal was observed. Snapshots of the HS-AFM images of the diffusion process are provided in Fig. 9A and the trajectories of two monovacancy defects are shown in Fig. 9B. The mobility of the monovacancy defects is obviously anisotropic with respect to the two axes of the lattice. A larger mobility of the defects was found along the b-axis (2.4 times larger) compared with the a-axis. The diffusion rate constants of migrating defects were $D_a = 20.5$ nm^2/s along the a-axis and $D_b = 48.8$ nm^2/s along the b-axis. The difference in intermolecular binding energy between the two types of subunits was estimated at ~ 0.52 kcal/mol. Furthermore, the monovacancy defects fused into bivacancy or even multivacancy defects. These multivacancy defects also diffused in the $C222$ crystals during successive imaging. Notably, the reverse reaction, fission of a large multivacancy into a smaller multivacancy defect or monovacancy defects, was rarely observed.

B. *Substrates for Visualizing Biomolecular Processes*

The flat substrate surface is important for the successful imaging of biological samples using AFM. Substrates used previously are not suitable for the immobilization and imaging of various biological samples.

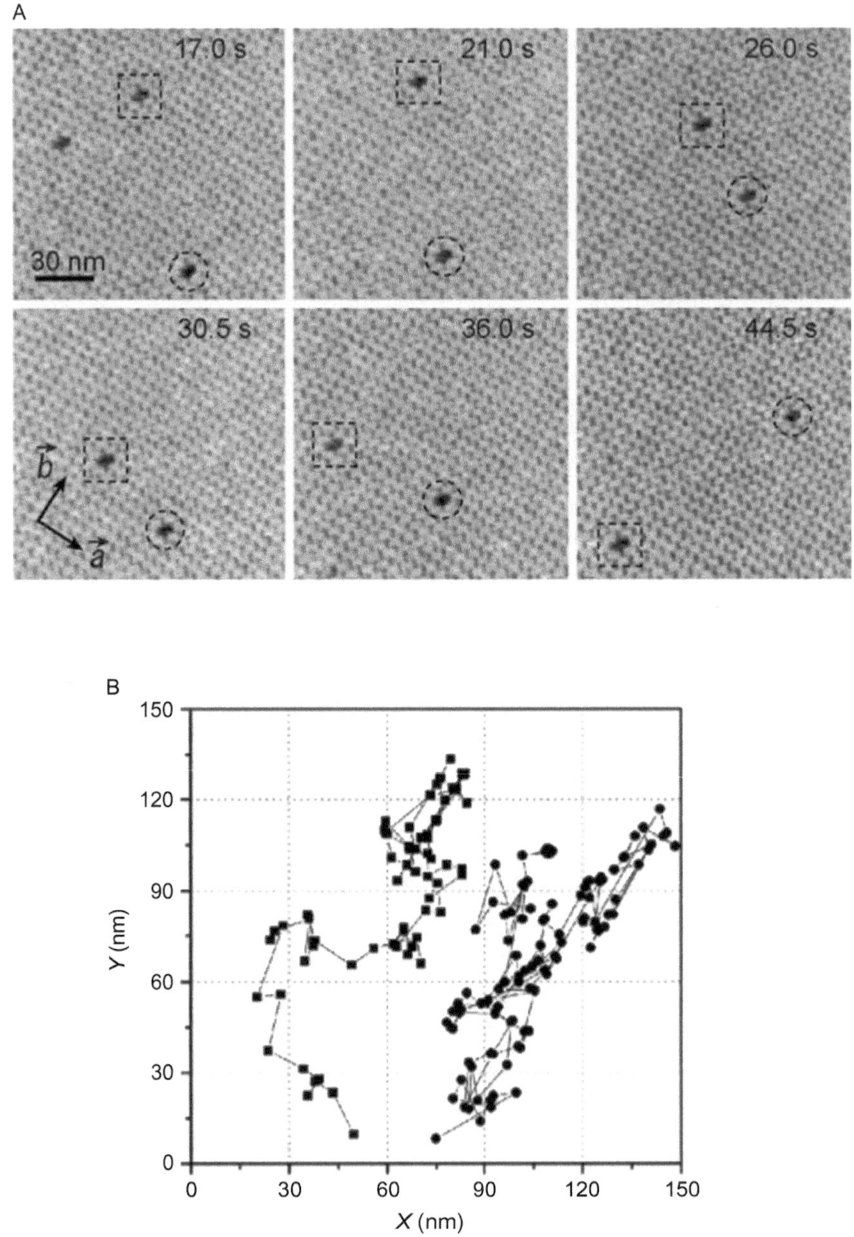

FIG. 9. (A) HS-AFM images of a streptavidin 2D crystal with monovacancy defects. The defects are marked by dashed squares and circles. The directions of the lattice

A pioneering study (Scheuring et al., 1999) used streptavidin 2D crystals formed on biotin-containing planar lipid layers. As mentioned before, in the 2D crystals, two biotin-binding sites face the free solution and, hence, can bind to the biotinylated samples to be imaged. In addition, as the biotinylated Ni-NTA is commercially available, the surface can also immobilize His-tag-conjugated proteins without lateral diffusion. Interestingly, streptavidin is resistant to nonspecific binding to many proteins, which possibly helps the specific immobilization of the target. Recently, the streptavidin 2D crystals were shown to be a novel surface for the HS-AFM analysis of biomolecular processes (Yamamoto et al., 2009). The three distinct crystalline arrangements of types 1–3 with *P*2 symmetry were characterized taking into consideration stability, degree of order, surface roughness, and nonspecific binding. Though the crystals met the required conditions for dynamic imaging of molecular processes, it was suggested that it would be better to consider their different properties depending on the sample to be imaged. Three different samples were analyzed to demonstrate the applicability of these crystal surfaces for improved AFM imaging: (i) calcium-induced structural changes on a calcium-modulated protein (calmodulin or CaM); (ii) immobilization of GroEL in a controlled orientation and binding of GroES to the immobilized GroEL; and (iii) the process of actin polymerization. Here, the Ca^{2+}-induced structural changes on CaM are described as a representative example to illustrate the ability of streptavidin as a novel substrate for visualizing biomolecular processes.

CaM (16.7 kDa) is a calcium-binding protein expressed in all eukaryotic cells. It can bind to and regulate various protein targets and affect many cellular functions. Many of the proteins that bind CaM are unable to bind Ca^{2+} themselves, and as such, they use CaM as a calcium sensor and signal transducer. The crystal structure of CaM (PDB: 1OSA) is given in Fig. 10A (Ban et al., 1994).

The type-3 *P*2 crystal is the lowest in roughness among the three types of crystals and was used for the analysis of CaM. CaM contains N-terminal and C-terminal globular domains connected by a central linker helix. Each domain contains two Ca^{2+}-binding sites, and CaM undergoes a conformational change after Ca^{2+} binding. The successive HS-AFM images

vectors are also indicated. (B) Trajectories of individual monovacancy defects. Closed squares and circles correspond to the defects marked by open squares and circles shown in (A), respectively. Figure adapted from Yamamoto et al. (2008). (For color version of this figure, the reader is referred to the online version of this chapter.)

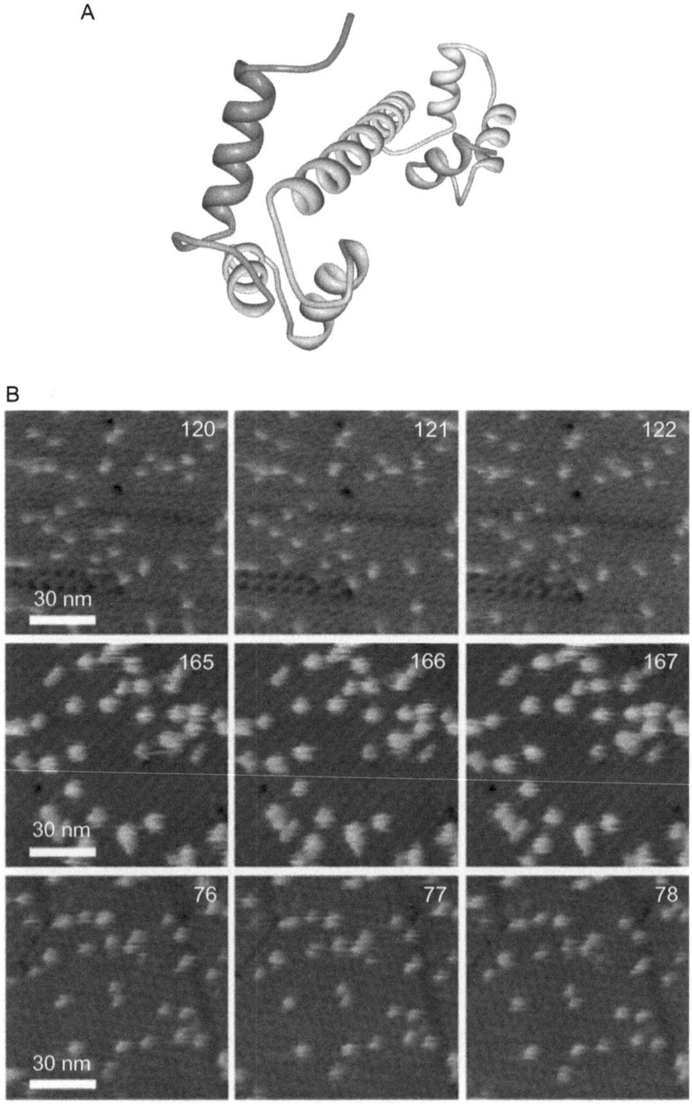

FIG. 10. (A) X-ray crystal structure of CaM (PDB: 1OSA). (B) Successive HS-AFM images of biotinylated CaM immobilized on the type-3 *P*2 crystal of streptavidin. The images were obtained in the absence (top panel) and presence (middle panel) of Ca^{2+}. The images in the bottom panel were acquired for the same sample after replacing the buffer with a solution that did not contain Ca^{2+}. The frame numbers are indicated in each image. HS-AFM images adapted from Yamamoto et al. (2009). (For color version of this figure, the reader is referred to the online version of this chapter.)

(captured at 1 frame/s) of biotinylated CaM specifically bound to the type-3 *P2* crystal in the absence of Ca^{2+} are shown in Fig. 10B (top). The immobilized CaM was observed as a single globe with an estimated height of 1.9 ± 0.3 nm. The addition of Ca^{2+} changed the appearance of CaM, resulting in a larger lateral dimension and a greater height of 2.6 ± 0.4 nm (Fig. 10B, middle). The average half-width was increased from 4.3 $\pm$ 1.0 nm to 5.9 $\pm$ 1.1 nm. This change in height and width was fully reversible in the same sample after the removal of Ca^{2+} (Fig. 10B, bottom), indicating the conformational change in CaM. The binding of Ca^{2+} induces an opening of the two EF-hand motifs in each globular domain. As a consequence, the distance between two helices in each EF-hand increases substantially after Ca^{2+} binding. These results indicate that HS-AFM imaging of streptavidin substrates enables the visualization of conformational changes of the immobilized proteins in the z direction at subnanometer resolution.

VI. Role of ClpX in the Regulation of FtsZ Polymer Dynamics

ClpX is a member of the Clp/Hsp100 class of ATPases associated with diverse cellular activities (AAA^+) family proteins. The highly conserved chaperone ClpX has been implicated in numerous cellular processes, including the turnover of regulatory proteins (Jenal and Fuchs, 1998; Zhou et al., 2001), the degradation of unfinished polypeptides (Wickner et al., 1999), and the remodeling of macromolecular complexes (Burton et al., 2001; Burton and Baker, 2005). It forms a complex with the tetradecameric protease component called ClpP, and the ClpXP complex degrades various cellular proteins (Flynn et al., 2003). ClpX consists of an N-terminal domain and an AAA^+ domain. The former recognizes specific substrate proteins, whereas the latter transmits energy from ATP to mechanical actions, such as unfolding and translocation of substrate proteins. The filamenting temperature-sensitive mutant Z (FtsZ; Fig. 11A) is a protein encoded by the *ftsZ* gene that assembles into a ring at the future site of the septum of bacterial cell division. FtsZ has the ability to bind guanosine triphosphate (GTP) and also exhibits a GTPase domain that allows it to hydrolyze GTP (Mukherjee and Lutkenhaus, 1994). *In vivo,* FtsZ forms filaments with a repeating arrangement of subunits, all arranged in a head-to-tail fashion. These filaments form a ring (Z-ring) around the longitudinal midpoint, or septum, of the cell. *In vitro,*

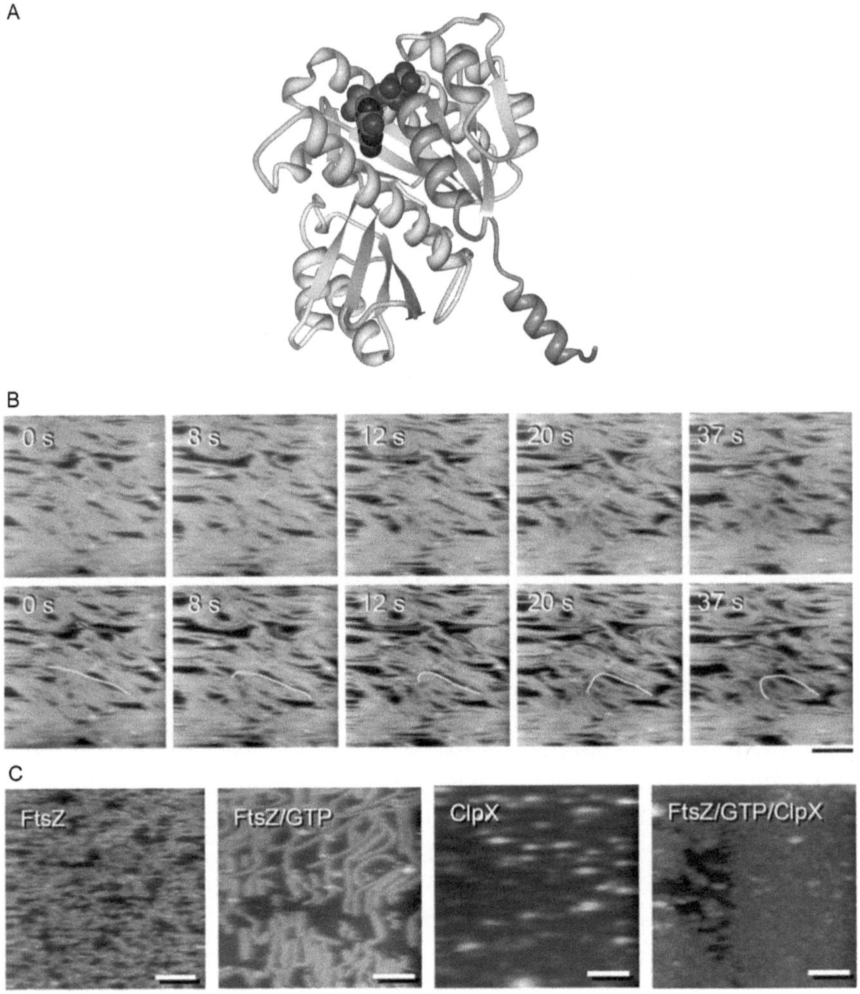

FIG. 11. (A) Crystal structure of FtsZ (ribbon) with bound GDP (sphere, PDB: 2VAP). (B) HS-AFM images of FtsZ polymers. Images taken at the indicated times explain the presence of many single, bundled, and curved filaments (top). To explain the dynamic nature of filaments, the same images are shown in gray scale (bottom) and selected filaments are colored in red and yellow. (C) Inhibition of FtsZ polymerization by ClpX. HS-AFM images of FtsZ in the absence and presence of GTP and ClpX. Scale bar, 100 nm. Images adapted from Sugimoto et al. (2010). (For interpretation of the references to color in this figure legend, the reader is referred to the online version of this chapter.)

depending on the experimental conditions, FtsZ assembles into several polymeric forms, such as single straight or curved filaments, bundles, and sheets (Mukherjee and Lutkenhaus, 1994; Beuria et al., 2003). To date, several proteins that regulate FtsZ dynamics have been identified, some of which stabilize the Z-ring (Ma and Margolin, 1999), while others inhibit FtsZ assembly (Cordell et al., 2003). In *Bacillus subtilis*, ClpX modulates the assembly of the FtsZ protein to help control the process of cell division. However, the mechanism of FtsZ recognition by ClpX and the role of ClpX in regulating the function of FtsZ are not clear. Moreover, the participation of ClpP in the inhibition of FtsZ polymerization by ClpX is controversial, as evaluated using different bacterial species and experimental conditions (Weart et al., 2005; Camberg et al., 2009). Further studies are required to address these differences.

Sugimoto et al. (2010) recently reported the HS-AFM analysis of the nature of FtsZ polymers and the role of ClpX in the regulation of FtsZ polymer dynamics using *E. coli* systems. FtsZ polymers are in highly dynamic equilibrium and the subunits within the polymer are rapidly exchanged on a time scale of 7 s. Thus, conventional AFM is unable to image such a fast dynamics, and HS-AFM analysis is highly desirable. First, GTP-dependent polymerization was confirmed using techniques such as sedimentation assay, light scattering analysis, SEM, and HS-AFM. The polymers were stable for at least 30 min under the conditions used. The height of the FtsZ filaments (5.2 nm in average) in the HS-AFM image was similar to that observed for monomers and dimers (4.4 nm in average), indicating that FtsZ monomers and dimers assemble into a single filament by longitudinal interaction. Next, the dynamic conformational changes in FtsZ polymers were captured with a scan rate of ~1 frame/s. As shown in Fig. 11B, some straight filaments were gradually curved, while others were laterally associated into bundles. Partial dissociation and reassembly of filaments were also observed.

The structure and size distribution of ClpX molecules were studied before the investigation of the effect of ClpX on FtsZ polymerization. ClpX was observed as heterogeneous globular particles with a size distribution ranging from 3 to 50 nm in height and 2 to 33 nm in width. This indicates that ClpX exists as heterogeneously sized particles, ranging from a monomer to a larger oligomer. The sedimentation assay indicated that ClpX specifically inhibits the polymerization of FtsZ. The same conclusion was also derived from the HS-AFM analysis, as shown in Fig. 11C. In the absence of GTP and ClpX, no polymerization was observed. The addition of GTP induced polymerization immediately. However, no filamentous structure was detected upon the simultaneous addition of GTP and ClpX, although small particles were observed. Note that, in this case, the ClpX dimers and FtsZ monomers or dimers were not distinguishable in the AFM image.

VII. Single-Molecular Analysis of DNA–Protein Interactions

The superior capability of HS-AFM over other methods can be best illustrated by the analysis of DNA–protein binding. In such binding interactions, multiple components (more than one part of the duplex DNA and protein either in a monomeric or multimeric form) participate and several conformational changes take place in the participating components. Monitoring these multicomponent interactions and multiple conformational changes simultaneously in real time is a challenging task that can be achieved by HS-AFM, whereas other techniques exhibit serious shortcomings.

The study of the interaction of proteins with DNA is important for gaining a deep understanding of how numerous biological processes, such as recombination, transcription, and repair, occur *in vivo*. This part of the review focuses on the mechanisms of DNA binding of restriction enzymes such as *Apa*I and *Eco*RII. Most bacteria produce at least one DNA restriction/modification protein system, which are thought to have evolved to provide a defense mechanism against invading viruses. Restriction enzymes cut the double-stranded DNA (dsDNA) at specific recognition sequences known as restriction sites. However, methylated host DNA is protected from endonuclease digestion, allowing the foreign virus DNA to be selectively and rapidly cleaved (Bickle and Kruger, 1993). The DNA-binding proteins, including restriction/modifying enzymes, first bind to a nonspecific site to begin the enzymatic reactions. Then, they slide on the DNA until the specific target is recognized, modify the corresponding site, and dissociate from the DNA (Pingoud and Jeltsch, 2001). Thus, the rate of DNA digestion can be strongly controlled by the rate at which it recognizes the target sequences. Several models, such as a 1D linear diffusion model (Terry et al., 1985) and a 3D hopping model (Gowers and Halford, 2003), have been proposed; however, the details of the searching mechanism and the reactions involved remained unclear.

A. *Apa*I Endonuclease

*Apa*I is a type IIP restriction endonuclease with a molecular weight of 38 kDa. The genomic source of this enzyme is *Acetobacter pasteurianus*. It specifically recognizes the sequence 5′-GGGCCC-3′. A volume distribution analysis of AFM images indicated that the free enzyme exists both as a monomer and as a dimer. The DNA–*Apa*I reaction mechanism was recently investigated based on a series of time-lapse sequential HS-AFM images (scanning rate of 1.56 frames/s) of circular plasmid DNA in the presence of *Apa*I (Yokokawa et al., 2006b). The results are summarized in Fig. 12. A circular plasmid DNA was seen in the center of the image (Fig. 12, 0 ms, white arrowhead). During successive scanning, the enzyme indicated by the white arrow moved closer to the DNA and bound to it at 640 ms. After specific recognition, it digested the target sequence (Fig. 12, 1360–2040 ms) and finally dissociated from the DNA, as shown in Fig. 12 (2720–4760 ms). The conversion of the circular DNA into a linear form after *Apa*I digestion is clearly seen in the AFM images. The volume analysis indicated that the dimer *Apa*I bound to the circular DNA (640 ms) and, immediately

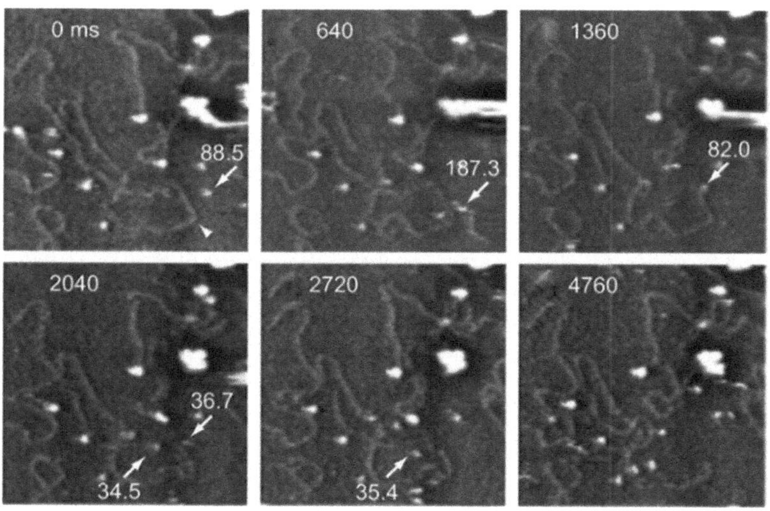

FIG. 12. Snapshots of the HS-AFM imaging of the *Apa*I-mediated DNA cleaving reaction. The numbers at the top indicate the time (ms) at which the images were taken, and the numbers indicated by arrows denote the particle volume. Circular plasmid DNA is indicated by a white arrowhead (0 ms). Images adapted from Yokokawa et al. (2006b).

after the digestion reaction, separated into the monomer units (1360 ms), each of which remained on one cleaved end. Finally, one of the monomers was dissociated from the digested DNA (2720 ms). Under these experimental conditions, the time constant of the binding of *Apa*I to DNA was ~0.8 s^{-1} and the average traveling distance of the enzyme on the DNA was ~80 nm, which corresponds to 240 bp.

B. *DNA-*Eco*RII Protein Complexes*

The restriction endonuclease (REase) *Eco*RII is an enzyme of the restriction/modification system that exists naturally in *E. coli.* It is a type IIE restriction enzyme with a molecular mass of 45.2 kDa that is composed of 402 amino acids. It is a dimer (Fig. 13A) and recognizes the 5′-CCWGG-3′ sequence. According to the definition of type IIE REases, it binds two DNA recognition sites and cleaves one of the sites (Roberts et al., 2003). However, a recent study suggested that the *Eco*RII protein actually requires three sites to cleave both strands of one recognition site concertedly (Tamulaitis et al., 2006). There is still much interest in how site-specific proteins search for their recognition site on DNA.

Recently, HS-AFM was used to image directly the single-molecule dynamics of the protein–DNA complexes formed by *Eco*RII (Gilmore et al., 2009). First, the three-site-binding behavior of the *Eco*RII was characterized using dry AFM imaging with fragments (Fig. 13B, top) reported previously (Shlyakhtenko et al., 2007). The single recognition site (i), looped structure with two recognition sites (ii), and a double-looped structure with three recognition sites (iii) were clearly seen in the AFM image (Fig. 13B, bottom). Complexes with one, two, and three sites had a frequency of 43%, 55%, and 2%, respectively.

The dynamics of the protein under aqueous buffer conditions was also imaged at a rate of 2 frames/s. As shown in Fig. 13C, two instances of protein translocation were found, one between 4 and 6 s and the other between 7 and 11.5 s. In the first instance, the protein tracked the DNA, whereas in the second case it translocated to another DNA fragment. The protein was located next to the DNA fragment in some frames (9–9.5 s), but completely overlapped the DNA in other frames (10–10.5 s). In addition, *Eco*RII appeared to interact briefly with two sites and form a transient loop, as observed at 3.5 and 6.5 s. In addition to the simple translocation of the protein interacting with a single DNA site, translocation was also

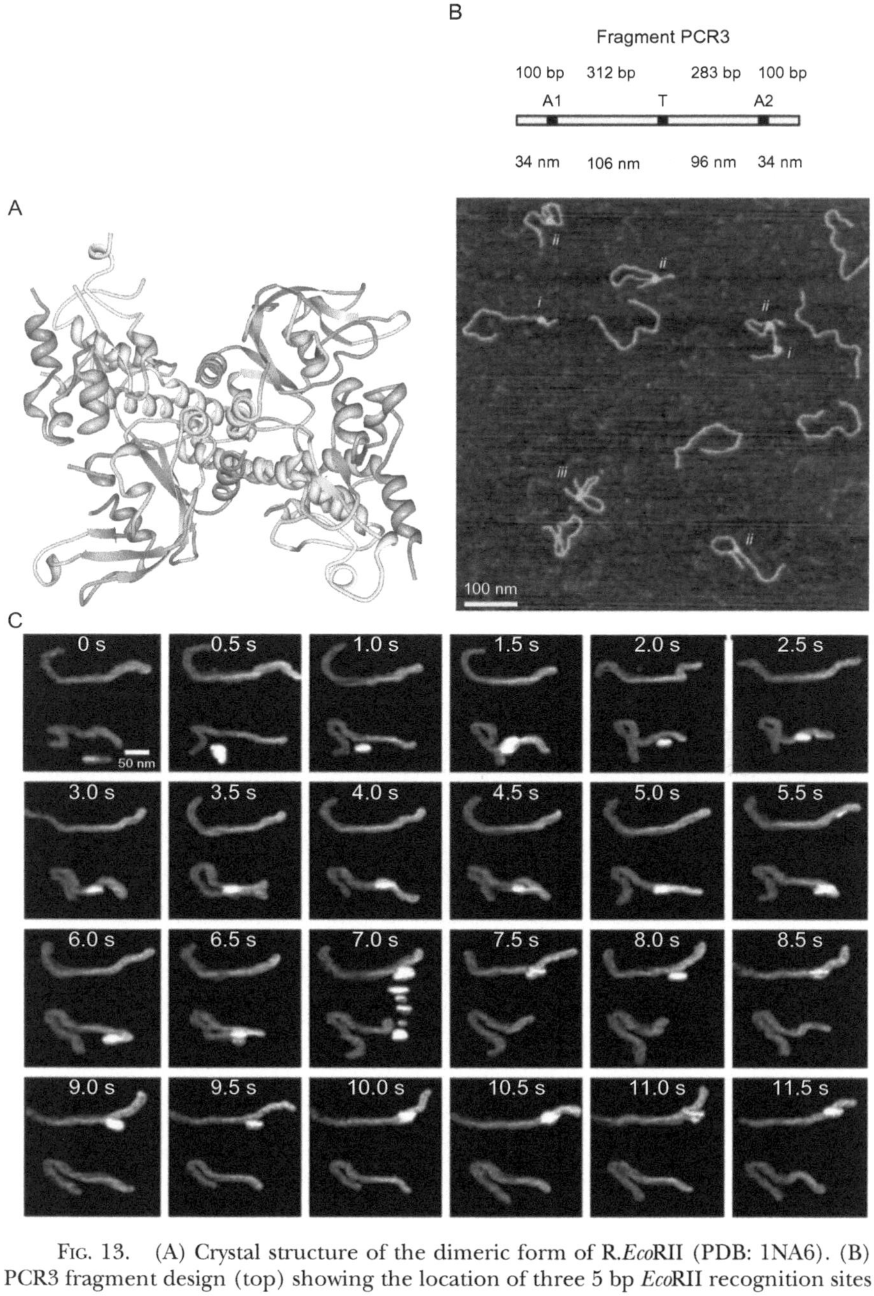

FIG. 13. (A) Crystal structure of the dimeric form of R.*Eco*RII (PDB: 1NA6). (B) PCR3 fragment design (top) showing the location of three 5 bp *Eco*RII recognition sites

observed on a looped DNA. In this case, the protein remained bound at one recognition site while the adjacent strand gradually translocated from one recognition site to another (data not shown). The estimated rate of translocation was $\sim$30 bp/s, which corresponds to a distance of 10.2 nm/s.

Another important issue to be addressed in the cases of protein–DNA interactions is how the proteins dissociate after the completion of the reaction. As *Eco*RII is a dimer, it either dissociates as a dimeric form or it separates into monomers and consequently dissociates from the DNA. The investigation of the single-looped complexes revealed that 58% dissociated to an intermediate complex that interacted with a single site (dimeric form), whereas 42% dissociated by breaking protein subunits (monomers) apart, resulting in a subunit that interacts with both sites. In addition to single-loop complexes, double-loop complexes also dissociated: 46% dissociated to a small loop, 31% dissociated to a large loop, and 15% dissociated spontaneously without a looped intermediate.

VIII. Action of Motor Proteins

The catalytic proteins, such as kinesins, dyneins, myosins, and DNA and RNA polymerases, that operate in biological cells by consuming chemical energy provided by ATP hydrolysis and converting it into mechanical work are termed motor proteins. They play crucial roles in several biological functions, such as cell division, cellular transport, muscle contraction, and genetic transcription. Single-molecule techniques and structural investigations have led to rapid progress in understanding how these biological motors operate. However, the main fundamental questions related to motor proteins—how the chemical energy is transformed into mechanical motion, how they generate force, how much fuel they consume, and their efficiency—remained unanswered. Furthermore, it is of interest to consider whether it is possible to use these biological nanomotors in artificial environments outside the cell to perform multiple actions aimed for our benefit.

along an 810 bp fragment. Dry AFM image (bottom) showing three different types of *Eco*RII–DNA complexes: (i) one-, (ii) two-, and (iii) three-site interactions. (C) Successive frames of *Eco*RII movement over DNA. Images adapted from Gilmore et al. (2009). (For color version of this figure, the reader is referred to the online version of this chapter.)

A. *Myosin V Translocation on Actin Filaments*

Myosin V is a dimeric molecular motor that consists of two heavy chains containing the myosin heads and a neck (Fig. 14A). It is a cargo-carrying processive motor that moves hand over hand, with the center of mass

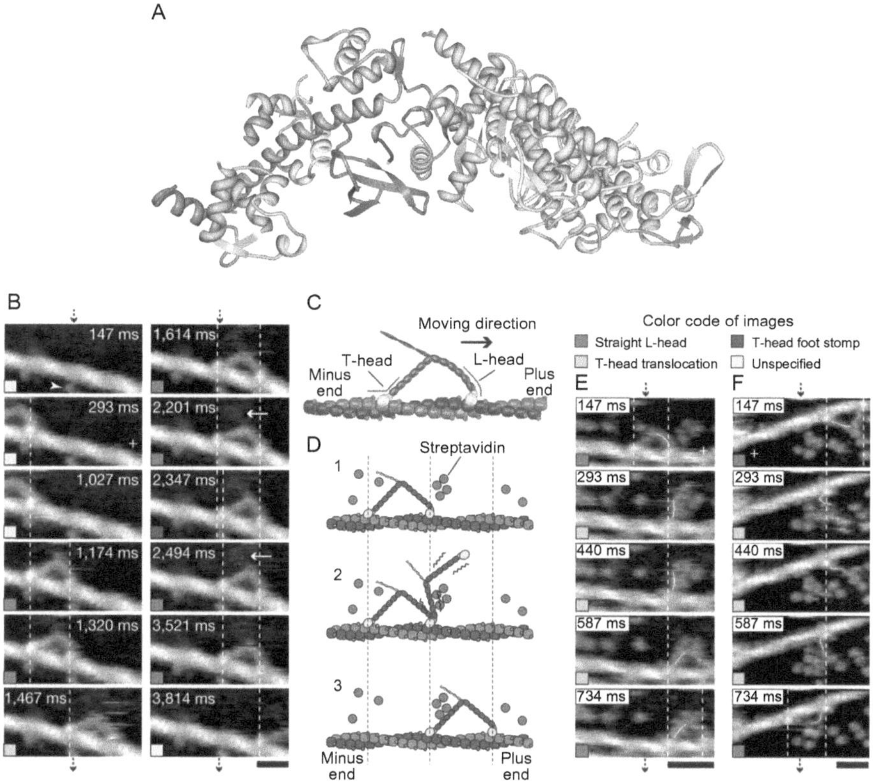

FIG. 14. (A) Crystal structure of myosin V (PDB: 1OE9). (B) Successive HS-AFM images showing processive movement of M5-HMM in 1 μM ATP. The arrowhead and arrows represent the streptavidin molecule and the coiled-coil tail of M5-HMM tilted toward the minus end of actin, respectively. (C) Drawing of the two-headed bound M5-HMM. (D) Schematic representation of the images shown in (E) and (F). Snapshots of HS-AFM images showing hand-over-hand movement in 1 μM (E) and 2 μM (F) ATP. The white lines, vertical dashed lines in the images, and plus sign indicate the swinging lever, centers of mass of the motor domains, and plus end of actin, respectively. Scale bars, 30 nm (B) and (F) and 50 nm (E). Images adapted from Kodera et al. (2010). (For color version of this figure, the reader is referred to the online version of this chapter.)

advancing about 36 nm along actin filaments for every ATP hydrolysis. The two heads of myosin V are held together by a coiled-coil stalk. Each head contains a catalytic domain (CD) that is responsible for actin binding and ATP hydrolysis, and a light chain-binding domain that likely acts as a lever arm to amplify small nucleotide-dependent conformational changes in the CD. Numerous single-molecule studies have demonstrated the movement of myosin V on actin filaments, and several walking hypotheses have been proposed (Yildiz et al., 2003); however, the dynamic behavior of the active molecule was not clearly understood.

A stunning HS-AFM video image of walking myosin V with unprecedented high resolution was recently published (Kodera et al., 2010). Those authors used biotin-containing lipid bilayers formed on a mica surface to immobilize partially biotinylated actin filaments using streptavidin. To facilitate the weak sideways adsorption of tail-truncated myosin V (M5-HMM) onto the bilayer surface, the positively charged lipid 1,2-dipalmitoyl-3-trimethylammonium–propane (DPTAP, 5%) was included in the bilayer. The processive movement of M5-HMM using discrete steps of $\sim$36 nm was imaged as shown in Fig. 14B and C. A processive run almost always continued to the full extent of each imaging range used. In the two-headed bound M5-HMM (in which "head" means the motor domain and neck domains), the neck–motor domain junction appeared smooth in the leading head (L-head) but was V shaped in the trailing head (T-head) without exception, because the neck regions emerge from different parts of the motor domain. The short coiled-coil tail was mostly tilted toward the minus end of actin, which is consistent with the results of previous studies.

Though the unidirectional processive movement was successfully imaged, the hand-over-hand movement was difficult to capture. This is because the movement of $\sim$36 nm was completed within a frame (time scale of 146.7 ms). Therefore, the process occurring within this time could not be imaged. However, the increase in the amount of streptavidin molecules placed on the substrate surface as moderate obstacles to the advance allowed the visualization of this process (Fig. 14D–F). After T-head detachment, the nearly straight leading neck swung from the reverse arrowhead (R-ARH) orientation to the arrowhead (ARH) orientation, confirming the swinging lever-arm motion initially proposed for muscle myosin. The detached T-head rotationally diffused around the advancing neck–neck junction and then bound to a forward site on the actin filament, completing one step. Thus, the hand-over-hand movement, including the intermediate process, was directly visualized at high resolution. Moreover, the unfolding of the coiled-coil tail, foot stomp in ATP, ADP, and nucleotide-free conditions, conformational transition in the L-head, and ADP release from the L-head were demonstrated in detail.

B. *Rotary Catalysis of Rotorless F_1-ATPase*

F_1-ATPase, a water-soluble portion of ATP synthase, is a rotary motor protein. It is composed of a hexameric headpiece assembled from alternating α and β subunits, denoted ($\alpha_3\beta_3$). The nucleotide-binding sites lie in the interface clefts between the subunits, but they are not all identical. Three catalytic sites are located mostly in the β subunits and alternate with three noncatalytic sites that lie mostly in the α subunits. The ($\alpha_3\beta_3$) hexamer surrounds an eccentric coiled-coil "shaft" (γ subunit), the rotation of which is driven by sequential ATP hydrolysis at the three catalytic sites. Some biochemical studies suggested that the $\alpha_3\beta_3$ ring alone does not possess intrinsic cooperativity and that the γ subunit mediates the interplay among the β subunits (Kaibara et al., 1996; García and Capaldi, 1998). In contrast, recent studies evidenced that F_1 retains catalytic power to rotate γ unidirectionally, even when most interaction sites between β and γ are abolished (Aloise et al., 1991; Furuike et al., 2008).

Recently, a detailed real time analysis was performed to clarify this issue (Uchihashi et al., 2011). The $\alpha_3\beta_3$ subcomplex was covalently immobilized on a mica surface and imaged using HS-AFM with a frame capture time of 80 ms (Fig. 15). In the absence of nucleotides, $\alpha_3\beta_3$ exhibited a pseudo sixfold symmetric ring in which three alternately arranged subunits were elevated relative to the other three (Fig. 15A, left). The simulated (Fig. 15A, right) and observed AFM images confirmed that the N-terminal side was selectively attached to the mica. The addition of the nonhydrolyzable ATP analog AMPPNP rendered the ring triangular and obscured the central hole (Fig. 15B, left). Although the three α subunits with lower protrusions retained the same conformation as those observed under nucleotide-free conditions, two of the three β subunits retracted toward the center and simultaneously lowered their protrusions. Consequently, the ring showed a single high protrusion. The simulated image (Fig. 15B, right) also showed an asymmetric ring similar to that of the observed image. This indicates that only two β subunits can assume the closed conformation, even in saturating AMPPNP conditions.

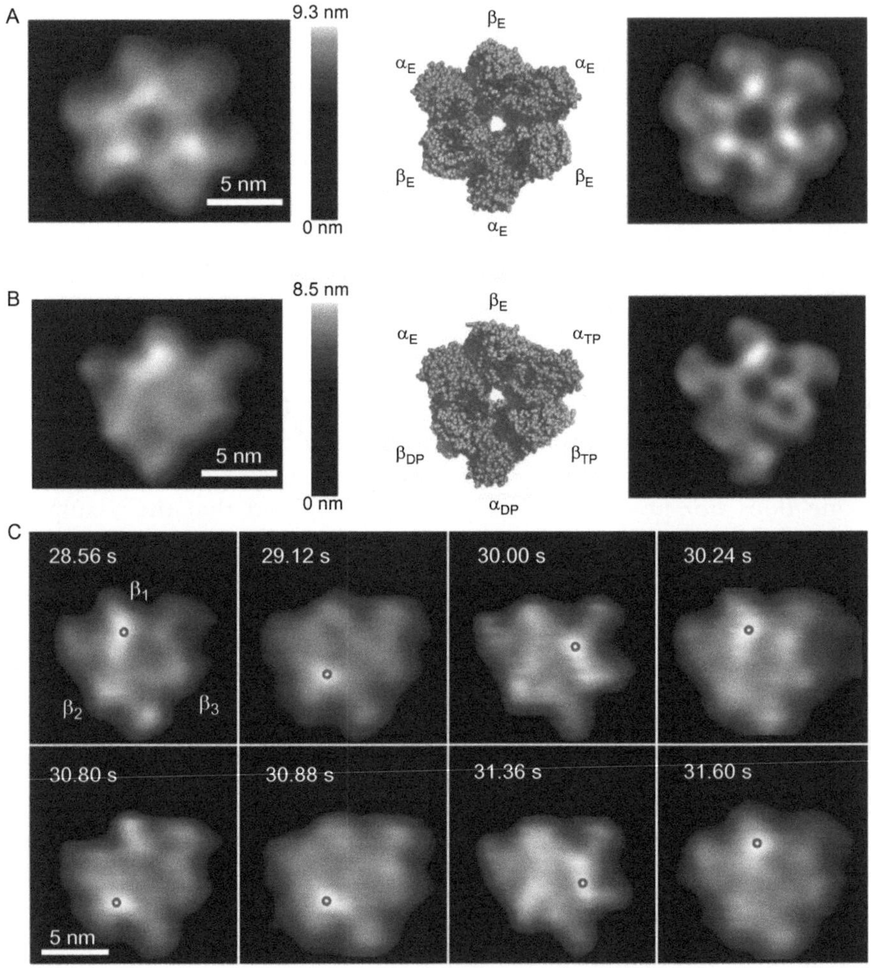

FIG. 15. (A) Averaged AFM image (left), crystal structure (middle, PDB: 1SKY), and simulated AFM image (right) of the C-terminal side of the $\alpha_3\beta_3$ subcomplex in the absence of nucleotides. (B) Averaged AFM image of the same in 1 mM AMPPNP (left). Atomic structure of the $\alpha_3\beta_3$ subcomplex with bound nucleotides obtained by removing γ from the crystal structure of F_1 (middle, PDB: 1BMF). Simulated AFM image (right). The α and β subunits, and the C-terminal DELSEED motif of β corresponding to the high protruding portions, are colored in cyan, pink, and red, respectively. (C) Successive HS-AFM images showing the conformational changes of β subunits in 2 μM ATP. The red circle in each image denotes the highest pixel. Frame rate, 12.5 frames/s. Figure adapted from Uchihashi et al. (2011). (For interpretation of the references to color in this figure legend, the reader is referred to the online version of this chapter.)

Next, the conformational changes of β subunits were observed in the presence of ATP at an imaging speed of 12.5 frames/s. After the addition of ATP, β subunits exhibited distinct conformational dynamics; each β subunit underwent a conformational transition between the outwardly extended high state (open) and the retracted low state (closed) (Fig. 15C). The most notable features were that only a single β subunit assumed the open state, as in the presence of AMPPNP, and that when the open-to-closed transition occurred at one β subunit, the opposite closed-to-open transition occurred simultaneously at its counterclockwise neighboring β subunit in most cases. Thus, the high and outwardly extended conformation propagates in the counterclockwise direction. These results proved that the stator $\alpha_3\beta_3$ ring alone possesses high cooperativity for sequential power stroking among the three catalytic β subunits. However, the ATP-binding rate and the efficiency of unidirectionality of the $\alpha_3\beta_3$ subcomplex were distinctly lower than those of F_1. Thus, the interaction with γ is dispensable but important for rapid and precise rotary catalysis.

IX. Movement of *Tr*Cel7A on Crystalline Cellulose Substrate

*Tr*Cel7A (formerly known as CBH I) is an enzyme that has catalytic activity on a crystalline cellulose substrate. Similar to most cellulolytic enzymes, it consists of two domains joined together by a glycosylated interdomain linker peptide: a small cellulose-binding domain (CBD) belonging to carbohydrate-binding module family 1 and a large CD from the glycoside hydrolase family 7 (Fig. 16A). The CD contains the active site, which carries out general acid-catalyzed hydrolysis of the β-1,4-glycosidic bonds of cellulose (Divne et al., 1998). The CBD improves the binding of the enzyme to cellulose and enhances its enzymatic degradation. The removal of the CBD causes a decrease in crystalline cellulose decomposition but has almost no effect on the hydrolysis of soluble or amorphous cellulose (Srisodsuk et al., 1993). *Tr*Cel7A is thought to hydrolyze the crystalline cellulose chain in a processive manner (Fig. 16A), making consecutive cuts without releasing the chain. The molecular process of crystalline cellulose degradation by *Tr*Cel7A was recently unraveled (Igarashi et al., 2009) following the development of the scanning speed of HS-AFM.

Crystalline cellulose prepared from *Cladophora* sp. was used in the analysis. As shown in Fig. 16B, crystalline cellulose was hydrophobically immobilized on a graphite grid and was observed using HS-AFM in the

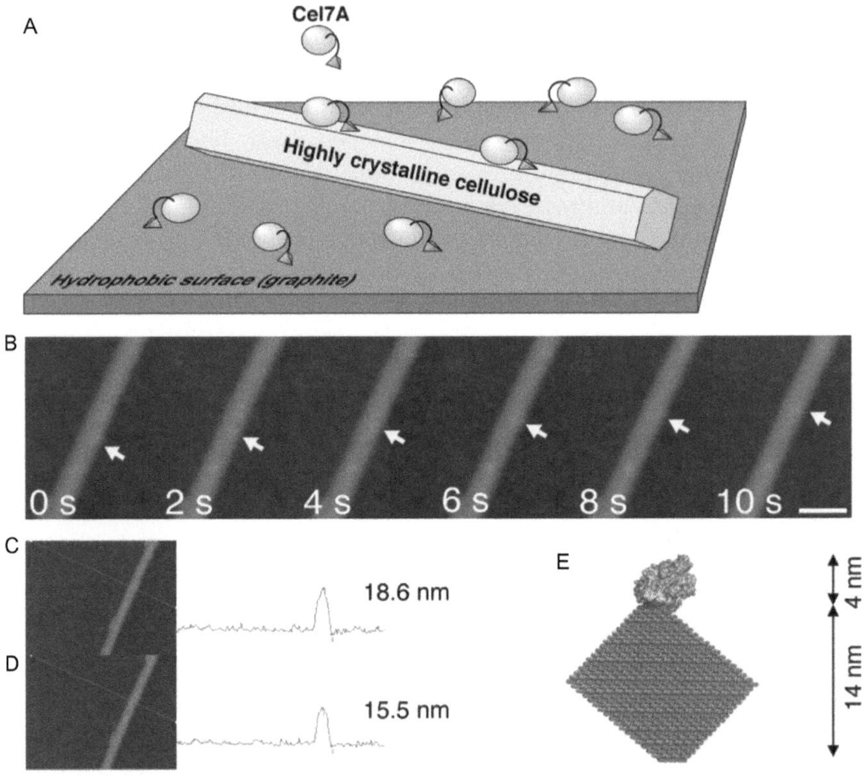

FIG. 16. (A) Schematic representation of the Cel7A reaction on crystalline cellulose fixed on the hydrophobic surface of graphite. CD, light gray; CBD, orange. (B) Time-lapse images of *Tr*Cel7A (white arrows) sliding on the substrate. Scale bar, 50 nm. Height analysis of the crystalline cellulose with (C) and without (D) enzyme. (E) 3D structure of the *Tr*Cel7A CD (PDB: 8CEL) and the CBD (PDB: 1CBH), and cellulose I_{α}. Figure adapted from Igarashi et al. (2009). (For interpretation of the references to color in this figure legend, the reader is referred to the online version of this chapter.)

presence of Cel7A. Small particles sliding on the surface of the cellulose substrate can be seen in the HS-AFM images. Furthermore, the particles slide in a single direction without becoming detached from the substrate. The size of the particles was 3.1 nm (Fig. 16C and D), which is consistent with the height estimated from the crystal structure of the catalytic module of *Tr*Cel7A (Fig. 16E). Moreover, these moving particles were not seen in the absence of the enzyme, which confirms that they are *Tr*Cel7A molecules sliding on crystalline cellulose. Cellulose crystals exhibited narrowing of the microcrystals after incubation with *Tr*Cel7A, which was also confirmed by HPLC analysis. This finding is consistent with the transmission electron microscopy observations of narrowing of the microcrystals at the reducing end side (Imai et al., 1998). Hence, it was concluded that the *Tr*Cel7A molecules move from the reducing end to the non-reducing end. Studies using CBH II types of enzymes further confirmed this conclusion.

The individual *Tr*Cel7A molecules all slid with a similar velocity on the crystalline cellulose surface, and the average velocity of the sliding movement was 3.5 ± 1.1 nm/s. It is noteworthy that a CD without the CBD moved with a velocity similar to that of the intact *Tr*Cel7A. However, sliding was not observed for a catalytically inactive enzyme (mutant E212Q) or for a variant lacking tryptophan at the entrance of the active site tunnel (mutant W40A). This clearly indicates that, in addition to the hydrolysis of glycosidic bonds, the loading of a cellulose chain into the active site tunnel is also essential for the movement of the enzyme.

X. Kinetics of Antimicrobial Peptide Activity

Antimicrobial peptides (Amps) are proteins with a small molecular weight that are found among all classes of life. They are an important class of antimicrobials and have been demonstrated to be novel therapeutic agents. These peptides possess activity against antibiotic-resistant bacteria, parasites, viruses, fungi, and even transformed or cancerous cells (Izadpanah and Gallo, 2005; Gottlieb et al., 2008). Though previous electron microscopy and AFM experiments provided some information on the activity of Amps on bacterial cells, early-stage kinetics has not been demonstrated at a nanometer-scale resolution and on a time scale of seconds. Recently, the kinetics of the predeath activity of a pore-forming, membrane-disrupting Amp (CM15) on individual *E. coli* cells was unambiguously described using HS-AFM with microfabricated small cantilevers (Fantner et al., 2010). In that study, bacteria were immobilized on poly-L-lysine-coated cover slides. Amp CM15 was added to the liquid droplet around the sample and images were acquired every 13 s. The HS-AFM phase images of two bacterial surfaces before and at several time points after the addition of CM15 are shown in Fig. 17. As can be seen from the images, the bacterial surface was smooth until the addition of CM15 and became corrugated after its addition. This observation of the changes in

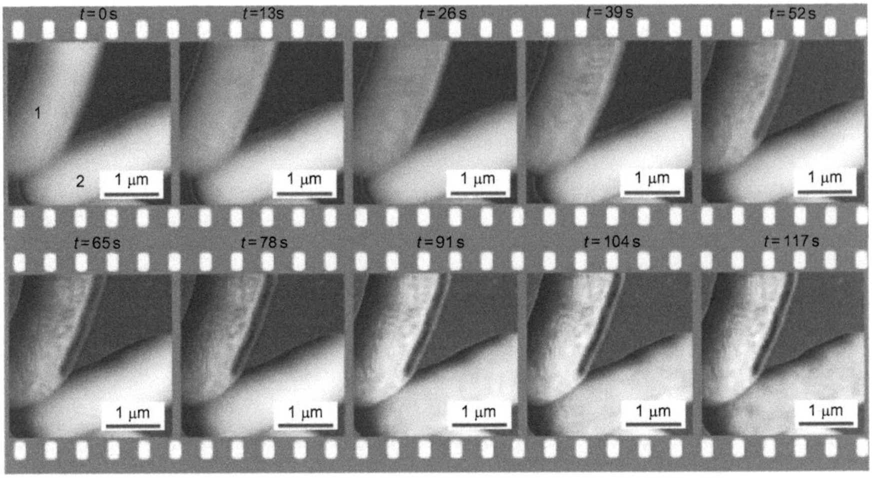

FIG. 17. Snapshots of HS-AFM imaging of live *Escherichia coli* cell disruption induced by CM15. CM15 was injected at $t = -6$ s and images were recorded every 13 s. The surface of the upper bacterium 1 started changing within 13 s. The lower bacterium 2 resisted until 78 s. Figure adapted from Fantner et al. (2010). (For color version of this figure, the reader is referred to the online version of this chapter.)

the bacterial surface is consistent with previous electron microscopy data (Kalfa et al., 2001; Mangoni et al., 2004; Meincken et al., 2005). Interestingly, the kinetics or the time of onset of the activity of the peptide varies significantly from one bacterial cell to another. For example, the surface change on bacterium 1 starts within 13 s of the addition of CM15 and is completed in ~60 s. However, bacterium 2 does not start changing until ~80 s and the change is not complete until ~120 s. The cell death caused by the cell surface corrugation was also investigated via the combined use of HS-AFM and fluorescence microscopy.

XI. DNA ORIGAMI: A SUBSTRATE FOR PROTEIN ANALYSIS

The nanomaterials prepared based on DNA served as novel substrates for biomolecular analysis, including the nanopatterning (Liu et al., 2005; Chhabra et al., 2007) and functional analysis of proteins. However, the nanostructures constructed were 10–20 nm in size (Lund et al., 2005; Park et al., 2006), which is not suitable for the high-throughput analysis of proteins, as these structures fall within the size range of several proteins.

Therefore, the construction of larger assemblies with a defined size was required. Furthermore, during enzymatic reactions, enzymes bend dsDNA. For example, *Eco*RI methyltransferase (M.*Eco*RI) bends the duplex DNA by 55–59° during the course of the reaction (Cheng, 1995). To monitor the selective methylation on one DNA duplex compared with another, the system of interest should be fixed in a DNA nanostructure with some relaxation so that enzymatic reaction can be feasible, as the DNA can undergo bending. In contrast, the control strand fixed with tension may be less affected by the enzyme, as bending will not be possible. Such structural control may be possible with the use of larger DNA nanostructures. In addition, distance-dependent studies can be performed with the aid of DNA nanostructures, to which the systems of interest can be fixed at a predetermined distance from the substrate. This type of analysis cannot be performed on the substrate-free bare surface of mica. The recent development of "scaffolded DNA origami" achieved the construction of larger nanostructures with a diameter of about 100 nm (Rothemund, 2006). Various 2D (Endo et al., 2011; Liu et al., 2011; Rajendran et al., 2011b, 2012a) and 3D (Andersen et al., 2009) structures were prepared using this method and were successfully applied to the analysis of various proteins (Rinker et al., 2008; Numajiri et al., 2010; Rajendran et al., 2012b).

*A. DNA-Methylation by M.*Eco*RI*

We have provided two examples of HS-AFM analysis of DNA–protein interactions using a DNA origami structure as a novel substrate. In the first report, we demonstrated the ability of DNA origami for the structural control of the selective methylation of a DNA duplex by M.*Eco*RI and further digestion with the restriction enzyme R.*Eco*RI (Endo et al., 2010a). DNA modification by enzymes often requires structural changes at the target strands, such as bending of the dsDNA. To control the reaction of M.*Eco*RI and examine its structural effect on methylation, a DNA origami scaffold termed "DNA origami frame" was designed and prepared (Fig. 18A). This frame accommodates two different lengths of the dsDNA fragments: tensed 64mer and relaxed 74mer strands. In the presence of *s*-adenosyl-L-methionine, this enzyme introduces a methyl group at the N6 position of the second adenine of the 5′-GGAATC-3′ sequence. After treatment with M.*Eco*RI and subsequent digestion with the restriction enzyme R.*Eco*RI, real time HS-AFM analysis revealed that the 74mer

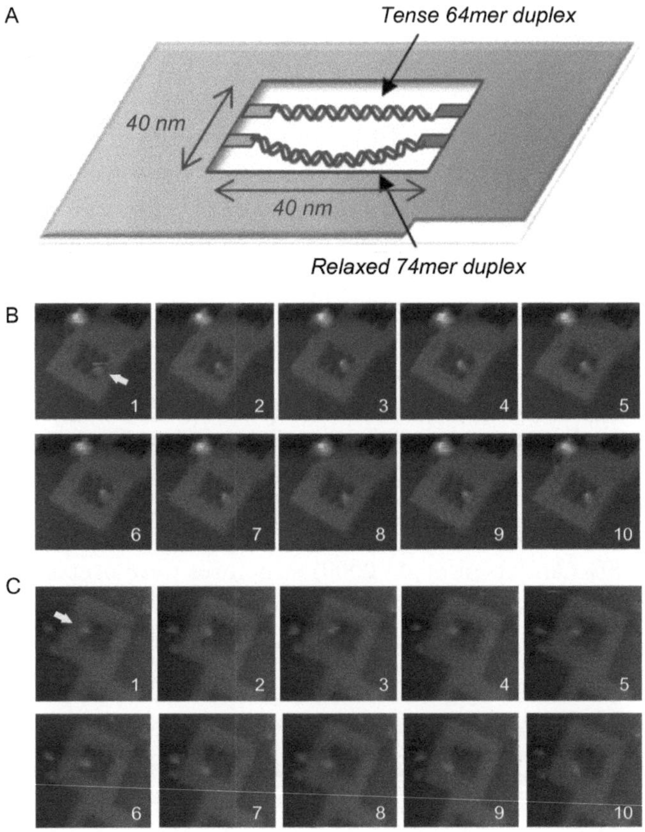

FIG. 18. Enzymatic reactions inside a DNA origami frame. (A) DNA frame structure designed for the incorporation of two different dsDNAs: tense 64mer and relaxed 74mer duplexes with the specific sequence for M.*Eco*RI at the center. Successive HS-AFM images of the DNA frame after treatment with M.*Eco*RI, which bound to the 64mer (B) and the 74mer (C) dsDNAs. The white arrow indicates the enzyme and numbers indicate the time (s) at which the images were taken. Scanning rate, 1 frame/s; image size, 200 × 200 nm. Figure adapted from Endo et al. (2010a). (For color version of this figure, the reader is referred to the online version of this chapter.)

DNA was not effectively cleaved compared with the 64mer. This indicates that the methylation occurred preferentially in the relaxed 74mer DNA (as it can undergo bending during the reaction) compared with that of the tense 64mer, which cannot undergo bending (Fig. 18B and C). This was further confirmed by the biochemical analysis of methylation and specific digestion using real time PCR. These results reveal the importance of structural flexibility for bending dsDNAs during the methyl transfer reaction mediated by M.*Eco*RI. Therefore, DNA methylation can be regulated using the structurally controlled dsDNAs constructed within a DNA nanostructure.

B. *Dynamics of Base-Excision Repair Enzymes*

In the second report, the reactivity of the DNA base-excision repair (BER) enzymes 8-oxoguanine glycosylase (hOGG1; Fig. 19A) and T4 pyrimidine dimer glycosylase (PDG; Fig. 19D) was studied using a similar strategy of structural control (Endo et al., 2010b). BER enzymes have glycosylase/AP-lyase activity for removing the damaged nucleobases and cleaving the DNA strand. Various dsDNAs with a damaged base were placed inside the origami frame and the repair reaction was analyzed at single-molecule level. Similar to the previous case, two substrate dsDNAs of different lengths, a tense

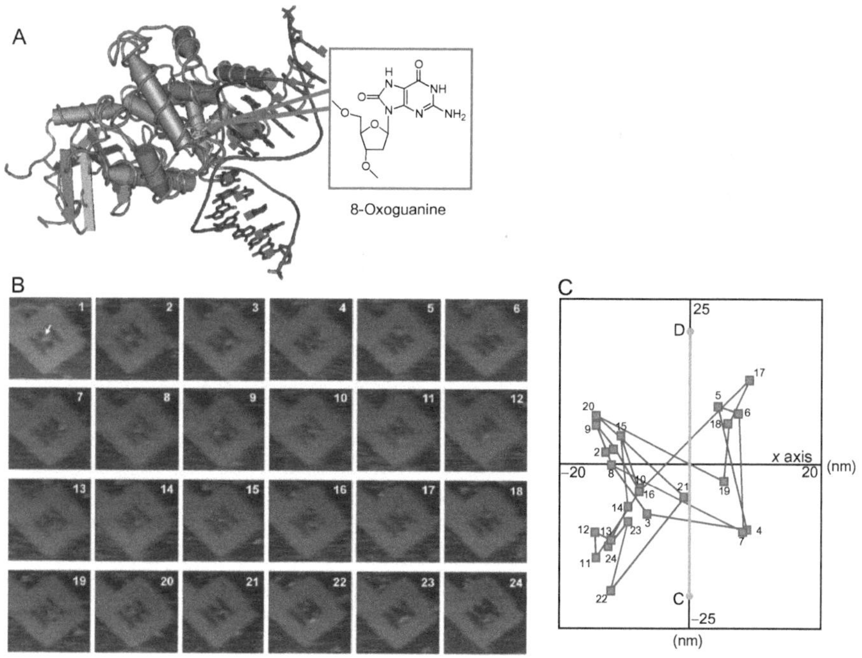

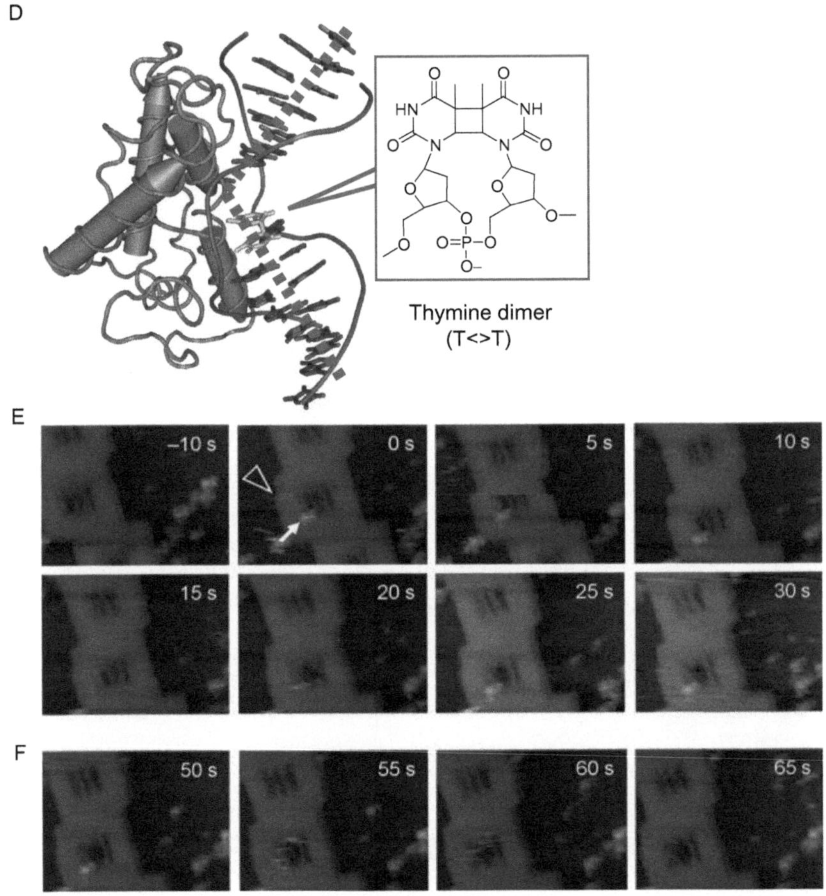

FIG. 19. Direct observation of the reactions of DNA repair enzymes. (A) Structure of the hOGG1–DNA complex (PDB: 1EBM). (B) Successive HS-AFM images of hOGG1 inside the DNA nanochip. Images were obtained every 1 s. Image size, 200 × 200 nm. (C) The trajectory of the movement of hOGG1. (D) Structure of the PDG–DNA complex (PDB: 1VAS). (E) Successive HS-AFM images of PDG binding to the dsDNA, and DNA cleavage. (F) Successive HS-AFM images of the dissociation of PDG from the cleaved dsDNA. Images were obtained every 5 s. Image size, 300 × 225 nm. Figure adapted from Endo et al. (2010b). (For color version of this figure, the reader is referred to the online version of this chapter.)

64mer and a relaxed 74mer, were placed inside the nanocavity of a DNA nanochip to examine the structural effect on glycosylase/AP-lyase activity, including the cleavage of the DNA strands and trapping of the reaction intermediates. The relaxed 74mer dsDNA was effectively cleaved and trapped using $NaBH_4$ reduction compared with the 64mer dsDNA. Furthermore, dynamic movement of the hOGG1 and the single DNA repair reaction, including sliding and DNA cleavage, were directly observed in the DNA nanochip using HS-AFM (Fig. 19B). The cavity of the DNA nanochip is associated with coordinates; thus, the position of the hOGG1 was plotted and the trajectory of the movement of the hOGG1 inside the DNA chip was analyzed (Fig. 19C). In addition, the whole reaction process of the DNA repair by PDG, that is, binding, sliding, catalytic reaction, and dissociation, was successfully observed using this method (Fig. 19E and F).

This method can also be used for other DNA-modifying and repair enzymes that bend the double-helix during the enzymatic reaction. In addition, the DNA origami frame structure may represent a versatile scaffold that can be used to observe various conformational changes of DNA, for example, G-quadruplex formation (Sannohe et al., 2010). The method can be extended to the direct observation of various enzymatic phenomena in the designed nanoscale space.

XII. Future Perspectives

Taking into consideration the novel applications of the HS-AFM method developed over the past decade, we have summarized various investigations of the structural and functional properties of several proteins in physiological solutions. The studies outlined here clearly describe the superior capability of the recent development of enhanced image acquisition frequency and its ability to document previously inaccessible data on various biological processes, including protein dynamics. The scan rate of HS-AFM has now reached a level where video-rate imaging is almost possible, which may shed light on the various biological mechanisms and kinetics of active proteins. HS-AFM is now the only viable technique that allows the simultaneous assessment of the structure and dynamics of single molecules. In addition to its advantages, the method has several limitations, and further developments are required before this technique can be applied efficiently to the study of several biological systems. The

HS-AFM method must be speeded up for the study of molecular events that occur in less than a millisecond. Such a development may help detect the hidden and unresolved intermediate states that proteins may adopt during their biological functions. As explained above, HS-AFM can be applied to cellular studies. However, further developments are required for the application of the method not only at the molecular scale but also at the cellular level. This may be achieved by incorporating a scanner stage that is suitable for larger scan areas, and by coupling the HS-AFM method with optical microscopy for simultaneous analyses. The substrate surface onto which the sample is placed is important for the precise imaging of the dynamic action of protein molecules under physiological conditions: it should be sufficiently flat to allow the easy identification of the deposited protein molecules. As described above, streptavidin 2D crystals and DNA origami structures were successfully employed as substrates; however, different choices of substrates should be explored. To date, only proof-of-concept experiments have been performed regarding the analysis of proteins using HS-AFM; we anticipate that a substantial amount of additional work is needed to answer several basic questions regarding protein dynamics *in vivo*.

Acknowledgments

We express our sincere thanks for the CREST grant from the Japan Science and Technology Corporation (JST), grants from the WPI program (iCeMS, Kyoto University), and for the global COE program from the Ministry of Education, Culture, Sports, Science and Technology (MEXT), Japan. A. R. expresses sincere thanks to the Japan Society for the Promotion of Science (JSPS) for the Postdoctoral fellowship. We thank Sekar Latha for her help with the graphics.

References

Aloise, P., Kagawa, Y., Coleman, P. S. (1991). Comparative Mg(2+)-dependent sequential covalent binding stoichiometries of 3'-O-(4-benzoyl)benzoyl adenosine 5'-diphosphate of MF1, TF1, and the alpha 3 beta 3 core complex of TF1. The binding change motif is independent of the F1 gamma delta epsilon subunits. *J. Biol. Chem.* **266**, 10368–10376.

Andersen, E. S., Dong, M., Nielsen, M. M., Jahn, K., Subramani, R., Mamdouh, W., et al. (2009). Self-assembly of a nanoscale DNA box with a controllable lid. *Nature* **459**, 73–76.

Ando, T., Kodera, N., Takai, E., Maruyama, D., Saito, K., Toda, A. (2001). A high-speed atomic force microscope for studying biological macromolecules. *Proc. Natl. Acad. Sci. USA* **98**, 12468–12472.

Ando, T., Uchihashi, T., Fukuma, T. (2008a). High-speed atomic force microscopy for nano-visualization of dynamic biomolecular processes. *Prog. Surf. Sci.* **83**, 337–437.

Ando, T., Uchihashi, T., Kodera, N., Yamamoto, D., Miyagi, A., Taniguchi, M., et al. (2008b). High-speed AFM and nano-visualization of biomolecular processes. *Pflugers Arch.* **456**, 211–225.

Ban, C., Ramakrishnan, B., Ling, K. Y., Kung, C., Sundaralingam, M. (1994). Structure of the recombinant Paramecium tetraurelia calmodulin at 1.68 Å resolution. *Acta Crystallogr. D Biol. Crystallogr.* **50**, 50–63.

Beuria, T. K., Krishnakumar, S. S., Sahar, S., Singh, N., Gupta, K., Meshram, M., et al. (2003). Glutamate-induced assembly of bacterial cell division protein FtsZ. *J. Biol. Chem.* **278**, 3735–3741.

Bickle, T. A., Kruger, D. H. (1993). Biology of DNA restriction. *Microbiol. Rev.* **57**, 434–450.

Binnig, G., Quate, C. F., Gerber, C. (1986). Atomic force microscope. *Phys. Rev. Lett.* **56**, 930–933.

Blankenburg, R., Meller, P., Ringsdorf, H., Salesse, C. (1989). Interaction between biotin lipids and streptavidin in monolayers: formation of oriented two-dimensional protein domains induced by surface recognition. *Biochemistry* **28**, 8214–8221.

Bucchianico, S. D., Poma, A. (2010). Atomic force microscope, a new tool for cytogenetic studies. In: Microscopy: Science, Technology, Applications and Education, Méndez Vilas, A. and Díaz, J. (Eds.), pp. 1287–1292. Formatex Research Center, Badajoz, Spain.

Burton, B. M., Baker, T. A. (2005). Remodeling protein complexes: insights from the AAA+ unfoldase ClpX and Mu transposase. *Protein Sci.* **14**, 1945–1954.

Burton, B. M., Williams, T. L., Baker, T. A. (2001). ClpX-mediated remodeling of Mu transpososomes: selective unfolding of subunits destabilizes the entire complex. *Mol. Cell* **8**, 449–454.

Butt, H. J., Downing, K. H., Hansma, P. K. (1990). Imaging the membrane protein bacteriorhodopsin with the atomic force microscope. *Biophys. J.* **58**, 1473–1480.

Camberg, J. L., Hoskins, J. R., Wickner, S. (2009). ClpXP protease degrades the cytoskeletal protein, FtsZ, and modulates FtsZ polymer dynamics. *Proc. Natl. Acad. Sci. USA* **106**, 10614–10619.

Casuso, I., Sens, P., Rico, F., Scheuring, S. (2010). Experimental evidence for membrane-mediated protein-protein interaction. *Biophys. J.* **99**, L47–L49.

Casuso, I., Rico, F., Scheuring, S. (2011). High-speed atomic force microscopy: structure and dynamics of single proteins. *Curr. Opin. Chem. Biol.* **15**, 704–709.

Chandrasekhar, G. N., Tilly, K., Woolford, C., Hendrix, R., Georgopoulos, C. (1986). Purification and properties of the groES morphogenetic protein of Escherichia coli. *J. Biol. Chem.* **261**, 12414–12419.

Chaudhry, C., Farr, G. W., Todd, M. J., Rye, H. S., Brunger, A. T., Adams, P. D., et al. (2003). Role of the [gamma]-phosphate of ATP in triggering protein folding by GroEL-GroES: function, structure and energetics. *EMBO J.* **22**, 4877–4887.

Chen, Z.-P., Levy, A., Lightman, S. L. (1995). Nucleotides as extracellular signalling molecules. *J. Neuroendocrinol.* **7**, 83–96.

Cheng, X. (1995). Structure and function of DNA methyltransferases. *Annu. Rev. Biophys. Biomol. Struct.* **24**, 293–318.

Chernov, A. A. (2003). Protein crystals and their growth. *J. Struct. Biol.* **142**, 3–21.

Chhabra, R., Sharma, J., Ke, Y., Liu, Y., Rinker, S., Lindsay, S., et al. (2007). Spatially addressable multiprotein nanoarrays templated by aptamer-tagged DNA nanoarchitectures. *J. Am. Chem. Soc.* **129**, 10304–10305.

Cockcroft, S., Gomperts, B. D. (1979). ATP induces nucleotide permeability in rat mast cells. *Nature* **279**, 541–542.

Cordell, S. C., Robinson, E. J. H., Löwe, J. (2003). Crystal structure of the SOS cell division inhibitor SulA and in complex with FtsZ. *Proc. Natl. Acad. Sci. USA* **100**, 7889–7894.

Divne, C., Ståhlberg, J., Teeri, T. T., Jones, T. A. (1998). High-resolution crystal structures reveal how a cellulose chain is bound in the 50 Å long tunnel of cellobiohydrolase I from Trichoderma reesei. *J. Mol. Biol.* **275**, 309–325.

Drenth, J. (1999). Principles of Protein X-ray Crystallography. 2nd edn. Springer-Verlag New York, Inc., New York.

Durbin, S. D., Carlson, W. E. (1992). Lysozyme crystal growth studied by atomic force microscopy. *J. Cryst. Growth* **122**, 71–79.

Endo, M., Katsuda, Y., Hidaka, K., Sugiyama, H. (2010a). Regulation of DNA methylation using different tensions of double strands constructed in a defined DNA nanostructure. *J. Am. Chem. Soc.* **132**, 1592–1597.

Endo, M., Katsuda, Y., Hidaka, K., Sugiyama, H. (2010b). A versatile DNA nanochip for direct analysis of DNA base-excision repair. *Angew. Chem. Int. Ed. Engl.* **49**, 9412–9416.

Endo, M., Sugita, T., Rajendran, A., Katsuda, Y., Emura, T., Hidaka, K., et al. (2011). Two-dimensional DNA origami assemblies using a four-way connector. *Chem. Commun.* **47**, 3213–3215.

Fantner, G. E., Barbero, R. J., Gray, D. S., Belcher, A. M. (2010). Kinetics of antimicrobial peptide activity measured on individual bacterial cells using high-speed atomic force microscopy. *Nat. Nanotechnol.* **5**, 280–285.

Flynn, J. M., Neher, S. B., Kim, Y.-I., Sauer, R. T., Baker, T. A. (2003). Proteomic discovery of cellular substrates of the ClpXP protease reveals five classes of ClpX-recognition signals. *Mol. Cell* **11**, 671–683.

Furuike, S., Hossain, M. D., Maki, Y., Adachi, K., Suzuki, T., Kohori, A., et al. (2008). Axle-less F1-ATPase rotates in the correct direction. *Science* **319**, 955–958.

García, J. J., Capaldi, R. A. (0301). Unisite catalysis without rotation of the γ-ε domain in Escherichia coli F1-ATPase. *J. Biol. Chem.* **273**, 15940–15945.

Gilmore, J. L., Suzuki, Y., Tamulaitis, G., Siksnys, V., Takeyasu, K., Lyubchenko, Y. L. (2009). Single-molecule dynamics of the DNA–EcoRII protein complexes revealed with high-speed atomic force microscopy. *Biochemistry* **48**, 10492–10498.

Gottlieb, C., Thomsen, L., Ingmer, H., Mygind, P., Kristensen, H.-H., Gram, L. (2008). Antimicrobial peptides effectively kill a broad spectrum of Listeria monocytogenes and Staphylococcus aureus strains independently of origin, sub-type, or virulence factor expression. *BMC Microbiol.* **8**, 205.

Gowers, D. M., Halford, S. E. (2003). Protein motion from non-specific to specific DNA by three-dimensional routes aided by supercoiling. *EMBO J.* **22**, 1410–1418.

Grallert, H., Buchner, J. (2001). Review: a structural view of the GroE chaperone cycle. *J. Struct. Biol.* **135**, 95–103.

Hansma, P. K., Schitter, G., Fantner, G. E., Prater, C. (2006). High-speed atomic force microscopy. *Science* **314**, 601–602.

Hartl, F. U., Hayer-Hartl, M. (2002). Molecular chaperones in the cytosol: from nascent chain to folded protein. *Science* **295**, 1852–1858.

Haupts, U., Tittor, J., Oesterhelt, D. (1999). Closing in on bacteriorhodopsin: progress in understanding the molecule. *Annu. Rev. Biophys. Biomol. Struct.* **28**, 367–399.

Hendrickson, W. A., Pähler, A., Smith, J. L., Satow, Y., Merritt, E. A., Phizackerley, R. P. (1989). Crystal structure of core streptavidin determined from multiwavelength anomalous diffraction of synchrotron radiation. *Proc. Natl. Acad. Sci. USA* **86**, 2190–2194.

Hightower, L. E. (1991). Heat shock, stress proteins, chaperones, and proteotoxicity. *Cell* **66**, 191–197.

Igarashi, K., Koivula, A., Wada, M., Kimura, S., Penttilä, M., Samejima, M. (2009). High speed atomic force microscopy visualizes processive movement of Trichoderma reesei cellobiohydrolase I on crystalline cellulose. *J. Biol. Chem.* **284**, 36186–36190.

Imai, T., Boisset, C., Samejima, M., Igarashi, K., Sugiyama, J. (1998). Unidirectional processive action of cellobiohydrolase Cel7A on Valonia cellulose microcrystals. *FEBS Lett.* **432**, 113–116.

Izadpanah, A., Gallo, R. L. (2005). Antimicrobial peptides. *J. Am. Acad. Dermatol.* **52**, 381–390.

Jenal, U., Fuchs, T. (1998). An essential protease involved in bacterial cell-cycle control. *EMBO J.* **17**, 5658–5669.

Joo, C., Balci, H., Ishitsuka, Y., Buranachai, C., Ha, T. (2008). Advances in single-molecule fluorescence methods for molecular biology. *Annu. Rev. Biochem.* **77**(1), 51–76.

Kaibara, C., Matsui, T., Hisabori, T., Yoshida, M. (1996). Structural asymmetry of F1-ATPase caused by the γ subunit generates a high affinity nucleotide binding site. *J. Biol. Chem.* **271**, 2433–2438.

Kalfa, V. C., Jia, H. P., Kunkle, R. A., McCray, P. B., Tack, B. F., Brogden, K. A. (2001). Congeners of SMAP29 kill ovine pathogens and induce ultrastructural damage in bacterial cells. *Antimicrob. Agents Chemother.* **45**, 3256–3261.

Karymov, M. A., Tomschik, M., Leuba, S. H., Caiafa, P., Zlatanova, J. (2001). DNA methylation-dependent chromatin fiber compaction in vivo and in vitro: requirement for linker histone. *FASEB J.* **15**, 2631–2641.

Khakh, B. S., Alan North, R. (2006). P2X receptors as cell-surface ATP sensors in health and disease. *Nature* **442**, 527–532.

Kodera, N., Yamamoto, D., Ishikawa, R., Ando, T. (2010). Video imaging of walking myosin V by high-speed atomic force microscopy. *Nature* **468**, 72–76.

Ku, A. C., Darst, S. A., Robertson, C. R., Gast, A. P., Kornberg, R. D. (1993). Molecular analysis of two-dimensional protein crystallization. *J. Phys. Chem.* **97**, 3013–3016.

Liu, Y., Lin, C., Li, H., Yan, H. (2005). Aptamer-directed self-assembly of protein arrays on a DNA nanostructure. *Angew. Chem. Int. Ed. Engl.* **44**, 4333–4338.

Liu, W., Zhong, H., Wang, R., Seeman, N. C. (2011). Crystalline two-dimensional DNA-origami arrays. *Angew. Chem. Int. Ed. Engl.* **50**, 264–267.

Lou, C., Wang, Z., Wang, S.-W. (2007). Two-dimensional protein crystals on a solid substrate: effect of surface ligand concentration. *Langmuir* **23**, 9752–9759.

Lund, K., Liu, Y., Lindsay, S., Yan, H. (2005). Self-assembling a molecular pegboard. *J. Am. Chem. Soc.* **127**, 17606–17607.

Ma, X., Margolin, W. (1999). Genetic and functional analyses of the conserved C-terminal core domain of Escherichia coli FtsZ. *J. Bacteriol.* **181**, 7531–7544.

Malkin, A. J., McPherson, A. (2002). Novel mechanisms for defect formation and surface molecular processes in virus crystallization. *J. Phys. Chem. B* **106**, 6718–6722.

Mangoni, M. L., Papo, N., Barra, D., Simmaco, M., Bozzi, A., Di Giulio, A., et al. (2004). Effects of the antimicrobial peptide temporin L on cell morphology, membrane permeability and viability of Escherichia coli. *Biochem. J.* **380**, 859–865.

Meincken, M., Holroyd, D. L., Rautenbach, M. (2005). Atomic force microscopy study of the effect of antimicrobial peptides on the cell envelope of Escherichia coli. *Antimicrob. Agents Chemother.* **49**, 4085–4092.

Miyagi, A., Ando, T., Lyubchenko, Y. L. (2011). Dynamics of nucleosomes assessed with time-lapse high-speed atomic force microscopy. *Biochemistry* **50**, 7901–7908.

Motojima, F., Yoshida, M. (2003). Discrimination of ATP, ADP, and AMPPNP by Chaperonin GroEL. *J. Biol. Chem.* **278**, 26648–26654.

Mukherjee, A., Lutkenhaus, J. (1994). Guanine nucleotide-dependent assembly of FtsZ into filaments. *J. Bacteriol.* **176**, 2754–2758.

Müller, D. J., Schabert, F. A., Büldt, G., Engel, A. (1995). Imaging purple membranes in aqueous solutions at sub-nanometer resolution by atomic force microscopy. *Biophys. J.* **68**, 1681–1686.

Numajiri, K., Kimura, M., Kuzuya, A., Komiyama, M. (2010). Stepwise and reversible nanopatterning of proteins on a DNA origami scaffold. *Chem. Commun.* **46**, 5127–5129.

Park, S. H., Pistol, C., Ahn, S. J., Reif, J. H., Lebeck, A. R., Dwyer, C., et al. (2006). Finite-size, fully addressable DNA tile lattices formed by hierarchical assembly procedures. *Angew. Chem. Int. Ed. Engl.* **45**, 735–739.

Park, H., Toprak, E., Selvin, P. R. (2007). Single-molecule fluorescence to study molecular motors. *Q. Rev. Biophys.* **40**, 87–111.

Picco, L. M., Bozec, L., Ulcinas, A., Engledew, D. J., Antognozzi, M., Horton, M. A., et al. (2007). Breaking the speed limit with atomic force microscopy. *Nanotechnology* **18**.

Pingoud, A., Jeltsch, A. (2001). Structure and function of type II restriction endonucleases. *Nucleic Acids Res.* **29**, 3705–3727.

Queisser, H. J., Haller, E. E. (1998). Defects in semiconductors: some fatal, some vital. *Science* **281**, 945–950.

Rajendran, A., Endo, M., Katsuda, Y., Hidaka, K., Sugiyama, H. (2011a). Photo-cross-linking-assisted thermal stability of DNA origami structures and its application for higher-temperature self-assembly. *J. Am. Chem. Soc.* **133**, 14488–14491.

Rajendran, A., Endo, M., Katsuda, Y., Hidaka, K., Sugiyama, H. (2011b). Programmed two-dimensional self-assembly of multiple DNA origami Jigsaw pieces. *ACS Nano* **5**, 665–671.

Rajendran, A., Endo, M., Sugiyama, H. (2012a). DNA origami: synthesis and self-assembly. *Curr. Protoc. Nucleic Acid Chem.* **48**, 12.9.1–12.9.18.
Rajendran, A., Endo, M., Sugiyama, H. (2012b). Single-molecule analysis using DNA origami. *Angew. Chem. Int. Ed. Engl.* **51**, 874–890.
Rinker, S., Ke, Y., Liu, Y., Chhabra, R., Yan, H. (2008). Self-assembled DNA nanostructures for distance-dependent multivalent ligand-protein binding. *Nat. Nanotechnol.* **3**, 418–422.
Roberts, G., Lian, L.-Y. (2011). Protein NMR Spectroscopy: Practical Techniques and Applications. John Wiley & Sons Ltd., West Sussex, UK.
Roberts, R. J., Belfort, M., Bestor, T., Bhagwat, A. S., Bickle, T. A., Bitinaite, J., et al. (2003). A nomenclature for restriction enzymes, DNA methyltransferases, homing endonucleases and their genes. *Nucleic Acids Res.* **31**, 1805–1812.
Rothemund, P. W. (2006). Folding DNA to create nanoscale shapes and patterns. *Nature* **440**, 297–302.
Sannohe, Y., Endo, M., Katsuda, Y., Hidaka, K., Sugiyama, H. (2010). Visualization of dynamic conformational switching of the G-quadruplex in a DNA nanostructure. *J. Am. Chem. Soc.* **132**, 16311–16313.
Scheuring, S., Müller, D. J., Ringler, P., Heymann, J. B., Engel, A. (1999). Imaging streptavidin 2D crystals on biotinylated lipid monolayers at high resolution with the atomic force microscope. *J. Microsc.* **193**, 28–35.
Schulz, G. E., Schirmer, R. H. (1979). Principles of protein structure. In: Springer Advanced Texts in Chemistry, Cantor, C. R. (Ed.). Springer-Verlag New York, Inc., New York.
Shibata, M., Yamashita, H., Uchihashi, T., Kandori, H., Ando, T. (2010). High-speed atomic force microscopy shows dynamic molecular processes in photoactivated bacteriorhodopsin. *Nat. Nanotechnol.* **5**, 208–212.
Shibata, M., Uchihashi, T., Yamashita, H., Kandori, H., Ando, T. (2011). Structural changes in bacteriorhodopsin in response to alternate illumination observed by high-speed atomic force microscopy. *Angew. Chem. Int. Ed. Engl.* **50**, 4410–4413.
Shinozaki, Y., Sumitomo, K., Tsuda, M., Koizumi, S., Inoue, K., Torimitsu, K. (2009). Direct observation of ATP induced conformational changes in single $P2X_4$ receptors. *PLoS Biol.* **7**, e1000103.
Shlyakhtenko, L. S., Gilmore, J., Portillo, A., Tamulaitis, G., Siksnys, V., Lyubchenko, Y. L. (2007). Direct visualization of the *Eco*RII–DNA triple synaptic complex by atomic force microscopy. *Biochemistry* **46**, 11128–11136.
Srisodsuk, M., Reinikainen, T., Penttilä, M., Teeri, T. T. (1993). Role of the interdomain linker peptide of Trichoderma reesei cellobiohydrolase I in its interaction with crystalline cellulose. *J. Biol. Chem.* **268**, 20756–20761.
Subramaniam, S., Gerstein, M., Oesterhelt, D., Henderson, R. (1993). Electron diffraction analysis of structural changes in the photocycle of bacteriorhodopsin. *EMBO J.* **12**, 1–8.
Sugimoto, S., Yamanaka, K., Nishikori, S., Miyagi, A., Ando, T., Ogura, T. (2010). AAA + chaperone ClpX regulates dynamics of prokaryotic cytoskeletal protein FtsZ. *J. Biol. Chem.* **285**, 6648–6657.
Tamulaitis, G., Sasnauskas, G., Mucke, M., Siksnys, V. (2006). Simultaneous binding of three recognition sites is necessary for a concerted plasmid DNA cleavage by EcoRII restriction endonuclease. *J. Mol. Biol.* **358**, 406–419.

Tatsuo, U., Masatsugu, S., Osamu, H. (2008). Techniques for imaging human metaphase chromosomes in liquid conditions by atomic force microscopy. *Nanotechnology* **19**, 384022.

Terry, B. J., Jack, W. E., Modrich, P. (1985). Facilitated diffusion during catalysis by EcoRI endonuclease. Nonspecific interactions in EcoRI catalysis. *J. Biol. Chem.* **260**, 13130–13137.

Thorgeirsson, T. E., Xiao, W., Brown, L. S., Needleman, R., Lanyi, J. K., Shin, Y.-K. (1997). Transient channel-opening in bacteriorhodopsin: an EPR study. *J. Mol. Biol.* **273**, 951–957.

Uchihashi, T., Iino, R., Ando, T., Noji, H. (2011). High-speed atomic force microscopy reveals rotary catalysis of rotorless F-ATPase. *Science* **333**, 755–758.

Viani, M. B., Schaffer, T. E., Paloczi, G. T., Pietrasanta, L. I., Smith, B. L., Thompson, J. B., et al. (1999). Fast imaging and fast force spectroscopy of single biopolymers with a new atomic force microscope designed for small cantilevers. *Rev. Sci. Instrum.* **70**, 4300–4303.

Viani, M. B., Pietrasanta, L. I., Thompson, J. B., Chand, A., Gebeshuber, I. C., Kindt, J. H., et al. (2000). Probing protein-protein interactions in real time. *Nat. Struct. Biol.* **7**, 644–647.

Wang, S.-W., Robertson, C. R., Gast, A. P. (1999). Molecular arrangement in two-dimensional streptavidin crystals. *Langmuir* **15**, 1541–1548.

Weart, R. B., Nakano, S., Lane, B. E., Zuber, P., Levin, P. A. (2005). The ClpX chaperone modulates assembly of the tubulin-like protein FtsZ. *Mol. Microbiol.* **57**, 238–249.

Wickner, S., Maurizi, M. R., Gottesman, S. (1999). Posttranslational quality control: folding, refolding, and degrading proteins. *Science* **286**, 1888–1893.

Xiao, W., Brown, L. S., Needleman, R., Lanyi, J. K., Shin, Y.-K. (2000). Light-induced rotation of a transmembrane α-helix in bacteriorhodopsin. *J. Mol. Biol.* **304**, 715–721.

Yamamoto, D., Uchihashi, T., Kodera, N., Ando, T. (2008). Anisotropic diffusion of point defects in a two-dimensional crystal of streptavidin observed by high-speed atomic force microscopy. *Nanotechnology* **19**, 1–9.

Yamamoto, D., Nagura, N., Omote, S., Taniguchi, M., Ando, T. (2009). Streptavidin 2D crystal substrates for visualizing biomolecular processes by atomic force microscopy. *Biophys. J.* **97**, 2358–2367.

Yatcilla, M. T., Robertson, C. R., Gast, A. P. (1998). Influence of pH on two-dimensional streptavidin crystals. *Langmuir* **14**, 497–503.

Yau, S. T., Petsev, D. N., Thomas, B. R., Vekilov, P. G. (2000). Molecular-level thermodynamic and kinetic parameters for the self-assembly of apoferritin molecules into crystals. *J. Mol. Biol.* **303**, 667–678.

Yildiz, A., Forkey, J. N., McKinney, S. A., Ha, T., Goldman, Y. E., Selvin, P. R. (2003). Myosin V walks hand-over-hand: single fluorophore imaging with 1.5-nm localization. *Science* **300**, 2061–2065.

Yokokawa, M., Wada, C., Ando, T., Sakai, N., Yagi, A., Yoshimura, S. H., et al. (2006a). Fast-scanning atomic force microscopy reveals the ATP/ADP-dependent conformational changes of GroEL. *EMBO J.* **25**, 4567–4576.

Yokokawa, M., Yoshimura, S. H., Naito, Y., Ando, T., Yagi, A., Sakai, N., et al. (2006b). Fast-scanning atomic force microscopy reveals the molecular mechanism of DNA cleavage by ApaI endonuclease. *IEE Proc. Nanobiotechnol.* **153**, 60–66.

Yuri, L. L. (2011). Preparation of DNA and nucleoprotein samples for AFM imaging. *Micron* **42**, 196–206.

Zhou, Y., Gottesman, S., Hoskins, J. R., Maurizi, M. R., Wickner, S. (2001). The RssB response regulator directly targets ςS for degradation by ClpXP. *Genes Dev.* **15**, 627–637.

REACTION INTERMEDIATES DISCOVERED IN CRYSTAL STRUCTURES OF ENZYMES

By IRENE T. WEBER,* JOHNSON AGNISWAMY,* GUOXING FU,* CHEN-HSIANG SHEN,* AND ROBERT W. HARRISON*,†

*Department of Biology, Georgia State University, Atlanta, Georgia, USA
†Department of Computer Science, Georgia State University, Atlanta, Georgia, USA

Abstract

Crystal structures of enzymes have provided valuable information for the reaction mechanisms. Structures of the enzyme complex with different reaction intermediates are particularly valuable. In several cases, these structures of intermediates were discovered accidently, presumably by trapping in the crystal during freezing prior to X-ray data collection. High to atomic resolution structures reveal the detailed geometry of the reaction intermediate and its interactions within the enzyme active site. In other cases, the protein can be crystallized with its substrate, including examples of protease precursors that represent their own substrates. Examples are described of an FAD-dependent dehydrogenase, HIV protease and caspases, where the structures provide snapshots of steps in the reaction and the conformational changes occurring during the reaction. Complementary techniques such as computational chemistry, neutron crystallography, Laue crystallography, and time-resolved spectroscopy can give a more complete picture of the reaction.

ADVANCES IN PROTEIN CHEMISTRY AND STRUCTURAL BIOLOGY, Vol. 87
DOI: 10.1016/B978-0-12-398312-1.00003-2

Copyright 2012, Elsevier Inc.
All rights reserved.

I. Introduction

X-ray crystallography is the predominant method for determining the structures of macromolecules with numerous successes for proteins and nucleic acids of various sizes from small Z-DNA hexamers to the huge protein–RNA complex of the ribosome (Yonath, 2011). Currently, almost 80,000 crystal structures of macromolecules have been deposited in the Protein Data Bank (PDB; Berman et al., 2000). These structures include several hundred identified with the search term of "reaction intermediate." Many of these structures exhibit inhibitors designed to mimic an intermediate of the catalytic reaction. Other enzyme structures reveal "natural" intermediates that were trapped during the preparation of the crystals. These structures provide important details for understanding the catalytic reaction and the geometry of the intermediate states. The high-resolution structure of an enzyme complex with its intermediate(s) has many applications in bioinformatics, computational biology, and medicine. Knowledge of the geometry of the reaction intermediate and its interactions within the enzyme active site has benefits in rational design of transition state analogs for development as effective therapeutic agents. This approach proved successful in targeting aspartic proteases for several diseases (Hoegl et al., 1999; Nguyen et al., 2008). Neuraminidase inhibitors for influenza has also benefited from knowledge of transition state analogs (Klumpp and Graves, 2006). Information on the enzyme residues that interact with intermediates is valuable to improve genomic annotation of the function of genes. Also, computational analysis of enzyme complexes with reaction intermediates can be applied in engineering enzymes of novel function. For example, cytochrome P450cam has been engineered to act on unnatural substrates with potential applications in chemical synthesis and bioremediation (Bell et al., 2003).

The enzyme reaction is expected to proceed from binding of one or more substrates, through intermediate steps in the reaction and finally formation and release of products, as indicated in the following scheme:

$$\text{Simple general scheme for an enzyme reaction:}$$
$$E + \mathrm{S} \leftrightarrow \mathrm{E}\cdot\mathrm{S} \leftrightarrow \mathrm{E}\cdot\mathrm{I} \leftrightarrow \mathrm{E}\cdot\mathrm{P} \leftrightarrow \mathrm{E} + \mathrm{P}.$$

The reaction commences when the enzyme (E) binds a substrate (S). One or more substrates may be required for the reaction to occur. The catalytic reaction proceeds via one or more steps involving enzyme-bound intermediates (E·I) toward the enzyme–product (E·P) complex. Finally, the product or products are released from the enzyme.

Generally, it is difficult to visualize the reaction intermediates since they are assumed to be transient, short-lived species. Also, it is difficult to crystallize an active enzyme together with its natural substrate or intermediates since the reaction can proceed even in the crystal and products may be released. For example, Mozzarelli et al. (1979) used single crystal microspectroscopy to compare reactions in the crystal and in solution for slowly reacting substrates of mitochondrial aspartate aminotransferase. The enzyme was catalytically competent in the crystal with similar rates to those in solution. The possibility of trapping intermediates will depend on their stability in the enzyme. It is presumed that critical slow steps in the reaction proceed via quasi-stable intermediates that can exist within the enzyme in the crystal environment. An exception, where intermediates can be observed more readily, is when a stable covalent intermediate is formed during the reaction. Otherwise, two common approaches are (1) to mutate the catalytic residue producing inactive enzyme for crystallizing with substrate, or (2) to synthesize a nonreactive analog of the substrate for cocrystallizing with the active enzyme. The product or products of a reaction can also be visible in the enzyme structure, again depending on the slow step(s) in the reaction. Further, many enzymes have flexible structural regions that alter conformation on the binding of substrate, or possibly during the reaction. Therefore, the crystals must accommodate the appropriate conformation of the enzyme, which means that not every crystal structure of an enzyme will be suitable for analysis of the catalytic reaction. Despite these challenges, many crystal structures have been solved of enzymes in complex with various reaction intermediates.

Here, we discuss the possible approaches to acquiring a crystal structure of an enzyme reaction intermediate as well as the advantages and disadvantages of high to atomic resolution X-ray or neutron diffraction data. Examples of successful identification of intermediates will be described focusing on our recent results. Our studies of the trapped reaction intermediates of two different proteases (Fang et al., 2006; Kovalevsky et al., 2007; Agniswamy et al., 2009), and a dehydrogenase (Fu et al., 2010) will be discussed. Finally, we describe several complementary techniques with high potential for studying reaction intermediates in the crystal.

II. Strategies for Trapping Intermediates

Reaction intermediates are, by definition, assumed to be transient stages in the catalytic process. Ideally, one or more intermediates would be revealed in different structures of an enzyme to cover multiple steps in the reaction. In several cases, intermediates have been discovered by chance. The intermediates can be identified as extra electron density visible in the active site of the enzyme during solution and refinement of the structure, as reported for streptogrisin B (Lee and James, 2008), HIV protease (PR) (Kovalevsky et al., 2007), bacterial carboxylesterase (Liu et al., 2004), and Dbv29, a flavin-containing oxidase (Liu et al., 2011). These intermediates presumably were captured in the active site during expression of the enzyme and retained during protein purification and growth of the crystals. Trapping of intermediates accumulated during the steady state of the reaction in enzyme crystals is favored by freezing, which has become a routine technique for synchrotron data collection (Wilmot and Pearson, 2002). The formation of semi-stable intermediates in the enzyme complex is promoted by using slow substrates, as reported for aromatic amine dehydrogenase (Roujeinikova et al., 2007) and HIV PR (Kovalevsky et al., 2007). Analogs of reaction intermediates can also be cocrystallized to study the enzyme mechanism, as reported for triosephosphate isomerase (Alahuhta and Wierenga, 2010). Recently, scanning different cryo-temperatures was used to capture different intermediates and light-induced conformational changes in bacteriophytochrome (Yang et al., 2011). These studies demonstrate the power of cryo-crystallography in revealing reaction states of enzymes.

III. Importance of High-Resolution Crystal Structures

High-resolution crystal structures are critical for revealing the detailed geometry of the bound reaction intermediates and catalytic residues. High-resolution diffraction data extend to better than 2.0 Å, while about 1.0 Å is considered to be atomic resolution. Several enzyme structures have been determined at subatomic resolutions. These structures provide highly accurate geometry for the active site, which can be critical for quantum mechanics (QM) calculations. The solvent structure is likely to be clearly visible, and water or ions are critical components of many enzyme reaction mechanisms. At high resolutions it will be possible to identify unusual features of the active site geometry, such as low barrier (unusually short) hydrogen bonds that can be important in enzyme catalysis (Cleland et al., 1998). Unusually short hydrogen bonds between the reaction intermediate and the catalytic residues have been identified in crystal structures of triosephosphate isomerase (Alahuhta and Wierenga, 2010) and HIV PR

(Das et al., 2006; Kovalevsky et al., 2007). In some enzymes, putative hydrogen atoms have been identified in atomic resolution crystal structures, such as for pancreatic elastase (Katona et al., 2002) or choline oxidase (Lyubimov et al., 2006). Another advantage is the ability to examine the protonation state of catalytic residues. In one model study, the protonation state of the catalytic histidine of RNase A was assessed using X-ray crystallography at different pH values (Berisio et al., 1999). The identification of hydrogen atoms in X-ray structures is debatable, however, even at atomic resolution, and neutron diffraction data are preferred for this purpose (Blakeley et al., 2004), as discussed later.

Despite the obvious advantages of atomic resolution crystal structures, there are several, less obvious, disadvantages. Clearly, not every enzyme crystal will be of suitable quality for X-ray data collection to atomic resolution. Moreover, the refinement of an atomic resolution structure is complicated by the necessity of modeling details such as a large amount of ordered solvent and potential disorder in the protein, the reaction intermediate, and the solvent. At one possible extreme, the highest resolution crystal structure of HIV PR to date (0.84 Å resolution) comprised two alternative conformations of the entire PR dimer that bound one or two inhibitor molecules, respectively (Kovalevsky et al., 2006). This type of disorder will preclude identification of protons and limits the accuracy of the active site geometry. Fortunately, this extreme disorder is not observed in most atomic or subatomic resolution enzyme structures.

IV. Selected Success Stories

The application of crystal structures of enzyme complexes with their reaction intermediates to understand enzyme mechanism is best illustrated by specific examples. We have achieved such structures, by serendipity in most cases, for a variety of enzymes. The earliest case was our discovery of a tetrahedral reaction intermediate in the newly determined crystal structure of the bacterial carboxylesterase Est30 (Liu et al., 2004). This enzyme is a member of the large family of serine hydrolases. We have applied substrate-based inhibitors to investigate the specificity of caspase-3 and -7 (Fang et al., 2006; Agniswamy et al., 2007). In addition, we have identified a precursor of the caspase, which is presumed to be activated by self-cleavage in the expression system (Agniswamy et al., 2009). Autoproteolytically processed precursors are common for all classes of proteases.

Transition state analog inhibitors bearing a hydroxyl group to mimic the gem-diol tetrahedral intermediate (TI) are used for HIV PR, which is a member of the aspartic protease family. Also, we and others have discovered the TIs of slowly processed substrates in the active site of HIV PR (Das et al., 2006, 2010; Kovalevsky et al., 2007; Prashar et al., 2009). Most recently, our crystal structure of bacterial D-arginine dehydrogenase (DADH) showed an unexpected bound product, and two different imino acid product complexes were obtained intentionally (Fu et al., 2010). These examples are described in the following sections.

A. *Redox Flavoenzymes: D-Arginine Dehydogenase*

The family comprising FAD-dependent oxidases and dehydrogenases plays a variety of important metabolic roles in all organisms. These enzymes contain a conserved FAD-binding domain and a substrate-binding domain with similar topology but variable sequence (Macheroux et al., 2001). A recently identified member of this family is the DADH. DADH is the catabolic enzyme involved in the first step of D-to-L inversion of D-arginine in *Pseudomonas aeruginosa,* which is a common opportunistic human pathogen (Li and Lu, 2009). This flavoenzyme converts D-arginine and other D-amino acids into imino acids, which dissociate from the enzyme, and are hydrolyzed to keto acids in a nonenzymatic reaction. Then, the keto acid is converted to the L-amino acid by the coupled anabolic enzyme, an NAD(P)H-dependent L-arginine dehydrogenase. The reaction is illustrated in Fig. 1 for D-arginine, which is the best substrate for DADH. DADH is characterized by broad substrate specificity, being able to oxidize basic and hydrophobic D-amino acids of various sizes, but not reacting with acidic residues (Fu et al., 2010; Li et al., 2010). The

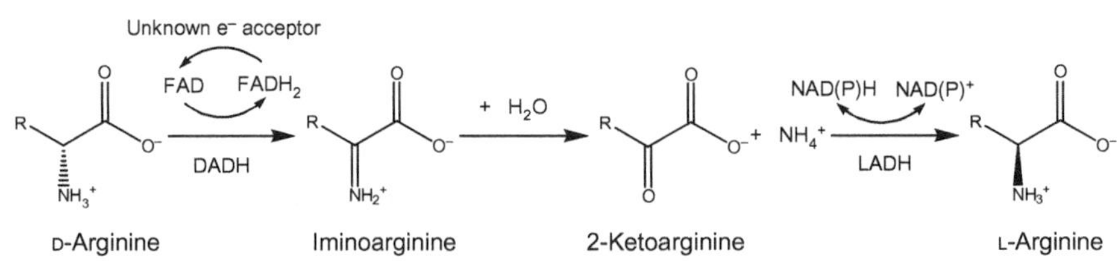

FIG. 1. Reaction scheme for conversion of D-amino acids to L-amino acids by DADH and LADH.

structures of DADH cocrystallized with D-arginine, D-histidine, or D-leucine were determined at the resolutions of 1.06–1.30 Å (Fu et al., 2010, 2011), which permitted detailed analysis of the conformational changes during the reaction and the substrate specificity of this enzyme.

DADH binds the cofactor FAD and its substrates bind near the flavin ring at the interface of the two domains (Fig. 2A). Well-defined electron density maps demonstrated that the imino acid products of the reaction were captured in the enzyme active site in the crystal structures. The first crystal structure unexpectedly revealed both ligand-free and iminoarginine-bound conformations (Fu et al., 2010). Comparison of structures of the ligand-free DADH and its complexes with imino acids showed that the protein undergoes significant conformational changes upon binding of the ligand. Two regions around the binding site showed two alternate conformations, which were well defined at the atomic resolution of this crystal structure. Residues 50–56 were designated as an active site lid controlling the substrate accessibility, similar to those reported in other

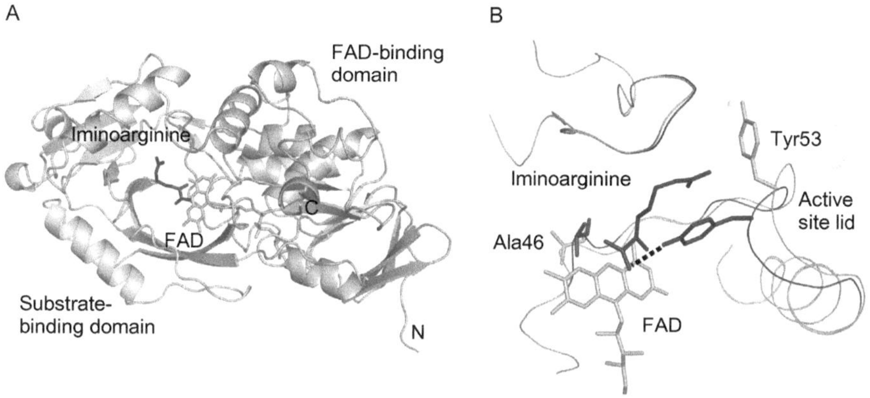

FIG. 2. (A) Overall structure of DADH shown in cartoon representation (PDB code: 3NYE; Fu et al., 2010). The enzyme is composed of an FAD-binding domain (gold) and a substrate-binding domain (cyan). Iminoarginine and cofactor FAD are shown as sticks and colored green and red, respectively. (B) Comparison of ligand-free conformation (cyan) and product-bound conformation (red) near the DADH active site (PDB code: 3NYC; Fu et al., 2010). Ala46, Tyr53, and iminoarginine are represented as sticks. A hydrogen bond (black dotted line) is formed between Tyr53 and iminoarginine in the product-bound conformation. (For interpretation of the references to color in this figure legend, the reader is referred to the online version of this chapter.)

flavin-dependent enzymes (Mattevi et al., 1996). The active site lid is flipped away from the active site in the ligand-free conformation and moves inward so that the hydroxyl of Tyr53 interacts with the bound imino acid in the product-bound conformation (Fig. 2B). Two conformations were also seen for a Ser45-Ala46 switch at the *si* face of the flavin ring, in which the hydrogen bond (O_{Ser45}—$N5_{FAD}$) in the ligand-free conformation is replaced by another polar interaction (N_{Ala46}—$N5_{FAD}$) in the product-bound conformation.

Iminoarginine binds DADH with the plane formed by its Cα atom, carboxyl group and imino group approximately parallel to the *re* face of the flavin, while the side chain points away from FAD (Fig. 3A). Several

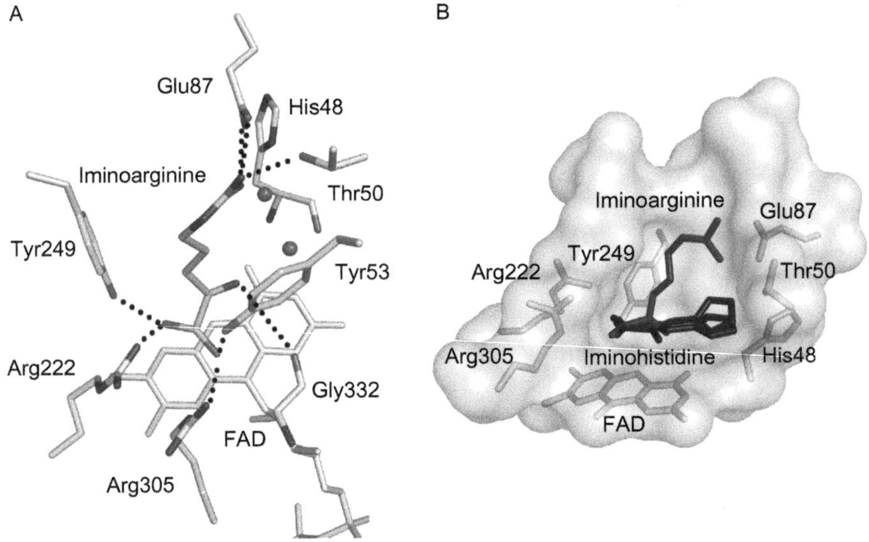

FIG. 3. (A) Interactions between DADH and iminoarginine product. Dotted lines represent hydrogen bonds and ionic interactions. (B) Comparison of DADH complexes with iminoarginine (PDB code: 3NYE) and iminohistidine (PDB code: 3NYF). DADH residues (light blue sticks) line the interior of the substrate binding pocket (Thr53, Gly332, and Gln336 were omitted for clarity). Glu87 (green) forms an ionic interaction with iminoarginine. Iminoarginine (red) and iminohistidine (blue) bind to the active site in very distinct conformations. FAD is represented by its isoalloxazine ring. (For interpretation of the references to color in this figure legend, the reader is referred to the online version of this chapter.)

DADH residues including Tyr53, Arg222, Tyr249, Arg305, and Gly332 form extensive polar interactions with iminoarginine and anchor the ligand in the active site. Two water molecules are located near the ligand imino group, and form a hydrogen bond network extending to the imidazole side chain of His48. An ionic interaction was observed between the guanidinium group of iminoarginine and the carboxylate group of Glu87, which was proposed to be responsible for the high specificity of DADH for positively charged D-amino acids. Similar residues determining substrate specificity have been identified in other enzymes: Asp189 in trypsin is responsible for its narrow selection for positively charged arginine and lysine (Lu et al., 1997), while Glu32 in choline oxidase provides an important ionic interaction with the positively charged choline (Quaye et al., 2008). The iminohistidine product of a weaker substrate binds to DADH in a distinct mode compared to iminoarginine. At least two conformations of iminohistidine were observed and nearly all its atoms lie parallel to the FAD isoalloxazine ring. Examination of the substrate-binding site revealed a triangular-like internal cavity with a narrow entrance at the top (Fig. 3B). The main chain atoms of the two ligands form conserved interactions with DADH residues, while their side chains display very different binding modes. The shape and volume of the binding site allows DADH to accommodate a variety of D-amino acids, while the substrate specificity depends on residues interacting with the ligand side chain. The binding site and different conformations of bound iminoarginine and iminohistidine are consistent with the kinetic studies on 19 D-amino acids (Fu et al., 2010) and detailed kinetic analysis (Yuan et al., 2011).

Interestingly, an acyl adduct rather than the imino product was identified in the DADH cocrystallized with D-leucine structure, which showed a covalent bond between the Cα atom of the decarboxylated ketoleucine product and the FAD N(5) atom (Fig. 4; Fu et al., 2011). The formation of this byproduct is likely due to a nucleophilic substitution in which the N (5) atom of the reduced flavin formed in the X-ray beam reacts with the Cα of ketoleucine, yielding 4-methyl-2-pentanone-FAD and CO_2. Photoreduction of DADH in the presence of 4-methyl-2-oxopentanoic acid (ketoleucine) successfully reproduced this acyl adduct in solution. Comparison of this structure with the DADH/iminoarginine complex indicated that ketoleucine has fewer interactions with DADH residues due to the covalent bond with FAD. However, a water molecule near the ketoleucine main chain was found to mediate a hydrogen bond network between the acyl

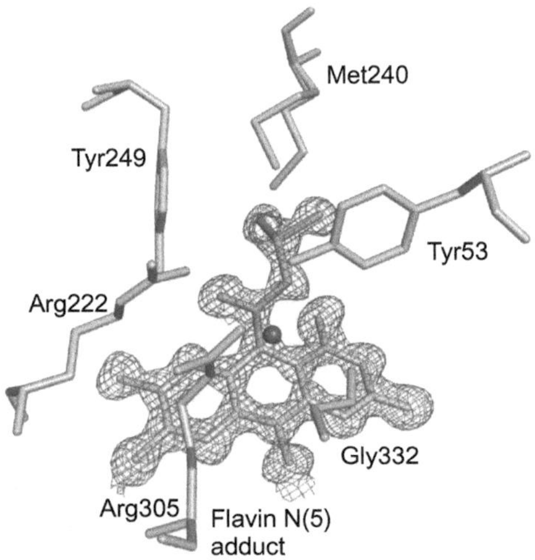

FIG. 4. Active site of the DADH/ketoleucine structure (PDB code: 3SM8; Fu et al., 2011). DADH active site residues and FAD are colored light orange and the ketoleucine part of the acyl adduct is colored green. Water molecule is shown as a red sphere. Fo–Fc omit map of the flavin N(5) adduct is contoured at 3σ and indicated as blue mesh. (For interpretation of the references to color in this figure legend, the reader is referred to the online version of this chapter.)

oxygen and the enzyme residues Tyr53, Arg305, and Gly332. A similar structure was described previously in the complex of D-amino acid oxidase with 3-methyl-2-oxobutyric acid (ketovaline) (PDB entry: 1DAO; Todone et al., 1997). Further, it has been reported that α-keto acids with small hydrophobic side chains are competitive inhibitors of DAAO and can react with flavins in the presence of light to form a covalent bond between the Cα atom and the flavin N(5) atom (Dixon and Kleppe, 1965; Todone et al., 1997).

The crystal structures shed light on several aspects of the reaction mechanism. The ping–pong bi–bi kinetic mechanism was established by steady-state kinetic studies with D-arginine or D-histidine as substrate (Yuan et al., 2011). A product release experiment with D-arginine indicates that the release of imino product is partially rate limiting for the overall turnover of the enzyme, which is in agreement with the observation that imino acids are trapped in the crystal structures. Also, DADH was proved to be a true dehydrogenase due to its lack of oxygen reactivity (Yuan et al.,

2011). Structural analysis suggested that Ala46 may block access of O_2 to the flavin C4a atom, while the lack of positive charge proximal to the flavin C4a and N1—C2 atoms contributes to its poor oxygen reactivity. In DAAO, Val substitution of Gly52, which is structurally equivalent to DADH Ala46, leads to the loss of reactivity to oxygen due to the steric hindrance blocking oxygen access (Saam et al., 2010). Further, substitution of the structurally equivalent Ala113 with Gly in L-galactono-γ-lactone dehydrogenase increased the reactivity of the reduced flavin with oxygen by about 400-fold (Leferink et al., 2009).

The structural studies on DADH have facilitated our understanding of the enzyme reaction with D-amino acids. The atomic resolution structure of DADH explicitly shows the open and closed forms of the active site lid that controls the binding of substrate and release of product, which has been proposed in other enzymes like DAAO without direct evidence. The ionic interaction of iminoarginine with Glu87 in DADH is supported by the kinetic study showing that D-arginine and D-lysine are the best substrates; the k_{cat}/K_{Arg} value is about 1000-fold higher than the k_{cat}/K_{His} value for the poorer substrate D-histidine. Moreover, comparison of DADH with related enzymes like DAAO and L-amino acid oxidase has revealed key components and structural arrangements that are likely responsible for their specificity toward different enantiomers and different amino acids (Fu et al., 2010). Overall, the high-resolution structures for DADH have provided valuable information on the substrate specificity and reaction pathways of flavin-dependent enzymes.

B. Aspartate Proteases: HIV Protease

Human immunodeficiency virus type 1 (HIV-1) PR has become one of the best characterized members of the aspartate protease family since HIV-1 was identified as the causative agent of AIDS. The enzymatically active form is a homodimer with 99 amino acids in each subunit and the active site composed of the two conserved catalytic triplets (Asp25-Thr26-Gly27) is located at the interface between the subunits (reviewed in Weber and Wang, 2010; Fig. 5). PR has an essential role in viral replication by precisely cleaving the viral precursor Gag and Gag–Pol polyproteins to liberate the functional proteins (Kohl et al., 1988). Thus, this enzyme has proved an excellent target for AIDS therapy and crystal structures are available for the ligand-free and inhibited enzyme conformations (Weber et al., 2009).

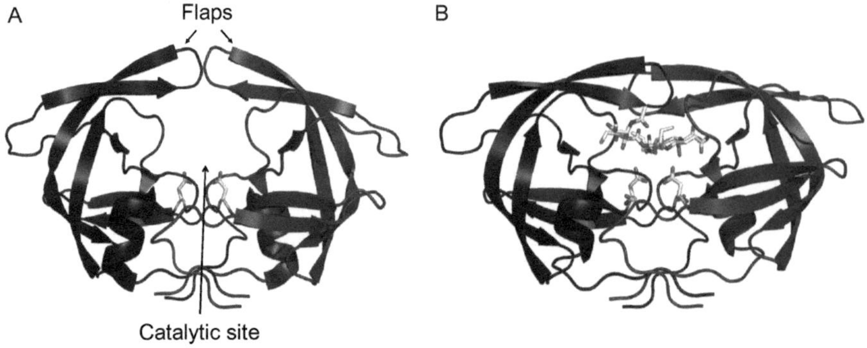

FIG. 5. The structure of HIV-1 PR. (A) Ligand-free PR dimer with flaps (residues 45–55) in the open conformation (PDB code: 1HHP; Spinelli et al., 1991). (B) PR dimer in complex with peptide analog p2/NC showing closed flaps (PDB code: 2AOD; Tie et al., 2005). The two subunits are colored red and blue; the catalytic Asp25, Asp25′, and peptide analog are in gray sticks. (For interpretation of the references to color in this figure legend, the reader is referred to the online version of this chapter.)

The reaction mechanism of HIV PR has been investigated using various experimental and theoretical techniques (see review by Brik and Wong, 2003). The reaction is considered to proceed via the general acid–base mechanism, as illustrated in Fig. 6. The two catalytic Asp25 and Asp25′ have an interaction of 2.5–3.0 Å between their carboxylate side chains, which implies one aspartate is protonated and the other has a negative charge. After substrate peptide is bound, the reaction proceeds via a tetrahedral gem-diol intermediate, which dissociates into the two product peptides. Interestingly, however, studies of the isotope effect and pH dependency have suggested two possible slow steps for the reaction with different peptide substrates (Hyland et al., 1991; Polgár et al., 1994). One rate-limiting step for substrate hydrolysis involves a chemical step and proton transfer, which is presumed to be the formation or dissociation of the TI. The slow step in reaction pathway for another substrate involved a physical change, which may correspond to closing of the flexible flaps when substrate binds (Fig. 5).

Current clinical inhibitors contain a hydroxyl group that mimics one hydroxyl of the tetrahedral reaction intermediate and interacts with the two catalytic aspartates (Weber and Wang, 2010). Therefore, the structures of the TI are of direct relevance to the design of novel and effective

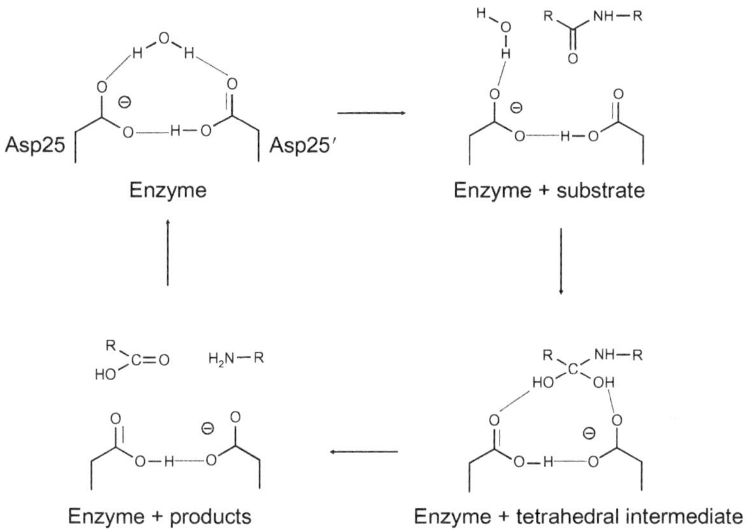

FIG. 6. Proposed reaction mechanism of peptide bond cleavage by HIV-1 PR.

inhibitors to combat the continued evolution of drug resistance in HIV. Further, high-resolution structural data exist for the other states of the enzyme as well, including unliganded, enzyme–substrate complexes, and enzyme–product complexes, giving an unparalleled experimental insight into the stages of the reaction pathway. The crystallographic evidence for the consecutive steps of the reaction will be described in the following sections.

The structure of PR in the absence of peptide substrate represents the first step in the reaction. Several recent structures of the ligand-free enzyme from different natural variants are described in Heaslet et al. (2007) and Robbins et al. (2010). The ligand-free PR structures often show a water molecule or metal ion interacting with the carboxylates of the catalytic aspartates. One obvious difference from the inhibitor-bound dimer is the flexible conformation of the flaps comprising residues 45–55 from each subunit (Fig. 5). The flaps are likely to alter their conformation during the proteolytic reaction. They must open outward to allow the large polyprotein substrate access to the active site. Subsequently, the flaps close down to bind the substrate and act to exclude unwanted solvent from the active site during the reaction. Finally, the flaps may need to move away to facilitate product release from the active site.

The structure of a substrate in complex with PR is the next stage in the reaction pathway. It has not proved possible, however, to capture the crystal structure of a bound substrate in the active enzyme. Instead, this stage is illustrated by the complex of either peptide substrate with the inactive enzyme containing the D25N mutation, or peptide analog with the active enzyme (see examples in Prabu-Jeyabalan et al., 2002; Tie et al., 2005). The structure of PR with the inactivating D25N mutation is very similar to that of wild-type PR (Sayer et al., 2008). The interactions seen in the structure of active PR crystallized with a reduced peptide analog of the p2/NC cleavage site are shown in Fig. 7 (Tie et al., 2005). The PR complex with peptides or peptide mimics is characterized by a series of conserved

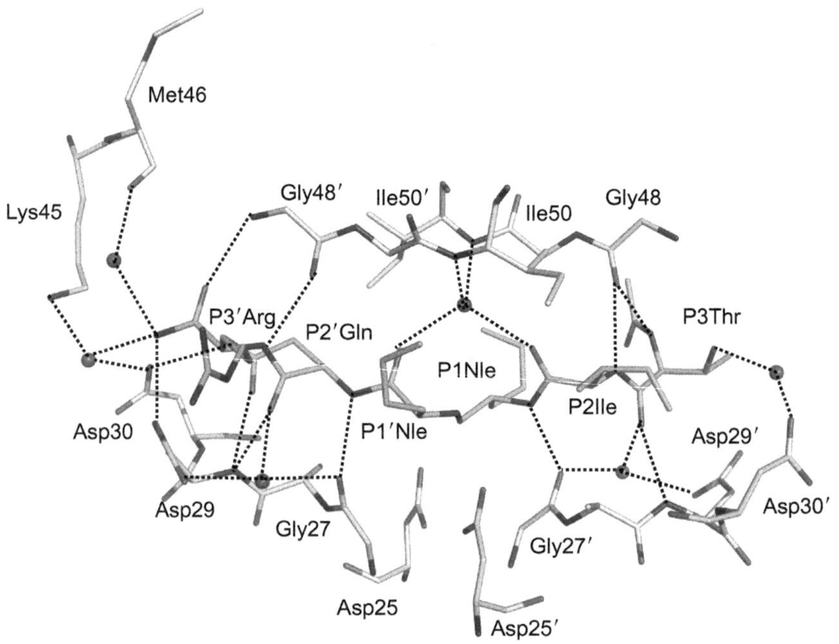

FIG. 7. Interactions of PR with a substrate analog of the p2–NC cleavage site (PDB code: 2AOD; Tie et al., 2005). The peptide analog is acetyl-Thr-Ile-Nle-r-Nle-Gln-Arg, where r is the reduced peptide bond and Nle is norleucine. Hydrogen bond interactions are indicated by dotted lines. Water molecules are shown as red spheres. (For interpretation of the references to color in this figure legend, the reader is referred to the online version of this chapter.)

hydrogen bond interactions connecting the main chain atoms of PR and its substrate. These structures provided the basis for the design of antiviral inhibitors targeting PR, which retain several of the typical hydrogen bond interactions shown by peptides.

In the next stage, the reaction is expected to proceed via formation of the gem-diol TI. The TI, having two hydroxyl groups attached to the carbonyl carbon at the scissile peptide bond, is a temporary species existing inside the PR active site cavity. We and other groups have identified a TI trapped in PR crystal structures (Kumar et al., 2005; Kovalevsky et al., 2007; Das et al., 2010). Although these intermediates represent different peptide sequences, the hydrogen bond interactions with PR were conserved in these structures. We describe our high-resolution structure (PDB code: 3B7V; Kovalevsky et al., 2007) of wild-type PR complexed with the peptide intermediate of autoproteolysis, NLL*TQI, where * indicates the gem-diol groups. Clear electron density was seen for the two hydroxyl oxygens of the TI (Fig. 8). The peptide intermediate showed the characteristic hydrogen bond interactions with the main chain of PR. One unusual feature is the short interaction between one hydroxyl of the intermediate and the outer carboxylate oxygen of Asp25 with a distance of 2.3 Å, which indicates a "low barrier" hydrogen bond. The other hydroxyl oxygen of the TI interacts with all four carboxylate oxygens of Asp25 and 25′ with more typical hydrogen bond distances of 2.6–3.2 Å (Fig. 9A).

In the later steps, dissociation of the TI results in two product peptides. Several PR–product complexes have been reported, including two product and single product complexes. A recent structure showed both the N-terminal and C-terminal peptide products bound in the active site cavity (Das et al., 2006). The termini of both products interact with the catalytic aspartates. An interesting feature is the 2.7 Å distance between the terminal nitrogen of one product and the carboxylate carbon of the other product peptide, which is longer than a regular peptide bond distance (Fig. 9B). This structure might represent the stage just after the cleavage of peptide bond and before the products have separated completely.

Different structures have been described of PR in complex with separate N-terminal or C-terminal products bound in the active site cavity (Rose et al., 1996; Prashar et al., 2009). In these structures, one half of the active site cavity contains only solvent molecules. The C-terminal product was observed to interact with one water molecule and the catalytic residues, as shown in Fig. 9C (Prashar et al., 2009). The water molecule might play a

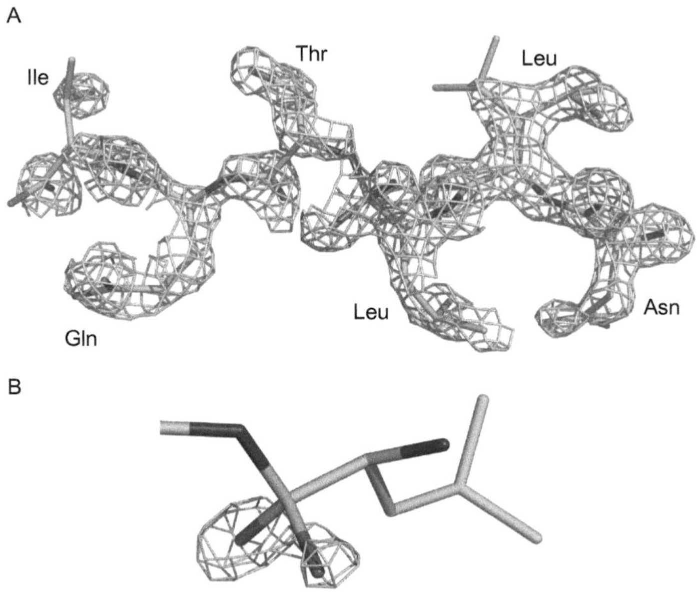

FIG. 8. Tetrahedral intermediate peptide trapped in HIV-1 PR crystal structure (PDB code: 3B7V; Kovalevsky et al., 2007). (A) Electron density map (2Fo–Fc) contoured at 1.0σ level for the TI fitted with sequence Asn-Leu-Leu-Thr-Gln-Ile from autoproteolysis. (B) Two hydroxyl oxygens of TI are clearly visible in the electron density (2Fo–Fc) contoured at 1.5σ level. (For color version of this figure, the reader is referred to the online version of this chapter.)

role in stabilizing the product complex, and has also been proposed to aid in recycling to the free enzyme. Single product complexes have been observed for either the N-terminal or C-terminal peptides, indicating that there is no preferred order of leaving for the products. Clearly, these HIV-1 PR crystal structures represent several steps in the reaction and reveal some unusual details, such as the short interaction seen for one hydroxyl of the TI.

C. Cysteine Hydrolases: Caspases

Cysteinyl aspartate specific proteases (caspases) are cysteine proteases that play a prominent role in apoptosis, or programmed cell death, and inflammation. Defects in modulation of caspase activity have been associated with

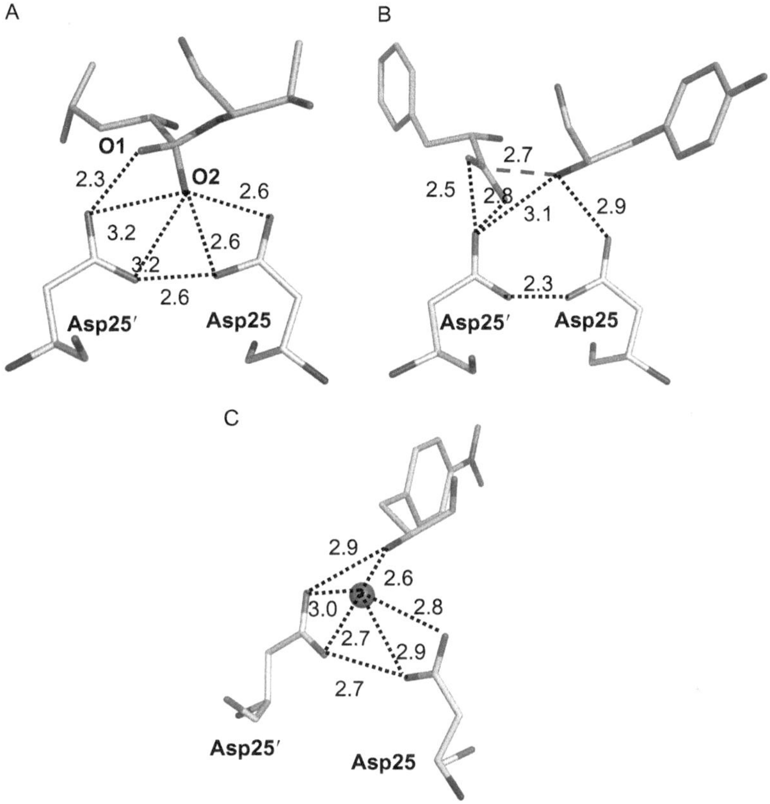

FIG. 9. Interactions of reaction intermediates with PR catalytic residues. (A) PR complex with TI (PDB code: 3B7V; Kovalevsky et al., 2007). (B) PR complex with both products (PDB code: 2NPH; Das et al., 2006). (C) PR complex with one product (PDB code: 2WHH; Prashar et al., 2009). Hydrogen bond interactions are indicated by dotted lines with interatomic distances in Å. Water molecules are shown as red spheres. (For interpretation of the references to color in this figure legend, the reader is referred to the online version of this chapter.)

numerous conditions including neurodegenerative, tumorigenic, and cardiovascular disorders (Li and Yuan, 2008). Caspases are synthesized as inactive procaspases or zymogens with an N-terminal prodomain of variable length in addition to a large p20 subunit and a small p10 subunit (Fig. 10). Activation by proteolysis and/or dimerization results in cleavage at one or more specific aspartic acid residues within the prodomain, p20 and p10

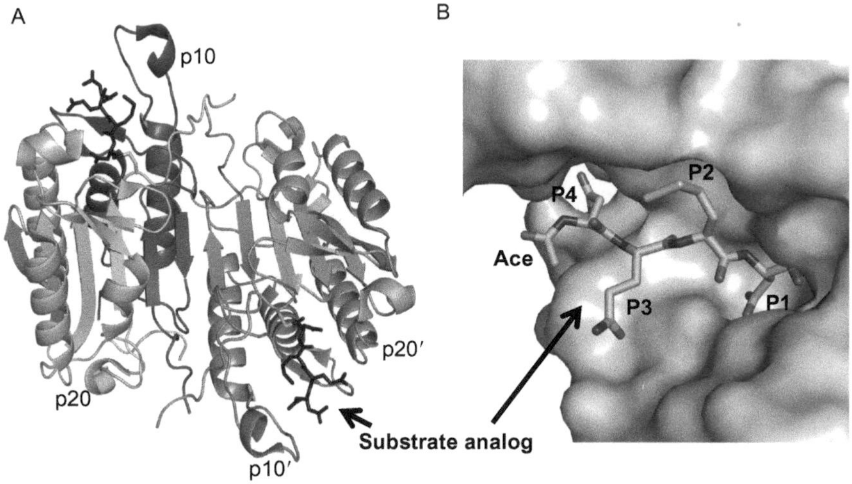

FIG. 10. (A) Caspase-7 tetrameric assembly in complex with aldehyde inhibitor Ac-DQMD-CHO (PDB code: 2QL5; Agniswamy et al., 2007). The p20 subunits are shown in two shades of green while the p10 subunits are colored in shades of red. The inhibitor is shown in blue sticks. (B) The aldehyde tetra-peptide inhibitor Ac-DQMD-CHO bound at the catalytic cleft of caspase-7. The caspase-7 is shown in surface representation and the inhibitor is shown as sticks colored by the element type. (For interpretation of the references to color in this figure legend, the reader is referred to the online version of this chapter.)

subunits to produce mature caspase. The active mature caspase is a heterotetramer with two p20 and two p10 subunits (Fig. 10). Caspases with longer prodomains are usually activated by dimerization, while those with short prodomains are activated by proteolytic cleavage (Pop and Salvesen, 2009). Human caspases are broadly classified into three groups. The inflammatory caspases (caspase-1, -4, -5, -12, and -13) are involved in inflammatory responses, while those responsible for apoptosis are known as apoptotic caspases (Lavrik et al., 2005). The apoptotic caspases are further divided into initiators (caspase-2, -8, -9, and -10) and executors (caspase-3, -6, and -7) of apoptosis. Pioneering studies with a positional scanning synthetic peptide library approach demonstrated that caspases generally recognize tetra-peptide sequences in their substrates, which are designated P4–P1 (Thornberry et al., 1997). Among the caspase family, members of the three groups have different optimal tetra-peptide sequences (Lavrik et al., 2005).

Structural studies by X-ray crystallography have provided the molecular basis for caspase specificity for peptides and assisted in structure-guided design of drugs for controlled cell death (Weber et al., 2008). Most of the peptides studied contain an aldehyde (—CHO) (Fig. 10B) or fluoromethylketone group at the C-terminal end (Rotonda et al., 1996; Mittl et al., 1997). These peptide analogs are classified as reversible inhibitors of caspases, however, they are observed in the crystal structures with a covalent bond between the aldehyde or fluoromethylketone group of the inhibitor and the mercapto (—SH) group of the catalytic cysteine (Fig. 11). The aldehydes form a thiohemiacetal bond with the side chain of the catalytic cysteine while ketones form a thioether bond. The aldehyde and ketone inhibitors bound to the catalytic cysteine mimic different states of TIs (Mittl et al., 1997; Fig. 11A). These structures played a vital role in understanding the caspase specificity and plasticity. The presence of alternate binding sites at S4 subsite for polar as well as small aliphatic P4 residues in caspase-3, -7, and -8 was discovered from the crystal structures (Blanchard et al., 2000; Agniswamy et al., 2007; Fang et al, 2009). Also, the presence or absence of an S5 subsite for P5 residues in substrates of caspase-2, -3, and -7 was revealed by structural studies (Schweizer et al.,

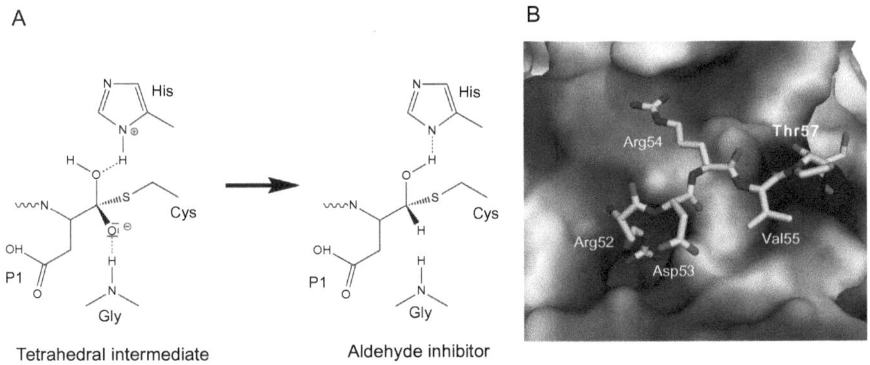

FIG. 11. (A) Reaction of the tetrahedral intermediate mimic of an aldehyde inhibitor in the active site of caspase. (B) Molecular surface of caspase-7 showing the N-terminal extension with residue Arg52 bound at the putative exosite (PDB code: 3IBF; Agniswamy et al., 2009). Blue, red, and white represent areas of positive, negative, and neutral electrostatic potential, respectively. (For interpretation of the references to color in this figure legend, the reader is referred to the online version of this chapter.)

2003; Fang et al., 2006; Fu et al., 2008). The molecular basis of caspase recognition of substrates defined by these mimics of a TI is valuable for the design of compounds for the pharmacological modification of caspase activity.

The removal of an N-terminal extension is important for the efficient activation of all caspases. For most inflammatory and initiator caspases, these extension domains are required for recruitment to respective activation complexes, but the precise role of the short N-terminal extensions in executioner caspases has not yet been determined. No N-terminal extension was visible in the crystal structures of procaspase-7 zymogen (Chai et al., 2001; Riedl et al., 2001). However, in the active unliganded caspase-7 structure, five residues of the N-terminal extension were visible with the terminal arginine buried inside a deep pocket of caspase-7 (Fig. 11B), which lead to the identification of an exosite of caspase-7 (Agniswamy et al., 2009). The N-terminal extension is not the physiological target of the exosite, but the proteolytic cleavage during expression and purification has resulted in a p20 subunit with N-terminal extension of convenient length to facilitate binding of the terminal arginine at the exosite. It is proposed that the N-terminal extension blocks the exosite and its removal results in exosite access and recruitment of potential binding partners of caspase-7 (Agniswamy et al., 2009).

Unlike many other proteases, caspases do not activate simply by the removal of the prodomain. The critical step in caspase activation, in addition to the removal of the N-terminal prodomain or peptide, is the cleavage of the linker peptide between the large p20 and the small p10 subunits. Structural studies by X-ray crystallography have trapped several intermediates in the activation pathway between procaspase and fully mature active caspase thereby shedding light on the conformational changes that caspases undergo during activation. Fig. 12 shows a scheme for the activation pathway based on these crystal structures. In crystal structures of the procaspase-7, the active site cavity composed of L1, L2, L2′, L3, and L4 loops is disordered and cannot bind substrate (Chai et al., 2001; Riedl et al., 2001). Particularly, the L2′ loop folds back into the central cavity of the homodimer and is physically restrained from forming the loop bundle as the linker peptide between the p20 and the p10 subunit is still intact. The L2, L2′, and L4 loops have to rotate by 90°, 180°, and 60°, respectively, to form the loop bundle necessary for substrate binding. The next stage in the activation pathway was revealed by the structure of the active isolated

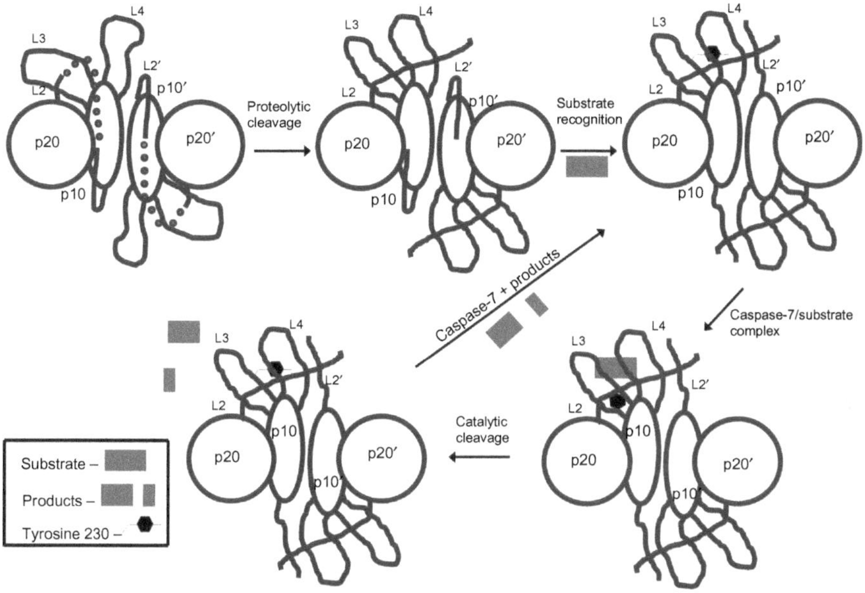

FIG. 12. Schematic diagram of caspase-7 activation and catalytic pathway. Prior to activation, the active site loop bundle in procaspase-7 is in a dissolved state. L2 and L2′ loops are physically restrained by the linker peptide (structure represented by PDB code: 1K88; Chai et al., 2001). Proteolytic cleavage of the linker peptide results in an intermediate form of the loop bundle but the L2′ loop is still inserted in the central cavity (PDB code: 1K86; Chai et al., 2001). Upon binding of substrate, the caspase-7 undergoes further conformational change to form the complete loop bundle with Tyr230 blocking the S2 subsite (PDB code: 3IBF; Agniswamy et al., 2009). Rotation of Tyr230 side chain by $\sim 90^{\circ}$ is required for formation of the enzyme substrate/inhibitor complex (PDB code: 2QL5; Agniswamy et al., 2007) and catalytic cleavage of the substrate. When the products leave, the protease is present in the tyrosine flipped state rather than pseudo zymogenic state.

caspase-7 with the cleaved linker peptide (Chai et al., 2001). Even after processing of the linker peptide, the L2′ loop is still inserted inside the central cavity in a closed conformation similar to that seen in procaspase-7. Though the L3 and L4 loops are more ordered than in procaspase-7, the loop bundle and the catalytic cleft are still not sufficiently well defined for the substrate to bind, and the caspase exists in a locked pseudo zymogenic intermediate state. Upon recognizing the substrate/inhibitor, the caspase-7 undergoes further conformational changes resulting in formation of an intact loop bundle suitable for the binding of substrate/inhibitor. The best evidence for this conformational state is revealed by the crystal structures of active unliganded caspase-3 and -7 with free catalytic site (Ni et al., 2003; Agniswamy et al., 2009). In these structures, the L2′ loop is flipped by 180° when compared to that arrangement in procaspse-7 and active isolated caspase-7 and interacts with the L2 loop. The loop bundle and the catalytic cleft are well defined with catalytic residues His144 and Cys186 in conformations similar to those of the mature inhibited structures (Agniswamy et al., 2009). The only difference is the conformation of Tyr230 which is rotated by ~90° around the Cα—Cβ bond and occupies the S2 subsite instead of the side chain of P2 residue of substrate or inhibitor. The Tyr230 must rotate away for the substrate or inhibitor to access the active site. It is likely that in the catalytic cycle of caspase-3 and -7, the tyrosine is flipped in between the substrate bound and ligand-free states (Agniswamy et al., 2009). Similar tyrosine flipping mechanism has been reported for other proteins, including DADH (Fu et al., 2010), as described earlier. For example, rotation of Tyr99 is reported for the protease Factor IXa which upon substrate binding opens up the S4 binding pocket (Hopfner et al., 1999). In Factor IXa, the premature onset of blood coagulation is prevented by this Tyr flipping mechanism. Thus, a set of trapped intermediate structures of caspases illuminate the activation pathway and substrate recognition of these enzymes.

V. Complementary Techniques

Structural data from X-ray crystallography can be further analyzed by computational techniques such as molecular mechanics (MM), QM, combined QM/MM, and molecular dynamics (MD) simulations. QM/MM calculations can exploit the structural data in the design of transition state analog drugs (De Vivo, 2011). Also, the calculations can model the complete reaction pathway with intermediates defined in crystal structures and short-lived species, as described for aromatic amine dehydrogenase (Pang et al., 2010) and MurD ligase (Perdih et al., 2009). Enzyme-intermediate structures can define functional residues for prediction of function in unknown gene sequences. Automated methods include prediction of function from enzyme structure using profiles for active site residues. Some examples are described in Pazos and Sternberg (2004) and Khurana et al. (2010).

Traditional X-ray crystallography can be complemented by other experimental techniques to study the reaction mechanism in crystals of enzymes. These complementary techniques such as neutron crystallography, Laue diffraction, and single crystal spectroscopy have a long history. Their more general application is limited by the requirement for specialized facilities that are less commonly available than the instruments or national laboratories serving macromolecular X-ray crystallography.

A. *Time-Resolved Single Crystal Spectroscopy*

Spectroscopy can be used to identify trapped intermediates, analyze the electronic structure, and follow the reaction in single crystals, even during X-ray data collection. More than a decade ago time-resolved studies with single crystal diffraction and spectroscopy were considered as "coming of age" in a review by Hajdu et al. (2000). An example of these studies used single crystal polarized absorption spectroscopy and freeze-trapping of reaction intermediates to follow the reaction of methylamine dehydrogenase with its redox partner amicyanin (Pearson and Wilmot, 2003). Recently, the techniques of combining single crystal spectroscopy and X-ray crystallography have been reviewed (Pearson and Owen, 2009; Orville et al., 2011; Stoner-Ma et al., 2011). Absorption, fluorescence, and Raman spectroscopic analysis have been applied. Fluorescence changes were observed on addition of photolabile GTP analog to H-ras p21 crystals using a fluorescence microspectrofluorometer in conjunction with synchrotron X-rays (Klink et al., 2006). Single crystal spectroscopic studies have an advantage over solution studies in that the protein molecules are spatially oriented, rather than the dynamic, random orientations of molecules in solution. In future, these techniques will become more easily available due to modern facilities under development at national synchrotron sources.

B. *Laue and Neutron Crystallography*

X-ray Laue diffraction has been applied to protein crystals in a limited number of studies over several decades (Moffat et al., 1984). Laue diffraction with polychromatic, rather than monochromatic synchrotron X-rays, permits time-resolved crystallography and generation of the protein structure in timescales of about 1 s. For example, recent studies have identified small structural changes of 1.3 Å occurring on photoactivation of a photosynthetic reaction center (Wöhri et al., 2010). Synchrotron X-ray diffraction studies can be complemented by neutron diffraction studies (Blakeley et al., 2004).

Neutron crystallography of proteins has a long history with critical contributions to the understanding of enzyme mechanisms (Schoenborn, 2010). Neutron diffraction experiments are vital in determining the position of hydrogen (or specifically deuterium) atoms and water molecules in the catalytic site of the enzyme in the crystal. Technological improvements in the optics, detectors, and use of Laue diffraction and spallation neutron sources with frozen protein crystals have greatly facilitated neutron crystallography (Meilleur et al., 2006; Schoenborn, 2010). These studies have proved valuable in elucidating protonation states in several enzymes (Blum et al., 2009). Different stages in the reaction of deuterated D-xylose isomerase were investigated by neutron diffraction (Glusker et al., 2010). Remarkable details for the hydrogen bond network and catalytic mechanism of human aldose reductase were obtained by combined QM, subatomic resolution X-ray crystallography (0.66 Å), and neutron Laue data (Blakeley et al., 2008). Clearly, a combination of techniques is extremely powerful for studying the reaction mechanism.

C. *Conclusions*

In conclusion, despite the technological advances and new national facilities for time-resolved spectroscopy and neutron crystallography, X-ray crystallography using frozen crystals and synchrotron sources is likely to remain the major technique for analyzing trapped reaction intermediates in enzyme crystals. The freezing process used in cryo-crystallography for synchrotron X-ray data collection gives a great advantage in trapping quasi-stable intermediates, coupled with the tendency of many enzymes to acquire substrates or other ligands from the expression media. The optimal analysis to explore the enzyme mechanism, when possible, will combine atomic resolution X-ray crystallography with time-resolved kinetics and neutron crystallography as well as computational techniques like QM/MM calculations and MD simulations.

Acknowledgments

This work was supported in part by the Georgia State University Research Program Enhancement Award in Bioinformatics, Georgia Research Alliance, and the United States National Institutes of Health grant GM062920. G. F. and C. H. S. were supported by Georgia State University Molecular Basis of Disease Fellowships.

References

Agniswamy, J., Fang, B., Weber, I. T. (2007). Plasticity of S2-S4 specificity pockets of executioner caspase-7 revealed by structural and kinetic analysis. *FEBS J.* **274**, 4752–4765.

Agniswamy, J., Fang, B., Weber, I. T. (2009). Conformational similarity in the activation of caspase-3 and -7 revealed by the unliganded and inhibited structures of caspase-7. *Apoptosis* **14**, 1135–1144.

Alahuhta, M., Wierenga, R. K. (2010). Atomic resolution crystallography of a complex of triosephosphate isomerase with a reaction-intermediate analog: new insight in the proton transfer reaction mechanism. *Proteins* **78**, 1878–1888.

Bell, S. G., Chen, X., Xu, F., Rao, Z., Wong, L. L. (2003). Engineering substrate recognition in catalysis by cytochrome P450cam. *Biochem. Soc. Trans.* **31**, 558–562.

Berisio, R., Lamzin, V. S., Sica, F., Wilson, K. S., Zagari, A., Mazzarella, L. (1999). Protein titration in the crystal state. *J. Mol. Biol.* **292**, 845–854.

Berman, H. M., Westbrook, J., Feng, Z., Gilliland, G., Bhat, T. N., Weissig, H., et al. (2000). The Protein Data Bank. *Nucleic Acids Res.* **28**, 235–242.

Blakeley, M. P., Cianci, M., Helliwell, J. R., Rizkallah, P. J. (2004). Synchrotron and neutron techniques in biological crystallography. *Chem. Soc. Rev.* **33**, 548–557.

Blakeley, M. P., Ruiz, F., Cachau, R., Hazemann, I., Meilleur, F., Mitschler, A., et al. (2008). Quantum model of catalysis based on a mobile proton revealed by subatomic X-ray and neutron diffraction studies of h-aldose reductase. *Proc. Natl. Acad. Sci. USA* **105**, 1844–1848.

Blanchard, H., Donepudi, M., Tschopp, M., Kodandapani, L., Wu, J. C., Grütter, M. G. (2000). Caspase-8 specificity probed at subsite S(4): crystal structure of the caspase-8-Z-DEVD-cho complex. *J. Mol. Biol.* **302**, 9–16.

Blum, M. M., Mustyakimov, M., Rüterjans, H., Kehe, K., Schoenborn, B. P., Langan, P., et al. (2009). Rapid determination of hydrogen positions and protonation states of diisopropyl fluorophosphatase by joint neutron and X-ray diffraction refinement. *Proc. Natl. Acad. Sci. USA* **106**, 713–718.

Brik, A., Wong, C. H. (2003). HIV-1 protease: mechanism and drug discovery. *Org. Biomol. Chem.* **1**, 5–14.

Chai, J., Wu, Q., Shiozaki, E., Srinivasula, S. M., Alnemri, E. S., Shi, Y. (2001). Crystal structure of a procaspase-7 zymogen: mechanisms of activation and substrate binding. *Cell* **107**, 399–407.

Cleland, W. W., Frey, P. A., Gerlt, J. A. (1998). The low barrier hydrogen bond in enzymatic catalysis. *J. Biol. Chem.* **273**, 25529–25532.

Das, A., Prashar, V., Mahale, S., Serre, L., Ferrer, J. L., Hosur, M. V. (2006). Crystal structure of HIV-1 protease in situ product complex and observation of a low-barrier hydrogen bond between catalytic aspartates. *Proc. Natl. Acad. Sci. USA* **103**, 18464–18469.

Das, A., Mahale, S., Prashar, V., Bihani, S., Ferrer, J. L., Hosur, M. V. (2010). X-ray snapshot of HIV-1 protease in action: observation of tetrahedral intermediate and short ionic hydrogen bond SIHB with catalytic aspartate. *J. Am. Chem. Soc.* **132**, 6366–6373.

De Vivo, M. (2011). Bridging quantum mechanics and structure-based drug design. *Front. Biosci.* **16**, 1619–1633.

Dixon, M., Kleppe, K. (1965). D-amino acid oxidase I. Dissociation and recombination of the holoenzyme. *Biochim. Biophys. Acta* **96**, 357–367.

Fang, B., Boross, P. I., Tozser, J., Weber, I. T. (2006). Structural insight into caspase-3 selectivity for substrate positions P2, P3, and P5. *J. Mol. Biol.* **360**, 654–666.

Fang, B., Fu, G., Agniswamy, J., Harrison, R. W., Weber, I. T. (2009). Caspase-3 binds diverse P4 residues in peptides as revealed by crystallography and structural modeling. *Apoptosis* **14**, 741–752.

Fu, G., Chumanevich, A. A., Agniswamy, J., Fang, B., Harrison, R. W., Weber, I. T. (2008). Structural basis for executioner caspase recognition of P5 position in substrates. *Apoptosis* **13**, 1291–1302.

Fu, G., Yuan, H., Li, C., Lu, C. D., Gadda, G., Weber, I. T. (2010). Conformational changes and substrate recognition in *Pseudomonas aeruginosa* D-arginine dehydrogenase. *Biochemistry* **49**, 8535–8545.

Fu, G., Yuan, H., Wang, S., Gadda, G., Weber, I. T. (2011). Atomic resolution structure of an N(5) flavin adduct in D-arginine dehydrogenase. *Biochemistry* **50**, 6292–6294.

Glusker, J. P., Carrell, H. L., Kovalevsky, A. Y., Hanson, L., Fisher, S. Z., Mustyakimov, M., et al. (2010). Using neutron protein crystallography to understand enzyme mechanisms. *Acta Crystallogr. D Biol. Crystallogr.* **66**, 1257–1261.

Hajdu, J., Neutze, R., Sjögren, T., Edman, K., Szöke, A., Wilmouth, R. C., et al. (2000). Analyzing protein functions in four dimensions. *Nat. Struct. Biol.* **7**, 1006–1012.

Heaslet, H., Rosenfeld, R., Giffin, M., Lin, Y. C., Tam, K., Torbett, B. E., et al. (2007). Conformational flexibility in the flap domains of ligand-free HIV protease. *Acta Crystallogr. D Biol. Crystallogr.* **63**, 866–875.

Hoegl, L., Korting, H. C., Klebe, G. (1999). Inhibitors of aspartic proteases in human diseases: molecular modeling comes of age. *Pharmazie* **54**, 319–329.

Hopfner, K. P., Lang, A., Karcher, A., Sichler, K., Kopetzki, E., Brandstetter, H., et al. (1999). Coagulation factor IXa: the relaxed conformation of Tyr99 blocks substrate binding. *Structure* **7**, 989–996.

Hyland, L. J., Tomaszek, T. A., Jr., Meek, T. D. (1991). Human immunodeficiency virus-1 protease. 2. Use of pH rate studies and solvent kinetic isotope effects to elucidate details of chemical mechanism. *Biochemistry* **30**, 8454–8463.

Katona, G., Wilmouth, R. C., Wright, P. A., Berglund, G. I., Hajdu, J., Neutze, R., et al. (2002). X-ray structure of a serine protease acyl-enzyme complex at 0.95-A resolution. *J. Biol. Chem.* **277**, 21962–21970.

Khurana, P., Gokhale, R. S., Mohanty, D. (2010). Genome scale prediction of substrate specificity for acyl adenylate superfamily of enzymes based on active site residue profiles. *BMC Bioinform.* **11**, 57.

Klink, B. U., Goody, R. S., Scheidig, A. J. (2006). A newly designed microspectrofluorometer for kinetic studies on protein crystals in combination with X-ray diffraction. *Biophys. J.* **91**, 981–992.

Klumpp, K., Graves, B. J. (2006). Optimization of small molecule drugs binding to highly polar target sites: lessons from the discovery and development of neuraminidase inhibitors. *Curr. Top. Med. Chem.* **6**, 423–434.

Kohl, N. E., Emini, E. A., Schleif, W. A., Davis, L. J., Heimbach, J. C., Dixon, R. A., et al. (1988). Active human immunodeficiency virus protease is required for viral infectivity. *Proc. Natl. Acad. Sci. USA* **85**, 4686–4690.

Kovalevsky, A. Y., Liu, F., Leshchenko, S., Ghosh, A. K., Louis, J. M., Harrison, R. W., et al. (2006). Ultra-high resolution crystal structure of HIV-1 protease mutant reveals two binding sites for clinical inhibitor TMC114. *J. Mol. Biol.* **363**, 161–173.

Kovalevsky, A. Y., Chumanevich, A. A., Liu, F., Weber, I. T. (2007). Caught in the act: 1.5 Å resolution crystal structures of the HIV-1 protease and the I54V mutant reveal a tetrahedral reaction intermediate. *Biochemistry* **46**, 14854–14864.

Kumar, M., Prashar, V., Mahale, S., Hosur, M. V. (2005). Observation of a tetrahedral reaction intermediate in the HIV-1 protease-substrate complex. *Biochem. J.* **389**, 365–371.

Lavrik, I. N., Golks, A., Krammer, P. H. (2005). Caspases: pharmacological manipulation of cell death. *J. Clin. Invest.* **115**, 2665–2672.

Lee, T. W., James, M. N. (2008). 1.2A-resolution crystal structures reveal the second tetrahedral intermediates of streptogrisin B (SGPB). *Biochim. Biophys. Acta* **1784**, 319–334.

Leferink, N. G., Fraaije, M. W., Joosten, H. J., Schaap, P. J., Mattevi, A., van Berkel, W. J. (2009). Identification of a gatekeeper residue that prevents dehydrogenases from acting as oxidases. *J. Biol. Chem.* **284**, 4392–4397.

Li, C., Lu, C. D. (2009). Arginine racemization by coupled catabolic and anabolic dehydrogenases. *Proc. Natl. Acad. Sci. USA* **106**, 906–911.

Li, J., Yuan, J. (2008). Caspases in apoptosis and beyond. *Oncogene* **27**, 6194–6206.

Li, C., Yao, X., Lu, C. D. (2010). Regulation of the dauBAR operon and characterization of D-amino acid dehydrogenase DauA in arginine and lysine catabolism of Pseudomonas aeruginosa PAO1. *Microbiology* **156**, 60–71.

Liu, P., Wang, Y.-F., Ewis, H., Abdelal, A., Lu, C. D., Harrison, R. W., et al. (2004). Covalent reaction intermediate revealed in crystal structure of the *Geobacillus stearothermophilus* carboxylesterase Est30. *J. Mol. Biol.* **342**, 551–556.

Liu, Y. C., Li, Y. S., Lyu, S. Y., Hsu, L. J., Chen, Y. H., Huang, Y. T., et al. (2011). Interception of teicoplanin oxidation intermediates yields new antimicrobial scaffolds. *Nat. Chem. Biol.* **7**, 304–309.

Lu, W., Apostol, I., Qasim, M. A., Warne, N., Wynn, R., Zhang, W. L., et al. (1997). Binding of amino acid side-chains to S1 cavities of serine proteinases. *J. Mol. Biol.* **266**, 441–461.

Lyubimov, A. Y., Lario, P. I., Moustafa, I., Vrielink, A. (2006). Atomic resolution crystallography reveals how changes in pH shape the protein microenvironment. *Nat. Chem. Biol.* **2**, 259–264.

Macheroux, P., Kappes, B., Ealick, S. E. (2001). Flavogenomics—a genomic and structural view of flavin-dependent proteins. *FEBS J.* **278**, 2625–2634.

Mattevi, A., Vanoni, M. A., Todone, F., Rizzi, M., Teplyakov, A., Coda, A., et al. (1996). Crystal structure of D-amino acid oxidase: a case of active site mirror-image convergent evolution with flavocytochrome b2. *Proc. Natl. Acad. Sci. USA* **93**, 7496–7501.

Meilleur, F., Myles, D. A., Blakeley, M. P. (2006). Neutron Laue macromolecular crystallography. *Eur. Biophys. J.* **35**, 611–620.

Mittl, P. R., Di Marco, S., Krebs, J. F., Bai, X., Karanewsky, D. S., Priestle, J. P., et al. (1997). Structure of recombinant human CPP32 in complex with the tetrapeptide acetyl-Asp-Val-Ala-Asp fluoromethyl ketone. *J. Biol. Chem.* **272**, 6539–6547.

Moffat, K., Szebenyl, D., Bilderback, D. (1984). X-ray Laue diffraction from protein crystals. *Science* **223**, 1423–1425.

Mozzarelli, A., Ottonello, S., Rossi, G. L., Fasella, P. (1979). Catalytic activity of aspartate aminotransferase in the crystal. Equilibrium and kinetic analysis. *Eur. J. Biochem.* **98**, 173–179.

Nguyen, J. T., Hamada, Y., Kimura, T., Kiso, Y. (2008). Design of potent aspartic protease inhibitors to treat various diseases. *Arch. Pharm. (Weinheim)* **341**, 523–535.

Ni, C. Z., Li, C., Wu, J. C., Spada, A. P., Ely, K. R. (2003). Conformational restrictions in the active site of unliganded human caspase-3. *J. Mol. Recognit.* **16**, 121–124.

Orville, A. M., Buono, R., Cowan, M., Héroux, A., Shea-McCarthy, G., Schneider, D. K., et al. (2011). Correlated single-crystal electronic absorption spectroscopy and X-ray crystallography at NSLS beamline X26-C. *J. Synchrotron Radiat.* **18**, 358–366.

Pang, J., Scrutton, N. S., de Visser, S. P., Sutcliffe, M. J. (2010). New insights into the multi-step reaction pathway of the reductive half-reaction catalysed by aromatic amine dehydrogenase: a QM/MM study. *Chem. Commun. (Camb.)* **46**, 3104–3106.

Pazos, F., Sternberg, M. J. (2004). Automated prediction of protein function and detection of functional sites from structure. *Proc. Natl. Acad. Sci. USA* **101**, 14754–14759.

Pearson, A. R., Owen, R. L. (2009). Combining X-ray crystallography and single-crystal spectroscopy to probe enzyme mechanisms. *Biochem. Soc. Trans.* **37**, 378–381.

Pearson, A. R., Wilmot, C. M. (2003). Catching catalysis in the act: using single crystal kinetics to trap methylamine dehydrogenase reaction intermediates. *Biochim. Biophys. Acta* **1647**, 381–389.

Perdih, A., Hodoscek, M., Solmajer, T. (2009). MurD ligase from E. coli: tetrahedral intermediate formation study by hybrid quantum mechanical/molecular mechanical replica path method. *Proteins* **74**, 744–759.

Polgár, L., Szeltner, Z., Boros, I. (1994). Substrate-dependent mechanisms in the catalysis of human immunodeficiency virus protease. *Biochemistry* **33**, 9351–9357.

Pop, C., Salvesen, G. S. (2009). Human caspases: activation, specificity, and regulation. *J. Biol. Chem.* **284**, 21777–21781.

Prabu-Jeyabalan, M., Nalivaika, E., Schiffer, C. A. (2002). Substrate shape determines specificity of recognition for HIV-1 protease: analysis of crystal structures of six substrate complexes. *Structure* **10**, 369–381.

Prashar, V., Bihani, S., Das, A., Ferrer, J. L., Hosur, M. (2009). Catalytic water co-existing with a product peptide in the active site of HIV-1 protease revealed by X-ray structure analysis. *PLoS One* **4**, e7860.

Quaye, O., Lountos, G. T., Fan, F., Orville, A. M., Gadda, G. (2008). Role of Glu312 in binding and positioning of the substrate for the hydride transfer reaction in choline oxidase. *Biochemistry* **47**, 243–256.

Riedl, S. J., Fuentes-Prior, P., Renatus, M., Kairies, N., Krapp, S., Huber, R., et al. (2001). Structural basis for the activation of human procaspase-7. *Proc. Natl. Acad. Sci.* **98**, 14790–14795.

Robbins, A. H., Coman, R. M., Bracho-Sanchez, E., Fernandez, M. A., Gilliland, C. T., Li, M., et al. (2010). Structure of the unbound form of HIV-1 subtype A protease: comparison with unbound forms of proteases from other HIV subtypes. *Acta Crystallogr. D Biol. Crystallogr.* **66**, 233–242.

Rose, R. B., Craik, C. S., Douglas, N. L., Stroud, R. M. (1996). Three-dimensional structures of HIV-1 and SIV protease product complexes. *Biochemistry* **35**, 12933–12944.

Rotonda, J., Nicholson, D. W., Fazil, K. M., Gallant, M., Gareau, Y., Labelle, M., et al. (1996). The three-dimensional structure of apopain/CPP32, a key mediator of apoptosis. *Nat. Struct. Biol.* **3**, 619–625.

Roujeinikova, A., Hothi, P., Masgrau, L., Sutcliffe, M. J., Scrutton, N. S., Leys, D. (2007). New insights into the reductive half-reaction mechanism of aromatic amine dehydrogenase revealed by reaction with carbinolamine substrates. *J. Biol. Chem.* **282**, 23766–23777.

Saam, J., Rosini, E., Molla, G., Schulten, K., Pollegioni, L., Ghisla, S. (2010). O2-reactivity of flavoproteins: dynamic access of dioxygen to the active site and role of a H^+ relay system in D-amino acid oxidase. *J. Biol. Chem.* **285**, 24439–24446.

Sayer, J. M., Liu, F., Ishima, R., Weber, I. T., Louis, J. M. (2008). Effect of the active site D25N mutation on the structure, stability, and ligand binding of the mature HIV-1 protease. *J. Biol. Chem.* **283**, 13459–13470.

Schoenborn, B. P. (2010). A history of neutrons in biology: the development of neutron protein crystallography at BNL and LANL. *Acta Crystallogr. D Biol. Crystallogr.* **66**, 1262–1268.

Schweizer, A., Briand, C., Grutter, M. G. (2003). Crystal structure of caspase-2, apical initiator of the intrinsic apoptotic pathway. *J. Biol. Chem.* **278**, 42441–42447.

Spinelli, S., Liu, Q. Z., Alzari, P. M., Hirel, P. H., Poljak, R. J. (1991). The three-dimensional structure of the aspartyl protease from the HIV-1 isolate BRU. *Biochimie.* **73**, 1391–1396.

Stoner-Ma, D., Skinner, J. M., Schneider, D. K., Cowan, M., Sweet, R. M., Orville, A. M. (2011). Single-crystal Raman spectroscopy and X-ray crystallography at beamline X26-C of the NSLS. *J. Synchrotron Radiat.* **18**, 37–40.

Thornberry, N. A., Rano, T. A., Peterson, E. P., Rasper, D. M., Timkey, T., Garcia-Calvo, M., et al. (1997). A combinatorial approach defines specificities of members of the caspase family and granzyme B. Functional relationships established for key mediators of apoptosis. *J. Biol. Chem.* **272**, 17907–17911.

Tie, Y., Boross, P. I., Wang, Y.-F., Gaddis, L., Liu, F., Chen, X., et al. (2005). Molecular basis for substrate recognition and drug resistance from 1.1 to 1.6 Å resolution crystal structures of HIV-1 protease mutants with substrate analogs. *FEBS J.* **272**, 5265–5277.

Todone, F., Vanoni, M. A., Mozzarelli, A., Bolognesi, M., Coda, A., Curti, B., et al. (1997). Active site plasticity in D-amino acid oxidase: a crystallographic analysis. *Biochemistry* **36**, 5853–5860.

Weber, I. T., Wang, Y.-F. (2010). HIV protease: role in viral replication, protein-ligand X-ray crystal structures and inhibitor design. In: Aspartic Proteases as Therapeutic Targets, Ghosh, A. K. (Ed.), vol. 45, pp. 109–137. Wiley-VCH Verlag GmbH & Co., Weinheim, Germany.

Weber, I. T., Fang, B., Agniswamy, J. (2008). Caspases: structure-guided design of drugs to control cell death. *Mini Rev. Med. Chem.* **8**, 1154–1162.

Weber, I. T., Zhang, Y., Tozser, J. (2009). HIV-1 protease and AIDS therapy. In: Proteases in Biology and Disease, Lendeckel, U. and Hooper, N. (Eds.), vol. 8, pp. 25–46. Springer-Verlag GmbH, Heidelberg.

Wilmot, C. M., Pearson, A. R. (2002). Cryocrystallography of metalloprotein reaction intermediates. *Curr. Opin. Chem. Biol.* **6**, 202–207.

Wöhri, A. B., Katona, G., Johansson, L. C., Fritz, E., Malmerberg, E., Andersson, M., et al. (2010). Light-induced structural changes in a photosynthetic reaction center caught by Laue diffraction. *Science* **328**, 630–633.

Yang, X., Ren, Z., Kuk, J., Moffat, K. (2011). Temperature-scan cryocrystallography reveals reaction intermediates in bacteriophytochrome. *Nature* **479**, 428–432.

Yonath, A. (2011). X-ray crystallography at the heart of life science. *Curr. Opin. Struct. Biol.* **21**, 622–626.

Yuan, H., Fu, G., Brooks, P. T., Weber, I. T., Gadda, G. (2011). Steady state kinetic mechanism and reductive half-reaction of D-arginine dehydrogenase from *Pseudomonas aeruginosa*. *Biochemistry* **49**, 9542–9550.

STRUCTURAL, FUNCTIONAL, AND MUTAGENESIS STUDIES OF UDP-GLYCOSYLTRANSFERASES

By VATSALA MALIK AND GARY W. BLACK

School of Life Sciences, Department of Biomedical Sciences, Northumbria University, Newcastle upon Tyne, United Kingdom

Abstract

The biosynthesis of the complex carbohydrates that govern many cellular functions requires the action of a diverse range of selective glycosyltransferases (GTs). Uridine diphosphate sugar-utilizing GTs (UGTs) account for the majority of characterized GTs. GTs have been classified into families (currently 92) based on amino-acid sequence similarity. However, as amino-acid sequence similarity cannot reliable predict catalytic mechanism, GTs have also been grouped into four clans based on catalytic mechanism and structural fold. GTs catalyze glycosidic bond formation with two possible stereochemical outcomes: inversion or retention of anomeric configuration. All UGTs also belong to one of two distinct structural folds, GT-A and GT-B. UGTs have conserved residues that are associated with nucleotide diphosphate sugar recognition and acceptor recognition. UGT diversification has been performed using *in vitro* DNA recombination, domain swapping, and random mutagenesis.

ADVANCES IN PROTEIN CHEMISTRY AND STRUCTURAL BIOLOGY, Vol. 87
DOI: 10.1016/B978-0-12-398312-1.00004-4

Copyright 2012, Elsevier Inc.
All rights reserved.

I. Natural Product Glycodiversification

Approximately 50% of prescription products sold in Europe and the USA are either natural products or derivatives thereof (Newman et al., 2003). Secondary metabolites from plants and microorganisms represent diverse as well as promising candidates for the development of new medicines apart from their agronomic relevance (Gachon et al., 2005; Ahn et al., 2009). For example, flavonoids, compounds with anti-inflammatory, antioxidant, and antitumor roles form a major part of medicinal research. Of these metabolites, the vital role of natural products harboring sugar moieties, predominantly glycoproteins and glycolipids, is well documented in eubacteria, eukaryotes, archaebacteria, and viruses. The occurrence of a myriad of complex carbohydrates such as polysaccharides, glycoproteins, and glycolipids governs a diverse range of defined cellular functions such as energy storage and cell wall structural integrity in the form of cellulose and starch, respectively, in plants (Zhang et al., 2007) and other processes such as cellular interactions, cell adhesion, and immune defense (Coutinho et al., 2003; Park et al., 2009).

Therefore, since drug-linked sugar moieties participate in interactions between drugs and their biological targets consequently increasing their pharmacokinetic properties (ADME) and promote drug solubility and stability, glycosylated natural products are being exploited by the pharmaceutical industry as potential leads in drug discovery and development (Kohara et al., 2007). Newman et al. (2003) have extensively reviewed the prospects of glycosylated products as sources for new drugs. For example, several glycosides produced by microorganisms such as anthracyclines, aminoglycosides, macrolides, and plants such as lignans and anthraquinones have been used as biologically active natural products in the treatment of various diseases. Some examples of these uses have been summarized in Table I. Biotechnology companies such as Neose (Horsham, PA), for example, have therefore invested billions of dollars with a view to improve glycosylation of medically useful glycoproteins.

Structure-based approach to drug design is at the forefront of the pharmaceutical industry, and therefore, structural modification of natural metabolites is an emerging research area. In addition, from a chemical standpoint, sugar conjugation results in both increased stability (through the protection of reactive nucleophilic groups) and water solubility (Luzhetskyy et al., 2008).

Table I

Glycosylated Plant- and Microorganism-Derived Natural Products and Their Pharmaceutical Uses (Adapted from Luzhetskyy et al., 2008)

Glycosylated natural product	Mode of action	Pharmaceutical use
Glycosylated natural products from microorganisms		
Anthracyclines	DNA intercalation	Acute lymphoblastic/myeloblastic leukemias; breast cancer, juvenile tumors, soft tissue, and aggressive sarcomas
Macrolides	Binding to 50S rRNA	Respiratory tract and soft tissue infections caused by streptococci, pneumococci, staphylococci, and enterococci
Coumarins	Gyrase inhibition	Infections caused by *Staphylococcus epidermidis*
Glycopeptides antibiotics	Binding to *N*-acyl-D-Ala-D-Ala termini of non-cross-linked lipid-PP-disaccharide-pentapeptides	Infections by Gram-positive bacteria
Glycosylated natural products from plants		
Anthraquinones	Absorption, inhibition of Na^+/K^+-ATPase	Laxative agents
Flavonoids	Radical-oxygen scavengers	Tone improvement in vein walls, treatment of congestive heart failure, and improvement in blood circulation
Etoposide	DNA-topoisomerase II inhibition	Anticancer therapy

Understanding the specific roles that glycans play in the disease process could offer new targets for specific therapeutic intervention of these events (Dove, 2001). For example, highly modified glycans on tumor cell surfaces that mediate carbohydrate–protein and carbohydrate–carbohydrate interactions are involved in critical events in the progression, dissemination, and invasion of cancer cells (Barchi, 2000; Gorelik et al., 2001). Antitumor drug design strategies therefore encompass targeting the production or interaction of tumor-associated glycans by carbohydrate-based drugs or vaccines that mimic the sugars involved in these interactions. For example, cytotoxicity tests with pluramycins, antibiotics consisting of an aryl-*C*-glycoside core, have demonstrated potency against several murine and human cell lines (Bililign et al., 2005).

Understanding the biochemical implications of these altered cell surface glycans has experienced a renaissance in the past decade. For example, cell wall proteomic profiling and direct mass spectrometric analysis of fungal wall glycoproteins could help identify suitable vaccine candidates (Yin et al., 2008).

In Nature, diversification of carbohydrate structure and function occurs in two ways: (i) novel sugar structures are generated through enzymatic modification of the functional groups of a common sugar precursor—this process is most elaborate in the prokaryotic cell wall and secondary metabolite biosynthesis (Klement et al., 2007) and (ii) a variety of glycoforms with differing regio- or stereochemistries are created by the attachment of sugar substrates to "acceptor" molecules that include natural product aglycones, proteins, lipids, and other sugars by the action of specific glycosyltransferases (GTs) (Thibodeaux et al., 2007). For example, glycoproteins that form the basis for drug design consist of oligosaccharide chains covalently linked to an amino acid backbone (N-linked or O-linked), the structural composition of which is defined by the expression of GTs and glycosidase genes in different cells and the availability of their sugar nucleotide cofactors ("donors") of a specific saccharide (Barchi, 2000).

In addition, despite their versatility and potential in drug development, there are concerns over using natural carbohydrates as drugs. This is because monovalent interactions between proteins and carbohydrates are relatively low compared to known drugs and their acceptors (Barchi, 2000). Moreover, most oligosaccharides are highly water soluble and therefore prone to degradation by digestive glycosidases or clearance by hepatic receptors, highlighting concerns over the overall drug bioavailability. Further, the chemical synthesis of oligosaccharides involves regio- and stereochemistry of bond formation to be controlled for the synthesis of well-defined products (Erb et al., 2009). This presents as a laborious and challenging process that requires additional protection and deprotection steps to obtain the required stereochemistry of the final product, and proving industrially unfeasible owing to decreased product yield, waste generation, and formation of reaction by-products (Palcic, 1999; Robertson and Steer, 2004). For example, the synthesis of the pharmaceutical compound quercetin containing five —OH groups, each of which can form glycosidic bonds, four of these must be protected to synthesize a specific desired monoglycoside (Lim and Bowles, 2004).

II. Glycosyltransferases

Due to the nature of some cellular functions, wherein carbohydrate moieties act as cellular language that depends on specific carbohydrate structures, the biosynthesis of complex sugars involves the action of a diverse range of selective GTs, generating an infinite number of glycolconjugates and polysaccharides (Aharoni et al., 2006). GTs catalyze the sequential transfer of the monosaccharide units of their activated donor (usually nucleotide donor) to the specific acceptor molecules such as lipids, saccharides, proteins resulting in the formation of a glycosidic bond (Hashimoto et al., 2009).

Activated sugar-dependent transferases account for the vast majority of glycosyl transfer reactions, and even though the range of acceptors is highly diverse, the donors are typically restricted to monosaccharides linked to a nucleoside monophosphate or a nucleoside diphosphate (NDP) leaving group such as uridine diphosphate (UDP), guanosine diphosphate, and cytidine monophosphate (Thorsoe et al., 2005). Of these, UDP-glycosyltransferases (UGTs) account for more than 60% of all GTs characterized so far (Hu and Walker, 2002).

A. GTs Across Genomes

In eukaryotes, only 10 different NDP-sugars are utilized in the majority of biosynthetic glycosylation reactions; however, prokaryotic NDP-sugars are much more diverse (Thorson et al., 2001). Bacterial GTs therefore serve as excellent targets for drug development since most of these sugars are absent in humans. Since these sugars are important constituents of the bacterial cell wall, their biosynthesis involves mechanistically novel enzymatic steps that have been actively targeted for inhibitor development. For example, CHLαGcT, a cholesterol glucosyltransferase from a peptic and gastric ulcer causing bacterium *Helicobacter pylori*, represents an excellent drug target due its vital role in the synthesis of cell wall lipids (Lebrun et al., 2006; Lee et al., 2006).

On the other hand, considering the vast number of glycosides in plants compared to other eukaryotes, it was believed that plant-derived GTs would be more promiscuous. Even though plants GTs are known for their highly regio- and stereoselective nature, they can recognize common features on a range of substrates including hormones, secondary metabolites, or xenobiotics. For example, glycosylation at the 3-OH position of certain plant pigments such as anthocyanins is crucial for the stability of the aromatic ring (Fukuchi-Mizutani et al., 2003; Gachon et al., 2005). This may imply that a wide range of substrates are accepted by plant GTs. For example, Offen et al. (2006) described the promiscuity of *Vv*GT1, a grape flavonoid UGT with respect to both the acceptor and the donor substrates. It is important to note, however, that even though *in vitro* studies suggest plasticity in GT reactions, low substrate specificity could lead to undesired side reactions within the natural environment unless activity is regulated under transcriptional control (Osmani et al., 2009).

In addition to prokaryotes and other eukaryotes, glycosylation is a well-established phenomenon in fungi where complex polysaccharides and glycolipids such as glucans and chitin form vital cell wall components (Gorelik et al., 2001; Hashimoto et al., 2009). Although glycolipids are known to contribute to the biophysical properties such as ion impermeability of membranes in these organisms, their specific roles have only recently being characterized. For example, membrane lipid–protein interactions are now known to govern the folding, localization, and function of membrane proteins. More recently, Warnecke and Heinz (2010) have suggested an alternative but complementary approach to the current methods being employed for deciphering the function of specific glycolipids. Instead of creating deletion mutants of GTs involved in lipid metabolism that result in phenotypes with incomplete or lack of membrane systems in many cases, this approach would involve an exchange of a lipid GT in the organism by a heterologous GT with different specificities, for example, substitution of a galactosyltransferase with a glucosyltransferase. This would help in creating mutants without the complete loss of general functions but would highlight the specific role of the lacking membrane lipid.

N- and O-linked glycoprotein synthesis is, however, a well-established process in yeasts such as *Saccharomyces cerevisiae*, *Pichia pastoris*, and *Candida albicans*, orchestrated by the action of specific GTs (Kim et al., 2006; Klutts et al., 2006; Mille et al., 2008). Cell wall glycoproteins involved in enhancing the stability and solubility of secretory proteins have functions in the maintenance of fungal morphology in filamentous yeasts such as the *Aspergillus* species and are involved in virulence, for example, in the fungal pathogen *Cryptococcus neoformans* that possesses a glycan-rich cell wall surrounded by an intricate polysaccharide capsule (Klutts et al., 2006; Goto, 2007). For example, an *Aspergillus nidulans* mutant lacking a protein *O*-mannosyltransferase showed abnormal hyphae development and absence of conidia, suggesting the vital role of protein *O*-mannosyltransferase in fungal development (Goto, 2007).

The majority of GTs involved in eukaryotic cell wall polysaccharide and glycoprotein synthesis are Golgi-resident type II-membrane-localized proteins consisting of a short N-terminal cytoplasmic domain followed by a transmembrane domain, a stem region, and the lumen facing C-terminal globular catalytic domain (Keegstra and Raikhel, 2001; Rabbani et al., 2006; Qasba et al., 2008).

Quite recently, proteomic approaches coupled with mass spectrometry have enabled the rapid and sensitive identification as well as the quantification of fungal wall glycoproteins and the associated GTs, which may pave the way to the development of novel diagnostic tools as well as vaccines (Yin et al., 2007; Lin et al., 2008).

B. Classification of GTs

The rapidly increasing number of genome projects has provided a wealth of sequence and biochemical information. GTs represent one of the most diverse groups of enzymes in biological systems, and their classification thus provides a useful insight into their divergent evolution. Over the years, a few different systems of GT classification have been proposed.

It was initially believed that the most logical way to classify GTs would be according to their reaction mechanisms. Based on this, all GTs would be grouped as either "inverting" or "retaining" (Kapitonov and Yu, 1999) (see reaction mechanisms described in Section II.B.1).

GTs may also be classified based on differences in the carbohydrate glycosyl donors involved in the reaction (Compain and Martin, 2001). Thus, they may either belong to the Leloir pathway, named after the 1970

Nobel Peace prize winner Luis Leloir following the discovery of UDP D-glucose and glycoside biosynthesis, or the non-Leloir pathway. The former comprises approximately 65% of all GTs, where donors may be activated by the addition of a mono- or diphospho-nucleotide, therefore forming a nucleotide sugar; and the non-Leloir pathway GTs utilize non-nucleotide donors as donors (Weijers et al., 2008).

However, the universally recognized system for GT classification based on amino acid sequence similarity (Campbell et al., 1997) is the Carbohydrate Active Enzymes (CAZy) database (http://www.cazy.org/) (Cantarel et al., 2009). This system comprises enzymes with roles in carbohydrate metabolism, biosynthesis, and modification and is composed of four enzyme classes—glycoside hydrolases, glycosyl transferases, polysaccharide lyases, and carbohydrate esterases. Each class is further divided into families that incorporate both structural and mechanistic (stereochemistry of the glycoside linkage) features with members of one family assuming similar 3D structural folds (GT-A, GT-B, or GT-C) and catalytic mechanism (Luzhetskyy et al., 2005). However, sequence comparison cannot reliably predict catalytic mechanism (Osmani et al., 2009), and therefore, Coutinho et al. (2003) proposed subgrouping of GTs into four clans as either inverting GT-A fold (clan I) or GT-B fold (clan II) or retaining GT-A fold (clan III) or GT-B fold (clan IV).

So far, GTs have been classified into 92 CAZy families with over 50,000 open reading frames, encoding putative GTs isolated from various plant, mammalian, fungal, and prokaryotic sources.

This classification is underpinned by the assumption that signature structural motifs can be deciphered on the basis of primary protein sequence information (Henrissat and Davies, 1997). Enzymes within a specific family may therefore have very different substrate specificities but be closely enough related to be considered divergent of one historical ancestor (i.e., homologous). It can also highlight convergent evolution whereby enzymes with very similar function and substrate specificity which would normally be grouped together by the IUBMB (International Union of Biochemistry and Molecular Biology) system of classification are shown to belong to several distinctly different families. For example, the large polyspecific families, such as GT family 2 (GT2), consist of more than 3500 sequences that vary in donor and/or acceptor specificity and therefore demonstrate distinct functions (Breton et al., 2006). GT classification based on catalytic mechanism and structural fold will be further described in the following sections.

1. Classification Based on Catalytic Mechanism

GT-catalyzed reactions proceed by the sequential binding of the metal ion (if required for activity as a cofactor) and sugar nucleotide to the enzyme, followed by the acceptor (Qasba et al., 2005; Seibel et al., 2006). The saccharide product is then ejected upon glycosyl transfer by "inversion" or "retention." Analogous with glycoside hydrolases, GTs catalyze glycosidic bond formation with two possible stereochemical outcomes—inversion of the anomeric configuration (e.g., UDP-glucose to β-glucoside) or retention of the anomeric configuration (e.g., UDP-glucose to α-glucoside). There is limited knowledge at the sequence level, indicating differences between the two reaction mechanisms even though differences in UDP-binding amino acids have been suggested (Rosen et al., 2004).

Inverting GTs catalyze via a single displacement mechanism with nucleophilic attack by the acceptors at the C-1 (anomeric carbon) of the sugar donor (Fig. 1). This mechanism requires an active site side chain to serve

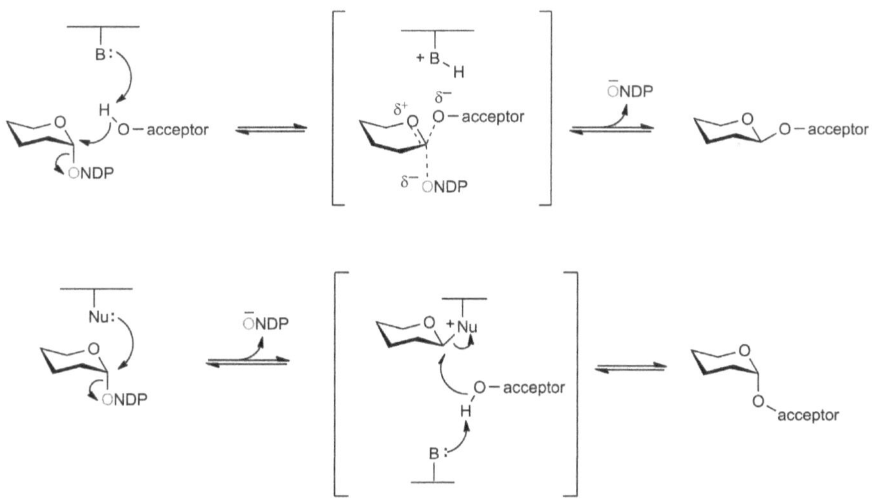

FIG. 1. *Reaction mechanism of GTs.* Inverting GTs-**1**: general base (B)-catalyzed S_N2 attack at C1 of the NDP-sugar donor by the acceptor forms an oxocarbenium-type transition state leading to the inversion of stereochemistry at the anomeric carbon; **2**: the double-displacement mechanism for retaining GTs involves the S_N2 attack of an active site nucleophile (Nu) at C1 of the NDP-sugar to form a covalent sugar–enzyme intermediate, followed by an S_N2 attack of the acceptor at C1 to form the glycosidic linkage with retention of anomeric configuration (For color version of this figure, the reader is referred to the online version of this chapter.).

as a base to activate the sugar acceptor for nucleophilic attack by deprotonation because sugar hydroxyls themselves are poor nucleophiles (Charnock et al., 2001; Lairson et al., 2008). In addition, most GT clan I enzymes characterized so far use a divalent cation (Mn^{2+} or Mg^{2+}) coordinated by the diphosphate moiety of the NDP-sugar. This ion binds with the conserved active site amino acid side chains to facilitate the departure of the NDP-leaving group by stabilizing the developing negative charge.

The inverting mechanism for GTs was first highlighted by the 3D X-ray structure of both native and UDP-complexed SpsA, a GT-2 family member from *Bacillus subtilis* (Charnock and Davies, 1999). Three side-chain aspartate residues are predominantly involved in binding the UDP-sugar. It showed that in the N-terminal, a nucleotide-binding domain, D39 coordinated with the uracil base of the sugar donor, whereas D98 and D99 (from a conserved DXD motif; described in Section II.B.2) were coordinated with ribose and UDP-bound Mn^{2+}, respectively. In addition, the side-chain carboxylate of D191 served as a base catalyst in the inverting mechanism.

Two reaction mechanism strategies have been proposed for retaining GTs. The earlier proposed strategy states the involvement of a double-displacement mechanism with the formation of a covalent intermediate. The S_N2 double-displacement mechanism involves a direct nucleophilic attack on the C1 of the UDP-sugar to form a β-linked glycosyl-enzyme intermediate. The leaving group phosphate itself plays the role of a base catalyst, activating the incoming acceptor hydroxyl for nucleophilic attack (Seibel et al., 2006; Lairson et al., 2008).

Another more likely adopted strategy suggests that the enzyme uses a S_N1 transition state in which the approach of the attacking donor and the leaving donor is on the same side of the sugar ring (Zhang et al., 2007). It involves the production of a nucleophile by the dissociation of the leaving group NDP and attack on the transition state by the acceptor from the same side of the sugar ring as the scissile glycosidic linkage to the leaving NDP.

Although both inverting and retaining catalytic mechanisms have been proposed, only the mechanism of inverting GTs has been well characterized so far (Qasba et al., 2005). This is due to the absence of any studies reporting the identification of the involvement of a catalytic nucleophile and/or a glycosyl intermediate as has been suggested in the reaction mechanism.

2. Classification Based on Structural Fold Topology

As compared with glycoside hydrolases, where structural representatives from a large number of the 122 families are available, there is a relatively limited structural database for GTs (Hu and Walker, 2002). Interestingly, in the former case, the highly diverse overall structural folds with conserved catalytic domains suggest evolutionary convergence during development of catalytic mechanisms. However, all NDP-sugar-utilizing GT structures so far fall into one of the two distinct structural fold families—the GT-A or GT-B. These observations indicate that a vast number of GTs may have evolved from a small number of progenitor sequences (Coutinho et al., 2003).

The two main GT folds (A and B) are well documented, with the overall topology of GT-A fold members consisting of two separate but closely associated β/α/β domains creating an active site in the center. Figure 2A depicts the structure of *Mycobacterium avium* GT (MAP2569c) that adopts this fold (Fulton et al., 2008). The N-terminal or NDP-sugar-binding domain possesses a common DXD or EXD motif (Asp-X-Asp or Glu-X-Asp) flanked by nonpolar residues. The DXD motif interacts with the phosphate group of the donor through the coordination of the divalent metal ion cofactor. The acceptor binds at the C-terminal. Most Leloir pathway Golgi and endoplasmic reticulum residing GTs, as well as most prokaryotic GTs, belong to this fold family (Hu and Walker, 2002).

All GT-A fold members characterized so far, such as the α-mannosyltransferases and galactosyltransferase, require the presence of a metal ion cofactor that coordinates with the DXD side-chain motif or its variants (e.g., EDD, TDD) and was therefore believed to be a characteristic feature of GT-A enzymes. However, one GT-A structure has been reported recently that does not possess this signature motif, and Arg and Lys side-chain residues were involved in stabilization of electrostatic charges (Pak et al., 2006). Differences are also observed in the function of the DXD conserved motif residues in inverting and retaining GT-A enzymes. In the former case, both Asp residues interact with the metal ion, whereas only the second Asp residue interacts with the metal ion in inverting members (Breton et al., 2006).

The second structural GT-B fold was first studied in the T4 β-glucosyltransferase, and members of this family consist of two separate Rossman-like β/α/β domains with a connecting linker region and a deep cleft

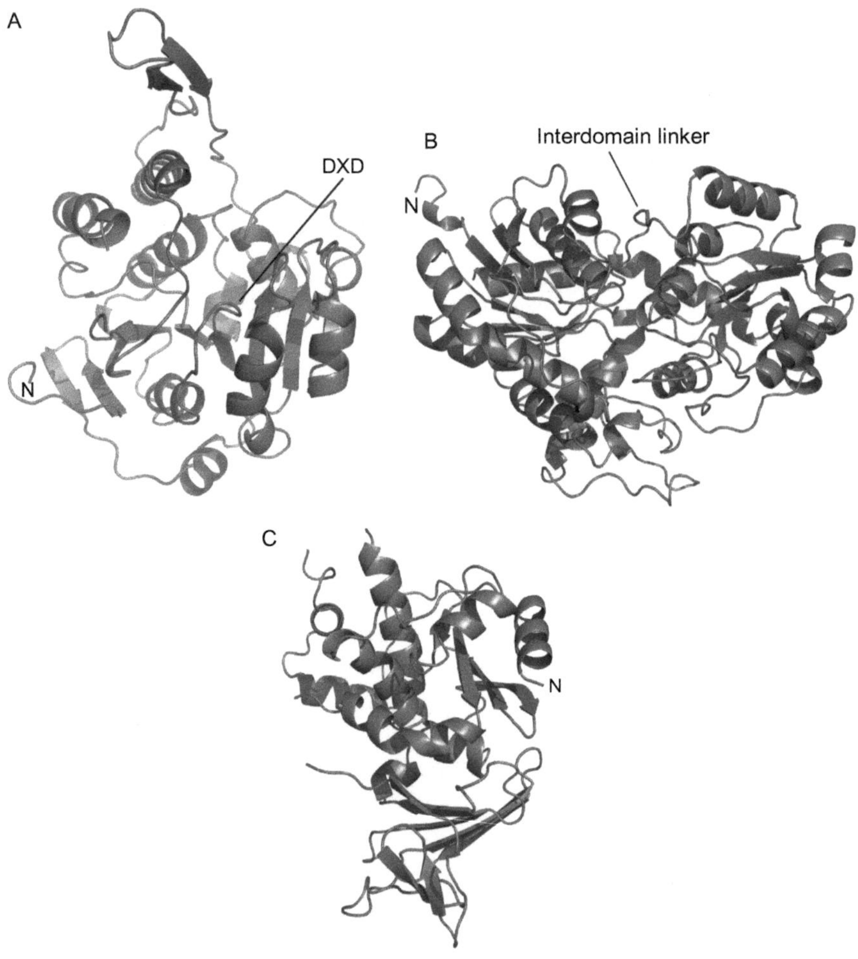

FIG. 2. Cartoon diagrams of GT structures depicting GT-A, GT-B, and GT-C folds. (A) GT-A fold of the MAP2569c structure showing two separate but closely associated β/α/β domains creating an active site in the center (Fulton et al., 2008); the catalytic DXD motif is indicated. (B) GT-B fold of the UGT71G1 structure also showing two separate β/α/β domains creating an active site in the center (Shao et al., 2005); the interdomain linker is indicated. (C) GT-C fold of the C-terminal domain of the PglB structure showing a 10-stranded β-sheet structure within the α-helical core domain (Maita et al., 2010). N, N-terminus. This figure was created using PyMOL (http://www.pymol.org/). (For interpretation of the references to color in this figure legend, the reader is referred to the online version of this chapter.)

between the domains which functions as the catalytic center. Figure 2B shows the GT-B fold of the crystal structure of a plant flavonoid GT (UGT71G1) (Shao et al., 2005). As opposed to the GT-A family, the C-terminal domain is the nucleotide-binding domain in this case. The N-terminal domains are less conserved and are evolved to accommodate different acceptors (Kapitonov and Yu, 1999). Members of this fold family comprise most of the prokaryotic GTs, such as those involved in glycosylation of natural secondary metabolites and cell wall synthesis (Thibodeaux et al., 2007), and are therefore more diverse compared to GT-A members.

As opposed to GT-A family enzymes, GT-B members catalyze in a metal ion-independent manner (Breton et al., 2006). An example of this can be demonstrated by Mulichak et al.'s (2003) study where the crystal structure of clan II GtfA from *Amycolatopsis orientalis* responsible for vancomycin biosynthesis showed that helix dipole and interactions with side-chain hydroxyl and imidazole groups were vital in stabilizing the negative charge on the leaving group and therefore did not require the presence of divalent cations.

Iterative searches and multivariate data analysis employing several computational models and sequence information of characterized and putative GTs have revealed the prediction of a third GT-C fold (Liu and Mushegian, 2003; Wimmerová et al., 2003; Rosen et al., 2004) now classified for 10 GT families based upon two crystal structures. The predicted structure of the GT-C fold had indicated a large hydrophobic integral membrane protein which consisted of 8–13 transmembrane helices with a catalytic signature motif in the extracellular loop (Igura et al., 2008; Lairson et al., 2008). Figure 2C shows the structure of the C-terminal domain of a GT66 oligosaccharyltransferase (OST) from *Campylobacter jejuni* (PglB) that adopts the GT-C fold (Maita et al., 2010). OSTs are involved in N-glycosylation by catalyzing oligosaccharide transfer from lipid donors to polypeptide chains and are responsible for the virulence of pathogens such as *C. jejuni* (Maita et al., 2010).

So far, all CAZy family members predicted to adopt this fold utilize lipid phosphate-activated donor sugar substrates (Lairson et al., 2008) as opposed to nucleotide-sugar donors or pyridoxal phosphate utilizing GT-A and GT-B fold enzymes. GT-C family members therefore belong to the non-Leloir pathway of GTs. Thus, GT-C families mainly comprise of α-mannosyltransferases that use dolichol pyrophosphate-mannose as the sugar donor for the O-linked mannosylation of membrane lipids. In addition, GT85 arabinofuranosyltransferases (Afts) are involved in the synthesis of *Mycobacterium tuberculosis* cell wall components as shown by Alderwick et al. (2006).

C. GT Family 1 UGTs

The GT1 family currently consists of over 3600 sequences with representatives from of all phylae. Members of this family adopt the GT-B fold and mechanistically are inverting in nature, classifying them within clan II according to the classification proposed by Coutinho et al. (2003). There is widespread interest in GT1 members as they are mainly recognized for the glycosylation of small molecules including important organic compounds such as terpenes, steroids, macrolides as well as are involved in detoxification pathways (Lim and Bowles, 2004; Lairson et al., 2008). For example, the pollutant 3,4-dichloroaniline was detoxified by glucosylation using an *Arabidopsis* GT (*N*-GT) (Loutre et al., 2003).

Of the total 93 3D structures of GTs solved so far, the GT1 family comprises 19, which is the maximum number of solved structures for any one GT family so far. This makes homology modeling an attractive tool for structure elucidation of novel GTs classified within this family.

GT1 includes UGTs as members where they represent the majority of all sequences within this family. Of these, plant-derived sequences are most prominent. For example, approximately 120 *Arabidopsis* sequences are included within the UGT superfamily (Hansen et al., 2010). Sequences sharing at least 40% amino acid sequence identity are further classified within the same UGT subfamily (Osmani et al., 2009). All characterized as well as putative plant UGTs identified so far lack the presence of a signal peptide sequence or membrane-spanning domains, suggesting that these enzymes mainly function within the cytosol but may be associated as peripheral components of the endomembrane system (Lim and Bowles, 2004).

In addition, all GT1 UGT sequences, regardless of their origin, share a conserved consensus sequence in the C-terminal domain denoted as the UGT signature sequence as originally proposed by Mackenzie et al. (1997). Among plant UGTs, this signature sequence is more conserved and is represented by a 44 amino acid PSPG (putative secondary product GT) box (Vogt and Jones, 2000). This region has been characterized as the NDP-sugar-binding domain (Mackenzie et al., 1997; Lim and Bowles, 2004) which will be described in the following section.

GT1 also includes a large number of glucuronosyltransferases (EC 2.4.1.17) involved in cellular homeostasis and phase II detoxification pathways (Offen et al., 2006). This class of GTs catalyzes glucuronic acid transfer to potentially toxic lipophilic compounds such as bilirubin and steroid hormones, thereby promoting their systemic clearance as highlighted earlier (Lim and Bowles, 2004; Meech and Mackenzie, 2010).

GT1 family UGTs contain a class of UDP-sugar:sterol glucosyltransferases (USGTs), also known as UDP-glucose:sterol 3-*O*-β-D-glucosyltransferases (EC 2.4.1.173). The first UGT of this class was purified from oat, following which USGTs from both prokaryotes and eukaryotes were characterized (Idnurm et al., 2003; Madina et al., 2007; Kim et al., 2010). However, reports of sterols from bacteria are rare, and only a few members of the *Mycoplasma* and *Helicobacter* species have shown to produce steryl glycosides from free sterols that they take up from their host organisms (Warnecke and Heinz, 2010). Within the last decade, the identification and cloning of several sterol GT genes have aided the use of genetic approaches to investigate the current limited understanding of the specific biological functions of steryl glycosides in eukaryotes (Grille et al., 2010).

D. *Conserved Amino Acid Residues are Involved in Substrate Recognition*

Until recently, the molecular basis for GT donor and acceptor selectivity was unclear. However, in the past decade, with a surge in the number of crystal structures that have become available from over 26 CAZy families within the GT 3D structure database (http://glyco3d.cermav.cnrs.fr/cgi-bin/rxgt/rxgt.cgi), it is now possible to better understand GT catalytic domains.

GTs are highly regio- and stereoselective with regard to substrate recognition *in vivo* (Weijers et al., 2008). However, as illustrated earlier in the case of plant GTs, modifications on the acceptor substrates can be tolerated *in vitro*. However, analysis of substrate specificity based on the contribution of individual functional groups for recognition and binding poses a great challenge due to the limited availability of unnatural oligosaccharides and high cost of NDP-sugar donors.

In the following sections (Sections II.D.1 and II.D.2), conserved amino acids and/or peptide motifs that play a major role in GT function will be discussed, with a special emphasis on GT-B fold UGTs as they are supported with better structural information.

1. *Conserved Residues in NDP-Sugar Recognition*

A general pattern for substrate recognition can be observed among characterized GT-A and GT-B fold members. An interdomain linker connects the N- and C-terminal domains, and the secondary structure elements are connected by less conserved, flexible loop regions associated with substrate binding. For example, in the GT-A family, these flexible loops are associated with NDP-donor-induced conformational change from open to closed where the loop acts as a lid covering the bound donor substrate creating an acceptor-binding site (Qasba et al., 2005). This is also suggested to prevent hydrolysis of the sugar donor (Unligil and Rini, 2000; Breton et al., 2006). The length and composition of the linker region may differ among different GTs and serve to define domain positioning and consequent substrate accommodation. For example, the interdomain linker shows both length and sequence variability in the crystal structures of four GT1 family plant UGTs—*Mt*UGT71GT, *Vv*GT1, *Mt*UGT85H2, and *At*UGT72B1 (Osmani et al., 2009). This region differs significantly in comparison with crystal structures of bacterial GT1 members—GtfA, GtfB, GtfD from *A. orientalis* where it is much shorter and is directly involved in NDP-sugar binding (Mulichak et al., 2001, 2003, 2004). No single motif is conserved across GT families classified within the CAZy database (Hashimoto et al., 2009). However, the presence of a few conserved residues in the NDP-binding pocket across GT members from different species determines nucleotide-sugar specificity and/or transfer to aglycones (Qasba et al., 2005). A classical example of this can be demonstrated by the presence of highly conserved His, Asp/Glu, Gln residues in the C-terminal PSPG motif (also known as the UGT signature sequence for UGTs, as described earlier) of the GT1 family members. A study by Potocka and Zimowski (2008) revealed that the Glu or Asp residue(s) in the *Solanum melongena* (eggplant) UGT were crucial for UDP D-glucose binding, whereas the His and Cys residues were important for both nucleotide and sterol binding.

The additional residues within the UGT signature sequence that do not directly interact with the sugar donor are believed to be involved in facilitating intramolecular interactions to stabilize the developing negative charge, as previously described (Section II.B.1).

As described earlier, GT-A fold members coordinate the essential metal cation via the DXD motif. Although GT-B fold members lack this motif, they do exhibit a pattern of Gly and Ser/Thr residues located within the donor-binding domain (Klutts et al., 2006). These residues have been shown to interact with the phosphate group of the NDP-donor. For example, the crystal structure of a GT-B family *N*-acetylglucosaminyltransferase (MurG) from *Escherichia coli* involved in peptidoglycan biosynthesis contains three glycine-rich loop motifs: one in the C-terminal that is suggested to bind the phosphate of the NDP-sugar, and the other two within the N-terminal domain proposed to interact with the phosphate groups of the lipid-linked acceptor (Ha et al., 2000).

Although, in the case of GT-B family UGTs, the majority of sugar substrate interactions occur in the structurally conserved C-terminal domain, some key N-terminal amino acids may also play a significant role in donor recognition and binding.

2. *Conserved Residues in Acceptor Recognition*

In both the GT-A and GT-B members, the acceptor-binding domain is generally less conserved, suggesting that these may have evolved in order to accommodate a diverse range of aglycones (Hansen et al., 2009). However, some key residues with charged side chains located within the acceptor-binding pocket act as the general base triggering deprotonation of the acceptor as described in the reaction mechanism of GTs. In a majority of GT-B inverting enzymes, this base is His, whereas the carboxylate group of the base catalyst Asp/Glu has been shown to be involved in deprotonation of the acceptor in some GT-A family members as illustrated earlier. The specific roles of these residues can be illustrated by the following examples. Mutation of the H19 and D117 resulted in complete loss of *At*UGT72B1 activity of the respective Ala variants, suggesting the vital role of these residues in the nucleophilic attack on the acceptor (Brazier-Hicks et al., 2007). It is also proposed that in a majority of inverting GTs, the histidine forms an acceptor-H-D triad where the Asp residue balances the charge on the His after proton abstraction. This was demonstrated by Shao et al. (2005) where substitution of the H22 and D121 with Ala in another plant UGT, UGT71G1, resulted in inactive variants suggesting that interaction of Asp with His was critical for catalytic activation by possibly forming an electron transfer chain.

E. *Bioinformatic Approaches for Structure and Function Prediction of GTs*

The number of families is continually increasing with the discovery and biochemical characterization of new GT genes. However, functional prediction of a putative GT based on sequence homology can be unreliable, as closely related sequences may express different catalytic activities. One such example can be presented by the UGTs of *Streptomyces fradie,* UrdGT1b and UrdGT1c, that are responsible for the synthesis of the antibiotic urdamycin. These GTs differ in acceptor specificity despite sharing 91% amino acid identity (Hoffmeister et al., 2001), thereby exemplifying the complex nature of substrate specificity.

For large polyspecific families such as the GT2, it is especially difficult to predict the specificity. Sequence similarities are generally observed for the entire catalytic domain in the case of monospecific GT families, as opposed to polyspecific families where they are mostly restricted to only a portion of the catalytic domain (Breton et al., 2006). These sequences therefore cannot be classified functionally using bioinformatics (Hansen et al., 2010). Thus, novel putative GTs with no characterized homologues can only be identified on the basis of catalytic signature motifs employing bioinformatics strategies. This was further highlighted in the case of the GT-C family classification as demonstrated by the following examples.

As outlined earlier, bacterial GTs represent excellent potential drug targets and with the genome sequence of human pathogens such as *M. tuberculosis* H37Rv available, cell wall-synthesizing GTs for this pathogen, for example, are both an attractive and unexplored area. For example, the *M. tuberculosis* cell wall-synthesizing enzymes EmbA, EmbB, and EmbC are currently one of the only few known drug targets for tuberculosis treatment and show no sequence similarity to homologues from other species (Berg et al., 2005).

Further, GT-C members characterized so far use lipid donors and are pivotal in archaeabacterial and eubacterial cell wall synthesis, clearly suggesting the exclusivity of the involvement of these GTs in the process. However, GT-C classification will need to be further established on the basis of structural data as so far, no characterized lipid phospho-sugar donor utilizing peptidoglycan-synthesizing GTs have been assigned the GT-C fold. For example, crystal structures of the soluble domains two prokaryotic peptidoglycan-synthesizing GTs grouped under GT51 have recently been determined and have interestingly shown to adopt a novel λ-lysozyme-type fold even though lysozymes degrade peptidoglycan chains (Lovering et al., 2007; Yuan et al., 2007).

On the other hand, the first 3D crystal structure of the GT-C assigned archaeal OST STT3 from *Pyrococcus furiosius* was determined recently (Igura et al., 2008). Similar to the *C. jejuni* OST, the *P. furiosius* OST adopts a novel α-helical fold surrounded by three β-sheet domains and is void of any characteristic Rossman-folds (Igura et al., 2008). Interestingly however, even though the C-terminal domain used for structural analysis consisted of the loop harboring the predicted active site, the truncated protein resulted in a loss of activity, suggesting the vital role of the

N-terminal transmembrane region in OST activity. On comparison with the only other GT-C family solved structure an OST from the prokaryote *C. jejuni*, it was observed that the two OSTs shared significant structural similarity despite low sequence similarity (Maita et al., 2010) and are now grouped under GT66.

It is interesting to note that both the GT-C structures available so far have been determined using only the truncated soluble C-terminal domains of OSTs. These domains consist of conserved catalytic motifs in the first extracellular loop after the N-terminal domain (Berg et al., 2005). The Asp in the conserved WWDYG motif acts as a catalytic base, whereas the DK and DXD motifs are involved in binding lipid-linked donors through a transiently bound cation (Igura et al., 2008). Based upon this, a novel Aft from *M. tuberculosis* (AftA) was identified and classified as a GT-C fold member (GT85) following structural sequence alignment with these conserved motifs (Alderwick et al., 2006).

The above studies therefore highlight the fact that computational methods based on motif-extraction, hidden Markov models (HMM) prediction, and structural comparison of sequenced genomes (Wimmerová et al., 2003; Rosen et al., 2004) may represent a more reliable approach in identifying novel GTs and predicting their structural folds, as primary sequence properties may be more conserved despite low sequence similarity.

III. Diversifying Functions of UGTs

The rational design of novel nonnatural glycosylated compounds with desired properties requires a detailed knowledge of the function and mechanism of GTs. As illustrated in the previous sections, GT substrate specificity cannot be predicted from sequence comparison alone (Osmani et al., 2009). Even though the number of GT families classified in the CAZy database has almost doubled within the last decade, structural elucidation studies have struggled to keep pace with the increasing amount of sequence information of gene clusters involved in glycosylated natural product synthesis. Structure- or sequence-guided rational GT engineering attempts have largely been limited to single site-directed mutagenesis to decipher the roles of specific residues and/or elicit desired changes to enzyme properties (Hancock et al., 2006; Kim et al., 2010).

Since GTs display a two-domain architecture with conserved 3D structural folds, yet varied substrate (donor and/acceptor) preferences, they represent ideal candidates for gene manipulation and engineering by methods such as directed evolution and combinatorial biosynthesis. These approaches would not only increase our present understanding of catalytic domains but would also help generate GTs with novel specificities (Hancock et al., 2006). Rational engineering and directed evolution approaches are not entirely exclusive of one another because even though the latter approach does not necessarily require knowledge of the enzyme structure, knowledge-based design of GT variant libraries may improve overall engineering strategies (Dalby, 2007).

The following sections will aim to review some of the approaches that have been employed to obtain novel enzymes with improved biochemical, physical, and catalytic properties as well as desired substrate specificities.

A. *GT Metabolic Pathway Engineering as a Means to Natural Product Glycodiversification*

Due to the limited availability of NDP-sugars as well as the difficulty in accessing chemical derivatives using chemical synthesis routes, whole-cell catalysis for the generation of glycosylated natural products is an evolving area. For example, in a recent study, a plant flavonoid GT was expressed in *S. cerevisiae* for the glycosylation of naringenin using endogenous UDP D-glucose (Werner and Morgan, 2009).

The metabolic pathway of interest of a natural product producing strain can be manipulated to generate novel natural products through gene disruption and heterologous expression of a gene encoding a GT catalyzing the synthesis of the required glycosylated product. For example, *dnrV*, the gene encoding 4-ketoreductase of the daunorubicin pathway of *Streptomyces peucetius* was disrupted and replaced by its homologue *eryBIV* that altered this biosynthetic pathway into producing epirubicin, an antitumor drug with improved therapeutic properties (Thibodeaux et al., 2007; Luzhetskyy and Bechthold, 2008).

Combinatorial biosynthesis is the systematic modification and interchange of genes involved in the biosynthesis of natural products leading to the production of unnatural or hybrid natural products *in vivo* (Rix et al., 2002; Wohlgemuth, 2005). In this emerging technology, whole cells are reengineered by introducing genes/chimeras for carbohydrate-processing pathway enzymes from other organisms. So far, this approach has been mainly adopted for the biosynthetic pathways of the polyketide group of antibiotics such as erythromycin and vancomycin (Rix et al.,

2002; Walsh et al., 2003) even though other examples have been reported. For example, in an attempt to produce a complex human *N*-glycan, the endogenous yeast glycosylation pathways of *P. pastoris* were replaced with chimeric GTs including mannosidases, *N*-acetylglucosaminyltransferases, and the UDP-*N*-acetylglucosamine (GlcNAc) transporter (Hancock et al., 2006). However, in some cases, the low transformation efficiency coupled with conjugational incompetence of host strains may pose limitations to the size of the target gene/insert to be expressed (Hoffmeister et al., 2001).

B. Directed Evolution of GTs

Directed evolution techniques mimic natural evolution processes such as random mutagenesis and sexual recombination (Dalby, 2007). It is the generation and selection of a molecular library of enzymes with sufficient diversity for the altered function to be represented. Evolved variants are then selected and used as templates for the next round of evolution (Cohen et al., 2001; Rubin-Pitel and Zhao, 2006).

In directed evolution studies, gene fragments of interest can be engineered *in vitro* or in their natural cellular environments for glycosylated product synthesis. However, in the case of GTs, such as those reported by Love et al. (2006), loss of enzyme activity does not result in a phenotypic change and therefore cannot form the basis of a cell-based selection method.

1. In Vitro *DNA Recombination and Domain Swapping*

GT engineering and domain-swapping studies have guided the identification of specific side-chain residues or peptide motifs that govern substrate specificity in addition to offering a greater understanding of intra- and interdomain interactions (Osmani et al., 2009). However, due to the complexity of substrate recognition where several regions are involved, chimeric GTs have been generated using homologous UGTs (Cartwright et al., 2008; Krauth et al., 2009). For example, based on previous knowledge of the specific protein region of 31 amino acids of the GTs, UrdGT1b and UrdGT1c, that defines both donor and acceptor recognition, Hoffmeister et al. (2002) created a chimeric gene library harboring codons for the 10 nonidentical residues in this region from both parental enzymes. In addition to exhibiting parental specificities for urdamycin synthesis, several variants displayed novel glycosylation patterns and exclusively catalyzed the synthesis of a derivative, urdamycin P that contained a branched saccharide appendage.

However, recently, Hansen et al. (2009) successfully applied the domain-swapping strategy too closely as well as distantly related family 1 plant UGTs. The chimeras were constructed based on sequence alignments and molecular modeling with known UGT structures and generated functionally active UGTs. For example, chimeras constructed with UGTs, *At*UGT71C1 and *At*UGT71C2, showed a threefold increase in K_{cat} of etoposide glycosylation compared with the wild-type enzymes. Increased catalytic efficiency toward etoposide and a novel regiospecificity toward *trans*-resveratrol were also observed for chimeras constructed with UGTs, *At*UGT71C1 and *Sr*UGT71E1, even though these sequences displayed only 38% amino acid sequence identity.

2. *Random Mutagenesis*

In random mutagenesis, libraries containing random combinations of single and multiple point mutations across entire gene/gene fragment can be generated by techniques such as error-prone PCR using a low-fidelity DNA polymerase. Thorson and coworkers (Williams et al., 2007; Williams and Thorson, 2008) evolved GTs involved in the glycosylation of specific macrolide antibiotics by random mutagenesis using error-prone PCR. In one such example, a triple mutant of the oleandomycin GT (OleD) from *Streptomyces antibioticus* showed a $\sim$30-fold higher specific activity value against the fluorescent acceptor 4-methylumbelliferone (4-Mu) as compared to wild-type OleD (Williams et al., 2007).

3. *Challenges for the Directed Evolution of GTs*

A hurdle to the scope of directed evolution experiments involving enzymatic and chemoenzymatic *in vitro* synthesis of glycoforms is created by the limited availability of NDP-sugar donors. However, recent work has suggested improvements in chemical methods of NDP-sugar synthesis, as well as enzyme-based approaches for the synthesis and regeneration of sugar nucleotides (Mao et al., 2006; Ruffing et al., 2006; Masada et al., 2007; Pesnot and Wagner, 2008). One such method for glycodiversification is glycorandomization which uses the inherent or engineered promiscuity of anomeric kinases and nucleotidyltransferases for the *in vitro* synthesis of accessible NDP-sugar donor libraries (Williams et al., 2007). However, the feasibility of the use of such methods requires further validation.

Also, most directed evolution and genetic engineering platforms require an efficient screening system for evolved variants. However, due to the magnitude of the library generated (10^3–10^7 variants), the exploitation of random mutagenesis studies is solely dependent on the availability of a robust high-throughput screening methodology based on a simple selection for glycosidic bond formation. Currently, for *in vitro* evolution studies, the lack of suitable and efficient high-throughput screening systems to select for the desired activity is well documented and represents the main focus of research for expanding the scope of GTs. The majority of the current screening methodologies involve either linking the aglycone to a surrogate fluorescent acceptor (Trubetskoy and Shaw, 1999) or derivitization of the acceptor (Warnecke et al., 1999; Cowan et al., 2008), which, in both cases, leads to a detectable change linked with product formation. A few recent studies have, however, reported the development of colorimetric screens for the rapid screening of GT activity (Deng and Chen, 2004), the wider scope of which needs to be further assessed.

Acknowledgment

We would like to thank Prof. Stephen Stanforth for his help in creating Fig. 1.

References

Aharoni, A., Thieme, K., et al. (2006). High-throughput screening methodology for the directed evolution of glycosyltransferases. *Nat. Methods* **3**(8), 609–614.

Ahn, B. Chan, et al. (2009). Formation of flavone di-O-glucosides using a glycosyltransferase from *Bacillus cereus*. *J. Microbiol. Biotechnol.* **19**(4), 387–390.

Alderwick, L. J., Seidel, M., et al. (2006). Identification of a novel arabinofuranosyltransferase (AftA) involved in cell wall arabinan biosynthesis in *Mycobacterium tuberculosis*. *J. Biol. Chem.* **281**, 15653–15661.

Barchi, J. J. (2000). Emerging roles of carbohydrates and glycomimetics in anticancer drug design. *Curr. Pharm. Des.* **6**(4), 485–501.

Berg, S., Starbuck, J., et al. (2005). Roles of conserved proline and glycosyltransferase motifs of EmbC in biosynthesis of lipoarabinomannan. *J. Biol. Chem.* **280**, 5651–5663.

Bililign, T., Griffith, B. R., et al. (2005). Structure, activity, synthesis and biosynthesis of aryl-C-glycosides. *Nat. Prod. Rep.* **6**, 742–760.

Brazier-Hicks, M., Offen, W. A., et al. (2007). Characterization and engineering of the bifunctional *N*- and *O*-glucosyltransferase involved in xenobiotic metabolism in plants. *Proc. Natl. Acad. Sci. USA* **104**, 20238–20243.

Breton, C., Šnajdrová, L., et al. (2006). Structures and mechanisms of glycosyltransferases. *Glycobiology* **16**(2), 29R–37R.

Campbell, J. A., Davies, G. J., et al. (1997). A classification of nucleotide-diphospho-sugar glycosyltransferases based on amino acid sequence similarities. *Biochem. J.* **326**, 929–942.

Cantarel, B. L., Coutinho, P. M., et al. (2009). The Carbohydrate-Active EnZymes database (CAZy): an expert resource for glycogenomics. *Nucleic Acids Res.* **37**, D-233–D-238.

Cartwright, A. M., Lim, E.-K., et al. (2008). A kinetic analysis of regiospecific glucosylation by two glycosyltransferases of *Arabidopsis thaliana*: domain swapping to introduce new activities. *J. Biol. Chem.* **283**, 15724–15731.

Charnock, S. J., Davies, G. J. (1999). Structure of the nucleotide-diphospho-sugar transferase, SpsA from *Bacillus subtilis*, in native and nucleotide-complexed forms. *Biochemistry* **38**, 6380–6385.

Charnock, S. J., Henrissat, B., et al. (2001). Three-dimensional structures of UDP-sugar glycosyltransferases illuminate the biosynthesis of plant polysaccharides. *Plant Physiol.* **125**(2), 527–531.

Cohen, N., Abramov, S., et al. (2001). *In vitro* enzyme evolution: the screening challenge of isolating the one in a million. *Trends Biotechnol.* **19**(12), 507–510.

Compain, P., Martin, O. R. (2001). Carbohydrate mimetics-based glycosyltransferase inhibitors. *Bioorg. Med. Chem.* **9**(12), 3077–3092.

Coutinho, P. M., Deleury, E., et al. (2003). An evolving hierarchical family classification for glycosyltransferases. *J. Mol. Biol.* **328**(2), 307–317.

Cowan, D. A., Kicman, A. T., et al. (2008). Ion trap MS/MS of intact testosterone and epitestosterone conjugates—adducts, fragile ions and the advantages of derivatisation. *Steroids* **73**(6), 621–628.

Dalby, P. A. (2007). Engineering enzymes for biocatalysis. *Recent Pat. Biotechnol.* **1**(1), 1–9.

Deng, C., Chen, R. R. (2004). A pH-sensitive assay for galactosyltransferase. *Anal. Biochem.* **330**(2), 219–226.

Dove, A. (2001). The bittersweet promise of glycobiology. *Nat. Biotechnol.* **19**(10), 913–917.

Erb, A., Weiss, H., et al. (2009). A bacterial glycosyltransferase gene toolbox: generation and applications. *Phytochemistry* **70**(15–16), 1812–1821.

Fukuchi-Mizutani, M., Okuhara, H., et al. (2003). Biochemical and molecular characterization of a novel UDP-glucose: anthocyanin 3′-O-glucosyltransferase, a key enzyme for blue anthocyanin biosynthesis, from gentian. *Plant Physiol.* **132**(3), 1652–1663.

Fulton, Z., McAlister, A., et al. (2008). Crystal structure of a UDP-glucose-specific glycosyltransferase from a *Mycobacterium* Species. *J. Biol. Chem.* **283**, 27881–27890.

Gachon, C. M. M., Langlois-Meurinne, M., et al. (2005). Plant secondary metabolism glycosyltransferases: the emerging functional analysis. *Trends Plant Sci.* **10**(11), 542–549.

Gorelik, E., Galili, U., et al. (2001). On the role of cell surface carbohydrates and their binding proteins (lectins) in tumor metastasis. *Cancer Metastasis Rev.* **20**(3–4), 245–277.

Goto, M. (2007). Protein O-glycosylation in fungi: diverse structures and multiple functions. *Biosci. Biotechnol. Biochem.* **71**(6), 1415–1427.

Grille, S., Zaslawski, A., et al. (2010). The functions of steryl glycosides come to those who wait: recent advances in plants, fungi, bacteria and animals. *Prog. Lipid Res.* **49** (3), 262–288.

Ha, S., Walker, D., et al. (2000). The 1.9 Å crystal structure of *Escherichia coli* MurG, a membrane-associated glycosyltransferase involved in peptidoglycan biosynthesis. *Protein Sci.* **9**, 1045–1052.

Hancock, S. M., Vaughan, M. D., et al. (2006). Engineering of glycosidases and glycosyltransferases. *Curr. Opin. Chem. Biol.* **10**(5), 509–519.

Hansen, E. S., Osmani, S. A., et al. (2009). Substrate specificities of family 1 UGTs gained by domain swapping. *Phytochemistry* **70**(4), 473–482.

Hansen, S. F., Bettler, E., et al. (2010). Exploring genomes for glycosyltransferases. *Mol. Biosyst.* **6**(10), 1773–1781.

Hashimoto, K., Tokimatsu, T., et al. (2009). Comprehensive analysis of glycosyltransferases in eukaryotic genomes for structural and functional characterization of glycans. *Carbohydr. Res.* **344**(7), 881–887.

Henrissat, B., Davies, G. (1997). Structural and sequence-based classification of glycoside hydrolases. *Curr. Opin. Struct. Biol.* **7**(5), 637–644.

Hoffmeister, D., Ichinose, K., et al. (2001). Two sequence elements of glycosyltransferases involved in urdamycin biosynthesis are responsible for substrate specificity and enzymatic activity. *Chem. Biol.* **8**(6), 557–567.

Hoffmeister, D., Wilkinson, B., Foster, G., Sidebottom, P. J., Ichinose, K., Bechthold, A. (2002). Engineered urdamycin glycosyltransferases are broadened and altered in substrate specificity. *Chem. Biol.* **9**(3), 287–295.

Hu, Y., Walker, S. (2002). Remarkable structural similarities between diverse glycosyltransferases. *Chem. Biol.* **9**(12), 1287–1296.

Idnurm, A., Warnecke, D. C., et al. (2003). Characterisation of neutral trehalase and UDP-glucose:sterol glucosyltransferase genes from the plant pathogenic fungus *Leptosphaeria maculans. Physiol. Mol. Plant Pathol.* **62**(5), 305.

Igura, M., Maita, M., et al. (2008). Structure-guided identification of a new catalytic motif of oligosaccharyltransferase. *EMBO J.* **27**, 234–243.

Kapitonov, D., Yu, R. K. (1999). Conserved domains of glycosyltransferases. *Glycobiology* **9**(10), 961–978.

Keegstra, K., Raikhel, N. (2001). Plant glycosyltransferases. *Curr. Opin. Plant Biol.* **4**(3), 219–224.

Kim, M. W., Kim, E. J., et al. (2006). Functional characterization of the Hansenula polymorpha HOC1, OCH1, and OCR1 genes as members of the yeast OCH1 mannosyltransferase family involved in protein glycosylation. *J. Biol. Chem.* **281** (10), 6261–6272.

Kim, B.-G., Sung, S. H., et al. (2010). Biological synthesis of isorhamnetin 3-*O*-glucoside using engineered glucosyltransferase. *J. Mol. Catal. B Enzym.* **63**(3), 194–199.

Klement, M.l., Ojemyr, L., et al. (2007). A processive lipid glycosyltransferase in the small human pathogen *Mycoplasma pneumoniae*: involvement in host immune response. *Mol. Microbiol.* **65**(6), 1444–1457.

Klutts, J. S., Yoneda, A., et al. (2006). Glycosyltransferases and their products: cryptococcal variations on fungal themes. *FEMS Yeast Res.* **6**, 499–512.

Kohara, A., Nakajima, C., et al. (2007). Characterization and engineering of glycosyltransferases responsible for steroid saponin biosynthesis in Solanaceous plants. *Phytochemistry* **68**(4), 478–486.

Krauth, C., Fedoryshyn, M., et al. (2009). Engineering a function into a glycosyltransferase. *Chem. Biol.* **16**(1), 28–35.

Lairson, L.l., Henrissat, B., et al. (2008). Glycosyltransferases: structures, functions, and mechanisms. *Annu. Rev. Biochem.* **77**, 521–555.

Lebrun, A. H., Wunder, C., et al. (2006). Cloning of a cholesterol-alpha-glucosyltransferase from *Helicobacter pylori*. *J. Biol. Chem.* **281**(38), 27765–27772.

Lee, H., Kobayashi, M., et al. (2006). Expression cloning of cholesterol alpha-glucosyltransferase, a unique enzyme that can be inhibited by natural antibiotic gastric mucin O-glycans, from *Helicobacter pylori*. *Biochem. Biophys. Res. Commun.* **349**(4), 1235–1241.

Lim, E.-K., Bowles, D. J. (2004). A class of plant glycosyltransferases involved in cellular homeostasis. *EMBO J.* **23**, 2915–2922.

Lin, C. H., Lin, C. W., et al. (2008). Proteomic identification of specific glycosyltransferases functionally implicated for the biosynthesis of a targeted glyco-epitope. *Proteomics* **8**(3), 475–483.

Liu, J., Mushegian, A. (2003). Three monophyletic superfamilies account for the majority of the known glycosyltransferases. *Protein Sci.* **12**, 1418–1431.

Loutre, C., Dixon, D. P., et al. (2003). Isolation of a glucosyltransferase from *Arabidopsis thaliana* active in the metabolism of the persistent pollutant 3,4-dichloroaniline. *Plant J.* **34**, 485–493.

Love, K. R., Swoboda, J. G., Noren, C. J., Walker, S. (2006). Enabling glycosyltransferase evolution: a facile substrate-attachment strategy for phage-display enzyme evolution. *ChemBioChem.* **7**(5), 753–756.

Lovering, A. L., de Castro, L. H., et al. (2007). Structural insight into the transglycosylation step of bacterial cell-wall biosynthesis. *Science* **315**, 1373–1374.

Luzhetskyy, A., Bechthold, A. (2008). Features and applications of bacterial glycosyltransferases: current state and prospects. *Appl. Microbiol. Biotechnol.* **80**, 945–952.

Luzhetskyy, A., Vente, A., et al. (2005). Glycosyltransferases involved in the biosynthesis of biologically active natural products that contain oligosaccharides. *Mol. Biosyst.* **1** (2), 117–126.

Luzhetskyy, A., Mendez, C., et al. (2008). Glycosyltransferases, important tools for drug design. *Curr. Top. Med. Chem.* **8**(8), 680–709.

Mackenzie, P. I., Owens, I. S., et al. (1997). The UDP glycosyltransferase gene superfamily: recommended nomenclature update based on evolutionary divergence. *Pharmacogenetics* **7**(255–269).

Madina, B. R., Sharma, L. K., et al. (2007). Purification and characterization of a novel glucosyltransferase specific to 27-hydroxy steroidal lactones from *Withania somnifera* and its role in stress responses. *Biochim. Biophys. Acta* **1774**(9), 1199–1207.

Maita, N., Nyirenda, J., et al. (2010). Comparative structural biology of eubacterial and archaeal oligosaccharyltransferases. *J. Biol. Chem.* **285**, 4941–4950.

Mao, Z., Shin, H. D., et al. (2006). Engineering the *E. coli* UDP-glucose synthesis pathway for oligosaccharide synthesis. *Biotechnol. Prog.* **22**(2), 369.

Masada, S., Kawase, Y., et al. (2007). An efficient chemoenzymatic production of small molecule glucosides with in situ UDP-glucose recycling. *FEBS Lett.* **581**, 2562–2566.

Meech, R. R., Mackenzie, P. I. P. I. (2010). UGT3A: novel UDP-glycosyltransferases of the UGT superfamily. *Drug Metab. Rev.* **42**(1), 45–54.

Mille, C., Bobrowicz, P., et al. (2008). Identification of a new family of genes involved in β-1,2-mannosylation of glycans in *Pichia pastoris* and *Candida albicans*. *J. Biol. Chem.* **283**, 9724–9736.

Mulichak, A. M., Losey, H. C., et al. (2001). Structure of the UDP-Glucosyltransferase GtfB that modifies the heptapeptide aglycone in the biosynthesis of vancomycin group antibiotics. *Structure* **9**(7), 547–557.

Mulichak, A. M., Losey, H. C., et al. (2003). Structure of the TDP-epi-vancosaminyl-transferase GtfA from the chloroeremomycin biosynthetic pathway. *Proc. Natl. Acad. Sci. USA* **100**(16), 9238–9243.

Mulichak, A. M., Lu, W., et al. (2004). Crystal Structure of vancosaminyltransferase GtfD from the vancomycin biosynthetic pathway: interactions with acceptor and nucleotide ligands. *Biochemistry* **43**(18), 5170–5180.

Newman, D. J., Cragg, G. M., et al. (2003). Natural products as sources of new drugs over the period 1981–2002. *J. Nat. Prod.* **66**, 1022–1037.

Offen, W., M-F, C., et al. (2006). Structure of a flavonoid glucosyltransferase reveals the basis for plant natural product modification. *EMBO J.* **25**(6), 1396–1405.

Osmani, S. A., Bak, S., et al. (2009). Substrate specificity of plant UDP-dependent glycosyltransferases predicted from crystal structures and homology modeling. *Phytochemistry* **70**(3), 325–347.

Pak, J. E., Arnoux, P., et al. (2006). X-ray crystal structure of leukocyte type core 2 β-1,6-N-acetylglucosaminyltransferase: evidence for a convergence of a metal-ion dependent glycosyltransferase mechanism. *J. Biol. Chem.* **281**, 26693–26701.

Palcic, M. M. (1999). Biocatalytic synthesis of oligosaccharides. *Curr. Opin. Biotechnol.* **10**(6), 616–624.

Park, S. H., Park, H. Y., et al. (2009). Reconstitution of antibiotics glycosylation by domain exchanged chimeric glycosyltransferase. *J. Mol. Catal. B Enzym.* **60**, 29–35.

Pesnot, T., Wagner, G. K. (2008). Novel derivatives of UDP-glucose: concise synthesis and fluorescent properties. *Org. Biomol. Chem.* **16**, 2884–2891.

Potocka A., Zimowski J. (2008). Metabolism of conjugated sterols in eggplant. Part 1. UDP-glucose:sterol glucosyltransferase. *Acta. Biochimica. Polonica.* **55**(1), 127-134.

Qasba, P. K., Ramakrishnan, B., et al. (2005). Substrate-induced conformational changes in glycosyltransferases. *Trends Biochem. Sci.* **30**(1), 53–62.

Qasba, P. K., Ramakrishnan, B., et al. (2008). Structure and function of beta-1,4-galactosyltransferase. *Curr. Drug Targets* **9**(4), 292–309.

Rabbani, S., Schwardt, O., et al. (2006). Glycosyltransferases: an efficient tool for the enzymatic synthesis of oligosaccharides and derivatives as well as mimetics thereof. *CHIMIA* **60**(1–2), 23–27.

Rix, U., Fischer, C., et al. (2002). Modification of post-PKS tailoring steps through combinatorial biosynthesis. *Nat. Prod. Rep.* **2002**(19), 542–580.

Robertson, D. E., Steer, B. A. (2004). Recent progress in biocatalyst discovery and optimization. *Curr. Opin. Chem. Biol.* **8**(2), 141–149.

Rosen, M.l., Edman, M., et al. (2004). Recognition of fold and sugar linkage for glycosyltransferases by multivariate sequence analysis. *J. Biol. Chem.* **279**(37), 38683–38692.

Rubin-Pitel, S. B., Zhao, H. (2006). Recent advances in biocatalysis by directed enzyme evolution. *Comb. Chem. High Throughput Screen.* **9**(4), 247–257.

Ruffing, A., Mao, Z., et al. (2006). Metabolic engineering of *Agrobacterium* sp. for UDP-galactose regeneration and oligosaccharide synthesis. *Metab. Eng.* **8**(5), 465–473.

Seibel, J., Jordening, H.-J., et al. (2006). Glycosylation with activated sugars using glycosyltransferases and transglycosidases. *Biocatal. Biotransformation* **24**(5), 311–342.

Shao, H., He, X., et al. (2005). Crystal structures of a multifunctional triterpene/flavonoid glycosyltransferase from *Medicago truncatula. Plant Cell* **17**, 3141–3154.

Thibodeaux, C. J., Melançon, C. E., Liu, H. W., et al. (2007). Unusual sugar biosynthesis and natural product glycodiversification. *Nature* **446**(7139), 1008–1016.

Thorsoe, K. S., Bak, S., et al. (2005). Determination of catalytic key amino acids and UDP sugar donor specificity of the cyanohydrin glycosyltransferase UGT85B1 from *Sorghum bicolor.* Molecular modeling substantiated by site-specific mutagenesis and biochemical analyses. *Plant Physiol.* **139**(2), 664–673.

Thorson, J. S., Hosted, T. J., et al. (2001). Nature's carbohydrate chemists: the enzymatic glycosylation of bioactive bacterial metabolites. *Curr. Org. Chem.* **5**, 139–167.

Trubetskoy, O. V., Shaw, P. M. (1999). A fluorescent assay amenable to measuring production of beta-d-glucuronides produced from recombinant UDP-glycosyl transferase enzymes. *Drug Metab. Dispos.* **27**(5), .

Unligil, U. M., Rini, J. M. (2000). Glycosyltransferase structure and mechanism. *Curr. Opin. Struct. Biol.* **10**(5), 510–517.

Vogt, T., Jones, P. (2000). Glycosyltransferases in plant natural product synthesis: characterization of a supergene family. *Trends Plant Sci.* **5**(9), 380–386.

Walsh, C. T., Losey, H. C., et al. (2003). Antibiotic glycosyltransferases. *Biochem. Soc. Trans.* **31**, 487–492.

Warnecke, D., Heinz, E. (2010). Glycolipid headgroup replacement: a new approach for the analysis of specific functions of glycolipids *in vivo. Eur. J. Cell Biol.* **89**(1), 53–61.

Warnecke, D., Erdmannn, R., et al. (1999). Cloning and functional expression of UGT genes encoding sterol glucosyltransferases from Saccharomyces cerevisiae, Candida albicans, Pichia pastoris, and Dictyostelium discoideum. *J. Biol. Chem.* **274**(19), 13048–13059.

Weijers, C. A. G. M., Franssen, M. C. R., et al. (2008). Glycosyltransferase-catalyzed synthesis of bioactive oligosaccharides. *Biotechnol. Adv.* **26**(5), 436–456.

Werner, S. R., Morgan, J. A. (2009). Expression of a *Dianthus* flavonoid glucosyltransferase in *Saccharomyces cerevisiae* for whole-cell biocatalysis. *J. Biotechnol.* **142**(3–4), 233–241.

Williams, G. J., Zhang, C., et al. (2007). Expanding the promiscuity of a natural-product glycosyltransferase by directed evolution. *Nat. Chem. Biol.* **3**(10), 657–662.

Williams, G. J., Thorson, J. S. (2008). A high-throughput fluorescence-based glycosyltransferase screen and its application in directed evolution. *Nat. Prot.* **3**(3), 357–362.

Wimmerová, M., Engelsen, S. B., et al. (2003). Combining fold recognition and exploratory data analysis for searching for glycosyltransferases in the genome of *Mycobacterium tuberculosis*. *Biochimie* **85**, 691–700.

Wohlgemuth, R. (2005). Development, production and application of recombinant yeast biocatalysts in organic synthesis. *CHIMIA* **59**(10), 735–740.

Yin, Q. Y., de Groot, P. W. J., et al. (2007). Mass spectrometry-based proteomics of fungal wall glycoproteins. *Trends Microbiol.* **16**(1), 20–26.

Yin, Q. Y., de Groot, P. W. J., et al. (2008). Mass spectrometry-based proteomics of fungal wall glycoproteins. *Trends Microbiol.* **16**(1), 20–26.

Yuan, Y., Barrett, D., et al. (2007). Crystal structure of a peptidoglycan glycosyltransferase suggests a model for processive glycan chain synthesis. *Proc. Natl. Acad. Sci. USA* **104**, 5348–5353.

Zhang, Y., Xiang, Y., et al. (2007). Structure and function of a chlorella virus-encoded glycosyltransferase. *Structure* **15**(9), 1031–1039.

VIPOXIN AND ITS COMPONENTS: STRUCTURE–FUNCTION RELATIONSHIP

By S.D. PETROVA,* V.N. ATANASOV,† AND K. BALASHEV‡

*Laboratory of Enzymology, Department of Biochemistry, Faculty of Biology, Sofia University, Sofia, Bulgaria
†Laboratory of Biocoordination & Bioanalytical Chemistry, Department of Analytical Chemistry, Faculty of Chemistry, Sofia University, Sofia, Bulgaria
‡Laboratory of Biophysical Chemistry, Department of Physical Chemistry, Faculty of Chemistry, Sofia University, Sofia, Bulgaria

Abstract

The neurotoxin Vipoxin has been of growing research interest since the time of its isolation from the venom of the Bulgarian viper *Vipera ammodytes meridionalis.* Vipoxin is a heterodimeric postsynaptic ionic complex composed of two protein subunits—a basic and strongly toxic His48 secretory phospholipase A_2 ($sPLA_2$) enzyme and an acidic, enzymatically inactive and nontoxic component, originally named Inhibitor. When separated, $sPLA_2$ enzyme loses its toxicity in 3–4 days and catalytic activity in 2 weeks. After the establishment of the high degree of sequence homology (62%)

DOI: 10.1016/B978-0-12-398312-1.00005-6
Copyright 2012, Elsevier Inc. All rights reserved.

and crystal structure of the subunits, Vipoxin was served as an example of molecular evolution of a toxic but unstable $sPLA_2$ into an inhibitor subunit which stabilizes the enzyme and preserves its pharmacological activity.

Beginning our research on Vipoxin, intrigued by the unique relationship—structure–function and based on the previous experience, we were more than surprised to establish the lack of so-called inhibitory function of the acidic subunit on the toxicity and catalytic activity of basic $sPLA_2$. On the contrary, the acidic subunit activated the $sPLA_2$ enzyme *in vitro.* Our studies undoubtedly proved that is more correctly to present Vipoxin as a heterodimeric complex composed of one basic catalytic subunit and one acidic *regulatory* subunit. Their interaction in a common quaternary protein structure is more than a noncovalent association between the two subunits. It allows pharmacological sites to be targeted and biological functions to be potentiated. We attempt to present the previous studies and new findings on Vipoxin and its components.

I. Introduction

Snake venoms are some of the most complex multifunctional mixtures of pharmacologically active proteins and polypeptides interfering in various physiological systems (Kini, 1997). They have been studied extensively for many years in search of the molecular basis of toxicity, have provided important biological tools to investigate vital physiological processes, and even have been used as therapeutic agents (Koh and Kini, 2012). It is inconceivable that some of the most toxic components of the snake venoms are phospholipase A_2 (PLA_2) enzymes, for they also have extremely important biochemical role in physiological processes such as maintenance of membrane homeostasis, membrane repair, cell proliferation, inflammation, signal transduction, etc. Snake PLA_2 enzymes provoke diverse pharmacological effects: neurotoxicity, myotoxicity, cardiotoxicity, anticoagulant effects, hemolytic activity, hemorrhage, organ, or tissue damage (Kini, 1997). There has been a burst in the available data on PLA_2 superfamily of enzymes, changing our understanding of the structure–function relationship by revealing completely new interactions, new effects, and new challenges to scientists (Kini, 1997, 2003, 2006; Six and Dennis, 2000; Murakami and Kudo, 2002, 2004; Wilton and Waite, 2002; Schaloske and Dennis, 2006; Dennis et al., 2011).

The neurotoxin Vipoxin was isolated from the snake venom of two subspecies *Vipera ammodytes*—*V. ammodytes* ssp. *meridionalis* and *V. ammodytes* ssp.

montandoni. V. ammodytes ssp. *meridionalis* is one of the most toxic snakes in Europe inhabiting only the Balkan peninsula. *V. ammodytes* ssp. *ammodytes,* which is considered to be evolutionary older, lacks Vipoxin but contains monomeric $sPLA_2$ enzymes. The intensive research on Bulgarian *V. ammodytes* ssp. *meridionalis* venom began in 1961, and 10 years later, Aleksiev et al. isolated the neurotoxin Vipoxin in a pure form (Shipolini et al., 1970). Vipoxin is a heterodimeric postsynaptic ionic complex composed of two protein subunits—a basic and strongly toxic His48 $sPLA_2$ enzyme and an acidic, enzymatically inactive and nontoxic component, originally named Inhibitor (Tchorbanov et al., 1978). Both subunits have the same polypeptide length (122 amino acids) and are closely related, sharing 62% sequence identity (Mancheva et al., 1987). Each subunit contains seven disulfide bridges that stabilize the native biologically active conformation.

We present herein our current understanding about the structure-function relationship of Vipoxin and its components, some of our new findings on the $sPLA_2$ enzymatic activity, acute toxicity, hemolytic and anticoagulant activities, and the role of the acidic subunit. We found that Vipoxin is more toxic than its isolated $sPLA_2$, reverting the existing view about the toxicity of the basic component. We also found that Vipoxin acidic component (VAC) was not an "inhibitor" of the toxic and catalytic activities. Further, we established that the isolated $sPLA_2$ and Vipoxin have one and same preferences to natural and synthetic substrates *in vitro.* Pure $sPLA_2$ possesses weak anticoagulant activity, whereas Vipoxin is anticoagulant ineffective, even though both exhibit comparable PLA_2 activity. Studying Vipoxin properties reveals remarkably complex structure–function relationship.

II. PLA_2—Structure and Function Relationship

PLA_2 enzymes (phosphatide *sn*-2 acylhydrolase, EC 3.1.1.4) catalyze the hydrolysis of the 2-acylester bond of 1,2-diacyl-3-*sn*-phosphoglycerides, releasing fatty acids and lysophospholipids (Six and Dennis, 2000). Precisely, these hydrolysis products are lipid-signaling molecules and play important roles in various biological processes, such as inflammation, signal and energy transduction, remodeling of membrane phospholipids giving rise to local changes, cell proliferation, host defense, etc. (Burke and Dennis, 2009; Zhao and Natarajan, 2009). PLA_2 enzymes occur ubiquitously in nature and are found intra- or extracellularly in all cell types. The general features of all PLA_2 enzymes are water solubility; interfacial catalysis; activity against a phospholipid substrate; preference to aggregated, organized substrate structures; a large number of cysteins in the forms of disulfide bonds in the $sPLA_2$; critical to life role for remodeling of cellular membranes; extreme diversity in biological functions (Kini, 1997; Murakami et al., 1997; Wilton and Waite, 2002).

According to the current classification, the superfamily of PLA_2 enzymes represents 15 groups and corresponding subgroups (Dennis, 1994; Six and Dennis, 2000; Wilton and Waite, 2002; Schaloske and Dennis, 2006; Burke and Dennis, 2009). All groups include the main types: $sPLA_2$, cytosolic PLA_2 ($cPLA_2$), calcium-independent intracellular PLA_2 ($iPLA_2$), lysosomal PLA_2, platelet-activating factor acetylhydrolases (acting specifically on short acyl chains at the *sn*-2 position (PAF-AH), as well as PLA_2 enzymes active on oxidized phospholipids (designated as lipoprotein-associated PLA_2 (Lp-PLA_2). A new type of PLA_2, isolated from adipose tissue, was designated as group XVI (GXVI) (Dennis et al., 2011).

Based on their catalytic mechanism, PLA_2 enzymes have been categorized into two types: PLA_2 enzymes using a His residue for the catalysis and PLA_2 enzymes that involve a catalytic Ser residue with subsequent formation of an acyl-serine intermediate (Six and Dennis, 2000). PAF-AH (GVII) PLA_2 enzymes use Ser/His/Asp triad and recently characterized AdPLA_2 uses His/Cys dyad (Duncan et al., 2008).

Based on their molecular mass, PLA_2 enzymes could be divided into high and low molecular mass enzymes, corresponding to their localization and biological function—into intracellular or secreted.

Two groups of high molecular PLA_2 enzymes (both of intracellular origin) have been characterized. The first group includes Ca^{2+}-independent enzymes ($iPLA_2$-GVI) with M_r 140–400 kDa that are found in various tissue types and many of them catalyze the hydrolysis of plasmalogens (Kini, 1997; 2003). The GVIA-1 PLA_2 (85 kDa) mechanism of action includes catalytically active Ser nucleophillic residue in a consensus sequence Gly-X-Ser465-X-Gly and plays a significant role in phospholipid remodeling and maintenance of homeostasis, myocardial ischemia, arachidonate release, insulin secretion, immunoglobulin G-mediated phagocytosis, lymphocyte proliferation, apoptosis, etc. (Reynolds et al., 1993; Balsinde et al., 1997; Tang et al., 1997; Barbour et al., 1999; Winstead et al., 2000).

The second group of high molecular iPLA$_2$ enzymes (M_r 85–100 kDa, GIV), with catalytically active Ser residue (Ser/Asp diad), have been characterized as Ca^{2+}-activated enzymes. Ba^{2+}, Sr^{2+}, Mn^{2+}, Mg^{2+}, as well as high NaCl concentrations can substitute for Ca^{2+}ions (Valentin et al., 1999; Berg et al., 2001). PLA$_2$ enzymes from this group not only catalyze preferentially the release of arachidonate from *sn*-2 position but also exhibit lysophospholipase and transacylase activities (Huang et al., 1997). Biological functions of GIV PLA$_2$ (inflammation, lung injury, intestinal ulceration, anaphylaxis, brain injury, etc.) are so diverse that these enzymes are distinctly different from all the remaining groups (Schaloske and Dennis, 2006).

Low molecular mass PLA$_2$ enzymes (M_r 13–18 kDa), isolated from different sources: synovial fluid, sperm, platelets, macrophages, pancreatic juice, snake venoms, plants, fungi, etc., are Ca^{2+}-dependent and extracellular secreted enzymes. They contain six to eight disulfide bridges that stabilize the structure, exhibit very high sequential and structural homology (evolutionary related), and utilize similar mechanism of action with catalytically important His/Asp diad (Reynolds et al., 1993; Six and Dennis, 2000; Schaloske and Dennis, 2006).

Mammalian sPLA$_2$ enzymes have a special place in pharmacology given their role in (patho)physiological processes such as membrane repair through acylation/deacylation pathways, biosynthesis of prostaglandins, fertilization and cell proliferation, smooth muscle contraction, allergy, and a variety of inflammatory conditions (Kini, 1997, 2003; Lambeau and Gelb, 2008).

Because of the apparent expansion of the PLA$_2$ superfamily of enzymes participating in so many diverse physiological processes, their classification criteria have been under a permanent reevaluation. sPLA$_2$ isolated from Vipoxin is a member of Group II (includes nonpancreatic mammalian and snake venom sPLA$_2$ enzymes isolated from *Viperidae* and *Crotalidae*), and its structure and catalytic activity obey to the common characteristics of the group.

A. *Structure*

GIIPLA$_2$ enzymes are very compact proteins, sharing common elements with strong three-dimensional structural homology and large proportion of conserved amino acid residues (AAs) occupying catalytically competent or structurally sensitive positions (Scott and Sigler, 1994; Scott, 1997). They have a core of three α-helices, a distinctive backbone loop that binds catalytically important Ca^{2+} ions, and two-stranded antipararlel β-sheets (β-wings) (Kini, 1997). All the members of this group lack the pancreatic and elapid loops but have additional C-terminal region, which forms an extra disulfide link with a Cys residue near the active site His48 (Kini, 1997). The C-terminal segment forms a semicircular "banister" around the Ca^{2+}-binding loop, particularly in viperid and crotalid $sPLA_2$ enzymes (Kini, 2006). In several $GIIBPLA_2$ enzymes, Asp49 is replaced by a Lys residue, leading to a noneffective coordination with Ca^{2+} (Yoshizumi et al., 1990; Arni et al., 1995; Ward et al., 2002; Sá et al., 2004). Lys49 PLA_2 homologues demonstrate membrane-damaging activity through a Ca^{2+}-independent, nonhydrolytic mechanism. In the case of ammodytin L from *V. ammodytes ammodytes*, Asp49 is replaced by Ser residue and this protein also exhibits low enzymatic activity (Pungercar et al., 1990). The existence of a hydrophobic channel (with invariable AA positions as Leu or Val for residue 2 and Phe and Ile for residues 5 and 9, respectively) entering the active site and occupied by the acyl chains of the bound substrate during the catalysis is also an essential characteristic (Scott, 1997).

As the calcium ion is required for both the substrate binding and catalysis, $sPLA_2$ enzymes coordinate this Ca^{2+} by a conserved and flexible loop of residues: Y25-G-C-Y/F-C-G-X-G-G33 (Bekkers et al., 1991; Scott and Sigler, 1994). Three backbone carbonyl oxygens—O28, O30, and O32—together with the oxygen atoms donated by the carboxylate of Asp49 and two water molecules form a tight pentagonal bipyramidal coordination cage for the primary Ca^{2+} in the absence of substrate (Scott et al., 1990; Scott, 1997; Steiner et al., 2001). The function of the Ca^{2+} ions is to provide an electrophilic stabilization of the oxyanion (by coordinating the carbonyl group) and to orientate and polarize the substrate and the sessile ester bond, resulting in lowering the activation barrier of the transition state. There is also evidence that the Ca^{2+} ion enhances protein–substrate interactions during the interfacial adsorption (Scott et al., 1990). A second, defined as a "supplemental electrophile," hepta-coordinated Ca^{2+}site, closely situated (11 Å) to the first, has been observed in several high-resolution crystal structures (White et al., 1990; Scott et al., 1990; Scott, 1997).

The structural models of $sPLA_2$ enzymes in the presence of inhibitors demonstrate the interaction between the enzyme and the phospholipid substrate based on a great number of weak bonds, important for the catalysis: (1) the oxygen atoms of the flexible glycerol backbone displace the Ca^{2+} associated water molecules; (2) *sn*-2 and *sn*-3 oxygens participate in a dual coordination with the primary Ca^{2+} ion, important for the natural "L" chirality selection at the *sn*-2 position; (3) interactions cause *sn*-2 ester bond to move out of the place and to expose to catalytic surface (Scott et al., 1990). According to the structural model, flexible alkyl chains at *sn*-1 and *sn*-2 positions lie parallel in a hydrophobic channel that extends (about 14 Å) from the catalytic site. It is suggested that only about 9–10 carbons of the *sn*-2 acyl chain interact with the enzyme and the rest of the chain (in the case of long-chain fatty acids) are buried in the lipid–water interface. The phenolic hydroxyl group of Tyr69 residue (important for the hydrophobic channel) transiently interacts with *sn*-3 phosphate of the substrate, and its substitution by Lys residue (in human nonpancreatic $sPLA_2$) greatly influences the hydrophobic channel structure (Scott et al., 1990).

$sPLA_2$ enzymes bind to membrane surfaces by relatively flat, exposed regions—IBS (interfacial binding surfaces/i-face), surrounding the entrance to the active site pocket (Pan et al., 2001; Winget et al., 2006). Jain et al. (1986a,b) suggested a tight desolvated contact between $sPLA_2$ and anionic phospholipid molecules and that this interaction altered the gel–fluid transition in bilayer without disruption of the bilayer organization. During this very important catalytic step, the water molecules around the interfacial anionic groups are replaced by the polarizable ligands from the enzyme's i-face. In this paradigm, the desolvation of $sPLA_2$ i-face is the most important factor and the "penetration" of $sPLA_2$ into the acyl chain region of the bilayer is discounted as energetically unfavorable. Verheij et al. (1980) defined the ability of $sPLA_2$s to penetrate the phospholipid membrane, which allows them to access and destroy the phospholipids even in tightly packed membranes. It was proposed that the penetrating ability is critical for exerting many pharmacological effects and is due to the aromatic residues localized on the IBS (Burke et al., 2008). Berg et al. (2004) discussed the electrostatic interactions between the basic AAs on the enzyme surface and anionic phospholipids as a factor playing an important role in interfacial binding.

The contribution of hydrophobic and electrostatic interactions is demonstrated especially in the cases of dimeric or trimeric $sPLA_2$ forms created upon the association with membrane. It was proposed that these forms (exposing more positively charged interfacial surface) represent another type of enzyme activation switching from monomeric to trimeric form at the membrane surface (Bahnson, 2005; Dennis et al., 2011)

It is known that the monomeric $sPLA_2$ possesses very rigid conformation (seven disulfide bridges) that is not predisposed to conformational changes similar to that of the allosteric enzymes with stable quaternary structure. Nevertheless, $sPLA_2$ enzyme is designed to bind and act on lamellar and micellar aggregates of phospholipids, and it is fully catalytically active as a monomer (Scott et al., 1990; Jain et al., 1991). The relationship structure–catalysis assumes that (1) interfacial binding does not induce a conformational transition in the protein and (2) a perfect preservation of the pentagonal bipyramidal coordination of the Ca^{2+} ion in the free and bound enzyme forms (Scott et al., 1990; White et al., 1990).

All $sPLA_2$ enyzmes possess signal sequences, which are cleaved proteolytically in the process of secretion of the mature protein (Six and Dennis, 2000). Some of them are activated by posttranslational modifications, catalyzed by transglutaminases (Fesus and Piacentini, 2002).

B. Catalysis

$sPLA_2$ enzymes act on phospholipids organized in different aggregates (monolayer, micellar, and bilayer phospholipid phases) according to their type—zwitterionic (phosphatidylcholine, PC), anionic (phosphatidylserine, PS; phosphatidate, PA; phosphatidylinositols, PI; phosphatidylglycerol, PG), neutral (phosphatidylethanolamine, PE; cardiolipins, CL), fatty acid composition, and biological milieu. The higher enzyme activity for most $sPLA_2$ enzymes is observed when micellar, anionic, or zwitterionic phospholipids are used (Kini, 1997; Berg et al., 2001; Dennis et al., 2011). $sPLA_2$ enzymes display distinct phospholipid headgroup preferences, but the catalytic efficiency is determined mainly by the physical properties of the aggregates: packing density, phase transition temperature, critical micellar concentration, some liquid crystalline characteristics, polymorphism of the aggregate, exchange of the substrate molecules between the phases (soluble and aggregated), presence of other molecules (ionic or nonionic detergents), and transfer of the substrate molecule from the ordered monolayer to the active site. All these factors and their integrity have an enormous impact on the enzyme and its catalytic competence.

The increase in activity upon changing of the substrate concentration from monomers to aggregates is designated as "interfacial activation," which is an inseparable prerequisite to $sPLA_2$ catalytic mechanism (Dennis et al., 2011). Many analytical models have been proposed in order to explain the consecutive steps of the interfacial catalysis, to enlighten the influence of all possible factors on the enzyme activity and to provide a closer look at the interface (Berg et al., 1991, 1997; Panaiotov and Verger, 2000; Cajal et al., 2004; Jain and Berg, 2006). All of them represent an extended but simplified enough Michaelis–Menten kinetic paradigm, applied for interfacial conditions (including a great number of reaction factors, separate reaction cycles, kinetic constants describing every single step, and their relations). The simplest kinetic model adapted for the catalysis of the short- and medium-chain fatty acid containing lipids considers the coupling between chemical Michaelis–Menten step and the process of enzyme penetration into the lipid monolayer, followed by the process of instantaneous solubilization of the reaction products into the water subphase (Fig. 1; Panaiotov et al., 1997). The generalized kinetic model has been enhanced by taking into account the processes of interfacial molecular reorganization and segregation of the insoluble lipolytic products, modifying the "interfacial quality" of the lipid structures (Panaiotov and Verger, 2000). Further, some kinetic parameters giving valuable information about the mechanism of enzyme action have been obtained from the proposed kinetic models, such as the "surface dilution model" for the catalysis of mixed micelles (Deems et al., 1975; Carman et al., 1995) and the "scooting and hopping modes" of $sPLA_2$ action on liposomal dispersions (Jain and Berg, 1989; Jain et al., 1995). Despite the benefits of the monolayer technique, the method has limited capacity for studying the enzyme–substrate interactions and the fine monolayer structure (e.g., monolayer heterogeneity).

Fluorescence microscopy has proven to be a well suited complementary technique in this respect as the first direct visualization of the action of $sPLA_2$ was performed in the late 1980s using wide-field microscopy on a lipid monolayer (Grainger et al., 1989). Using this method, Grainger et al. (1990) provided the existence of liquid–solid lipid domains in the phospholipid monolayers, which $sPLA_2$ can recognize and hydrolyze.

The method which brought the visualization of the molecular events at the nanometer scale was the atomic force microscopy (AFM). The very first AFM application for visualization of the enzyme lipolysis of supported

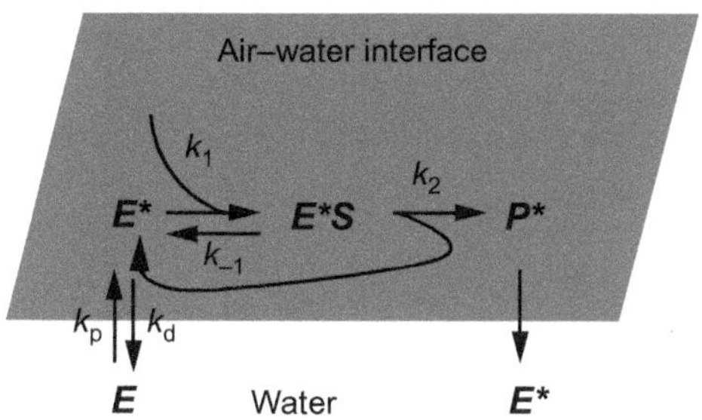

FIG. 1. Adaptation of the Michaelis–Menten kinetic scheme describing interfacial catalysis of short- and medium-chain lipids with soluble reaction products (according to Panaiotov and Verger, 2000). (For color version of this figure, the reader is referred to the online version of this chapter.)

1,2-dipalmitoyl-*sn*-glycero-3-phosphocholine (DPPC) bilayers was demonstrated by Gaub's group analyzing the enzyme behavior on a gel-phase membrane (Grandbois et al., 1998).

Nielsen et al. (1998) reported the presence of a latency period (also known as a "lag phase") in $sPLA_2$ degradation of supported phospholipid bilayers. Lag phase is defined as a period of a low enzyme activity where no substrate hydrolysis is observed. The duration of the lag phase is determined by the bilayer properties and its fine molecular structure, that is, microheterogeneity, product accumulation, etc. Combining kinetic and structural studies of the hydrolysis of supported DPPC bilayers by AFM, Nielsen et al. (1998) showed that the burst in enzymatic activity was located around the product domains (seen as small depressions) created in the early stages of hydrolysis and was followed by a disruption of the lipid bilayer in the vicinity of these depressions. The $sPLA_2$ subunit of Vipoxin reflects all these common principles in structure and function but still remains some characteristic, particular properties making it unique in its specialized pharmacological effects .

By analogy with the serine proteases, Verheij et al. (1980) and later Scott (1990) proposed a catalytic mechanism for $sPLA_2$ involving Asp99, His48 and a water molecule (instead of Ser in proteases), forming a catalytic "triad" in which water is activated by His48. NMR and IR studies of the co-crystals of $sPLA_2$ with the substrate analog - (R)-(2-dodecanamidoisohexyl)-phosphocholine (DAHPC) established that the carbonyl bond undergoes significant polarization in the enzyme-ligand complex, facilitated by the enzyme-bound Ca^{2+} ion (Slaich et al., 1992). Using of 1-hexadecyl-3-(trifluoroethyl)-*sn*-glycero-2-phosphomethanol (MJ33) as a transition state analog, revealed that a second, previously not observed water molecule, liganded by Ca^{2+}, is actually the attacking nucleophile (Sekar et al., 1997).

Yu et al. (1998) proposed alternative catalytic mechanism in which Asp-His-water catalytic triad and the Ca^{2+} ion both activate the second water molecule (*w5*) to attack the scissile and polarized carbonyl carbon (Fig.2).

III. Vipoxin and Its Components

A. *Purification of the Components*

Vipoxin was isolated from a crude venom of *V. ammodytes meridionalis* (Thracian Herpetological Society and National Centre of Infectious and Parasitic Diseases, Bulgaria) using ion-exchange chromatography on

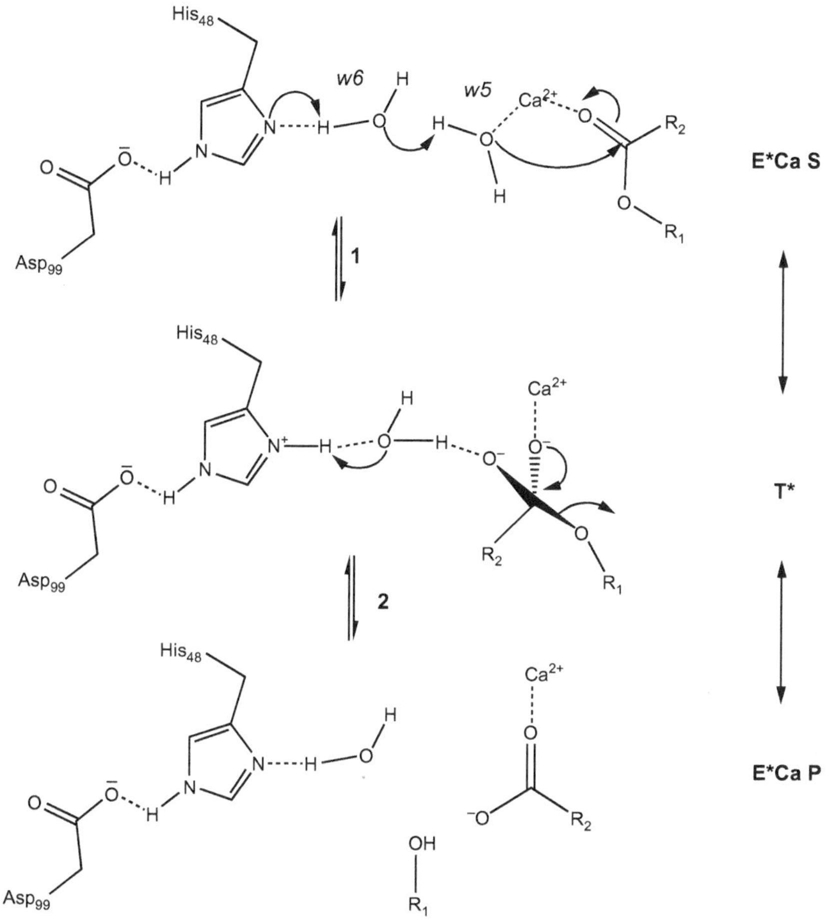

FIG. 2. Calcium-coordinated oxyanion mechanism for the $sPLA_2$ enzymes. The catalytic triad Asp99-His48-*w6* and Ca^{2+} cofactor, coordinated to *w5* and to the *sn*-2-ester carbonyl are presented. Formation of the tetrahedral intermediate occurs by the nucleophilic attack of a water molecule *w5*, activated by His48 as a general-base catalyst and Ca^{2+} ion polarizing the scissile bond. Ca^{2+} electrostatically stabilizes the resulting tetrahedral intermediate (Berg et al., 2001).

SP-Sephadex C-50 (Pharmacia, Sweden) according to the procedure described previously (Tchorbanov and Aleksiev, 1981). The separation of Vipoxin components was modified and optimized by using cation-exchange FPLC on Mono S HR 5/5 column (Pharmacia, Sweden),

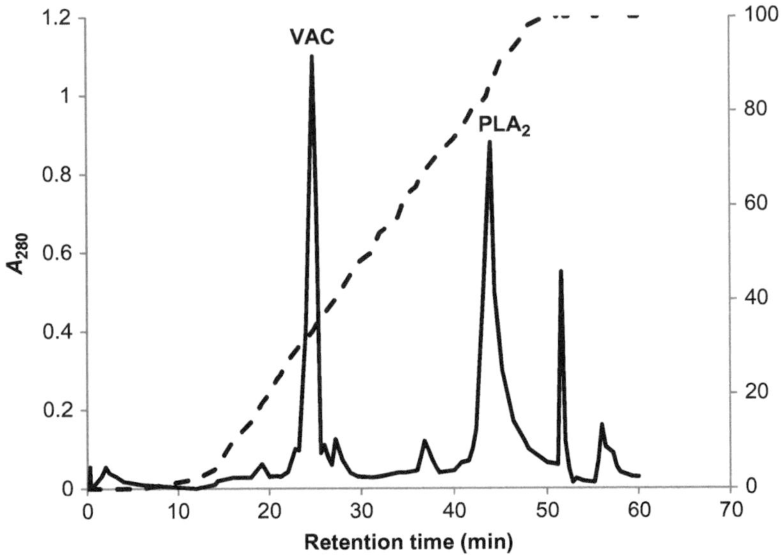

FIG. 3. Purification of Vipoxin components by cation-exchange chromatography on Mono S HR 5/5. The column (Pharmacia, Sweden) was pre-equilibrated with 0.1 M acetate buffer (pH 4.5) in the presence of 6 M carbamide and eluted at a flow rate of 0.4 ml/min with a linear gradient of 0.5 M NaCl in the same buffer.

pre-equilibrated with 0.1 M acetate buffer (pH 4.5) in the presence of 6 M carbamide (Atanasov et al., 2009a; Fig. 3).

B. *Vipoxin Structure*

1. *sPLA_2 Component*

The Vipoxin's sPLA_2 is a strongly Ca^{2+}-dependent enzyme (Atanasov et al., 2005). Although all the crystallographic data obtained so far are in the absence of Ca^{2+} in the buffer solutions, Georgieva et al. (2004) found that both components of Vipoxin—sPLA_2 and the acidic subunit (VAC)—have Ca^{2+}-binding loops structurally different from similar loops in other sPLA_2 enzymes and that binding of Ca^{2+} ions provokes conformational change and transition of the complex from nonspherical shape to more compact and symmetric one. Reduction of the disulfide bonds in both components results in a drastic change in the far UV circular dichroism spectra, demonstrating a loss of α-helical structure (Georgieva et al., 2004). In the absence of the metal ion, Ca^{2+}-binding loop is conformationally more flexible compared to the structures of similar $sPLA_2$ enzymes (Murakami et al., 2006). The carbonyl O atoms of Tyr28 and Gly32 from the $sPLA_2$ subunit are involved in intermolecular contacts with the VAC, and the carbonyl O of Gly32 is 5.5 Å away from the Ca^{2+}-binding site (Devedjiev et al., 1997; Banumathi et al., 2001). The unusual salt bridge between Asp49 and Lys69 stabilizes the region which in other $sPLA_2$ enzymes is occupied by a Ca^{2+} ion as well as the whole molecule.

The X-ray structure of Vipoxin, established in 1997 (Perbandt et al., 1997), showed the presence of His48 in the toxic subunit (similar to $sPLA_2$ from *Crotalus atrox*) and Gln48 in the nontoxic (VAC) subunit (Fig. 4). His48 is most important for the exhibition of phospholipase activity in the $sPLA_2$ enzymes (His48, Asp99, Tyr52 are in the catalytic site). The replacement of His with Gln in the VAC explains one of the reasons for the lack of enzymatic activity (Perbandt et al., 1997; Banumathi et al., 2001; Georgieva

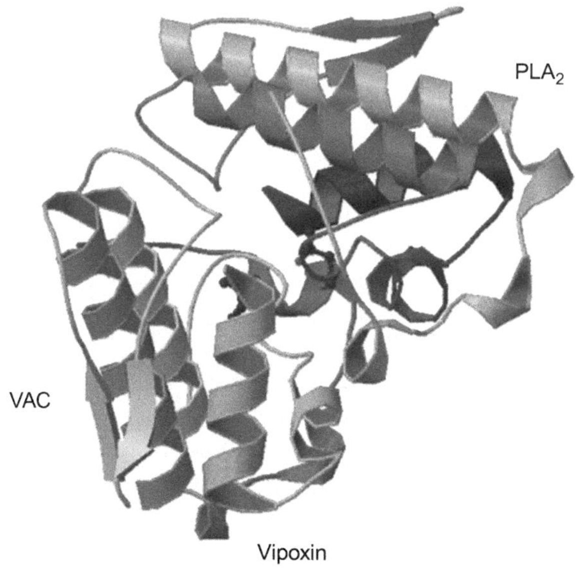

FIG. 4. Ribbon representation of heterodimeric Vipoxin complex, (according to Banumathi et al., 2001). (For interpretation of the references to color in this figure legend, the reader is referred to the online version of this chapter.)

et al., 2003b). Hydrophobic surfaces forming the hydrophobic channel and recognized in both Vipoxin subunits consist of Leu2, Phe5, Met8, Ile9, Tyr22, Cys29, Cys45, Ala102, Ala103, and Phe106 (Leu106 in VAC) (Banumathi et al., 2001).

Recently, Faure et al. (2011) established completely new characteristics of the crystal structure of the presynaptic neurotoxin Crotoxin isolated from *Crotalus durissus terificus*, not present in the reported so far structures of other heterodimeric snake venom β-neurotoxins such as viperotoxin (*Vipera russelli formosensis*) and Vipoxin (*V. ammodytes meridionalis*) (Perbandt et al., 1997; Perbandt et al., 2003). Crotoxin is analogous to Vipoxin in its quaternary structure, composed of two noncovalently associated subunits—a basic and toxic $GIIAPLA_2$ (CB, 122 AAs and structurally similar to the $sPLA_2$ from Vipoxin) and an acidic, nonenzymatic, and nontoxic one CA (functionally active CA consists of three polypeptide chains—α, β, and γ linked by disulfide bonds, derived posttranslationally from a PLA_2-like precursor). The unique structural scaffold of the crotoxin showed a second Na^+ site (stabilizing the Ca^{2+}-binding loop in CB) and a novel Cl^--binding site indicating that posttranslational cleavage of the pro-CA precursor is a prerequisite for the assembly of the CA–CB complex. These findings have thrown a new light on the possible role of the crotoxin acidic subunit which is also the main question in our investigations on Vipoxin.

2. *Vipoxin Acidic Component*

According to the investigations up to now, as VAC has no enzyme activity, it stabilizes the unstable PLA_2 and preserves the catalytically and physiologically active conformation (Tchorbanov et al., 1978; Betzel et al., 1999; Georgieva et al., 2003a). The observed space for calcium in the structure of the acidic component supports the hypothesis, suggesting a common origin with the sPLAs. The role of the Ca^{2+} ion here is to stabilize the conformation of the loop, which is important for recognition of the basic $sPLA_2$ (Devedjiev et al., 1997). The acidic component and basic $sPLA_2$ associate with their hydrophobic sides by complementary, well-fitted hydrophobic interactions (Trp20 and Trp31 from $sPLA_2$ and Phe3, Trp31, and Tyr119 from the nontoxic component and the conservative hydrophobic regions on the recognition surface of the $sPLA_2$), causing a loss of 13% of the whole accessible surface area upon the formation of Vipoxin (Devedjiev et al., 1993; Devedjiev et al., 1997; Georgieva et al., 2003a). The complex is additionally stabilized by electrostatic interactions between the positively charged $sPLA_2$ (pI 10.4) and the negatively charged acidic component (pI 4.6) (Tchorbanov et al., 1978; Banumathi et al., 2001). Additionally, 13 solvent water molecules are buried inside the interface between the subunits, providing a hydrogen-bonding network, which enhances the stabilization of the complex (Devedjiev et al., 1997; Banumathi et al., 2001). Further, it was proposed that in the complex, the acidic component shields the access of the substrate molecules to the active site of the enzyme, thus lowering the enzyme activity, that is, Vipoxin is enzymatically active and allows the access of small substrates to the $sPLA_2$ active site.

Different roles were attributed to VAC, but none has been proved yet. Nevertheless, the diversity of pharmacological effects manifested by the toxic $sPLA_2$ component is strongly dependent on the intramolecular interactions between the subunits. Blinov et al. (1979) proposed that Vipoxin exerts its toxicity through dissociation to components. The neurotoxic component binds to the target membrane, and the nontoxic component remains in solution. Similar to other multicomponent $sPLA_2$, VAC is involved in the specific binding of the neurotoxin to the membrane acting as a "chaperone" directing the $sPLA_2$ subunit to its target acceptor (Betzel et al., 1999). Such a synergism between enzymatic and nonenzymatic subunits has been described for crotoxin (Faure et al., 1991; Bon, 1997).

For the first time, we establish heterogeneicity (Fig. 5) of the VAC that imposes further purification and analysis of the structure and eventual functional role of different separated acidic forms. We fractionated VAC by anion-exchange chromatography on Mono Q HR 5/5 column and three of the purified acidic fractions, displaying different molecular masses, were further purified by RP-HPLC. The origin of their existence could be explained with availability of different isoforms (pIs), tendency of association, impurities from total venom purification step, acidic deamination of Asn and Gln to Asp and Glu, or posttranslational modifications. Whether the existence of different VAC forms will be associated with different degree of stability and activity of the complex, similar to the crotoxin isoforms (Faure et al., 1993), or with a blocking of different pharmacological sites upon association with target sites is a matter of further investigation, but they definitely have a specific role in the pharmacological and biochemical properties.

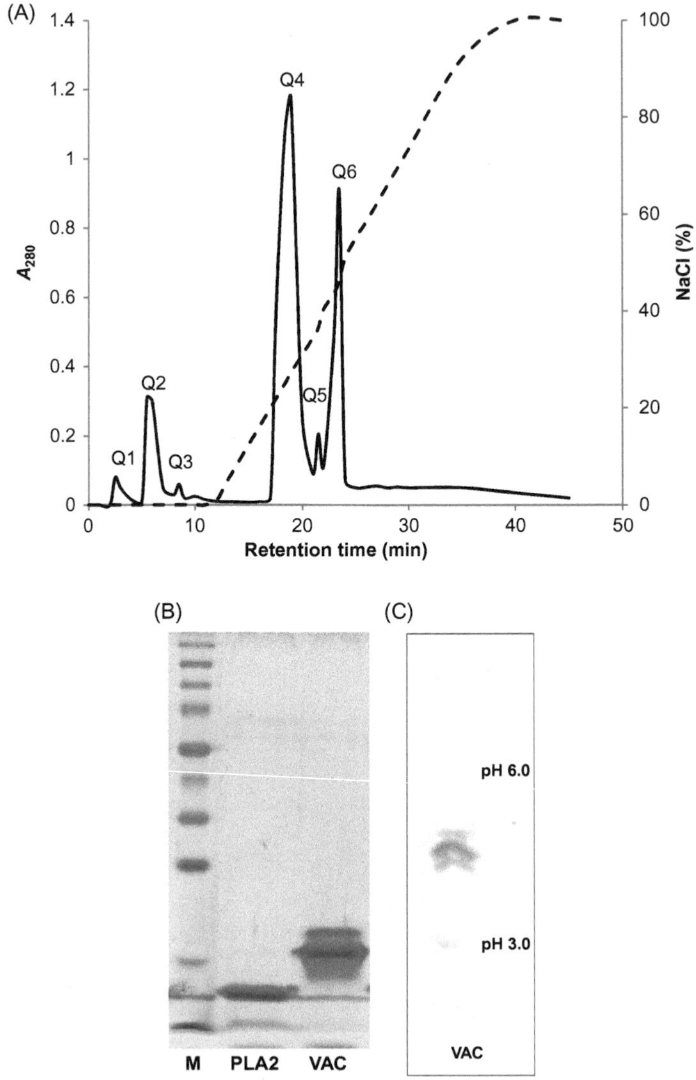

FIG. 5. (A) Separation of VAC by anion-exchange chromatography on Mono QHR 5/5. The column was equilibrated with 0.1 M acetate buffer (pH 5.5), containing 6 M carbamide and eluted at a flow rate of 0.4 ml/min with a linear gradient of 0.5 M NaCl in the same buffer (from 0% to 100%). Different fractions named Q1–6 were isolated. (B) SDS–PAGE of Vipoxin and its components (silver staining); (C) IEF of VAC (silver staining). (For color version of this figure, the reader is referred to the online version of this chapter.)

C. *Enzyme Activity*

1. *Catalytic Properties*

We assay PLA_2 activity by two analytical methods: (1) HPLC measurement of the free fatty acids released from phospholipid substrates by an analytical procedure described previously (Atanasov et al., 2009a) using DPPC and 1-palmitoyl-2-oleoyl-*sn*-glycero-3-phosphocholine (POPC) as substrates and (2) a spectrophotometric method described by Cho and Kézdy (1991) and Holzer and Mackessy (1996), using a synthetic substrate (4-nitro-3-octanoyloxy-benzoic acid, ONBA), modified for 96-well plates (Ponce-Soto et al., 2002). In all enzymatic activity experiments, we analyze the activity of Vipoxin complex, pure $sPLA_2$, isolated from Vipoxin and the reconstituted system of the purified Vipoxin subunits: $sPLA_2$ and VAC preincubated for different time intervals prior to reaction.

We evaluate the interactions between phospholipid monolayers and pure $sPLA_2$/Vipoxin (enzyme activity as well as the penetrating ability) using model membranes, composed of different phospholipids in different proportions by AFM (Balashev et al., 2010) and surface tension γ (mN/m) determination using monolayer films made in a multi-well plate (3×5) MicroTrough X (Jordanova et al., 2009).

It is well known that $GIIPLA_2$ enzymes display different phospholipid headgroup preferences (choline, ethanolamine, serine, inositol) and hence different substrate specificities. These differences in the enzymatic activity could be related to the various pharmacological effects, etc., a relatively weak ability to hydrolyze PE might be a characteristic of presynaptically acting $sPLA_2$ toxins; higher affinity to anionic PS supports mainly the anticoagulant acting $sPLA_2$ enzymes (Kini, 1997, 2006). Some $sPLA_2$-inducing neurotoxicity release more docosahexanoic acid (22:6) or oleic acid (18:1) (Fletcher and Jiang, 1995). Comparing large number of $sPLA_2$ enzymes, Rosenberg (1997) concluded that there is no clear correlation between the rates of the enzymatic activities measured *in vitro* and the potency of the biological actions.

Our results on the interaction between Vipoxin's PLA_2 and different spread phospholipid monolayers (in 50 mM Tris–HCl, pH 8.0, containing 100 mM NaCl and 10 mM $CaCl_2$) in equilibrium and dynamic conditions showed that maximum reaction rate was reached when monolayers composed of PLs with unsaturated fatty acids at *sn*-2 position were used:

palmitoleoylphosphatidylglycerol (POPG), dioleoylphosphatidylethamolamine (DOPE), and palmitoleoylphosphatidylcholine (POPC) (Table I). The effect of POPG concentration on the rate of the enzymatic activity showed maximum at concentrations above 0.2 mM at which the enzyme was fully saturated (Fig. 6). Zhao et al. (2007) established the counterion effect of Na^+ on POPG lipids in overcoming the electrostatic repulsion between negative charges of PG headgroups. This behavior results in a

TABLE I
The Effect of Pure $sPLA_2$ Isolated from Vipoxin on the Surface Tension of Monolayers from Different PLs

Lipids	$\Delta\gamma$ (mN/m)
DOPE	16.89
DPPC:POPG (7:3)	4.43
DPPS	1.92
POPC	22.4
POPG	45.73

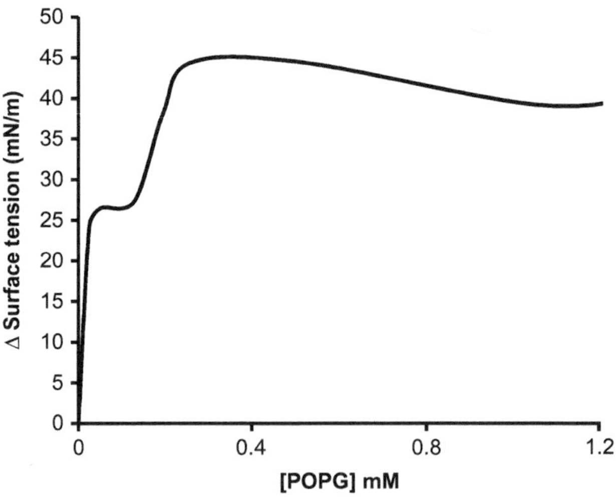

FIG. 6. The effect of POPG (1-palmitoyl-2-oleoyl-*sn*-glycero-3-phosphoglycerol) concentration on the enzymatic activity of $sPLA_2$ (measured as the change of the surface tension at the 10th minute from the beginning of the reaction).

complexation of PG molecules and formation of large PG clusters (including salt bridges and strong hydrogen bonding), and this probably activate $sPLA_2$.

Using HPLC assay of the enzymatic activity and POPC as a substrate, we established a more complex effect of the substrate concentration on the reaction rate associated with the different states of the substrate arrangement at concentrations above the CMC (Fig. 7). The course of the curve exhibits different parts indicating different substrate arrangements. At a concentrations above 0.08 mM POPC, pure $sPLA_2$ showed a burst in activity which could be explained with the ordered arrangement of the substrate under the conditions of the reaction mixture. Many investigations on molecular dynamics of different phospholipids have proved the complex intermolecular interactions between PL molecules, strongly dependent on concentration, hydrogen bonding, and the composition of the reaction milieu (Boggs, 1987; Böckmann and Grubmüller, 2004; Sachs et al., 2004; Zhao et al., 2007) with great impact on structural organization of the substrate and hence on $sPLA_2$ activity. Under the same reaction conditions and concentrations lower than 0.1 mM, Vipoxin and $sPLA_2$+VAC

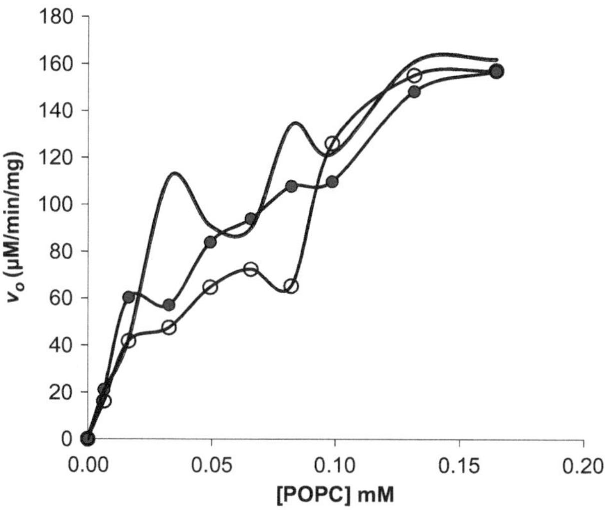

FIG. 7. The effect of the POPC (1-palmitoyl-2-oleoyl-*sn*-glycero-3-phosphocholine) concentration on the rate of enzymatic reaction catalyzed by Vipoxin (solid line): purified PLA2, –o–; reconstitued PLA2+VAC (1:1), –●–.

(1:1 molar ratio) reconstituted mixture demonstrated higher $sPLA_2$ activity. At concentrations higher than 0.1 mM, the $sPLA_2$ activity is practically the same for the three systems used, although maximal activity for Vipoxin was reached at higher concentrations of POPC. Interestingly, we found that the pure $sPLA_2$ has lower activity compared to the reconstituted mixture PLA_2 and VAC (1:1 ratio) against the synthetic substrate ONBA also.

The mechanism of VAC "inhibition" was proven (Aleksiev and Tchorbanov, 1976) as pH independent (at least in the range of pH 7–10) and specific (maximal effect of inhibition was achieved at equimolar ratio of the components) furthermore, the free and inhibited $sPLA_2$ component had identical Michaelis constants. This suggests that the inhibitors do not compete with the substrate for binding in the active site.

Analyzing the role of the VAC on the enzymatic and pharmacological activities, we were more than surprised by the lack of its "inhibitory activity" on the $sPLA_2$ component. Even more, VAC activated it. Our studies demonstrate that enzyme activity of the reconstituted mixture of the purified subunits preincubated for 1 h (4 °C) at a molar ratio VAC: $sPLA_2$ of 0.8 was even higher against ONBA than that of the pure $sPLA_2$ in the control reaction (Fig. 8). Both methods used (HPLC and spectrophotometric) revealed that VAC acts as an "inhibitor" only at molar ratios VAC:$sPLA_2$ below 0.6:1. Increasing of the VAC: $sPLA_2$ molar ratio up to 1:1 (which is postulated to be the case in the natural Vipoxin), restores the initial $sPLA_2$ activity, measured without VAC in the reaction mixture. The preliminary experiments (Aleksiev and Tchorbanov, 1976; Tchorbanov et al., 1977) on the incubation of the separated and purified Vipoxin components have shown an increasing of the complex stability and reducing of the enzyme activity.

Having analyzed our results on the enzymatic activity of Vipoxin and its $sPLA_2$, we come round to the opinion that is more correctly to present Vipoxin as a heterodimeric complex composed of one basic *catalytic* subunit and one acidic *regulatory* subunit. Their interaction in a common quaternary protein structure is more than a noncovalent association between two components. The existing and recurring view presented Vipoxin more in the light of two associated individual components which can display their functions independently. Our view is that Vipoxin is a heterodimer of two subunits which act in concert with each other and both can be destabilized upon separation. The fact that their separation is

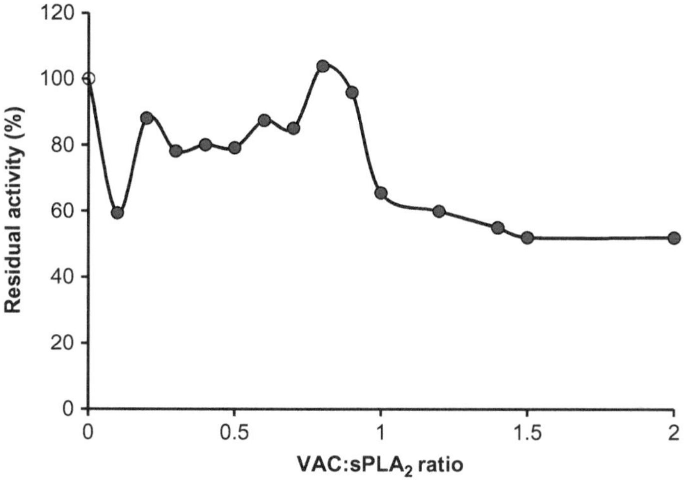

FIG. 8. $sPLA_2$ activity assayed in the presence of VAC after 1 h preincubation of the subunits at 4 °C using ONBA as a substrate. The activity is presented as a percentage of the initial $sPLA_2$ activity (100%) measured in the absence of VAC in the reaction mixture under the same conditions.

possible only in the presence of 6 M carbamide proves the biological importance of the whole complex. Jan et al. (2002) proposed that the acidic subunits in the heterodimeric toxins (Vaspin, Vipoxin) may be a product of a random evolution rather than a direct evolution, conferring an advantage to the selective acting on both subunits, that is, the chaperone function of the acidic subunits may be an acquired function, allowing postsynaptic membranes of the neuromuscular junction to be targeted.

For the first time, we applied ATM as an analytical tool to visualize the degradation of a phospholipid bilayer composed of DPPC by Vipoxin's $sPLA_2$. Figure 9 shows a typical example of time course set of images from an experiment prior and after injection of $sPLA_2$ (Balashev et al., 2010). Few bilayer defects are easily distinguishable at the first image (Fig. 8A) obtained prior to the enzyme injection. The subsequent images (Fig. 8B–F) reveal progressive changes in the surface topology of the bilayer as a result of the enzymatic action. With the $sPLA_2$ present, holes in the bilayer start enlarging as a result of hydrolysis of the lipid molecules from the adjacent to the liquid phase upper layer, followed by a desorption of the products: LysoPPC and palmitic acid molecules. Simultaneously, DPPC molecules

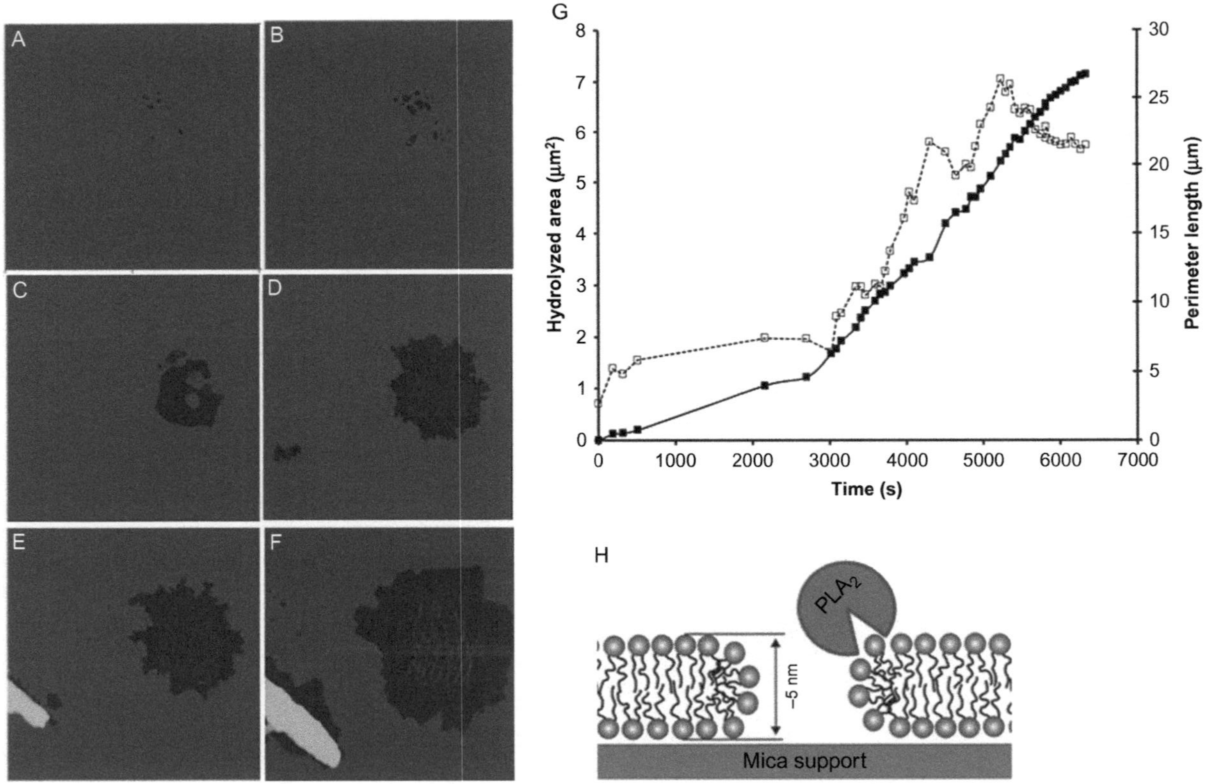

FIG. 9. AFM images of a time sequence of the Vipoxin sPLA$_2$ hydrolysis of a supported DPPC bilayer on mica. (A) Image prior to injection of Vipoxin PLA$_2$ into the AFM liquid cell; (B–F) images are taken in the time interval 150–6400 s after the

from the bottom layer adjacent to the mica support also desorb: as soon as the adjoining upper layer "disappears," their hydrophobic tails are exposed unfavorably toward the liquid phase. A typical result of the kinetics analysis of the growing holes is presented at Fig. 8G. We assume that if ever the enzyme binds to the bilayer, it is activated and prefers to "scoot" through the lipid bilayer. This assumption favors the "scooting mode" as most realistic molecular model of $sPLA_2$ action (Jain et al., 1986a). The kinetic curve of the hydrolyzed area versus time shows two distinct slopes which corresponded to the two regimes of the enzyme hydrolysis. The first one, prior to the appearance of new bilayer defects, is a slow-rate hydrolysis, while the second one is a burst of the enzymatic activity. This kinetic behavior supports the hypothesis for the existence of a lag phase characterized by low enzymatic activity, although the lag period measured in the present experiments is between 1.5 and 2.5 times longer than the one reported in Grandbois et al. (1998). The first stage of bilayer hydrolysis occurs preferably at the rims of the existing structural defects. The kinetic curves (see Fig. 8G) prove the correlation between the hydrolysis rate and the length of the edge of the bilayer defect. Figure 8H presents the molecular model of hydrolysis of the supported bilayer. The model shows the lipid organization at the rim of the bilayer defects where the enzyme attack occurs.

2. *The Effect of Metal Ions*

The investigated $sPLA_2$ is a Ca^{2+}-dependent enzyme. The coordination of calcium ions has not been established in any previous crystal structures of Vipoxin or its components. In the case of Vipoxin, it has been proposed an unusual topology, caused by the intermolecular contacts between the

Vipoxin PLA_2 injection. Lighter areas correspond to intact bilayer, whereas dark areas correspond to the mica support. Image size is 5 × 5 μm. Images have been corrected for baseline tilt and bow. (G) Analysis of the complete image series of (A)–(F) (including frames not shown in the figure). The graph shows the hydrolyzed area (filled squares) and the perimeter length (open squares) of the bilayer defects as functions of time. (H) Schematic representation of the model of hydrolysis of the supported bilayer by $sPLA_2$. The model shows the lipid organization at the rim of the bilayer defects where the enzyme attack occurs. With some changes, figures are adapted from Balashev et al. (2010). (The data points obtained from the time series images are presented in a format: desorbed area as a primary ordinate vs. time and perimeter length as a secondary *Y*-axis.) (For color version of this figure, the reader is referred to the online version of this chapter.)

subunits (Mancheva et al., 1987; Devedjiev et al., 1997; Perbandt et al., 1997; Banumathi et al., 2001; Georgieva et al., 2004). Ca^{2+}-binding loop of Vipoxin is considerably destabilized in the absence of cofactor (Banumathi et al., 2001; Murakami et al., 2006).

Aleksiev and Tchorbanov (1976) found that the optimal concentration of Ca^{2+} ions was $1-3\times10^{-3}$ M and that Mg^{2+} ions activated slightly the initial velocity of the hydrolysis at a concentration of 2×10^{-3} M but inhibited the reaction at higher concentrations. The other divalent cations (Cu^{2+}, Zn^{2+}, Ba^{2+}, Sr^{2+} in 1.5 mM concentration) suppressed the enzymatic activity against egg yolk lecithin as a substrate (Aleksiev and Tchorbanov, 1976; Atanasov et al., 2005). Further, we analyzed the effect of different Me^{2+} ions (Ca^{2+}, Mg^{2+}, Cd^{2+}, Sr^{2+} Ba^{2+} in 5 mM concentration) on the $sPLA_2$ activity toward synthetic (ONBA) and natural substrates (DPPC and POPC). Our experimental data showed that the metal ions having ionic radii larger than Ca^{2+} (Cd^{2+}, Sr^{2+}, Ba^{2+}) influence more effectively the ONBA hydrolysis than those possessing smaller (Mg^{2+}) one. Therefore, the size of the Me^{2+} ionic radius is crucial for the formation of $sPLA_2$–Me^{2+}–ONBA complex prior to the subsequent ONBA hydrolytic reaction. In the case of DPPC and POPC as natural substrates, Ca^{2+} and Cd^{2+} ions activate the hydrolytic activity of $sPLA_2$ at a comparable level, the smaller (Mg^{2+}) and the larger (Sr^{2+}) ions are less effective, while Ba^{2+} acts as an enzyme inhibitor. The Me^{2+} ions form the corresponding $sPLA_2$–Me^{2+} complexes, but the further formation of $sPLA_2$–Me^{2+} substrate complexes does or does not take place, depending on the size of ionic radii as well as on the structure of the natural substrates. We also assume that Ca^{2+} ions have definite effect on enzyme's *sn*-2 specificity, determined by the A2/A1 ratio of the final reaction products and play an important role in the substrate structure and organization (Scott and Sigler, 1994; Scott, 1997; Böckmann and Grubmüller, 2004).

3. *The Effect of pH and Temperature*

Aleksiev and Tchorbanov (1976) established a pH optimum at pH 9.9–10 for both $sPLA_2$ component and Vipoxin complex using egg yolk lecithin as a substrate, proposing a complicated character of pH activation, due to the ionization of more than one group. Using DPPC, POPC, and synthetic ONBA as substrates, we found maximal activity at a pH range 8.0–9.2.

The effect of temperature on the initial velocity of the lecithin hydrolysis was measured in the range of 0–50 °C and a maximum velocity was observed at 23–26 °C, which was identical throughout the pH range 7–10 (Aleksiev and Tchorbanov, 1976).

IV. Pharmacological Effects of Vipoxin

The enzymatic activity of $sPLA_2$ toxins results in a number of effects such as: membrane damage leading to a change in the environment of the membrane proteins; the products of the reaction—lysophospholipids and free fatty acids are lipid-signaling molecules with their own effect; binding to the target proteins can block their specific interactions with other ligands; binding to the target proteins may activate or inhibit enzymatic activity, etc. (Kini, 1997). The basic factor for exhibiting a diversity of pharmacological effects is considered to be the penetratability of $sPLA_2$ enzymes, by which enzymes insert themselves into the phospholipid membrane of the target cell. As most basic $sPLA_2$ enzymes have higher penetratability compared to the acidic and neutral isoenzymes, it is accepted that positively charged residues flanking the hydrophobic segments play a significant role in pharmacological manifestations (Verheij et al., 1980; Kini, 1997). However, the pharmacological affinity is in the low nanomolar range, whereas the binding between the $sPLA_2$ enzymes and phospholipids is in the high micromolar range, which requires searching for other mechanisms as well.

Some snake venom $sPLA_2$ enzymes require additional protein factors to express their specific pharmacological effect at full potency (Kini, 1997). Kini and Evans (1997) and Kini (2006) proposed a target model, that is, a presence of specific "target sites" on the surface of the target cell complementary (in terms of charges, hydrophobicity, and van der Waals contact surfaces) to the eventual pharmacological sites of $sPLA_2$. Some target proteins such as membrane-bound receptors are identified using radiolabeled PLA_2 enzymes and photoaffinity-labeling techniques (Lambeau et al., 1997).

Comparing the regions responsible for the toxicity, Wang et al. (1992) showed drastic changes in the electrostatic charge, mainly in the β-wing and the segment including residues 115 ± 128. The β-wing of Vipoxin's $sPLA_2$ subunit contains positively charged residues, while the same loop of the double-stranded antiparallel β-sheet in the VAC is negatively charged (Glu instead of Lys). Certain exposed aromatic residues in this region could also be very important for the neurotoxicity. Pungercar et al. (1999) established more than 10-fold higher toxicity of AtxA from *V. ammodytes ammodytes* compared with AtxC due to a substitution of Ile124 by Phe.

A. *Acute Toxicity*

Initially, the toxicity of Vipoxin and its PLA_2 subunit was examined in 1973 by measuring LD_{100} values on mice (Aleksiev and Shipolini, 1973; Aleksiev and Tchorbanov, 1976). For the separated $sPLA_2$, LD_{100} was found to be—1–3 μg per 20 g mouse (i.v.) and on this basis PLA_2 component of Vipoxin was classified to the group of the most toxic $sPLA_2$ toxins, which toxicity increases fivefold after separation from the acidic subunit named "inhibitor." Further, the $sPLA_2$ toxicity was lost 3–4 days after separation from the VAC (Aleksiev and Shipolini, 1973). The toxicity of Vipoxin complex was established to be 8 μg per 20 g mouse (i.v.), indicating lower toxicity for the whole complex (Tchorbanov and Aleksiev, 1981). These first experiments on the acute toxicity of Vipoxin and its $sPLA_2$ were not as detailed, for the results were presented as LD_{100} values and no data on methodology were published. Aleksiev and Tchorbanov (1976) and Tchorbanov et al. (1978) proposed a postsynaptic neurotoxicity of Vipoxin, whereas the separated $sPLA_2$ manifested a presynaptic action due to the enzymatic activity. Aleksiev and Tchorbanov (1976) demonstrated that *in vitro*, the acidic subunit reduces the catalytic activity and toxicity of the PLA_2 subunit up to 60%. In contrast, in other multichain neurotoxins, the toxic subunit has a lower lethality than the whole complex (Bon, 1997).

Our recent studies (Atanasov et al., 2011a) on the acute toxicity expressed as LD_{50} after intraperitoneal and intravenous administration of purified $sPLA_2$ component and Vipoxin showed completely different results. Vipoxin LD_{50} (mice, i.p. and i.v.) value was found to be 1.1 mg/kg b.w. (i.p./i.v.). The LD_{50} values of $sPLA_2$ are much higher—11 mg/kg b.w. (i.p.) and 2.6 (i.v.), respectively. When these values are presented in mol/kg b.w., Vipoxin turns out to be more toxic than the $sPLA_2$ (30 vs. 15 kDa). In this way, our results present a new insight about the role of VAC, which is crucial for the toxicity of both the $sPLA_2$ and the whole complex. Typical neurotoxic symptoms were observed in the experimental animals when Vipoxin is applied: agitation, ataxia, tremor and convulsions, paresis of back limbs, seizure activity, and usually death 2–5 min after generalized seizure. The same toxicity signs were observed when the reconstituted Vipoxin complex (PLA_2:VAC, in 1:1 molar ratio) was used. Surprisingly, in the case of the pure $sPLA_2$, these symptoms were not observed. The animals were apathetic with any lack of neurotoxic symptoms (only transitional paresis in isolated cases followed by a complete animal recovery).

These findings are in agreement with our understanding of the unique regulatory mechanism by which VAC potentiates the $sPLA_2$ toxicity in Vipoxin. A ''cheparonic'' function was proposed and discussed previously in the light of the stability of the complex (Betzel et al., 1999). The molecular synergism by which the nontoxic acidic subunit potentiates the toxicity, and activity of basic $sPLA_2$ has already been postulated for the crotoxin acidic component A (Faure and Bon, 1988; Faure et al., 1991, 1993, 1994; Bon, 1997).

B. *Hemolytic Activity*

For the first time, we demonstrate that Vipoxin's hemolytic activity is directly related to its enzymatic activity (Atanasov et al., 2009a). The hemolytic activity of the PLA_2 enzyme could be a result of direct disruption of the membranes of red blood cells (similar to cardiotoxins) (Jiang et al., 1989), or indirect—the result of released hydrolytic products disrupting the red blood cells (Nikai et al., 1993; Adhikari et al., 2007; Ibbara-Alvarado et al., 2007). We tested the direct and indirect hemolytic activity of Vipoxin, purified $sPLA_2$, and $sPLA_2$:VAC reconstituted complex (at 1:1 molar ratio) (Atanasov et al., 2009b). The effect of $sPLA_2$ isolated from Vipoxin on the red blood cells is associated with the indirect hemolysis, that is, the effect of lysophospholipid formation as a reaction product and is Ca^{2+} dependent. The hemolytic activity measured with pure $sPLA_2$ remains constant during 24 h of incubation. However, after 24 h at 4 °C, $sPLA_2$ enzyme activity decreases up to 77%, while the enzyme activity of the reconstituted $sPLA_2$:VAC (1:1 ratio) complex measured under the same conditions decreases only up to 29%, indicating for the stabilizing role of the acidic component.

We tested the direct and indirect hemolytic activity of the purified PLA_2 and PLA_2:VAC complex at different pH values. On the one hand, the results showed maximal values of direct and indirect hemolysis for pure $sPLA_2$ to be at pH 6.5, while the $sPLA_2$:VAC mixture showed minimum of direct and indirect hemolysis at the same pH value. It can be proposed that at pH 6.5, the association between Vipoxin's components is optimal. This pH dependence of the subunit interactions could also be a factor regulating the pharmacological activity.

On the other hand, at pH 7.5, the enzymatic activity of pure $sPLA_2$ (against ONBA) is optimal, while the hemolytic activity of $sPLA_2$ and reconstituted PLA_2:VAC (1:1) mixture is practically indistinguishable under the same conditions. The differences in enzymatic and hemolytic activity of $sPLA_2$ at different pH values suggest involving different ionizable groups important for both the enzymatic activity and interactions with the red blood cell membrane as a natural target.

Stoykova et al. (2011) established that direct $sPLA_2$ activity on the red blood cell membrane caused formation of spherocytes and spherocytic echinocytes. This morphological change of the red cell membrane (decreasing its effective surface) could be explained with the need of a protective mechanism of the cell toward direct hemolysis. The formation of echinocytes is a typical phenomenon for *in vivo* envenomation with snake toxins.

C. *Anticoagulant Activity*

In general, two main mechanisms are proposed for the anticoagulant action of $sPLA_2$ enzymes—hydrolysis of/or binding to procoagulant phospholipids and binding to factor Xa (inhibiting formation of prothrombinase complex) (Verheij et al., 1980; Mounier et al., 2001; Zhong et al., 2001; Kini, 2005). By comparison of the amino acid sequences, Kini and Evans (1987) differentiate the strong anticoagulants by weak and noncoagulant $sPLA_2$ enzymes. Strongly anticoagulant $sPLA_2$ enzymes inhibit blood coagulation at concentrations below 2 μg/ml, weakly anticoagulant PLA_2 enzymes are effective up to 10 μg/ml. The third group of $sPLA_2$ enzymes acts in concentrations higher than 15 μg/ml and has little or no effect on the clotting times (Kini and Evans, 1997). In the strong anticoagulant enzymes, the region between 54 and 77 residues is basic, whereas in weakly and nonanticoagulant enzymes is neutral or negatively charged (Glu54-Lys55-Gly57 and Lys75-Lys77 couple).

The weak anticoagulants exhibit their action through hydrolysis of procoagulant phospholipids or inhibit activation of factors X–Xa by extrinsic tenase complex. In contrast, the strong anticoagulants act as affecting target proteins or interfered with the prothrombinase complex. Similar residues are not presented in the structure of $sPLA_2$ subunit of Vipoxin, which characterized it as weak anticoagulant or noncoagulant protein (Banumathi et al., 2001).

Measuring both the characteristic coagulation times and the enzymatic activity, we concluded that the anticoagulant activity of the $sPLA_2$ from Vipoxin is a result of its enzymatic activity on the procoagulant phospholipids rather than using nonenzymatic mechanisms (binding to specific coagulation factors). The activated partial thromboplastin time (aPTT) was affected to the highest degree (contact activation, intrinsic, pathway). The prothrombin time and thrombin time were not changed compared to the initial control sample.

VAC displays no anticoagulant effect. More interestingly, the whole Vipoxin complex is also anticoagulant *ineffective*, although it exhibits PLA_2 activity. This effect could be explained with the occupation of the positively charged $sPLA_2$ contact sites upon association of the subunits, which results in the absence and/or weakening the contact with negatively charged phospholipids of the coagulation complex.

The alkylated with 4-bromophenacyl bromide $sPLA_2$ enzyme does not affect the coagulation system. The modified enzyme lost catalytic, hemolytic, and anticoagulant activities which confirmed enzyme-dependent mechanism of its anticoagulant activity.

V. Future Prospects

A. *Immunological Studies*

Vipoxin and its components have not been investigated immunologically, and the results from our preliminary experiments are promising (Stoyanova et al., 2011). We obtained numerous clones of recombinant scFv antibodies using Human Single Fold scFv Libraries named Tomlinson I+J (Cambridge, UK). The monoclonal antibodies against $sPLA_2$, Vipoxin, and VAC have been tested for their effectiveness to inhibit the catalytic activity of $sPLA_2$ and Vipoxin, and some of the results show a radical difference between them. Further, our investigations will be directed also to more detailed analysis of the regulatory role of the VAC using the obtained purified recombinant antibodies, and some of them will be tested against other $sPLA_2$ enzymes displaying proinflammatory effects. Immunological studies are crucial for obtaining the effective antivenoms, for determining the pharmacological sites of the $sPLA_2$, for distinguishing the biological effects and enzymatic activity of $sPLA_2$, and therefore for enlightening the role of the catalytic activity for the toxicity of venom $sPLA_2$.

B. *Neurotoxicity*

Many of the obtained results on the neurotoxicity earlier have to be revised, and neurotoxicity has to be discussed at the molecular level. The main questions about the modulation of the postsynaptic activity and the transformation from postsynaptic (Vipoxin) to presynaptic ($sPLA_2$) activity still remain.

C. *Inhibition by Natural and Chemical Inhibitors*

The inhibition of the $sPLA_2$ enzymatic activity has always been of great pharmacological and medical importance (Dennis et al., 2011). Noetzel et al. (2002) showed that aristolochic acid, vitamin E (D-α-tocopherol), and oleyloxyethylphosphocholine suppressed the enzymatic activity of Vipoxin and its isolated $sPLA_2$ component. Another field of research for us is designing and synthesizing of new inhibitors of $sPLA_2$ and Vipoxin, which will be used for mechanistic studies, for the investigation of the inflammation pathways on a molecular basis, and as pharmaceutical agents.

D. *$sPLA_2$ Targeting and Cell Signaling Pathways of Vipoxin and Its Components*

One of the most interesting fields of the $sPLA_2$ research is related to its interactions with functionally active molecules, driving specific cell-signaling pathways. This essential study of Vipoxin and its subunits has not been addressed yet.

Acknowledgments

This work was supported by the project of the Bulgarian National Fund of Scientific Research, Grant DO-02-83/2008 (''Biochemical, Immunological, and Pharmacological properties of Vipoxin and its components'').

References

Adhikari, D., Samanta, S. K., Dutta, A., Roy, A., Vedasiromoni, J. R., Sen, T. (2007). *In vitro* hemolysis and lipid peroxidation-inducing activity of the tentacle extract of the sea anemone (*Paracondylactis indicus* Dave) in rat erythrocytes. *Ind. J. Pharmacol.* **39**, 155–159.

Aleksiev, B., Shipolini, R. (1973). On the isolation and the structure of the toxic proteins from the venom of Bulgarian viper (*Vipera ammodytes ammodytes*). *In Proceedings of the International Symposium on Polymers* Varna, Bulgariavol. 1, pp. 417–420.

Aleksiev, B., Tchorbanov, B. (1976). Action on phosphatidylcholine of the toxic phospholipase A_2 from the venom of Bulgarian viper (*Vipera ammodytes ammodytes*). *Toxicon* **14**, 477–485.

Arni, R. K., Ward, R. J., Gutierrez, J. M., Tulinski, A. (1995). Structure of a calcium-independent phospholipase-like myotoxic protein from *Bothrops asper* venom. *Acta Crystallogr. D* **51**, 311–317.

Atanasov, V., Bardarov, V., Aleksiev, B., Mitewa, M. J. (2005). Vipoxin specificity studied by gas chromatographic determination of enzymatic reaction products. Influence of Ca^{2+}, Mg^{2+} and Sr^{2+}. *J. Venom. Anim. Toxins incl. Trop. Dis.* **11**, 350–360.

Atanasov, V., Petrova, S., Mitewa, M. (2009a). HPLC assay of phospholipase A_2 activity using low-temperature derivatization of fatty acids. *Anal. Lett.* **42**, 1341–1351.

Atanasov, V. N., Danchev, D., Mitewa, M., Petrova, S. (2009b). Hemolytic and anticoagulant study of the neurotoxin vipoxin and its components—basic phospholipase A_2 and an acidic Inhibitor. *Biochemistry (Moscow)* **74**, 339–345.

Atanasov, Stoykova, S., Goranova, Y., Gluhcheva, Y., Petrova, S. (2011a). *In vivo* toxicity of vipoxin and its phospholipase A2. 17th Congress of the European Section of the International Society on Toxinology, Valencia (Spain), Abstract book, P77, p. 168.

Bahnson, B. J. (2005). Structure, function and interfacial allosterism in phospholipase A2: insight from the anion-assisted dimer. *Arch. Biochem. Biophys.* **433**, 96–106.

Balashev, K., Atanasov, V., Mitewa, M., Petrova, S., Bjørnholm, T. (2010). Kinetics of degradation of dipalmitoylphosphatidylcholine (DPPC) bilayers as a result of vipoxin phospholipase A_2 activity: an atomic force microscopy (AFM) approach. *Biochim. Biophys. Acta* **1808**, 191–198.

Balsinde, J., Balboa, M. A., Dennis, E. A. (1997). Inhibition of group VI Ca^{2+}-independent phospholipase A_2 blocks phospholipid fatty acid remodelling in murine $P388D_1$ macrophages. *J. Biol. Chem.* **272**, 29317–29321.

Banumathi, S., Rajashankar, K. R., Noetzel, C., Aleksiev, B., Singh, T. P., Genov, N., et al. (2001). Structure of the neurotoxic complex vipoxin at 1.4 Å resolution. *Acta Crystallogr. D* **57**, 1552–1559.

Barbour, S. E., Kapur, A., Deal, C. L. (1999). Regulation of phosphatidylcholine homeostasis by calcium-independent phospholipase A_2. *Biochem. Biophys. Acta* **1439**, 77–88.

Bekkers, A. C. A. P. A., Franken, P. A., Toxopeus, E., Verheij, H. M., Haas, G. H. (1991). The importance of glycine-30 for enzymatic activity of phospholipase A_2. *Biochem. Biophys. Acta* **1076**, 374–378.

Berg, O. G., Yu, B. Z., Rogers, J., Jain, M. K. (1991). Interfacial catalysis by phospholipase A2: determination of the interfacial kinetic rate constants. *Biochemistry* **30**, 7283–7297.

Berg, O. G., Rogers, J., Yu, B. Z., Yao, J., Romsted, L. S., Jain, M. K. (1997). Thermodynamic and kinetic basis of interfacial activation: resolution of binding and allosteric effects on pancreatic phospholipase A_2 at zwitterionic interfaces. *Biochemistry* **36**, 14512–14530.

Berg, O. G., Tsai, M. D., Gelb, M. H., Jain, M. K. (2001). Interfacial Enzymology: the secreted phospholipase A2-paradigm. *Chem. Rev.* **101**, 2613–2653.

Berg, O. G., Yu, B. Z., Chang, C., Koehler, K. A., Jain, M. K. (2004). Cooperative binding of monodisperse anionic amphiphiles to the i-face: phospholipase A2-paradigm for interfacial binding. *Biochemistry* **43**, 7999–8013.

Betzel, C., Genov, N., Rajashankar, K. R., Singh, T. P. (1999). Modulation of phospholipase A2 activity generated by molecular evolution. *Cell. Mol. Life Sci.* **56**, 384–397.

Blinov, N. O., Tscorbanov, B. P., Grishin, E. V., Aleksiev, B. V. (1979). Neurotoxic action of a protein complex from the venom of the Bulgarian viper (*Vipera ammodytes ammodytes*). *C. R. Acad. Bulg. Sci.* **32**, 663–666.

Böckmann, R. A., Grubmüller, H. (2004). Multistep binding of divalent cations to phospholipid bilayers: a molecular dynamics study. *Angew. Chem. Int. Ed.* **43**, 1021–1024.

Boggs, J. M. (1987). Lipid intermolecular hydrogen bonding: influence on structural organization and membrane function. *Biochim. Biophys. Acta* **906**, 353–404.

Bon, C. (1997). Multicomponent neurotoxic phospholipases A_2. In: Venom Phospholipase A2 Enzymes: Structure, Function and Mechanism, Kini, R. M. (Ed.), pp. 269–286. John Wiley & Sons, Chichester.

Burke, J. E., Dennis, E. A. (2009). Phospholipase A_2 structure/function, mechanism, and signaling. *J. Lipid Res.* **50**, S237–S242.

Burke, J. E., Karbarz, M. J., Deems, R. A., Li, S., Woods, V. L., Jr., Dennis, E. A. (2008). Interaction of group IA phospholipase A2 with metal ions and phospholipid vesicles probed with deuterium exchange mass spectrometry. *Biochemistry* **47**, 6451–6459.

Cajal, Y., Berg, O. G., Jain, M. K. (2004). Origins of delays in monolayer kinetics: phospholipase A_2 paradigm. *Biochemistry* **43**, 9256–9264.

Carman, G. M., Deems, R. A., Dennis, E. A. (1995). Lipid signaling enzymes and surface dilution kinetics. *J. Biol. Chem.* **270**, 18711–18881.

Cho, W., Kézdy, F. J. (1991). Chromogenic substrates and assay of phospholipases A_2. *Methods Enzymol.* **197**, 75–79.

Deems, R. A., Eaton, B. R., Dennis, E. A. (1975). Kinetic of phospholipase A_2 activity toward mixed micelles and its implications for study of lipolytic enzymes. *J. Biol. Chem.* **250**, 9013–9020.

Dennis, E. A. (1994). Diversity of group types, regulation, and function of phospholipase A_2. *J. Biol. Chem.* **269**, 13057–13060.

Dennis, E. A., Cao, J., Hsu, Y.-H., Magrioti, V., Kokotos, G. (2011). Phospholipase A_2 enzymes: physical structure, biological function, disease implication, chemical inhibition, and therapeutic intervention. *Chem. Rev.* **111**, 6130–6185.

Devedjiev, Y., Atanasov, B., Mancheva, I., Aleksiev, B. (1993). Crystals of phospholipase A_2 inhibitor: the non-toxic component of vipoxin from the venom of Bulgarian Viper (*Vipera ammodytes*). *J. Mol. Biol.* **229**, 1147–1149.

Devedjiev, Y., Popov, A., Atanasov, B., Bartunik, H.-D. (1997). X-ray Structure at 1.76 Å. Resolution of a polypeptide phospholipase A_2 inhibitor. *J. Mol. Biol.* **266**, 160–172.

Duncan, R. E., Sarkadi-Nagy, E., Jaworski, K., Ahmadian, M., Sul, H. S. (2008). Identification and functional characterization of adipose-specific phospholipase A_2 (AdPLA). *J. Biol. Chem.* **283**, 25428–25435.

Faure, G., Bon, C. (1988). Crotoxin, a phospholipase A_2 neurotoxin from the south american rattlesnake *Crotalus durissus terrificus*: purification of several isoforms and comparison of their molecular structure and of their biological activities. *Biochemistry* **27**, 730–738.

Faure, G., Guillaume, J. L., Camoin, L., Saliou, B., Bon, C. (1991). Multiplicity of acidic subunit isoforms of crotoxin, the phospholipase A_2 neurotoxin from *Crotalus durissus terrificus* venom, results from posttranslational modifications. *Biochemistry* **30**, 8074–8083.

Faure, G., Harvey, A. L., Thomson, E., Saliou, B., Fadvanyi, F., Bon, C. (1993). Comparison of crotoxin isoforms reveals that stability of the complex plays a major role in its pharmacological action. *Eur. J. Biochem.* **214**, 491–496.

Faure, G., Choumet, V., Bouchier, C., Camoin, L., Guillaume, J. L., Monegier, B., et al. (1994). The origin of the diversity of the crotoxin isoforms in the venom of *Crotalus durissus terrificus. Eur. J. Biochem.* **223**, 161–164.

Faure, G., Xu, H., Saul, F. A. (2011). Crystal structure of Crotoxin reveals key residues involved in the stability and toxicity of this potent heterodimeric β-neurotoxin. *J. Mol. Biol.* **412**, 176–191.

Fesus, L., Piacentini, M. (2002). Transglutaminase 2: an enigmatic enzyme with diverse functions. *Trends Biochem. Sci.* **27**, 534–539.

Fletcher, J. E., Jiang, M. S. (1995). Presynaptically acting snake venom phospholipase A_2 enzymes attack unique substrates. *Toxicon* **33**, 1565–1576.

Georgieva, D. N., Genov, N., Nikolov, P., Aleksiew, B., Rajashankar, K. R., Voelter, W., et al. (2003a). Structure-function relationships in the neurotoxin Vipoxin from the venom of *Vipera ammodytes meridionalis. Specrochim. Acta A* **59**, 617–627.

Georgieva, D. N., Rypniewski, W., Perbrandt, M., Jain, M. K., Genov, N., Betzel, C. (2003b). Crystallization and preliminary X-ray diffraction studies of a toxic phospholipase A2 from the venom of *Vipera ammodytes meridionalis* complexed to a synthetic inhibitor. *Biochem. Biophys. Acta* **1650**, 1–3.

Georgieva, D. N., Genov, N., Hristov, K., Dierks, K., Betzel, C. (2004). Interactions of the neurotoxin vipoxin in solution studied by dynamic light scattering. *Biophys. J.* **86**, 461–466.

Grainger, D. W., Reichert, A., Ringsdorf, H., Salesse, C. (1989). An enzyme caught in action—direct imaging of hydrolytic function and domain formation of phospholipase A_2 in phosphatidylcholine monolayers. *FEBS Lett.* **252**, 73–82.

Grainger, D. W., Reichert, A., Ringsdorf, H., Salesse, C. (1990). Hydrolytic action of phospholipase A2 in monolayer phase transition region: direct observation of enzyme domain formation using fluorescence microscopy. *Biochem. Biophys. Acta* **1023**, 365–379.

Grandbois, M., Clausea-Schaurnann, H., Gaub, H. (1998). Atomic force microscope imaging of phospholipid bilayer degradation by phospholipase A_2. *Biophys. J.* **74**, 2398–2404.

Holzer, M., Mackessy, S. P. (1996). An aqueous endpoint assay of snake venom hospholipase A_2. *Toxicon* **34**, 1149–1155.

Huang, M. Z., Gopalakrishnakone, P., Chung, M. C. M., Kini, R. M. (1997). Complete amino acid sequence of an acidic, cardiotoxic phospholipase A_2 from the venom of *Ophiophagus hannah* (King cobra): a novel cobra venom enzyme with "pancreatic loop. *Arch. Biochem. Biophys.* **338**, 150–156.

Ibarra-Alvarado, C., Alejandro García, J., Aguilar, M. B., Rojas, A., Falcón, A., Heimer de la Cotera, E. P. (2007). Biochemical and pharmacological characterization of toxins obtained from the fire coral Millepora complanata. *Comp. Biochem. Physiol. Part C* **146**, 511–518.

Jain, M. K., Berg, O. G. (1989). The kinetics of interfacial catalysis by phospholipase A_2 and regulation of interfacial activation: hopping versus scooting. *Biochim. Biophys. Acta* **1002**, 127–156.

Jain, M. K., Berg, O. G. (2006). Coupling of the i-face and the active site of phospholipase A_2 for interfacial activation. *Curr. Opin. Chem. Biol.* **10**, 473–479.

Jain, M. K., Rogers, J., Jahagirdar, D. V., Marecek, J. F., Ramirez, F. (1986a). Kinetics of interfacial catalysis by phospholipase A_2 in intravesicle scooting mode, and hetero-fusion of anionic and zwitterionic vesicles. *Biochim. Biophys. Acta* **860**, 435–447.

Jain, M. K., Maliwal, B. P., De Haas, G. H., Slotboom, A. J. (1986b). Anchoring of phospholipase A2: the effect of anions and deuterated water, and the role of N-terminus region. *Biochim. Biophys. Acta* **860**, 448–461.

Jain, M. K., Ranadive, G., Yu, B. Z., Verheij, H. M. (1991). Interfacial catalysis by phospholipase A2: monomeric enzyme is fully catalytically active at the bilayer interface. *Biochemistry* **30**(29), 7330–7340.

Jain, M. K., Gelb, M. H., Rogers, J., Berg, O. G. (1995). Kinetic basis for interfacial catalysis by phospholipase A_2. *Methods Enzymol.* **249**, 567–614.

Jan, V., Maroun, R. C., Robbe-Vincent, A., De Harob, L., Choumet, V. (2002). Toxicity evolution of Vipera aspis aspis venom: identification and molecular modeling of a novel phospholipase A_2 heterodimer neurotoxin. *FEBS Lett.* **527**, 263–268.

Jiang, M. S., Fletcher, J. E., Smith, L. A. (1989). Factors influencing the hemolysis of human erythrocytes by cardiotoxins from *Naja naja kaouthia* and *Naja naja atra* venoms and a phospholipase A2 with cardiotoxin-like activities from *Bungarus fasciatus* venom. *Toxicon*, **27**, 247–257.

Jordanova, A., Georgiev, G. A., Alexandrov, S., Todorov, R., Lalchev, Z. (2009). Influence of surfactant protein C on the interfacial behavior of phosphatidylethanolamine monolayers. *Eur. Biophys. J.* **38**, 369–379.

Kini, R. M. (1997). Phospholipase A_2—a complex multifunctional protein puzzle. In: Venom Phospholipase A_2 Enzymes: Structure, Function and Mechanism, Kini, R. M. (Ed.), pp. 1–28. John Wiley & Sons, Chichester.

Kini, R. M. (2003). Excitement ahead: structure, function and mechanism of snake venom phospholipase A_2 enzymes. *Toxicon* **42**, 827–840.

Kini, R. M. (2005). Structure–function relationships and mechanism of anticoagulant phospholipase A_2 enzymes from snake venoms. *Toxicon* **45**, 1147–1161.

Kini, R. M. (2006). Anticoagulant proteins from snake venoms: structure, function and mechanism. *Biochem. J.* **397**, 377–387.

Kini, R. M., Evans, H. J. (1987). Structure-function relationships of phospholipases. The anticoagulant region of phospholipases A_2. *J. Biol. Chem.* **262**, 14402–14407.

Kini, R. M., Evans, H. J. (1997). The anticoagulant effects of snake venom phospholipases A_2. In: Venom Phospholipase A_2 Enzymes: Structure, Function and Mechanism, Kini, R. M. (Ed.), pp. 353–368. John Wiley & Sons, Chichester.

Koh, C. Y., Kini, R. M. (2012). From snake venom toxins to therapeutics—cardiovascular examples. *Toxicon* **59**, 497–506.

Lambeau, G., Gelb, M. H. (2008). Biochmistry and physiology of mammalian secreted phospholipases A_2. *Annu. Rev. Biochem.* **77**, 495–520.

Lambeau, G., Cupillard, L., Lazdunski, M. (1997). Membrane receptors for venom phospholipases A_2. In: Venom Phospholipase A2 Enzymes: Structure, Function and Mechanism, Kini, R. M. (Ed.), pp. 389–412. John Wiley & Sons, Chichester.

Mancheva, I., Kleinschmidt, T., Aleksiev, B., Braunitzer, G. (1987). Sequence homology between phospholipase and its inhibitor in snake venom. The primary structure of phospholipase A_2 of vipoxin from the venom of the Bulgarian viper (*Vipera ammodytes ammodytes, Serpentes*). *Biol. Chem. Hoppe Seyler* **368**, 343–352.

Mounier, C., Bon, C., Kini, R. M. (2001). Anticoagulant venom and mammaliansecreted phospholipases A2: protein- versus phospholipid-dependent mechanism of action. *Haemostasis* **31**, 279–287.

Murakami, M., Kudo, I. (2002). Phospholipase A_2. *J. Biochem. (Tokyo)* **131**, 285–292.

Murakami, M., Kudo, I. (2004). Secretory phospholipase A_2. *Biol. Pharm. Bull.* **27**, 1158–1164.

Murakami, M., Nakatani, Y., Atsumi, G., Inoue, K., Kudo, I. (1997). Regulatory functions of phospholipase A_2. *Crit. Rev. Immunol.* **17**, 225–283.

Murakami, M. T., Gabdoulkhakov, A., Genov, N., Cintra, A. C. O., Betzel, C., Arni, R. K. (2006). Insights into metal ion binding in phospholipases A2: ultra-high resolution crystal structures of an acidic phospholipase A_2 in the Ca^{2+} free and bound states. *Biochimie* **88**, 543–549.

Nielsen, L. K., Risbo, J., Callisen, T. H., Bjørnholm, T. (1998). Lag-burst kinetics in phospholipase A_2 hydrolysis of DPPC bilayers visualized by atomic force microscopy. *Biochem. Biophys. Acta* **1420**, 266–271.

Nikai, T., Komori, Y., Yagihashi, S., Ohara, A., Ohizumi, Y., Sugihara, H. (1993). Isolation and characterization of phospholipase A_2, from *Agkistrodon bilineatus* (common cantil) venom. *Int. J. Biochem.* **25**, 879–884.

Noetzel, C., Chandra, V., Perbandt, M., Rajashankar, K., Singh, T., Aleksiev, B., et al. (2002). Enzymatic activity and inhibition of the neurotoxic complex vipoxin from the venom of Vipera ammodytes meridionalis. *Z. Naturforsch.* **57c**, 1078–1083.

Pan, Y. H., Epstein, T. M., Jain, M. K., Bahnson, B. J. (2001). Five coplanar anion binding sites on one face of phospholipase A_2: relationship to interface binding. *Biochemistry* **40**, 609–617.

Panaiotov, I., Verger, R. (2000). Enzymatic reactions at interfaces: interfacial and temporal organization of enzymatic lipolysis. In: Physical Chemistry of Biological Interfaces, Bazkin, A. W. N. W. (Ed.), pp. 359–400. Marcel Dekker, New York.

Panaiotov, I., Ivanova, M., Verger, R. (1997). Interfacial organization of enzymatic lipolysis. *Curr. Opin. Colloid Interface Sci.* **2**, 517–525.

Perbandt, M., Wilson, J. C., Eschenburg, S., Mancheva, I., Aleksiev, B., Genov, N., et al. (1997). Crystal structure of vipoxin at 2.0 A: an example of regulation of a toxic function generated by molecular evolution. *FEBS Lett.* **412**, 573–577.

Perbandt, M., Tsai, I. H., Fuchs, A., Banumathi, S., Rajashankar, K. R., Georgieva, D., et al. (2003). Structure of hetero dimeric neurotoxic complex viperotoxin F (RV-4/RV-7) from the venom of *Vipera russelli formosensis* at 1.9 Å resolution. *Acta crystallogr. D* **59**, 1679–1687.

Ponce-Soto, L. A., Toyama, M. H., Hyslop, S., Novello, J. C., Marangoni, S. (2002). Isolation and preliminary enzymatic characterization of a novel PLA_2 from *Crotalus durissus collilineatus* venom. *J. Protein Chem.* **21**, 131–136.

Pungercar, J., Liang, N., Strukelj, B., Gubenshek, F. (1990). Nucleotide sequence of a cDNA encoding ammodytin L. *Nucleic Acids Res.* **18**, 4601–4646.

Pungercar, J., Krizaj, I., Liang, N.-S., Gubenshek, F. (1999). An aromatic, but not a basic, residue is involved in the toxicity of group-II phospholipase A_2 neurotoxins. *Biochem. J.* **341**, 139–145.

Reynolds, L. J., Hughes, L. L., Louis, A. I., Kramer, R. M., Dennis, E. A. (1993). Metal ion and salt effects on the phospholipase A.sub.2, lysophospholipase, and transacylase activities of human cytosolic phospholipase A.sub.2. *Biochem. Biophys. Acta* **1167**, 272–280.

Rosenberg, P. (1997). Pitfalls to avoid in the study of correlations between enzymatic activity and pharmacological properties of phospholipase A_2 enzymes. In: Venom Phospholipase A_2 Enzymes: Structure, Function and Mechanism, Kini, R. M. (Ed.), pp. 155–183. John Wiley & Sons, Chichester.

Sá, J. M., Chioato, L., Ferreira, T. L., De Oliveura, A. H. C., Ruller, R., Rosa, J. C., et al. (2004). Topology of the substrate-binding site of a Lys49-phospholipase A_2 influences Ca^{2+}-independent membrane-damaging activity. *Biochem. J.* **382**, 191–198.

Sachs, J. N., Nanda, H., Petrache, H. I., Woolf, T. B. (2004). Changes in phosphatidylcholine headgroup tilt and water order induced by monovalent salts: molecular dynamics simulations. *Biophys. J.* **86**, 3772–3782.

Schaloske, R. H., Dennis, E. A. (2006). The phospholipase A2 superfamily and its group numbering system. *Biochim. Biophys. Acta* **1761**, 1246–1259.

Scott, J. K. (1997). Phospholipase A_2: structure and catalytic properties. In: Venom Phospholipase A_2 Enzymes: Structure, Function and Mechanism, Kini, R. M. (Ed.), pp. 97–128. John Wiley & Sons, Chichester.

Scott, D. L., Sigler, P. B. (1994). Structure and catalytic mechanism of secretory phospholipases A_2. *Adv. Protein Chem.* **45**, 53–88.

Scott, D. L., White, S. P., Otwinowski, Z., Yuan, W., Gelb, M. H., Siegler, P. B. (1990). Interfacial catalysis: the mechanism of phospholipase A2. *Science* **250**, 1541–1546.

Shipolini, R., Ivanov, C., Aleksiev, B., Ganchev, K. (1970). Preliminary studies on the isolation and physiological action of fractions from the venom of *Vipera ammodytes ammodytes*. *C. R. Acad. Bulg. Sci.* **23**, 279–282.

Six, D. A., Dennis, E. A. (2000). The expanding superfamily of phospholipase A2 enzymes: classification and characterization. *Biochim. Biophys. Acta* **1488**, 1–19.

Steiner, R. A., Rozeboom, H. J., de Vries, A., Kalk, K. H., Murshudiv, G. N., Wilson, K. S., et al. (2001). X-ray structure of bovine pancreatic phospholipase A_2 at atomic resolution. *Acta Crystallogr. D* **57**, 516–526.

Stoyanova, V., Atanasov, V., Petrova, S. (2011). Recognition of *Vipera ammodytes meridionalis* neurotoxin vipoxin and its components using phage-displayed scFv and polyclonal antivenom sera. 17th Congress of the European Section of the International Society on Toxinology, Valencia (Spain). Abstract book, P125, p. 237.

Stoykova, S., Atanasov, V., Goranova, Y., Petrova, S. (2011). Effect of free fatty acids and heparin on vipoxin's PLA2-induced hemolysis. 17th Congress of the European Section of the International Society on Toxinology, Valencia (Spain), Abstract book, P94, p. 190.

Tang, J., Kriz, R. W., Wolfman, N., Shaffer, M., Seehra, J., Jones, S. S. (1997). A novel cytosolic Calcium-independent phospholipase A_2 contains eight ankyrin motifs. *J. Biol. Chem.* **272**, 8567–8575.

Tchorbanov, B., Aleksiev, B. (1981). A simple procedure for the isolation of Vipoxin P a neurotoxin with phospholipase A2 activity from the venom of the Bulgarian viper (Vipera ammodytes). *J. Appl. Biochem.* **3**, 558–561.

Tchorbanov, B., Aleksiev, B., Bukolova-Orlova, T., Burstein, E., Atanasov, B. (1977). Subfractionation and recombination of a neurotoxic complex from the venom of the bulgarian viper (*Vipera ammodytes ammodytes*). *FEBS Lett.* **76**, 266–268.

Tchorbanov, B., Grishin, E., Aleksiev, B., Ovchinnikov, Yu. (1978). A neurotoxic complex from the venom of the Bulgarian viper (Vipera ammodytes ammodytes) and a partial amino acid sequence of the toxic phospholipase A_2. *Toxicon* **16**, 37–44.

Valentin, E., Ghomashchi, F., Gelb, M. H., Lazdunski, M., Lambeau, G. (1999). Cloning and recombinant expression of a novel mouse-secreted phospholipase A2. *J. Biol. Chem.* **274**, 31195–31202.

Verheij, H. M., Boffa, M. C., Rothen, C., Verger, M. C., de Haas, G. H. (1980). Correlation of enzymatic activity and anticoagulant properties of phospholipase A_2. *Eur. J. Biochem.* **112**, 25–32.

Wang, Y. M., Lu, P. J., Ho, C. L., Tsai, I. H. (1992). Characterization and molecular cloning of neurotoxic phospholipases A_2 from Taiwan viper (*Vipera russelli formosensis*). *Eur. J. Biochem.* **209**, 635–641.

Ward, R. J., Chioato, L., de Oliveira, A. H. C., Ruller, R., S´A, J. M. (2002). Active-site mutagenesis of a Lys49-phospholipase A2: biological and membrane-disrupting activities in the absence of catalysis. *Biochem. J.* **362**, 89–96.

White, S. P., Scott, D. L., Otwinowski, Z., Gelb, M. H., Sigler, P. B. (1990). Crystal structure of cobra-venom phospholipase A2 in a complex with a transition-state analogue. *Science* **250**, 1560–1563.

Wilton, D. C., Waite, M. (2002). Phospholipases. *New Compr. Biochem.* **36**, 291–314.

Winget, J. M., Pan, Y. H., Bahnson, B. J. (2006). The interfacial binding surface of phospholipase A_2s. *Biochim. Biophys. Acta* **1761**, 1260–1269.

Winstead, M. V., Balsinde, J., Dennis, E. A. (2000). Calcium-independent phospholipase A_2: structure and function. *Biochem. Biophys. Acta* **1488**, 28–39.

Yoshizumi, K., Liu, S. Y., Miyata, T. (1990). Purification and amino acid sequence of basic protein I, a lysine-49-phospholipase A_2 with low activity, from the venom of *Trimeresurus flavoviridis* (Habu snake). *Toxicon* **28**, 43–54.

Zhao, Y., Natarajan, V. (2009). Lysophosphatidic acid signaling in airway epithelium: role in airway inflammation and remodeling. *Cell. Signal.* **21**, 367–377.

Zhao, W., Rorg, T., Gurtovenko, A. A., Vattulainen, I., Karttunen, M. (2007). Atomic-scale structure and electrostatics of anionic palmitoyloleoylphosphatidylglycerol lipid bilayers with Na^+ counterions. *Biophys. J.* **92**, 1114–1124.

Zhong, X., Liu, J., Wu, X., Zhou, Y. (2001). Expression, purification and biochemical characterization of a recombinant phospholipase A_2, with anticoagulant activity from *Agkistrodon halys* Pallas. *J. Nat. Toxins* **10**, 17–25.

EXPERIMENTAL AND THEORETICAL STUDIES OF ENZYME-CATALYZED HYDROGEN-TRANSFER REACTIONS

By ZHEN WANG,[1] DANIEL ROSTON,[1] AND AMNON KOHEN

Department of Chemistry, The University of Iowa, Iowa City, Iowa, USA

Abstract

The mechanisms of enzyme-catalyzed reactions are medicinally important and present a fascinating intellectual challenge. Many experimental and theoretical techniques can shed light on these mechanisms, and here, we shall focus on the utility of kinetic isotope effects (KIEs) to study enzymatic reactions that involve hydrogen transfers. We will provide a short background on the prevailing models to interpret KIEs and then present more detailed reviews of two model enzymes: alcohol dehydrogenase and thymidylate synthase. These two examples provide a context to describe the types of experiments and theoretical calculations that drive this field forward and the kind of information each can furnish. We emphasize the importance of cooperation between experimentalists and theoreticians to continue the progress toward a comprehensive theory of enzyme catalysis.

[1]These authors contributed equally to this work.

ADVANCES IN PROTEIN CHEMISTRY AND STRUCTURAL BIOLOGY, Vol. 87
DOI: 10.1016/B978-0-12-398312-1.00006-8

Copyright 2012, Elsevier Inc.
All rights reserved.

I. Introduction

As of yet, there exists no comprehensive theory of enzyme catalysis that is broadly accepted in the enzymology community. Both computational and experimental enzymologists continue to pursue important open questions on the mechanisms of reaction in enzymes and how the catalytic effects arise. Pauling's theory that enzymes lower the energy of the transition state (TS) is certainly correct, but questions remain, for example, on how electrostatic interactions lower the free-energy barrier, the role played by quantum mechanical tunneling through the TS, and the possibility that motions throughout the enzyme enable efficient tunneling at the active site. A collaborative relationship where theoreticians propose new models that can explain experimental outcomes, and experimentalists push the bounds of those theories with new experimental results, is important in addressing these questions.

The focus of this review will be to demonstrate a combination of different approaches in studying reactions that involve the transfer of hydrogen (hydrogen atom, hydride, or proton), primarily through kinetic isotope effects (KIEs), both experimentally and theoretically. KIEs are the ratio of rates between reactants that differ only in isotopic composition. In previous reviews and book chapters (Sen and Kohen, 2010; Kohen et al., 2011), we provided a historical background to demonstrate how KIE experiments can probe information in the TS of a reaction, and KIEs in enzymes cannot be explained by semiclassical (no tunneling) transition state theory (TST) because of the quantum nature of the hydrogen atom. Here, we shall assume *a priori* that the hydrogen atom is light enough that full quantum mechanical treatments—particularly quantum tunneling—are necessary to adequately describe the reactions. The prevailing model of reactivity, then, is an extension of Marcus theory of electron tunneling (Marcus and Sutin, 1985), where the tunneling particle is a hydrogen rather than an electron. This model, which has fallen under different names (Marcus-like models (Sen and Kohen, 2010; Kohen et al., 2011), environmentally coupled tunneling (Nagel and Klinman, 2010), vibrationally enhanced tunneling (Schwartz, 2006), etc.), proposes a mechanism where heavy-atom reorganization leads to a "tunneling ready state" (TRS) where the potential surfaces of the reactants and products are degenerate and thus the hydrogen can transfer to products by tunneling (Fig. 1A). Several groups have independently developed the functional form of the rate constant (k) for a

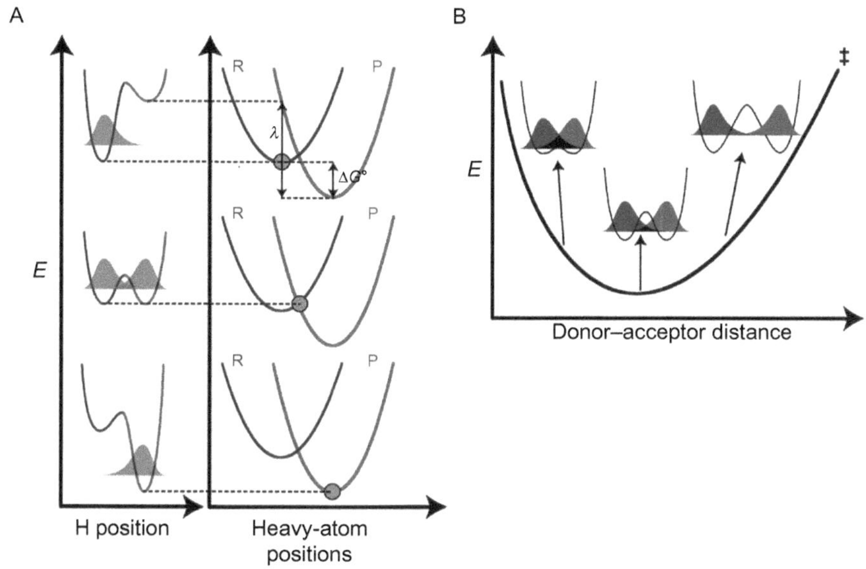

FIG. 1. *Marcus-like models of hydrogen tunneling.* (A) Three slices along the collective reaction coordinate, showing the effect of heavy-atom motions on the hydrogen's zero-point energy in the reactant potential energy surface (PES) (blue) and the product PES (red). The left panel presents the H-transfer coordinate and the right panel presents the heavy-atom coordinate. In the top panel, the hydrogen is localized in the reactant well, and the energy of the product state is much higher than that of the reactant state. Heavy-atom reorganization brings the system to the TRS (middle panel), where the zero-point energy levels of the reactant and product surfaces are equal (degenerate) and the hydrogen can tunnel, if the donor–acceptor distance (DAD) is short enough (see B). Further, heavy-atom reorganization breaks the transient degeneracy and traps the hydrogen in the product state (bottom panel). The rate of reaching the TRS depends on the reorganization energy (λ) and driving force (ΔG°), which are indicated in the top panel. (B) The effect of DAD sampling on tunneling probability at the TRS. Tunneling probability is proportional to the overlap between the hydrogen wavefunctions of the reactant (blue) and product (red) states, which depends on the DAD. (For interpretation of the references to color in this figure legend, the reader is referred to the online version of this chapter.)

H-transfer reaction (Borgis et al., 1989; Kuznetsov and Ulstrup, 1999; Knapp et al., 2002; Hammes-Schiffer, 2006; Pudney et al., 2010):

$$k = C(T)\frac{|V|^2}{\hbar}\sqrt{\frac{\pi}{\lambda k_B T}}e^{-(\Delta G^{\circ}+\lambda)^2/4\lambda k_B T}\int_0^{\infty} F(m,\mathrm{DAD})e^{-E(\mathrm{DAD})/k_B T}\,\mathrm{dDAD} \quad (1)$$

In this equation, the terms in front of the integral give the standard Marcus theory (Marcus and Sutin, 1985) rate of reaching a TRS (Fig. 1A) based on the electronic coupling (V), the reorganization energy (λ), the reaction's driving force (ΔG°), and the absolute temperature (T). The additional term $C(T)$ is the fraction of reactive complexes (Sen and Kohen, 2010) to account for the effect of other steps, which cannot be separated from the H-transfer step in experimental rate measurements with a multistep enzyme reaction. The terms in front of the integral are essentially insensitive to the mass of the transferring particle, so the isotope effects are determined by factors within the integral. The integral computes the probability of tunneling to products once the system reaches the TRS (Fig. 1B). The first term in the integral is the probability of a particle with mass (m) tunneling as a function of donor–acceptor distance (DAD). This probability function is multiplied by the second term, a Boltzmann factor, which gives the probability of finding the system at any given DAD, and the product is integrated over all DADs to yield the total tunneling probability. An important characteristic of this model is that the entire system is assumed to be in equilibrium. In many previous publications, the use of the word "dynamics" simply meant "motions" without distinguishing between motions that are in thermal equilibrium and "ballistic dynamics," which are motions not in thermal equilibrium (Kohen et al., 1999; Boehr et al., 2006; Klinman, 2006; Nagel and Klinman, 2006; Bhabha et al., 2011). This imprecise language subsequently led some to believe that a controversy exists over whether true nonequilibrium dynamics contribute to enzyme-catalyzed reactions (Pisliakov et al., 2009; Adamczyk et al., 2011). While several such proposals do exist (Welch, 1986; Liebl et al., 1999; Boekelheide et al., 2011), the Marcus-like models do not require and thus do not invoke nonequilibrium dynamics, and in this context, we prefer to use the word "motion" instead of "dynamics" to avoid confusion.

The appeal of the Marcus-like models is that they seem to explain the wide range of kinetic data observed in enzyme reactions. Since the thermal activation to reach the TRS is unconnected to the tunneling probability, this model can account for temperature-dependent rates with temperature-independent KIEs, which is not interpretable with simple tunneling corrections to TST (Bell, 1980). Further, since the distribution of DADs at the TRS is thermally regulated, this model can also explain temperature-dependent KIEs even when tunneling is known to be the primary mode of H transfer. The great flexibility of this model has allowed experimentalists to concisely interpret their results in qualitative terms, but as will be discussed below, the model has not achieved widespread use to quantitatively model enzyme kinetics, likely because of the difficulty in accurately calculating the reorganization energy.

While some theoreticians employ an empirical valence bond approach that contains many clear analogies to the Marcus-like models (Hammes-Schiffer, 2006; Warshel et al., 2006), most theoreticians rely on various corrections to the conventional TST to model the behavior of enzyme-catalyzed reactions (Pu et al., 2006; Major et al., 2009; Truhlar, 2010; Kanaan et al., 2011). Correcting for the effects of barrier recrossing and quantum tunneling has allowed for high-level simulations to give great insights into experimental results and even to successfully predict the results ahead of experiments. The remainder of the chapter will use the examples of two well-studied enzymes to explore how experiments and theoretical calculations can be used together to probe the nature of enzymatic H transfers. The first enzyme, alcohol dehydrogenase (ADH, E.C. 1.1.1.1), involves a very simple chemical transformation, while the second example, thymidylate synthase (TSase, E.C. 2.1.1.45), encompasses a complex chemical mechanism with certain steps that are still not well understood. As will be demonstrated in the context of TSase, the computational models for quantum mechanics/molecular mechanics (QM/MM) simulations naturally include all the designated degrees of freedom (such as DAD and protein motions) in the Marcus-like models.

Given the many mechanistic steps of enzymes (binding of substrates, conformational changes, release of products, etc.), specialized kinetic methods are necessary to experimentally extract information on the chemical step of interest from the observed data. Particularly when measuring KIEs, one has to bear in mind that KIEs can be hidden by "kinetic complexity," where partially rate-limiting steps that are not isotopically sensitive diminish the observed KIE from the value of the "intrinsic" KIE on the step of interest (Cook, 1991). The observed KIE ${}^{i}(V/K)_{H}$ is then (i is another isotope of hydrogen):

$$ {}^{i}(V/K)_{H} = \frac{(k_{H}/k_{i})_{int} + C_{f} + C_{r} \cdot EIE}{1 + C_{f} + C_{r}} \tag{2} $$

where $(k_{H}/k_{i})_{int}$ is the intrinsic KIE and the equilibrium isotope effect (EIE) is the isotope effect on the equilibrium constant of the reaction.

C_f and C_r are the forward and reverse kinetic commitments to catalysis, respectively (Cook and Cleland, 2007). C_f reflects the ratio of the forward rate of the isotopically sensitive step (the step of interest) to the net reverse rate of the nonisotopically sensitive steps that precede this step. Similarly, C_r reflects the reverse rate of the isotopically sensitive step to the net forward rate of the nonisotopically sensitive steps that follow this step. For an irreversible step, $C_r = 0$ and all the kinetic complexity is caused by C_f from the preceding steps. Analyzing the kinetic commitments can sometimes provide information on the reaction mechanism, as will be seen in the context of TSase.

The most preferable method to expose the intrinsic KIEs is to find a set of reaction conditions where the step of interest is completely rate limiting. Unfortunately, most mature enzymes using natural substrates and near physiological conditions are not rate-limited by the chemical step. Different experimental approaches have been developed to address this limitation. For example, for yeast ADH (yADH), the chemical step has been "exposed" by using nonnatural substrate, and for horse liver ADH (hlADH), an active site mutation was required in addition to using nonnatural substrate (Bahnson et al., 1993). For systems where these solutions are not feasible, an instrumental procedure is to use Northrop's method for finding intrinsic KIEs based on a combination of deuterium and tritium KIEs (Northrop, 1991; Sen et al., 2011):

$$\frac{{}^{D}(V/K)_{H} - 1}{{}^{T}(V/K)_{H} - 1} = \frac{(k_H/k_D)_{int} - 1}{(k_H/k_D)_{int}^{1.43} - 1} \tag{3}$$

In this equation, ${}^{D}(V/K)_{H}$ and ${}^{T}(V/K)_{H}$ are the observed H/D and H/T isotope effects on the second-order rate constant k_{cat}/K_M, respectively, and $(k_H/k_D)_{int}$ is the intrinsic H/D KIE on the chemical step of interest. A variation of Eq. (3) is to calculate intrinsic KIEs with a combination of H/T and D/T isotope effects:

$$\frac{{}^{T}(V/K)_{D}^{-1} - 1}{{}^{T}(V/K)_{H}^{-1} - 1} = \frac{(k_H/k_T)_{int}^{-0.2994} - 1}{(k_H/k_T)_{int}^{-1} - 1} \tag{4}$$

Since Eqs. (3) and (4) use different "reference isotopes" (H for Eq. (3) and T for Eq. (4)), the question arises that which combination of observed KIEs provides more accurate assessments of the intrinsic KIEs. A recent triple-labeling KIE experiment (Sen et al., 2011) has discussed this question and demonstrated the versatility of this method with the enzyme dihydrofolate reductase.

An unfortunate aspect of Northrop's method is its reliance on the Swain–Schaad exponent (SSE) (Swain et al., 1958), which was derived from semiclassical (no tunneling) TST. Its applicability to systems where tunneling is known to be important appears somewhat questionable (Kohen and Jensen, 2002; Hirschi and Singleton, 2005; Smedarchina and Siebrand, 2005). Deviations from Swain–Schaad behavior will be discussed in greater detail in Section II.

II. Alcohol Dehydrogenase

ADH has served as a model for H-transfer reactions for many years owing to a number of convenient characteristics of the reaction. For one, following the activation of the alcohol as zinc-alkoxide, the reaction only involves one chemical conversion: the transfer of the hydride from the α-carbon of the alcohol to C4 of the NAD^+ cofactor (Scheme 1). That hydride-transfer step is rate limiting in yADH when benzyl alcohol is the substrate (Klinman, 1972). Thus, KIEs on that step are well exposed and not significantly masked by the "kinetic complexity" mentioned above. Another advantage of ADH is that both the forward and reverse reactions can be studied under relatively similar conditions, providing twice as much data as available for irreversible reactions (Klinman, 1976).

Some of the intriguing work on ADH has centered on secondary KIEs (2° KIEs), where a hydrogen not being transferred is isotopically labeled.

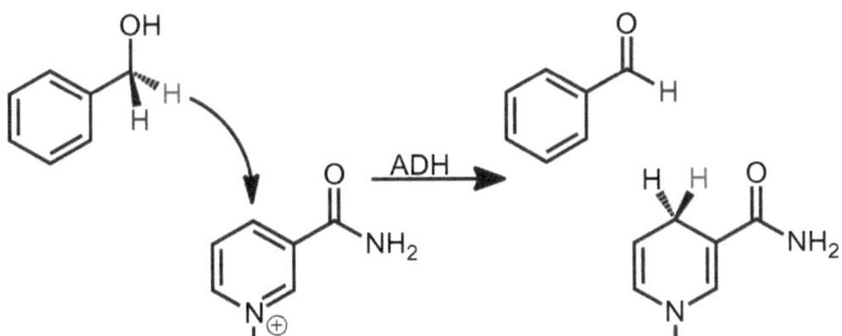

Scheme 1. The reaction catalyzed by ADH. The 1° H is red and the 2° H is blue. (For interpretation of the references to color in this figure legend, the reader is referred to the online version of this chapter.)

The traditional interpretation is that 2° KIEs can indicate the position of a TS because they reflect the extent to which a reactive center has rehybridized at that state (Streitwieser et al., 1958). In some of the earliest experiments to probe the nature of the hydride transfer, Klinman measured 2° KIEs on the forward reaction and found a value close to the EIE, which was believed to serve as an upper limit for the KIE (complete rehybridization from sp^2 to sp^3 at the TS as in the product) and therefore indicated a very late (aldehyde like) TS (Klinman, 1976). In contrast, measurements of linear free-energy relationships using *para*-substituted benzyl alcohols and aldehydes indicated an early (alcohol-like) TS (Klinman, 1972, 1976). This contradiction was perhaps the first indication that conventional TST would not suffice to describe the reaction, but it was not until Cleland and coworkers found that 2° KIEs on the NAD^+ cofactor exceeded the relevant EIEs that quantum mechanical tunneling was proposed to have an important role in this reaction (Cook et al., 1981). Conventional TST requires that the KIE should fall somewhere between unity (no isotope effect for an early TS) and the EIE (largest KIE for a late TS), depending on the extent to which the H-donating or accepting carbon has rehybridized at the TS. In accordance with its EIE, the sp^2 to sp^3 transition of NAD^+ to NADH was expected to have an inverse KIE (<1) because of the loss of vibrational freedom in the 2° hydrogens, but the measured value was significantly larger than unity. Similarly, in the reverse reaction, $NADH \rightarrow NAD^+$, the KIE was greater than the EIE. The most probable explanation for the inflated KIEs was that motion of the 2° hydrogen was involved in the reaction coordinate mode and that this coupled mode tunneled through the TS (referred to as "tunneling and coupled motion," Fig. 2).

Following these initial findings, Klinman and coworkers tested the notion of "tunneling and coupled motion" by measuring the SSE (Swain et al., 1958; Cha et al., 1989) using what has come to be known as a mixed-labeling experiment (Kohen, 2006). The SSE relates tritium and deuterium isotope effects and with semiclassical TST (no tunneling) it should be 3.3:

$$\mathrm{SSE} = \frac{\ln\left(\frac{k_\mathrm{H}}{k_\mathrm{T}}\right)}{\ln\left(\frac{k_\mathrm{D}}{k_\mathrm{T}}\right)} = 3.3 \qquad (5)$$

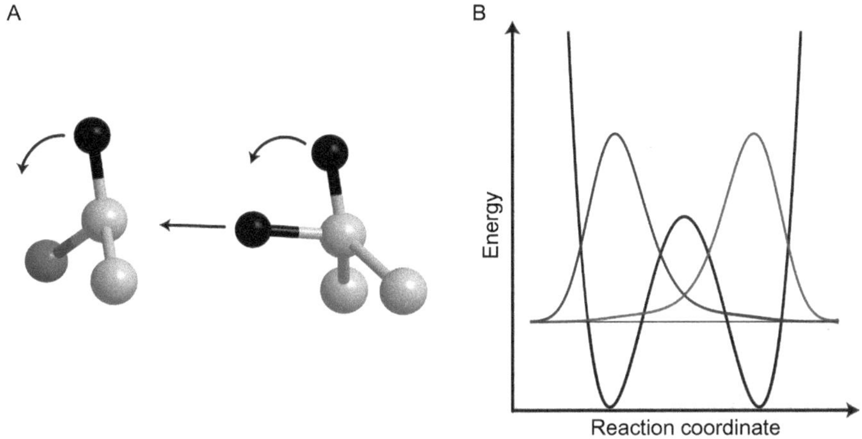

FIG. 2. *The model of tunneling and coupled motion in ADH.* (A) The reaction coordinate involves motions of the transferring hydrogen as well as the two 2° hydrogens that are involved in rehybridization of the donor and acceptor carbons. Since motions of all these three atoms define the reaction coordinate (B), tunneling affects all the three atoms. The blue curve represents the wavefunction of this mode in the reactant state and the red in the product state; the probability of tunneling is proportional to the overlap of these two wavefunctions. (For interpretation of the references to color in this figure legend, the reader is referred to the online version of this chapter.)

k_i indicates the rate with isotope i. The mixed-labeling experiment tests the SSE by measuring the 2° H/T KIE with H at the 1° position and the D/T KIE with D at the 1° position, yielding the mixed-labeling SSE (mSSE). Conventional TST predicts no isotope effects on isotope effects (the rule of the geometrical mean) (Bigeleisen, 1958), so the mSSE will be equal to the normal SSE so long as conventional TST applies

$$\text{mSSE} = \text{SSE} = \frac{\ln\left(\frac{k_H^H}{k_T^H}\right)}{\ln\left(\frac{k_D^D}{k_T^D}\right)} = 3.3 \quad (6)$$

In this case, k_i^j indicates the rate with isotope i at the 2° position and j at the 1° position. The mixed-labeling experiments on yADH found a 2° mSSE that was inflated to a value over 10, which provided very strong evidence that nonclassical behavior was important in this reaction (Cha et al., 1989). Subsequent studies have confirmed the inflated mSSE in a variety of ADHs

(Rucker et al., 1992; Bahnson et al., 1993; Kohen et al., 1999) but surprisingly few other reactions (Alston et al., 1996).

Since the publication of these violations of the mSSE, researchers have spent a great deal of effort attempting to reproduce the findings. Certain phenomenological models based on Huskey and Schowen (1983), for example, have found plausible reaction coordinate couplings which, along with tunneling corrections, yield the inflated mSSE (Rucker and Klinman, 1999; Kohen and Jensen, 2002). A hindrance to conducting higher level QM/MM calculations on this enzyme has been that the crystal structure of yADH was very difficult to solve and has only recently been deposited in the PDB (Pal et al., 2009). A crystal structure of hlADH has been available for some time (Ramaswamy et al., 1994), but that enzyme has kinetic complexity that masks the KIEs and shows only mildly inflated mSSEs, even in mutants where the H-transfer step is more exposed (Bahnson et al., 1993, 1997). While QM/MM simulations that addressed KIEs in hlADH reproduced the observed rather than intrinsic KIEs, it is no small feat as these calculations provided insight into the atomic-level view of the reaction (Agarwal et al., 2000; Alhambra et al., 2000, 2001; Webb et al., 2000; Cui et al., 2002; Warshel et al., 2006). These high-level QM/MM simulations provided detailed structures of the TS of the reaction and some explanation for the inflated mSSEs. Unfortunately, no simulations addressed the KIEs on the NAD^+ cofactor that fell outside the range of the EIE to unity—an effect that was found in hlADH as well as in yADH (Cook et al., 1981)—or the contradiction in the linear free-energy relationships.

Another major open challenge for theoreticians is to explain the temperature dependence of KIEs (or lack thereof) observed in certain ADHs. A thermophillic ADH from *Bacillus stearothermophilus* (bsADH), for which H transfer is rate limiting across the temperature range studied, showed an inflated mSSE within its physiological temperature range and demonstrated temperature-independent 1° KIEs, despite the fact that the overall rate was temperature dependent (Kohen et al., 1999). Temperature-independent KIEs with a temperature-dependent rate can be rationalized by the Marcus-like models described above, where isotopically insensitive heavy-atom motions compose the reaction coordinate, or as stated by Kiefer and Hynes "the solvent coordinate is the reaction coordinate" (Kiefer and Hynes, 2003). Interestingly, below its physiological temperature range, bsADH showed temperature-dependent KIEs. This indicated a sort of phase transition at low temperatures, which alters the DAD sampling at the TRS (Fig. 1B). This phenomenon was used to propose that certain protein motions bring the donor and acceptor close enough for hydrogen tunneling, but those motions are constrained below the physiological temperature range. No high-level calculations have addressed this behavior in bsADH, and only a few attempts have examined the temperature dependence of KIEs in other enzymes (Pu et al., 2005; Liu and Warshel, 2007; Kanaan et al., 2011).

Noticing that the inflated mSSEs from many different ADHs actually resulted from similar H/T 2° KIEs with H transfer but deflated D/T 2° KIEs with D transfer, Klinman proposed an alternative explanation (Klinman, 2006; Nagel and Klinman, 2006). Instead of having *inflated* H/T KIEs due to "tunneling and coupled motion," this model suggests that the 2° D/T KIEs in the mixed-labeling experiment are *deflated* because they are measured with D at the 1° position and D transfer requires a shorter DAD than H transfer. The steric interactions of the shorter DAD hinder the rehybridization of the 2° hydrogens, thus reducing the observed KIEs.

This explanation is intuitively satisfying, and building on the approach of Huskey and Schowen (1983), we recently found quantitative success in modeling the yADH reaction using these ideas (Roston and Kohen, 2010). We were able to parameterize the geometry of the substrates at the TRS in order to replicate the many 2° KIEs measured for the reaction. Additionally, the geometry that reproduced the KIEs (and mSSE) also reproduced the results of the linear free-energy relationships, demonstrating that in the context of Marcus-like models these data are not contradictory. Important questions remain, though, as to why inflated mSSEs are observed only in the ADH reactions. Other reactions where Marcus-like models have been successful in explaining certain results do not show inflated mSSE (e.g., Sikorski et al., 2004), despite the fact that the model predicts that D transfer will occur from shorter DADs in all reactions.

III. Thymidylate Synthase

TSase catalyzes the last step of *de novo* synthesis of 2′-deoxythymidine-5′-monophosphate (dTMP, one of the four DNA bases), in which the substrate 2′-deoxyuridine-5′-monophosphate (dUMP) is methylated and reduced by the cofactor N^5,N^{10}-methylene-5,6,7,8-tetrahydrofolate (CH_2H_4folate) (Scheme 2; Carreras and Santi, 1995). The catalytic mechanism of TSase

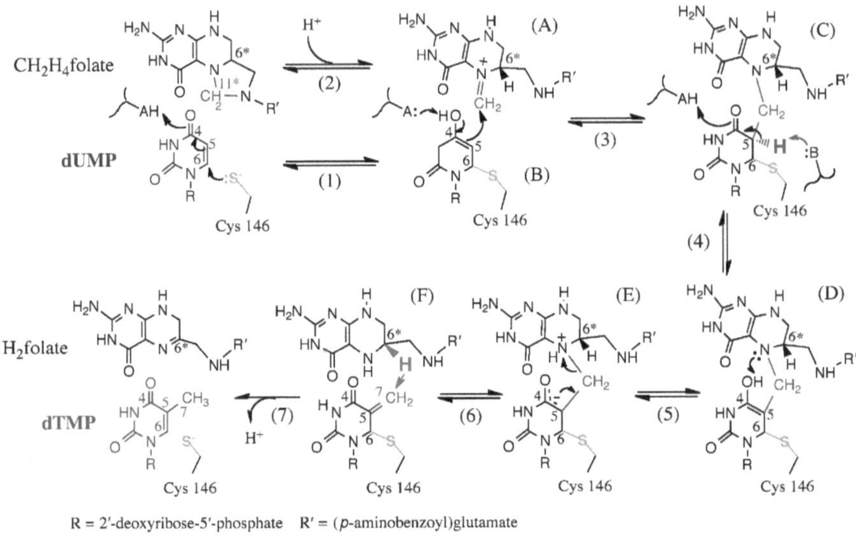

SCHEME 2. The traditionally proposed chemical mechanism of TSase, modified from Finer-Moore et al. (2003). The numbering of carbons on the CH_2H_4folate is denoted with an asterisk (*) to discriminate from those on the dUMP/dTMP. (For color version of this figure, the reader is referred to the online version of this chapter.)

comprises a series of bond cleavages and formations that include two different C—H bond activations: a reversible proton transfer (step 4 in Scheme 2) and an irreversible rate-limiting hydride transfer (step 7 in Scheme 2). Early mechanistic studies of TSase included steady-state and pre-steady-state kinetics, crystallography, 2° α-hydrogen isotope effects, etc., which have been summarized in a few excellent reviews (Bruice and Santi, 1991; Carreras and Santi, 1995; Finer-Moore et al., 2003; Stroud and Finer-Moore, 2003). In this section, we focus on investigations on the two C—H activation steps (steps 4 and 7 in Scheme 2) with both experimental and computational methods and highlight the integration of different approaches in addressing specific issues (Agrawal et al., 2004; Hong et al., 2006, 2007; Newby et al., 2006; Kanaan et al., 2007, 2009, 2010a,b, 2011; Wang and Kohen, 2010). In this section, KIEs refer to 1° KIEs, and TSase refers to the enzyme from *Escherichia coli*, unless otherwise specified.

A. *The Hydride Transfer*

Spencer et al. measured both 1° and 2° KIEs on the hydride transfer with steady-state kinetic experiments of wild-type TSase (wt TSase) at 20 °C (Spencer et al., 1997). The authors observed a large 1° KIE (3.72±0.12) on k_{cat}, suggesting that the hydride transfer step was rate limiting for the catalysis (i.e., lack of "kinetic complexity"). They also found a large normal 2° H/D KIE, which was independent of the 1° isotope. The observed 2° KIE was explained as a result of rapid equilibrium between two or more reaction intermediates (i.e., complexes A–F in Scheme 2) prior to the hydride transfer. In other words, all the steps that precede the hydride transfer seem to be reversible during the TSase-catalyzed reaction.

We measured intrinsic 1° KIEs on the hydride-transfer step in the physiological temperature range (5–40 °C) of wt TSase (Agrawal et al., 2004). The small kinetic commitment (Eq. (2)) observed on the hydride transfer (Fig. 3A) supports the conclusions of Spencer et al. that this step

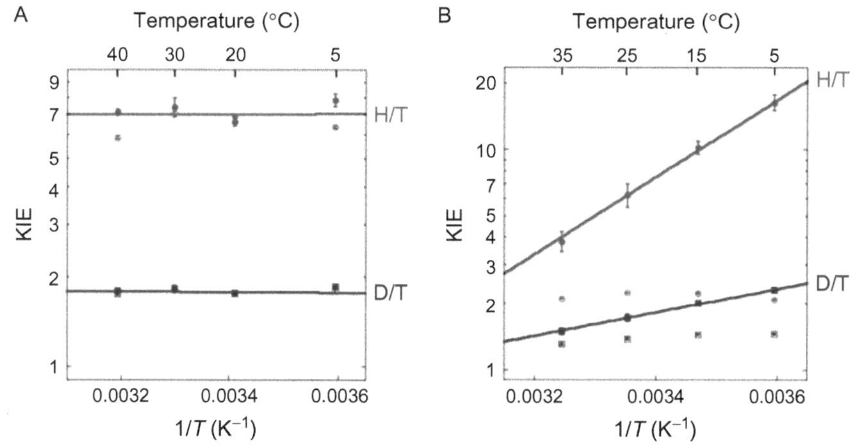

FIG. 3. Observed (empty structures) and intrinsic (filled structures) primary KIEs on (A) the hydride transfer (step 7 in Scheme 2/4) and (B) the proton transfer (step 4 in Scheme 2/4) catalyzed by wt ecTSase. The hydride transfer (rate limiting) presents temperature-independent KIEs, while the proton transfer presents temperature-dependent KIEs. This figure is modified from Agrawal et al. (2004) and Wang and Kohen (2010). (For color version of this figure, the reader is referred to the online version of this chapter.)

is rate limiting. Within the experimental temperature range (5–40 °C), the intrinsic KIEs are temperature independent (Fig. 3A), and the isotope effects on Arrhenius pre-exponential factors exceed the high limit of semiclassical prediction (0.5–1.6 for H/T and 0.9–1.2 for D/T; Kohen et al., 2011). We also measured initial velocities of the reaction over the same temperature range and observed a significant activation energy on k_{cat}. This is another example of an enzyme-catalyzed H transfer with a temperature-dependent rate but temperature-independent KIEs, which is in accordance with Marcus-like models.

B. The Proton Transfer

Compared with the rate-limiting hydride transfer, the proton transfer is a fast and reversible step, and therefore, its KIEs are greatly masked by the kinetic complexity. Previous studies used a saturating concentration of CH_2H_4folate and measured a 1° KIE of unity on the proton transfer due to the large kinetic commitment (Smith Lomax and Greenberg, 1967; Finer-Moore et al., 2003). We found that the observed KIE for this step varies with the concentration of CH_2H_4folate, owing to the sequential binding order of dUMP and CH_2H_4folate (Hong et al., 2007). Thus, we measured the intrinsic KIEs on the proton transfer with a low concentration of CH_2H_4folate (yet high enough to ensure sufficient conversion of dUMP to dTMP) (Wang and Kohen, 2010), following the Northrop's method (Eq. (4)). In contrast to the hydride-transfer step, the intrinsic KIEs on the proton transfer are temperature dependent (Fig. 3B). The different temperature dependences of KIEs on the two hydrogen-transfer steps suggest that the enzyme employs different strategies in the sequential C—H bond activations (Wang and Kohen, 2010). Similar observations have been reported for a few other enzymes, such as morphinone reductase (Basran et al., 2003) and protochlorophyllide oxidoreductase (Heyes et al., 2009).

C. Mutagenesis Studies

KIEs have been measured for several TSase mutants to elucidate the molecular mechanisms of both the hydride and proton transfers. Two distinct mechanisms have been proposed for the hydride-transfer step (Scheme 3). Barrett et al. suggested a two-step radical mechanism where W80 stabilizes the cationic radical intermediate of H_4folate (Barrett et al.,

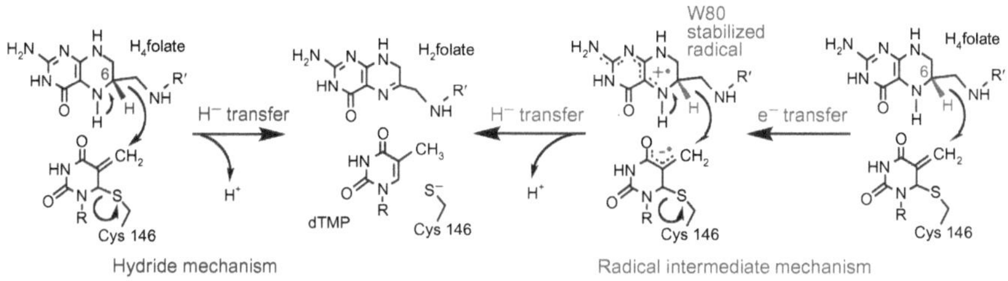

SCHEME 3. Two proposed mechanisms for the hydride transfer catalyzed by TSase, modified from Hong et al. (2006). (For color version of this figure, the reader is referred to the online version of this chapter.)

1999). Liu et al., on the other hand, proposed a one-step hydride-transfer mechanism where W80 reorients L143 to ideally position H_4folate and to protect the active site cavity (Fritz et al., 2002). We measured steady-state kinetics as well as KIEs on W80M TSase to differentiate the two proposed mechanisms (Hong et al., 2006). As methionine cannot stabilize the cation radical as well as tryptophan (Scheme 3 right panel), the two-step radical intermediate mechanism predicts a direct change in the hydride-transfer step of W80M, which would lead to different KIE values. We found that, although the hydride-transfer step of the W80M mutant is one to two orders of magnitude slower than that of wt TSase, the intrinsic KIEs of the two enzymes are similar and both are temperature independent within 5–40 °C (Fig. 4A). These findings suggest that, while the W80M mutation alters the parameters affecting the rate (e.g., reorientation of protein residues, substrates, and/or water molecules), it does not change the nature of the H transfer, which supports the one-step hydride-transfer mechanism.

Another residue that has been examined by KIE experiments is a strictly conserved tyrosine residue (Y94 in TSase from *E. coli*). In the crystal structure of the ternary complex TSase–(5-fluoro)dUMP–CH_2H_4folate, the Y94 hydroxyl is in van der Waals contact with the fluorine in (5-fluoro) dUMP (Hyatt et al., 1997). A series of Y94 mutants showed greatly deteriorated k_{cat} (500-fold decrease for Y94F), and accumulation of the ternary complex with dUMP and CH_2H_4folate during the reaction (complexes C, D, or E in Scheme 2) (Liu et al., 1999; Saxl et al., 2003). These observations suggested that Y94 assists the proton abstraction from C5 of dUMP (step 4 in Scheme 2), which is critical for breakdown of the ternary complex. We measured steady-state rates and KIEs with Y94F to examine

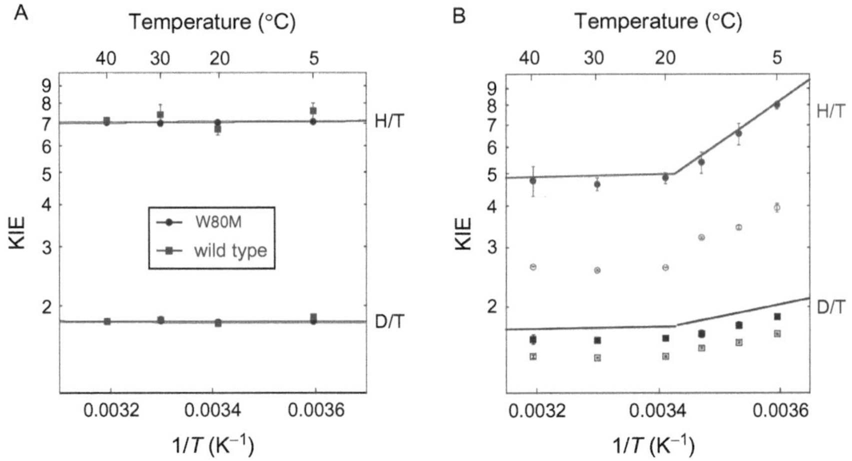

FIG. 4. Observed (empty structures) and intrinsic (filled structures) primary KIEs on the hydride-transfer step (step 7 in Schemes 2 and 4) of (A) wt ecTSase and W80M and (B) Y94F in the physiological temperature range (5–40 °C). The W80M mutant does not affect the magnitude or the temperature dependence of intrinsic KIEs on the hydride-transfer step, while the Y94F mutant affects both. This figure is modified from Hong et al. (2006, 2007). (For color version of this figure, the reader is referred to the online version of this chapter.)

this hypothesis (Hong et al., 2007). The observed KIEs on the proton-transfer step revealed that Y94F alters the binding mechanism of dUMP and CH_2H_4folate. At 25 °C, the intrinsic KIEs on the proton transfer are similar for both wt and Y94F, but the kinetic complexity is larger for Y94F. These observations suggest that Y94 may not be the general base for the proton abstraction but rather participates in other kinetic steps. However, the KIE studies with dihydrofolate reductase have shown mutations that affect the hydride-transfer step present the same intrinsic KIEs as the wt enzyme at one temperature, but different temperature dependence of the KIEs (Wang et al., 2006). Therefore, the effect of Y94F on the proton-transfer step needs to be more carefully examined by measuring the temperature dependence of its intrinsic KIEs.

In order to study the role of Y94 in later steps during the TSase-catalyzed reaction, we measured the KIEs on the hydride transfer of Y94F (Hong et al., 2007) and found a larger kinetic commitment than that in the wt (Fig. 4B). Since all the kinetic complexity is caused by C_f (Eq. (2)) for the irreversible hydride-transfer step, these data suggest that one or more steps prior to the hydride transfer are rate limiting for Y94F. This finding agrees with the experiments mentioned above, suggesting that the breakdown of the ternary complex becomes rate limiting for Y94F (which involves steps 4, 5, and 6 in Scheme 2). Since the hydride-transfer step is no longer rate limiting for Y94F, the kinetic parameters of the initial velocities (e.g., k_{cat}, K_M) cannot provide information on this step. Therefore, we used the method developed by Klinman and coworkers (Miller and Klinman, 1985; Francisco et al., 2002) to assess the activation parameters of the hydride-transfer step and found that the Y94F mutation has a major effect on the entropy of activation. The intrinsic KIEs on the hydride transfer are found to be temperature dependent within 5–20 °C but temperature independent within 20–40 °C. These observations are similar to the case of bsADH (see Section II) and suggest that Y94F affects the DAD sampling for the hydride-transfer step. Taken together, the experimental data with Y94F indicate that Y94 is involved in the enzyme stability and play a role in multiple physical and chemical steps.

D. High-Level Simulations

We performed QM/MM calculations on the potential energy surfaces (PESs) of TSase-catalyzed reaction (Kanaan et al., 2007) and proposed a mechanism (Scheme 4) that is different from the traditionally proposed one (Scheme 2). In the computed mechanism, the cleavage of the C6—S bond replaces the enol formation at the C4 carbonyl and serves as a mechanistic feature throughout the reaction. For example, the C4-enol/ketone formation is concerted with steps 1, 3, 4, 5, and 6 in Scheme 2, while C6—S formation/cleavage is concerted with steps 3, 4, 6, and 7 in Scheme 4. The proposed lability of the C6—S bond agrees with the diffuse electron density between the C6 and S atoms in the crystal structure of the ternary complex TSase–(5-fluoro)dUMP–CH_2H_4folate (Hyatt et al., 1997). Another feature of the computed mechanism is that a water molecule acts as a proton donor and acceptor in multiple steps (including step 4). This agrees with experimental observations that water molecules at the TSase active site serve as proton donors/acceptors and even play a role in substrate/product discrimination (Fauman et al., 1994; Sage et al., 1996; Huang and Santi, 1997). This computational work also revealed that the hydride transfer (step 7 in Scheme 4) presents the highest energy barrier along the minimum potential energy path, which agrees with the

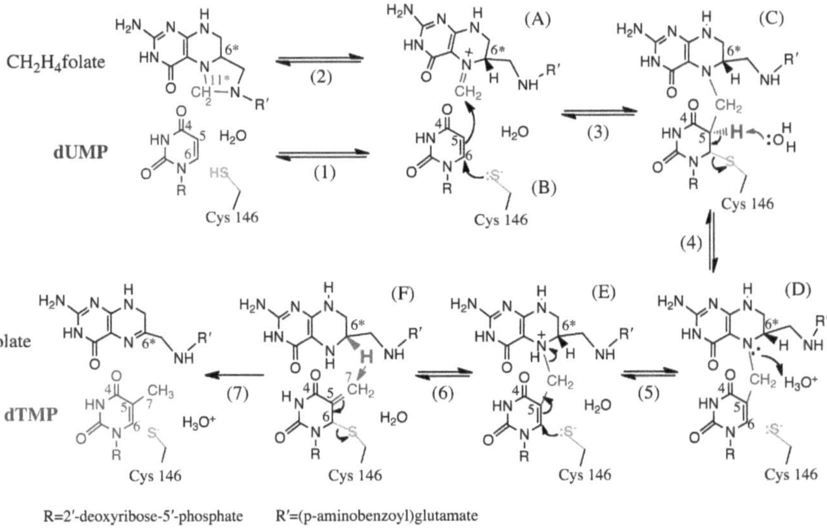

SCHEME 4. The recently proposed chemical mechanism of ecTSase based on QM/MM PESs exploration. This scheme is a simplified version of the complete mechanism proposed in Kanaan et al. (2007). The mechanistic feature (cleavage/formation of C6—S bond) is substantially different from that of Scheme 2 (enol/ketone formation). (For color version of this figure, the reader is referred to the online version of this chapter.)

experimental conclusion (Spencer et al., 1997; Agrawal et al., 2004) that the hydride transfer is rate limiting.

To further examine the hydride-transfer mechanism, especially the role of protein motions in this step, we performed the potential of mean force (PMF) calculations along the reaction coordinate. We first computed the 1D PMF at 20 °C using the antisymmetric combination of the lengths of the breaking and forming bonds as the distinguished reaction coordinate (DRC, e.g., $d_{C6^*-H}-d_{H-C7}$, see complex F in Scheme 4) and found that C6—S bond cleavage is concerted with the hydride transfer (Kanaan et al., 2007). Thus, we used this C6—S distance (d_{C6-S}) as an additional dimension of the DRC and computed the 2D PMF within 5–40 °C. The 2D PMFs gave a more precise description of the TS with a lower energy barrier compared with the 1D PMFs (Kanaan et al., 2009) and suggested that the C6—S cleavage and the hydride transfer are concerted but not synchronized. The 2D PMFs also allowed us to analyze the key geometrical parameters of the reactants, TS, and products, as well as the interaction energies between the protein and substrates along the DRC. The results suggested that movements of multiple active site residues are coupled to the change in DAD (d_{C6^*-C7} in this case), which leads to more preferable interactions between the protein and substrates at the TS than in the reactants. Further studies with both rare event trajectories (Kanaan et al., 2010b) and Grote–Hynes theory (Kanaan et al., 2010a) also revealed those protein motions accommodate the geometric and electrostatic characteristics of the substrates at the TS, thereby reducing the free-energy barrier. Most importantly, the positively charged R166 approaches C146 as the reaction proceeds and stabilizes the TS by polarizing the electron density around the S atom. The TS structures, and particularly the DAD for the hydride transfer, are invariant within the examined temperature range, which corroborates the experimental findings of temperature-independent KIEs.

Based on the better-defined mechanism for the hydride transfer, we computed the rate and KIEs of this step in the same temperature range using the ensemble-averaged variational TST with multidimensional tunneling and Grote–Hynes theory (Kanaan et al., 2011). The results suggested that, while the nuclear vibrational terms make the major contributions in reducing free-energy barriers, quantum mechanical tunneling significantly increases the reaction rate at all temperatures (90–94% rate increase for protium and 77–82% for tritium). These computations yielded temperature-independent KIEs that agree with previous experimental data (Agrawal et al., 2004). We also observed promoting modes of the enzyme that compress the DAD, reducing both the height and width of the barrier. This leads to DADs that are invariant with temperatures and also enhances the probability of hydride transfer by both classical and tunneling mechanisms. These findings are consistent with the phenomenological Marcus-like models and corroborate previous computations for another enzymatic system that concluded the temperature dependence of DAD sampling is the source for temperature dependence of KIEs (Liu and Warshel, 2007).

IV. Concluding Remarks

Numerous efforts have been devoted to identifying the mechanisms of enzyme-catalyzed reactions, and the focus of research constantly shifts with accumulated knowledge in this field. One of the attractive topics is the role of protein motions in the chemical transformations catalyzed by enzymes, starting from the general notion of "induced fit" (Fersht, 1999) and continuing to more specific molecular hypotheses. In this chapter, we reviewed recent works on ADH and TSase to illustrate studies in this area, focusing on experimental techniques and computer-based calculations recently used by us and others. Many other examples, both from our group and from other groups, were not discussed here but have also made substantial contributions to the examination of the role of protein motions in enzymatic functions. KIEs have been measured in a wide variety of enzymes as well as noncatalyzed reactions and proved to be a powerful tool in probing TS structures and motions toward the TS (or TRS as discussed above). Recent developments in KIE experiments and theories allow us to evaluate the involvement of quantum mechanical tunneling and protein motions in H-transfer reactions. It is not likely to be a coincidence that various wt enzymes present temperature-independent 1° KIEs with their natural substrates under physiological conditions (Kohen et al., 1999; Agrawal et al., 2004; Sikorski et al., 2004; Bandaria et al., 2009), and the prevailing rationale for these data among experimentalists is the phenomenological Marcus-like models, which are largely in accordance with relevant computer simulations. The Marcus-like models, for example, have seen notable success in quantitatively interpreting many of the seemingly contradictory experimental observations in ADH (Roston and Kohen, 2010). We are currently using similar methods to study several other enzyme systems to expand the applicability of this phenomenological model. The goal of computations, from our point of view, is not only to rationalize the experimental findings but also to make experimentally testable predictions that allow examinations of the generality of a theory's underlying rationale. The QM/MM calculations with TSase demonstrated this ability of high-level simulations. Particularly, the newly proposed C6—S lability may be examined by 2° α-H isotope effects on C6 of dUMP (if coupled to labeled proton or hydride transfers), and mutation of Arg 166. We are currently studying the proton transfer step of TSase with the QM/MM calculations, and preliminary results seem to rationalize many ambiguous experimental data, including the role of Y94 in this step. We believe that an interactive combination of experimental and theoretical approaches is the key to resolving the dilemma between different theoretical models and apparently contradictory experimental findings. Such interactions may enhance our understanding of specific enzymatic systems and develop models that address the nature of enzyme catalysis in general.

Acknowledgments

This work was supported by NIH R01 GM065368 and NSF CHE 496 0715448 to A. K., the Iowa CBB to Z. W., and the predoctoral fellowship from the NIH (T32 GM008365) to D.R.

References

Adamczyk, A. J., Cao, J., Kamerlin, S. C. L., Warshel, A. (2011). Catalysis by dihydrofolate reductase and other enzymes arises from electrostatic preorganization, not conformational motions. *Proc. Natl. Acad. Sci. USA* **108**, 14115–14120.

Agarwal, P. K., Webb, S. P., Hammes-Schiffer, S. (2000). Computational studies of the mechanism for proton and hydride transfer in liver alcohol dehydrogenase. *J. Am. Chem. Soc.* **122**, 4803–4812.

Agrawal, N., Hong, B., Mihai, C., Kohen, A. (2004). Vibrationally enhanced hydrogen tunneling in the *E. coli* thymidylate synthase catalyzed reaction. *Biochemistry* **43**, 1998–2006.

Alhambra, C., Corchado, J. C., Sanchez, M. L., Gao, J. L., Truhlar, D. G. (2000). Quantum dynamics of hydride transfer in enzyme catalysis. *J. Am. Chem. Soc.* **122**, 8197–8203.

Alhambra, C., Corchado, J., Sanchez, M. L., Garcia-Viloca, M., Gao, J., Truhlar, D. G. (2001). Canonical variational theory for enzyme kinetics with the protein mean force and multidimensional quantum mechanical tunneling dynamics. Theory and application to liver alcohol dehydrogenase. *J. Phys. Chem. B* **105**, 11326–11340.

Alston, W. C., Kanska, M., Murray, C. J. (1996). Secondary H/T and D/T isotope effects in enzymatic enolization reactions. Coupled motion and tunneling in the triosephosphate isomerase reaction. *Biochemistry* **35**, 12873–12881.

Bahnson, B. J., Park, D. H., Kim, K., Plapp, B. V., Klinman, J. P. (1993). Unmasking of hydrogen tunneling in the horse liver alcohol-dehydrogenase reaction by site-directed mutagenesis. *Biochemistry* **32**, 5503–5507.

Bahnson, B. J., Colby, T. D., Chin, J. K., Goldstein, B. M., Klinman, J. P. (1997). A link between protein structure and enzyme catalyzed hydrogen tunneling. *Proc. Natl. Acad. Sci. USA* **94**, 12797–12802.

Bandaria, J. N., Cheatum, C. M., Kohen, A. (2009). Examination of enzymatic H-tunneling through kinetics and dynamics. *J. Am. Chem. Soc.* **131**, 10151–10155.

Barrett, J. E., Lucero, C. M., Schultz, P. G. (1999). A model for hydride transfer in thymidylate synthase based on unnatural amino acid mutagenesis. *J. Am. Chem. Soc.* **121**, 7965–7966.

Basran, J., Harris, R. J., Sutcliffe, M. J., Scrutton, N. S. (2003). H-tunneling in the multiple H-transfers of the catalytic cycle of morphinone reductase and in the reductive half-reaction of the homologous pentaerythritol tetranitrate reductase. *J. Biol. Chem.* **278**, 43973–43982.

Bell, R. P. (1980). The Tunnel Effect in Chemistry. Chapman & Hall, New York.

Bhabha, G., Lee, J., Ekiert, D. C., Gam, J., Wilson, I. A., Dyson, H. J., et al. (2011). A dynamic knockout reveals that conformational fluctuations influence the chemical step of enzyme catalysis. *Science* **332**, 234–238.

Bigeleisen, J. (1958). Second-order sum rule for the vibrations of isotopic molecules and the second rule of the mean. *J. Chem. Phys.* **28**, 694–699.

Boehr, D. D., McElheny, D., Dyson, H. J., Wright, P. E. (2006). The dynamic energy landscape of dihydrofolate reductase catalysis. *Science* **313**, 1638–1642.

Boekelheide, N., Salomon-Ferrer, R., Miller, T. F., 3rd (2011). Dynamics and dissipation in enzyme catalysis. *Proc. Natl. Acad. Sci. USA* **108**, 16159–16163.

Borgis, D. C., Lee, S. Y., Hynes, J. T. (1989). A dynamical theory of nonadiabatic proton and hydrogen-atom transfer-reaction rates in solution. *Chem. Phys. Lett.* **162**, 19–26.

Bruice, T. W., Santi, D. V. (1991). Isotope effects in reactions catalyzed by thymidylate synthase. In: Enzyme Mechanism from Isotope Effects, Cook, P. F. (Ed.). vol. 457–479. CRC Press, Boca Raton.

Carreras, C. W., Santi, D. V. (1995). The catalytic mechanism and structure of thymidylate synthase. *Annu. Rev. Biochem.* **64**, 721–762.

Cha, Y., Murray, C. J., Klinman, J. P. (1989). Hydrogen tunneling in enzyme-reactions. *Science* **243**, 1325–1330.

Cook, P. F. (1991). Enzyme Mechanism from Isotope Effects. CRC Press, Boca Raton.

Cook, P. F., Cleland, W. W. (2007). Isotope effects as a probe of mechanism. In: Enzyme Kinetics and Mechanism, Cook, P. F. and Cleland, W. W. (Eds.). Garland Science, London; New York.

Cook, P. F., Oppenheimer, N. J., Cleland, W. W. (1981). Secondary deuterium and N-15 isotope effects in enzyme-catalyzed reactions—chemical mechanism of liver alcohol-dehydrogenase. *Biochemistry* **20**, 1817–1825.

Cui, Q., Elstner, M., Karplus, M. (2002). A theoretical analysis of the proton and hydride transfer in liver alcohol dehydrogenase (LADH). *J. Phys. Chem. B* **106**, 2721–2740.

Fauman, E. B., Rutenber, E. E., Maley, G. F., Maley, F., Stroud, R. M. (1994). Water-mediated substrate/product discrimination: the product complex of thymidylate synthase at 1.83 Å. *Biochemistry* **33**, 1502–1511.

Fersht, A. (1999). Structure and Mechanism in Protein Science : A Guide to Enzyme Catalysis and Protein Folding. W.H. Freeman, New York.

Finer-Moore, J. S., Santi, D. V., Stroud, R. M. (2003). Lessons and conclusions from dissecting the mechanism of a bisubstrate enzyme: thymidylate synthase mutagenesis, function, and structure. *Biochemistry* **42**, 248–256.

Francisco, W. A., Knapp, M. J., Blackburn, N. J., Klinman, J. P. (2002). Hydrogen tunneling in peptidylglycine alpha-hydroxylating monooxygenase. *J. Am. Chem. Soc.* **124**, 8194–8195.

Fritz, T. A., Liu, L., Finer-Moore, J. S., Stroud, R. M. (2002). Tryptophan 80 and leucine 143 are critical for the hydride transfer step of thymidylate synthase by controlling active site access. *Biochemistry* **41**, 7021–7029.

Hammes-Schiffer, S. (2006). Hydrogen tunneling and protein motion in enzyme reactions. *Acc. Chem. Res.* **39**, 93–100.

Heyes, D. J., Sakuma, M., de Visser, S. P., Scrutton, N. S. (2009). Nuclear quantum tunneling in the light-activated enzyme protochlorophyllide oxidoreductase. *J. Biol. Chem.* **284**, 3762–3767.

Hirschi, J., Singleton, D. A. (2005). The normal range for secondary Swain-Schaad exponents without tunneling or kinetic complexity. *J. Am. Chem. Soc.* **127**, 3294–3295.

Hong, B., Haddad, M., Maley, F., Jensen, J. H., Kohen, A. (2006). Hydride transfer versus hydrogen radical transfer in thymidylate synthase. *J. Am. Chem. Soc.* **128**, 5636–5637.

Hong, B., Maley, F., Kohen, A. (2007). Role of Y94 in proton and hydride transfers catalyzed by thymidylate synthase. *Biochemistry* **46**, 14188–14197.

Huang, W., Santi, D. V. (1997). Active site general catalysts are not necessary for some proton transfer reactions of thymidylate synthase. *Biochemistry* **36**, 1869–1873.

Huskey, W. P., Schowen, R. L. (1983). Reaction-coordinate tunneling in hydride-transfer reactions. *J. Am. Chem. Soc.* **105**, 5704–5706.

Hyatt, D. C., Maley, F., Monfort, W. R. (1997). Use of strain in a stereospecific catalytic mechanism: crystal structures of *Escherichia coli* thymidylate synthase bound to FdUMP and methylenetetrahydrofolate. *Biochemistry* **36**, 4585–4594.

Kanaan, N., Martí, S., Moliner, V., Kohen, A. (2007). A quantum mechanics/molecular mechanics study of the catalytic mechanism of the thymidylate synthase. *Biochemistry* **46**, 3704–3713.

Kanaan, N., Marti, S., Moliner, V., Kohen, A. (2009). QM/MM study of thymidylate synthase: enzymatic motions and the temperature dependence of the rate limiting step. *J. Phys. Chem. A* **113**, 2176–2182.

Kanaan, N., Roca, M., Tunon, I., Marti, S., Moliner, V. (2010a). Application of Grote-Hynes theory to the reaction catalyzed by thymidylate synthase. *J. Phys. Chem. B* **114**, 13593–13600.

Kanaan, N., Roca, M., Tunon, I., Marti, S., Moliner, V. (2010b). Theoretical study of the temperature dependence of dynamic effects in thymidylate synthase. *Phys. Chem. Chem. Phys.* **12**, 11657–11664.

Kanaan, N., Ferrer, S., MartiÃÅ, S., Garcia-Viloca, M., Kohen, A., Moliner, V. (2011). Temperature dependence of the kinetic isotope effects in thymidylate synthase. A theoretical study. *J. Am. Chem. Soc.* **133**, 6692–6702.

Kiefer, P. M., Hynes, J. T. (2003). Kinetic isotope effects for adiabatic proton transfer reactions in a polar environment. *J. Phys. Chem. A* **107**, 9022–9039.

Klinman, J. P. (1972). The mechanism of enzyme-catalyzed reduced nicotinamide adenine dinucleotide-dependent reductions. Substituent and isotope effects in the yeast alcohol dehydrogenase reaction. *J. Biol. Chem.* **247**, 7977–7987.

Klinman, J. P. (1976). Isotope effects and structure-reactivity correlations in the yeast alcohol dehydrogenase reaction. A study of the enzyme-catalyzed oxidation of aromatic alcohols. *Biochemistry* **15**, 2018–2026.

Klinman, J. P. (2006). Linking protein structure and dynamics to catalysis: the role of hydrogen tunnelling. *Philos. Trans. R. Soc. B* **361**, 1323–1331.

Knapp, M. J., Rickert, K., Klinman, J. P. (2002). Temperature-dependent isotope effects in soybean lipoxygenase-1: correlating hydrogen tunneling with protein dynamics. *J. Am. Chem. Soc.* **124**, 3865–3874.

Kohen, A. (2006). Kinetic isotope effects as probes for hydrogen tunneling in enzyme catalysis. In: Isotope Effects in Chemistry and Biology, Kohen, A. and Limbach, H. H. (Eds.), pp. 743–764. Taylor & Francis, CRC Press, Boca Raton, FL(Chapter 28).

Kohen, A., Jensen, J. H. (2002). Boundary conditions for the Swain-Schaad relationship as a criterion for hydrogen tunneling. *J. Am. Chem. Soc.* **124**, 3858–3864.

Kohen, A., Cannio, R., Bartolucci, S., Klinman, J. P. (1999). Enzyme dynamics and hydrogen tunneling in a thermophilic alcohol dehydrogenase. *Nature* **399**, 496–499.

Kohen, A., Roston, D., Stojković, V., Wang, Z. (2011). Kinetic isotope effects in enzymes. In: Encyclopedia of Analytical Chemistry, Meyers, R. A. (Ed.), vol. S1–S3, pp. 77–99. John Wiley & Sons, Ltd, Chichester, UK.

Kuznetsov, A. M., Ulstrup, J. (1999). Proton and hydrogen atom tunnelling in hydrolytic and redox enzyme catalysis. *Can. J. Chem.* **77**, 1085–1096.

Liebl, U., Lipowski, G., Negrerie, M., Lambry, J. C., Martin, J. L., Vos, M. H. (1999). Coherent reaction dynamics in a bacterial cytochrome *c* oxidase. *Nature* **401**, 181–184.

Liu, H., Warshel, A. (2007). Origin of the temperature dependence of isotope effects in enzymatic reactions: the case of dihydrofolate reductase. *J. Phys. Chem. B* **111**, 7852–7861.

Liu, Y., Barrett, J. E., Schultz, P. G., Santi, D. V. (1999). Tyrosine 146 of thymidylate synthase assists proton abstraction from the 5-position of 2′-deoxyuridine 5′-monophosphate. *Biochemistry* **38**, 848–852.

Major, D. T., Heroux, A., Orville, A. M., Valley, M. P., Fitzpatrick, P. F., Gao, J. (2009). Differential quantum tunneling contributions in nitroalkane oxidase catalyzed and the uncatalyzed proton transfer reaction. *Proc. Natl. Acad. Sci. USA* **106**, 20734–20739.

Marcus, R. A., Sutin, N. (1985). Electron transfers in chemistry and biology. *Biochim. Biophys. Acta* **811**, 265–322.

Miller, S. M., Klinman, J. P. (1985). Secondary isotope effects and structure-reactivity correlations in the dopamine β-monooxygenase reaction: evidence for a chemical mechanism. *Biochemistry* **24**, 2114–2127.

Nagel, Z. D., Klinman, J. P. (2006). Tunneling and dynamics in enzymatic hydride transfer. *Chem. Rev.* **106**, 3095–3118.

Nagel, Z. D., Klinman, J. P. (2010). Update 1 of: tunneling and dynamics in enzymatic hydride transfer. *Chem. Rev.* **110**, PR41–PR67.

Newby, Z., Lee, T. T., Morse, R. J., Liu, Y., Liu, L., Venkatraman, P., et al. (2006). The role of protein dynamics in thymidylate synthase catalysis: variants of conserved 2′-deoxyuridine 5′-monophosphate (dUMP)-binding Tyr-261. *Biochemistry* **45**, 7415–7428.

Northrop, D. B. (1991). Intrinsic isotope effects in enzyme catalyzed reactions. In: Enzyme Mechanism from Isotope Effects, Cook, P. F. (Ed.). vol. 181–202. CRC Press, Boca Raton, FL.

Pal, S., Park, D.-H., Plapp, B. V. (2009). Activity of yeast alcohol dehydrogenases on benzyl alcohols and benzaldehydes: characterization of ADH1 from Saccharomyces carlsbergensis and transition state analysis. *Chem. Biol. Interact.* **178**, 16–23.

Pisliakov, A. V., Cao, J., Kamerlin, S. C. L., Warshel, A. (2009). Enzyme millisecond conformational dynamics do not catalyze the chemical step. *Proc. Natl. Acad. Sci. USA* **106**, 17359–17364.

Pu, J., Ma, S., Gao, J., Truhlar, D. G. (2005). Small temperature dependence of the kinetic isotope effect for the hydride transfer reaction catalyzed by Escherichia coli dihydrofolate reductase. *J. Phys. Chem. B* **109**, 8551–8556.

Pu, J., Gao, J., Truhlar, D. G. (2006). Multidimensional tunneling, recrossing, and the transmission coefficient for enzymatic reactions. *Chem. Rev.* **106**, 3140–3169.

Pudney, C. R., Johannissen, L. O., Sutcliffe, M. J., Hay, S., Scrutton, N. S. (2010). Direct analysis of donor-acceptor distance and relationship to isotope effects and the force constant for barrier compression in enzymatic H-tunneling reactions. *J. Am. Chem. Soc.* **132**, 11329–11335.

Ramaswamy, S., Eklund, H., Plapp, B. V. (1994). Structures of horse liver alcohol dehydrogenase complexed with NAD+ and substituted benzyl alcohols. *Biochemistry* **33**, 5230–5237.

Roston, D., Kohen, A. (2010). Elusive transition state of alcohol dehydrogenase unveiled. *Proc. Natl. Acad. Sci. USA* **107**, 9572–9577.

Rucker, J., Klinman, J. P. (1999). Computational study of tunneling and coupled motion in alcohol dehydrogenase-catalyzed reactions: implication for measured hydrogen and carbon isotope effects. *J. Am. Chem. Soc.* **121**, 1997–2006.

Rucker, J., Cha, Y., Jonsson, T., Grant, K. L., Klinman, J. P. (1992). Role of internal thermodynamics in determining hydrogen tunneling in enzyme-catalyzed hydrogen transfer reactions. *Biochemistry* **31**, 11489–11499.

Sage, C. R., Rutenber, E. E., Stout, T. J., Stroud, R. M. (1996). An essential role for water in an enzyme reaction mechanism: the crystal structure of the thymidylate synthase mutant E58Q. *Biochemistry* **35**, 16270–16281.

Saxl, R. L., Reston, J., Nie, Z., Kalman, T. I., Maley, F. (2003). Modification of *Escherichia coli* thymidylate synthase at tyrosine-94 by 5-Imidazolylpropynyl-2′-deoxyuridine 5′-monophosphate. *Biochemistry* **42**, 4544–4551.

Schwartz, S. D. (2006). Vibrationally enhanced tunneling from the temperature dependence of KIE. In: Isotope Effects in Chemistry and Biology, Kohen, A. and Limbach, H. H. (Eds.), pp. 475–498. Taylor & Francis, CRC Press, Boca Raton, FL(Chapter 18).

Sen, A., Kohen, A. (2010). Enzymatic tunneling and kinetic isotope effects: chemistry at the crossroads. *J. Phys. Org. Chem.* **23**, 613–619.

Sen, A., Yahashiri, A., Kohen, A. (2011). Triple isotopic labeling and kinetic isotope effects: exposing H-transfer steps in enzymatic systems. *Biochemistry* **50**, 6462–6468.

Sikorski, R. S., Wang, L., Markham, K. A., Rajagopalan, P. T., Benkovic, S. J., Kohen, A. (2004). Tunneling and coupled motion in the Escherichia coli dihydrofolate reductase catalysis. *J. Am. Chem. Soc.* **126**, 4778–4779.

Smedarchina, Z., Siebrand, W. (2005). Generalized Swain-Schaad relations including tunneling and temperature dependence. *J. Chem. Phys. Lett.* **410**, 370–376.

Smith Lomax, M. I., Greenberg, G. R. (1967). A new assay of thymidylate synthetase activity based on the release of tritium from deoxyuridylate-5-3H. *J. Biol. Chem.* **242**, 109–113.

Spencer, H. T., Villafranca, J. E., Appleman, J. R. (1997). Kinetic scheme for thymidylate synthase from *Escherichia coli*: determination from measurements of ligand binding, primary and secondary isotope effects and pre-steady-state catalysis. *Biochemistry* **36**, 4212–4222.

Streitwieser, A., Jagow, R. H., Fahey, R. C., Suzuki, S. (1958). Kinetic isotope effects in the acetolysis of deuterated cyclopentyl tosylates. *J. Am. Chem. Soc.* **80**, 2326–2332.

Stroud, R. M., Finer-Moore, J. S. (2003). Conformational dynamics along an enzymatic reaction pathway: thymidyate synthase, "the movie". *Biochemistry* **42**, 239–247.

Swain, C. G., Stivers, E. C., Reuwer, J. F., Schaad, L. J. (1958). Use of hydrogen isotope effects to identify the attacking nucleophile in the enolization of ketones catalyzed by acetic acid. *J. Am. Chem. Soc.* **80**, 5885–5893.

Truhlar, D. G. (2010). Tunneling in enzymatic and nonenzymatic hydrogen transfer reactions. *J. Phys. Org. Chem.* **23**, 660–676.

Wang, Z., Kohen, A. (2010). Thymidylate synthase catalyzed H-transfers: two chapters in one tale. *J. Am. Chem. Soc.* **132**, 9820–9825.

Wang, L., Goodey, N. M., Benkovic, S. J., Kohen, A. (2006). Coordinated effects of distal mutations on environmentally coupled tunneling in dihydrofolate reductase. *Proc. Natl. Acad. Sci. USA* **103**, 15753–15758.

Warshel, A., Sharma, P. K., Kato, M., Xiang, Y., Liu, H. B., Olsson, M. H. M. (2006). Electrostatic basis for enzyme catalysis. *Chem. Rev.* **106**, 3210–3235.

Webb, S. P., Agarwal, P. K., Hammes-Schiffer, S. (2000). Combining electronic structure methods with the calculation of hydrogen vibrational wavefunctions: application to hydride transfer in liver alcohol dehydrogenase. *J. Phys. Chem. B* **104**, 8884–8894.

Welch, G. R. (1986). The Fluctuating Enzyme. Wiley, New York.

PROTEIN FLEXIBILITY AND ENZYMATIC CATALYSIS

By M. KOKKINIDIS,[*,†] N.M. GLYKOS,[‡] AND V.E. FADOULOGLOU[*,‡]

[*]Department of Biology, University of Crete, Heraklion, Crete, Greece
[†]Institute of Molecular Biology and Biotechnology, Heraklion, Crete, Greece
[‡]Department of Molecular Biology and Genetics, Democritus University of Thrace, Alexandroupolis, Greece

Abbreviations

1-CPI	1-(4-chlorophenyl)imidazole
3D	three dimensional
4-BP	4-benzylpyridine
4-CPI	4-(4-chlorophenyl)imidazole
4-NBP	4-(4-nitrobenzyl)pyridine
(S/R)-GOP	*(S/R)*-glycidol phosphate
TIM	Triosephosphate Isomerase
UDX	UDP-α-D-xylose
NMR	Nuclear Magnetic Reasonance
NSE	Neutron Spin Echo
CYP	Cytochrome P450

ADVANCES IN PROTEIN CHEMISTRY AND STRUCTURAL BIOLOGY, Vol. 87
DOI: 10.1016/B978-0-12-398312-1.00007-X

Copyright 2012, Elsevier Inc.
All rights reserved.

Abstract

The dynamic nature of protein structures has been recognized, established, and accepted as an intrinsic fundamental property with major consequences to their function. Nowadays, proteins are considered as networks of continuous motions, which reflect local flexibility and a propensity for global structural plasticity. Protein–protein and protein–small ligand interactions, signal transduction and assembly of macromolecular machines, allosteric regulation and thermal enzymatic adaptation are processes which require structural flexibility. In general, enzymes represent an attractive class among proteins in the study of protein flexibility and they can be used as model systems for understanding the implications of protein fluctuations to biological function. Flexibility of the active site is considered as a requirement for reduction of free energy barrier and acceleration of the enzymatic reaction while there is growing evidence which concerns the connection between flexibility and substrate turnover rate. Moreover, the role of conformational flexibility has been well established in connection with the accessibility of the active site, the binding of substrates and ligands, and release of products, stabilization and trapping of intermediates, orientation of the substrate into the binding cleft, adjustment of the reaction environment, etc.

I. Introduction

Long before the appearance of any experimental evidence supporting the dynamic character of protein structures, their view as static and rigid objects was proving increasingly insufficient to explain a large body of biochemical data, which were gathered mainly in the field of enzymes. The "key-lock" hypothesis, proposed at the end of the nineteenth century by Fisher (1894) and used for the first half of the twentieth century to explain enzymatic activity was based on the concept of fixed structures, that is, the substrate had to fit into a structurally well-defined active site of fixed shape which was also complementary to the substrate's shape. However, the predictions of the "key-lock" theory were not always sufficient to explain the experimental data. Thus, the need for revision led Koshland in 1958 to propose the "induced fit" theory which introduced the concept of an enzyme active site which undergoes structural changes induced by the binding of the substrate. This theory is based on the idea that protein structures or at least their active sites possess a flexibility which enables them to adopt more than one conformation. This was proposed even before the determination of the first protein three-dimensional (3D) structure (myoglobin; Kendrew et al., 1958). One year later, Linderstrom-Lang and Schellmann (1959) proposed a breath-like continuous movement of the protein structures, an idea which was experimentally probed during the following decades. Evidence came from the analysis of crystallographic temperature factors which had lower values for buried/semi-buried areas and systematically higher for solvent-exposed residues. In some cases, loops and side chains exposed to the solvent have very high values indicating high mobility. Moreover, relatively high temperature factors also characterize regions of the active site of enzymes. The same conclusions were drawn by comparing the apo- and holo-states of a protein complexed with different compounds or by comparing homologous proteins from different organisms. Beginning with the analysis of crystallographic structures of lysozyme, the importance of conformational flexibility for catalytic efficiency was widely accepted. In the 1970s, Karplus and coworkers studied the molecular dynamics of folded proteins by solving the equations of motion for the atoms, applying empirical potential energy functions, and the mobile character of protein structures was confirmed for once more. In the 1980s, the development of nuclear magnetic resonance (NMR) provided an additional experimental tool to explore the dynamical aspects of protein structures. Nowadays, the dynamic nature of structures is well established and proteins are considered as a network of continuous motions, which reflect local flexibility and a propensity for global structural plasticity; an implication of the latter is a mechanism for the establishment of new folds as proposed by Glykos et al. (1999). Internal motions have different amplitudes and their frequencies range over different time scales. The motions which affect the protein structure include bond vibrations at the time scale of femto- to pico-seconds, side-chain rotations at the time scale of nanoseconds, and more complicated motions of larger time scales such as motion of flexible termini and loops, large concerted domain motions, and conformational adjustment upon substrate binding.

A growing variety of methods was used and/or developed to study the protein mobility at time scales ranging from femtoseconds to seconds. Next to classical macromolecular crystallography, NMR and molecular dynamics simulations, the protein motions can be studied by a variety of time-resolved techniques and a wide range of spectroscopies such as time-resolved Laue diffraction and intermediate trapping, single-molecule fluorescence resonance energy, neutron spin echo (NSE), etc.

Some types of protein functions can be easily related to structural flexibility. For example, almost any interaction between a protein and another molecule is associated with conformational changes. Ligand recognition and binding, protein–protein interactions, and processes as complicated as signal transduction and assembly of multiprotein machines are based on protein flexibility (Gazi et al., 2009; Teilum et al., 2011). Allosteric regulation is achieved through conformationally flexible regulatory proteins. Even the evolutionary process of thermophilic, mesophilic, or psychrophilic adaptations of protein structures, affect protein sequences such that structural flexibility is directly affected with implications for protein stability and enzymatic catalysis. In general, enzymes represent an attractive class among proteins in the study of protein flexibility and they can be used as model systems for understanding the implications of protein fluctuations to biological function.

Enzymes catalyze a variety of biological processes and may accelerate them by as many as 20 orders of magnitude in comparison with the corresponding uncatalyzed reaction (Wolfenden and Snider, 2001). One of the most intriguing questions in biochemistry is how the enzymes achieve this high rate enhancement. Numerous studies have been reported aiming to elucidate dynamic effects associated with catalysis. These studies demonstrate that among other factors, the ability of enzymes to adopt different conformations is of high importance for the efficiency of the catalytic processes (Kamerlin and Warshel, 2010). Flexibility of the active site is considered as a requirement for reduction of free energy barrier and acceleration of the enzymatic reaction (Hammes-Schiffer and Benkovic, 2006; Henzler-Wildman and Kern, 2007; Nashine et al., 2010). During the catalytic cycle, the enzyme molecule passes through different states and each state is associated with a different active site conformation. Rapid transition between the different conformational states is therefore mandatory for the maximum enzyme activity. Some aspects of the role of this conformational flexibility have been established in connection with the accessibility of the active site, the binding of substrates and ligands and release of products, stabilization and trapping of intermediates, orientation of the substrate into the binding cleft, adjustment of the reaction environment, etc. Furthermore, growing evidence concerns the connection between flexibility and the substrate turnover rate (reviewed in Yon et al., 1998; Hammes, 2002; Agarwal, 2006).

Extreme cases of protein flexibility are the molten globule state and the intrinsically disordered proteins. These exhibit significant conformational flexibility which is frequently necessary for the establishment of intermolecular interaction networks (Ohgushi and Wada, 1983; Gazi et al., 2009; Schlessinger et al., 2011). In the partially folded molten globule state, the polypeptide chain adopts a nearly native-like secondary structure, and a dynamic tertiary structure due to the partial absence of native intramolecular packing interactions.

During the last decade, many proteins have been described that fail to adopt a stable tertiary structure under physiological conditions and yet display biological activity (Dunker et al., 2008; Uversky and Dunker, 2010). This state of the proteins, defined as intrinsic disorder, has been found to be rather widespread; disordered regions lacking stable secondary and tertiary structure are often a prerequisite for biological activity, suggesting that structure–function relationships can be frequently only understood in a dynamic context in which function arises from conformational freedom. Fully or partly nonstructured proteins are described as intrinsically disordered or intrinsically unstructured proteins. The term natively unfolded proteins indicates that protein function is associated with a dynamic ensemble of different conformations (Gazi et al., 2008, 2009).

Here, we will focus on the flexibility of natively folded proteins and, in particular, of enzymes. We will review the different ways in which flexibility is linked to enzymatic catalysis through conformational flexibility of loops surrounding the active site, and of domain and active site residue dynamics. Specific examples from the literature will be reviewed. In addition, an overview of widely used specific methodologies for studying protein flexibility will be presented, as well as some more recently developed ones. The issues of allostericity, thermal adaptation, and ligand specificity in relation to flexibility will be reviewed in detail. The topic is presented from the structural point of view with an emphasis on X-ray crystallography results.

II. Selected Methods Used for Studying Protein Flexibility

A summary of selected methods used today to investigate and analyze protein flexibility is presented below. A detailed application of X-ray crystallography, mutagenesis, and molecular dynamics simulations to a specific example is described in a following section.

A. *Crystallography and Time-Resolved X-ray Methods*

X-ray crystallography is the method of choice for obtaining the molecular structure of proteins at atomic resolution. The X-ray diffraction pattern of the crystalline specimen is recorded and the electron density map of the molecule under study is calculated. The molecular model is built into the electron density map and refined and thus the position of the atoms is determined with high accuracy. In fact, crystallography provides a static model of a dynamic molecule; therefore this model represents an average, in space and time, of 3D structure of the molecule. However, even the static crystal structure contains useful information for the dynamic nature of the molecule, as it is explained below.

The refined atomic model contains information for the degree of thermal motion of the atoms. The mean-squared atomic displacement is commonly expressed as the temperature factor (B-factor). Temperature factors express not only the static disorder, which is an ensemble of substates present in solution and trapped in the crystal but also the dynamic disorder, which represents fluctuations in the crystal or otherwise crystal defects. Thus, B-factors cannot be interpreted simply as the amplitude of atomic fluctuations because both true intramolecular motion and lattice disorder contribute to them. Moreover, crystal contacts affect B-factors. In the context of protein structures, the B-factors can be taken as indicating the relative vibrational motion of different parts of the structure and comparing them along the structure may allow one to draw conclusions about which are the most flexible parts. Atoms with low B-factors belong to a part of the structure that is well ordered while atoms with large B-factors generally belong to a part that is very flexible.

It is quite common that some protein segments, that is, amino and carboxyl termini, loops or even other protein segments yield weak or nondetectable electron density. A common reason (apart from crystal lattice defects) for missing electron density is that the unobserved region fails to scatter X-rays coherently due to variation in position of the atoms of this particular segment, that is, the unobserved atoms are disordered or highly flexible.

Sometimes there are residues in a structure which present clear electron density for two positions, or they present what we call multiple distinct conformations. This is a direct evidence that the side chain spend time in more than one conformations or otherwise that present a relative flexibility. In other cases however, the movement from one conformation to the other occurs so quickly that there is no definite electron density for some of the atoms of the side chain.

Several time-resolved X-ray structural methods (for a review see Westenhoff et al., 2010) have been developed in the recent years aiming to follow in real time the conformational changes of a protein while performing its action. Time-resolved Laue diffraction, intermediate trapping, time-resolved wide-angle X-ray scattering, and time-resolved X-ray absorption are the most common of them. Time-resolved Laue diffraction (Schottea et al., 2004) provides conformational snapshots of the whole protein in high resolution and allow us to visualize in real time and with atomic detail the conformational evolution of a protein. Detection of structural changes as small as 0.2–0.3 Å with a time resolution of 100 ps is possible. The time-resolved Laue diffraction experiments are of a pump-probe type. The reaction is triggered within the protein crystal by photolysis caused by ultra-short laser pulses which play the role of the pump for the initiation of the reaction. Diffraction patterns are collected at specific time delays after triggering. This cycle must be repeated many times for each spatial rotation of the crystal and many times for the same time delay. The applicability of the method depends on whether the molecule retains its biological activity in the crystalline state, whether the molecule is inherently photosensitive or if it could be engineered as so, whether the system is reversible, whether the concentration of the produced intermediate is sufficiently high to be detectable because at any given time during reaction several intermediates are likely to be present unless they are well separated in time. The method has been successfully applied to heme-containing proteins (Srajer and Royer, 2008).

As an alternative, the intermediate trapping method could be used. In this method, the lifetime of intermediates is sufficiently extended so that they can be studied by static crystallography. Freezing, changes in pH, chemical modifications, or solvent modifications are some ways which are used to trap the intermediate. The successful application of an alternative approach of this method has been used to reveal the dynamics of the active site of the small guanosine nucleotide-binding protein H-Ras-p21 (Klink et al., 2006; Klink and Scheidig, 2010).

In time-resolved wide-angle X-ray scattering diffraction, data are recorded as a function of time from the molecule in solution (Fischetti et al., 2003). Because the structural information obtained is averaged over all orientations of the randomly oriented molecules in the solution, the atomic detail is poor. This technique is used to characterize large-scale global conformational changes. On the other hand, the time-resolved X-ray absorption spectroscopy is used to characterize the geometry and the changes of the geometry of the coordination structure in the active sites of metalloproteins.

B. X-ray Free Electron Lasers

The X-ray free electron laser (XFEL) is a novel radiation source, which combines a particle accelerator with laser physics. Bunches of electrons are first brought to high energies in a superconducting accelerator. Then they fly through a special arrangement of magnets, called undulator, in which they emit laser-like flashes of radiation. By this way, X-ray flashes of high energy are generated and they can be directed to the sample whose diffraction pattern is recorded just before its explosion. The special characteristics of the generated flashes, that is, coherent radiation, wavelengths at the level of Angstroms, and short pulses at the level of hundreds of femtoseconds allows for atomic resolution snapshots of the molecule under study. XFEL may allow imaging of single particles/molecules and a large variety of different applications such as investigation of protein folding and dynamics, the actions of catalysts, and the splitting of chemical bonds can be afforded (Neutze et al., 2004). Applicability of the method with tiny crystals of photosystem I has been demonstrated (Chapman et al., 2009).

C. Spectroscopies, NMR, NSE, and Hydrogen–Deuterium Exchange

Spectroscopy is the study of the interaction of electromagnetic radiation with matter. An advantage of the spectroscopic methods is that they can give a dynamic picture of the selected part of the molecule, for example, time-resolved intrinsic fluorescence of Trp reflects internal mobility of this amino acid. Real-time information on structural changes for macromolecules containing a chromophore, such as heme proteins, is provided by time-resolved spectroscopic studies including absorption, resonance Raman, and infrared spectroscopy. In those cases, we observe structural changes limited to the chromophore environment.

The single-molecule fluorescence resonance energy transfer (FRET) reports proximity of moieties within a molecule for distances in the range of 1–10 nm. FRET is measured between two dyes, donor and acceptor. If the dyes are separated by a large distance (larger than 10 nm), then there is little interaction between them and even if the donor emits photons upon its excitation by laser, acceptor will not interact with this energy. However, if the two dyes are brought closer, the acceptor takes the energy from the donor and emits photons of different colors. So, we can use FRET to measure distance changes in the nanometer scale. A recent application of the single-molecule FRET is described by Kahra et al. (2011) for the investigation of conformational plasticity and dynamics of the peptidyl-prolyl *cis–trans* isomerase SlyD.

NMR spectroscopy is used to obtain detailed structural information for the whole system and, in addition, can yield information on the dynamics of specific parts of the structure. The spin of protons is the property which causes the nucleus to produce NMR signal. Specific nuclei, called NMR active, that is, ^{15}N and ^{13}C, when they are found under a magnetic field absorb radiation at a characteristic frequency and migrate to a higher spin energy level. When spin returns to the basic level, energy is emitted and this signal can be recorded and processed to generate the (NMR) spectrum. Protein NMR spectroscopy is the use of NMR phenomenon to extract structural information. The protein structure determination by this method is a process which results in a convergent ensemble of structures.

Molecular motions generate local fluctuating magnetic fields which have, as a consequence, NMR relaxation. Relaxation times can be measured and used to determine parameters as correlation times and chemical exchange rates which yield information on the dynamics of protein parts as the backbone or side chains. Motions which can be detected occur on the time scale of 10 ps to 10 ns but even slower motions between 10 μs to 100 ms can be studied.

NSE spectroscopy is an inelastic neutron-scattering technique with recent applications to the study of protein dynamics (Mezei, 1980). It is a technique of high effective energy resolution which determines the time–space correlation function at the temporal scales from nanoseconds to microseconds and at the spatial scale from several Angstroms to some hundreds. NSE is used to determine relaxation processes in a macromolecule, that is, internal dynamic modes and it has the potential to determine the global shape fluctuations and domain motions. NSE has been applied to the NHERF1 multidomain protein to distinguish and characterize couple domain motions which are involved in the dynamic propagation of allosteric signals at the nanoseconds timescale (Farago et al., 2010). It is shown that NSE can be used to determine the domain mobility tensor which determines the degree of dynamical coupling between domains. Bu et al. (2005) have used the method to determine internal coupled domain motions within DNA polymerase I from *Thermus aquatius.*

The hydrogen–deuterium (H/D) exchange is a chemical reaction. A covalently bonded hydrogen atom is replaced by a deuterium atom from the solution. Usually the examined protons are the amides in the backbone of a protein. The method gives information about the solvent accessibility of various parts of the molecule and detects global or local unfolding on timescales of milliseconds and longer. It is monitored by NMR and/or mass spectrometry. The conformational changes that occur in factor XIII due to monovalent and divalent ion binding has been recently reported by Woofter and Maurer (2011). The H/D exchange effects observed in the presence of a wide range of ions in different conditions and the deuterium incorporation was analyzed by MALDI-TOF MS. Moreover, Oyeyemi et al. (2011) have applied the technique to elucidate the relationship between flexibility and thermal adaptation in the case of dihydrofolate reductases.

D. Computational Methods

Biomolecular simulations are an important technique for characterizing protein conformational changes (Klepeis et al., 2009). The advantage of the computational methods is that one can follow the details of the protein dynamics in the pico-second timescale and examine the structural features in the atomic level. The disadvantage is that without experimental validation of the potentials used in the simulations, the predictions are on the risk of questioning. The starting point is a high-resolution structure determined by X-rays or NMR. A prerequisite is also the availability of computational power. Then, an empirical force field, which is a set of parameters describing the potential energy of the system together with the equations of motion, is applied to the system and the successive positions of all the atoms of the system, and thus the progressive motion of the molecule, can be watched. The computational demands prevent the method from reaching timescales greater than milliseconds. To achieve longer timescales, simplified models can be used as the implicit solvent or the coarse-graining models.

Molecular dynamics simulations apply empirical molecular mechanics potential energy functions which are suitable for studying conformational changes and dynamics as the conformational changes during the catalytic cycle, associated with substrate binding and product release as well as fluctuations around an average structure. However, they cannot be applied to model chemical reactions, so cannot investigate enzyme reaction mechanisms directly. To study the connection between conformational and chemical changes as well as to investigate the hypothesis that protein motions accelerate the reaction rates of catalysis, the quantum mechanics/molecular mechanics (QM/MM) methods are used. An analytic overview of how the computational methods have been used to increase our understanding of the dynamical aspects of enzymatic catalysis has been presented by McGeagh et al. (2011).

III. Determinants of Active Site Flexibility

Enzymes are generally characterized by conservation of their functional groups through evolution. On the other hand, the flexibility requirements of many active sites frequently impose a relaxation of strict conservation so that different functional residues are not conserved during evolution while converging toward the same mechanistic role. The resulting plasticity of active sites has been reviewed by Todd et al. (2002). In principle, upon catalysis, residues in and around the active site must undergo conformational changes associated with the binding and release of substrates and cofactors, the protection of the reaction space from the aqueous environment, stabilization of reaction intermediates, and interactions between catalytic residues and binding subsites with various groups of reactants upon completion of catalysis. In some cases, the structural requirements for conformational flexibility are limited to individual amino acids especially to side chains capable of adopting different rotamers. In other cases, the interactions of the macromolecule with substrate and cofactors require conformational changes mainly involving the external loops located in the periphery of the active site. On the other hand, several times the required structural reorganization associated with catalysis involves large global movements and rearrangements of whole protein domains.

A. *Flexible Active Site Loops*

Loop regions belong to the most flexible parts of protein structures. In crystal structures, they are characterized by higher than the average temperature factors and display the highest conformational variability among equivalent regions of homologous protein structures. In some cases, they correspond to segments of weak or missing density in electron density maps because they fail to scatter X-rays coherently due to their pronounced disorder. The conformational variations of loop regions become evident in NMR structural ensembles, which show a multitude of different conformations for the same loop. The high mobility of loops has been frequently confirmed by trajectories of molecular dynamics simulations.

The extreme flexibility of loops which reflects the absence of conformational constraints is consistent with their low sequence conservation, an indicator of low evolutionary pressure. A pronounced exception to this conservation rule, are the loops participating in the active sites of enzymes. The critical role of loops for the enzymatic function was recognized early and their conformational transitions are now accepted as key events in catalytic processes (Malaban et al., 2010). Such loops, commonly called "lids" or "flaps", have been the subject of investigation by many research groups. Their location at the entry of active site plays a major role in substrate selectivity and recognition and facilitation of substrate binding into the binding cleft. The structural comparison of apo- and holo-states of various enzymes highlights that the main structural difference between the two states is the different conformations of active/binding site loops. At the unbound state, the flaps adopt an open conformation which leaves the binding cleft and active site accessible to solvent. In the bound state, the flaps adopt a closed conformation which blocks the entry to the active site. A characteristic example is the 60/70 loop of the zinc-dependent endopeptidase BoNT/A LC (Thompson et al., 2011). Binding of three different inhibitors induces a consistent change in the conformation of this loop as a result of the hydrophobic interaction of a loop residue (Pro69) with an aromatic moiety of the inhibitor (Fig. 1A). Compared to the uncomplexed enzyme, ligand binding induces a more compact conformation because the loop is pulled into the active site enclosing more tightly specific subsites of the inhibitor and inducing a more effective inhibition. Thus a disordered, solvent-exposed loop adopts upon substrate binding a more compact and ordered conformation-making interactions with subsites of the ligand and/or other residues of the protein. The closing of the flaps traps the ligand into the cleft and shields fully or partially the ligand molecule from the aqueous environment, thereby stabilizing the bound state. Moreover, the access of the active site to other molecules is prevented and the reaction intermediates are protected and stabilized (Kember, 1993). This mechanism provides an effective way

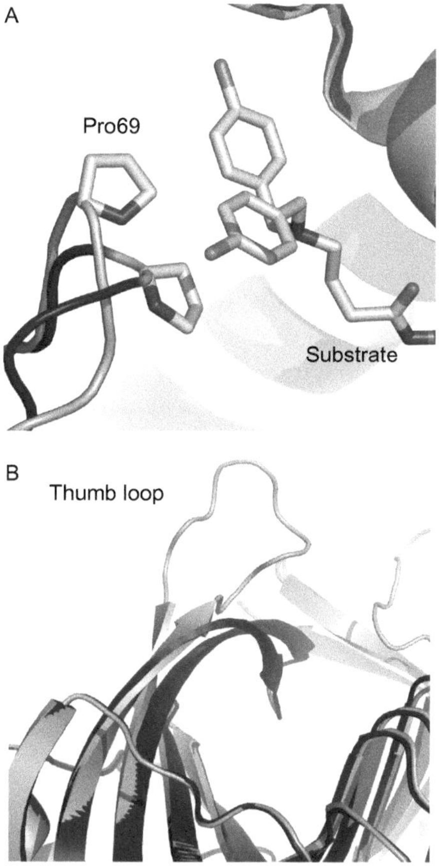

FIG. 1. (A) Structural comparison of the active site, 60/70 loop of the zinc-dependent endopeptidase BoNT/ALC at the apo- (green) and holoenzyme states (blue). Hydrophobic interactions of Pro69 with the substrate, which is also shown by stick representation inside the active site, induce a more compact protein conformation by pulling the 60/70 loop toward the active site (PDB entries 3QIX and 3QIY). All figures have been prepared by PYMOL. (B) The "thumb" loop is the most flexible part of the structure of glycoside hydrolase family 11 xylanases. Here, two conformations of this loop, the open (green) and closed one (blue) are shown (PDB entries 3EXU and 2B46). (For interpretation of the references to color in this figure legend, the reader is referred to the Web version of this chapter.)

to control the accessibility to the active site. After the end of the reaction, the opening of the loops permits release of the product and the beginning of a new catalytic cycle. For example, in the glycoside hydrolase family 11 xylanases, a highly conserved "thumb loop" in the proximity of the active site has been found in three conformational states, a closed, a loose, and an extended open one (Pollet et al., 2009; Fig. 1B). Molecular dynamics simulations suggest that the thumb is the most flexible part of the xylanase structure: a dynamic catalytic cycle has been proposed based on these three different conformations of the thumb. These conformational changes are directly associated with the binding of the substrate and the release of the product. Moreover, a mutation which hinders the thumb movement results in a fourfold decrease of turnover number thus suggesting a direct relation between thumb movement and catalytic properties.

Flaps are not the only flexible structural elements around active sites. Quite often, the rearrangement of the active site geometry is accompanied or facilitated by movements and conformational changes of secondary structural elements directly adjacent to the active site, for example, helices which frequently precede or follow the flaps. Examples are presented at the following paragraphs for the cases of EryK P450 cytochrome and *Bc*ZBP deacetylase.

At least two models are used to describe and mechanistically explain the transition from the open to the closed conformation. The first of them, the well-known "induced fit" model, postulates that large conformational changes at a local level, and the transition to the closed conformation are induced by substrate binding into the active site cleft. In a number of examples, the unbound enzyme state has as a favored conformation the open one and the bound enzyme state, the closed one. The extensively studied loop 6 of triosephosphate isomerase (TIM) is a typical case interpreted through the "induced fit" model (Lolis and Petsko, 1990). TIM is a homodimer in which the subunits catalyze the interconversion of D-glyceraldehyde 3-phosphate and dihydroxyacetone phosphate. Each subunit is a classical $(\beta\alpha)_8$ barrel (Branden and Tooze, 1999) with the active site being formed by the loops located at the C-termini of the eight β-strands (loops 1–8) of the barrel. A detailed structural comparison of the unbound and bound forms revealed that the only significant difference is the movement of the 10-residue loop 6 resulting in the blockage of the active site entry at the bound state. Some residues of the active site loop are moved up to 7 Å from their original position.

However, in some cases, experimental evidence supports as an alternative a "conformational selection" scenario in which the enzyme preexists in two alternative and stable conformations which differ in their ability to bind ligands. This concept has been reported for the monomeric cytochrome P450 (CYP) from *Saccharopolyspora erythraea*, commonly called EryK (Savino et al., 2009). X-ray crystallography studies revealed two different structures, an open and a closed one, which were determined for the unbound state of EryK in different ionic strength conditions. Characterization of binding kinetics of the ErD substrate to EryK could be sufficiently explained by the "conformational selection" model and allowed for the exclusion of the "induced fit" model. Comparison of the two structures revealed conformational shift involving among others a movement of an active site loop of about 11 Å and a shift of the N-terminus of an α-helix of about 10 Å toward this active site loop. This reorganization closes the access channel for the active site.

The rigid body movement observed for the TIM active site loop 6 is the most common type of loop motion in general, and is termed "hinge-bending motion" because the loop is linked to the protein through two regions known as hinge regions and moves around them as a rigid body. For α-helices, a common movement is rotation around one of their ends which moves the other end away from its original position, occasionally up to several Angstroms. The high flexibility of the loops is often associated with the presence of glycine residues while in hinges alanines and prolines are also quite common. Mutagenesis has been frequently used to probe the significance of loop flexibility for catalysis with hinge residues being quite often the targets of such experiments. In the case of glutathione synthetase (Tanaka et al., 1993), replacement of Gly and Pro in the hinge by Val significantly impaired the enzymatic activity. Structural studies show that the mutations caused no structural alterations to the rest of the structure. Absence of electron density indicated that a high degree of flexibility has been retained for the mutant loop as well as for the native one. The observed loss of catalytic activity could reflect problems in conformational adaption of the mutant loops in the closed state which results in disruptive interactions with the bound substrate. A study by Kursula and coworkers (2004) demonstrated by mutagenesis the importance of small residues at position 3 of the C-terminal three-residue hinge (LysThrAla) of TIM loop 6. When the small Ala residue at position 3 was replaced by bulky residues (Gln, Leu, or Lys), the unliganded form of the mutants resembled the closed conformation. The results suggest that these mutations shift the equilibrium of the oscillation motion of loop in favor of the closed conformation. Kinetic data implies that in these mutants substrate binding is the rate-limiting step.

Except from hinge residues, the importance of residues in the middle of loops for active site flexibility has also been investigated. When Gly76 of the active site loop (residues 71–80) of pepsin was replaced by Ala, Val, and Ser, a lower catalytic efficiency was measured and interpreted as the result of lower flap flexibility. A lower flexibility could alter the pattern of interactions (hydrogen bonds) which are responsible for substrate alignment in the active site (Okoniewska et al., 2000).

In conclusion, mobile loops in the proximity of active sites frequently play various roles in catalysis and it is widely accepted that their pronounced conformational flexibility may contribute to their functions. Mutagenesis experiments support the concept of this flexibility being essential for catalytic efficiency and activation processes.

B. *Active Sites Located at Domain Interfaces*

A significant portion of oligomeric enzymes form their catalytic and binding sites at the interface between subunits while, in the case of monomeric enzymes, active and binding sites often lie between domains (Ali and Imperiali, 2005). Among the most well-studied examples of enzymes with their active/binding sites formed between domains are the alcohol and glutamate dehydrogenases, citrate synthase and hexokinase, glutathione transferases and ribonuclease A. An obvious consequence from the location of active sites in the interdomain or intersubunit interface is that even subtle local conformational changes between the domains, which are commonly called lobes, could affect catalysis and enzyme interaction with substrate and products. A model consistent with many experiments and with the Koshland's (1958) "induced fit" model implies that the ligand is first bound to one lobe; subsequent movement of the other lobe brings it near to the ligand permitting additional interactions which stabilize the structure (Stillman et al., 1993; Hayward, 2004).

Lesk and Chothia (1984) classified domain motions as hinge or shear domain motions. In the former case, the two domains are connected through a short extended and flexible peptide fragment, the hinge region. Hinge permits rotations while the lobes move as rigid bodies with their internal structure being preserved and all the deformations restricted to the hinge region. As a result, the binding cleft may exist in many different intermediate states between the open and the closed ones. On the other hand, the shear motion is generated by small rearrangement motions, that is, small changes in torsion angles and side-chain motions causing small shifts of secondary structure elements with respect to the original position. The shear domain closure is thus the cumulative effect of relatively small shifts of loosely packed secondary structures. This type of motion is more likely to occur when there is an extended surface between domains. Domain motions around the active sites of citrate synthase and hexokinase have been grouped with this type of motions.

Large-scale domain motions that can be classified as hinge motions were observed for the homodimeric restriction endonuclease PvuII in connection with the binding of cognate oligonucleotides. The U-shaped dimer exists in an open (Athanasiadis et al., 1994) and a closed, DNA-bound conformation (Cheng et al., 1994). Each PvuII monomer consists of a DNA-binding subdomain which also carries the catalytic site and a helical dimerization subdomain. Although the structures of the individual subdomains are highly conserved in the open and the closed conformation, their relative orientation changes upon DNA binding, with the opening of the DNA-binding cleft being reduced by nearly 28 Å upon transition from the open to the closed conformation. This large conformational change results mainly from small changes ($<10°$) in the φ,ψ backbone conformational angles of two Gly residues located in a loop region at the interface between the dimerization and DNA-binding subdomains.

Alcohol dehydrogenases are present in many different organisms and catalyze the conversion of alcohols to aldehydes functioning as dimers or tetramers (Lesk and Chothia, 1984). Each subunit consists of a catalytic and a coenzyme-binding domain which are linked together through two α-helices (termed α2 and α3). Comparing the crystal structures of the apo- and holo-liver alcohol dehydrogenase, Eklund et al. (1981) described the conformational changes between these states as a rigid body rotation of one domain relative to the other one. The two hinge regions were determined at the α2, α3 helices and was found to consist of three and four residues, respectively. Colonna-Cesari and co-workers (1986) applied empirical energy functions to simulate the domain rotation. The analysis showed that most of the hinge residues undergo small changes at their main chain, φ/ψ angles ($<15°$) while motion of the side chains with variations in χ torsion angles of about 50° was observed. The internal dynamics of the enzyme was investigated by small angle neutron scattering and NSE spectroscopy and confirmed a large-scale correlated domain motion (Biehl et al., 2008). By normal mode analysis, this motion can be understood as the result of the movement of the outside catalytic domain with respect to the rigid core. Moreover, this motion results in a cleft opening which is much larger than the relatively narrow pocket observed in the crystal structures.

RNase A is a model system which has been used in numerous experiments for the study of dynamical properties of proteins and it is represented in PDB with more than a hundred deposited structures (e.g., see Wlodawer et al., 1988; Santoro et al., 1993; Toiron et al., 1996; Leonidas et al., 1997). It is a stable monomeric enzyme which catalyzes the cleavage of the P—O^5 bond in single-stranded RNA. Its structure consists of two all-beta domains which form a V shape and are linked by a hinge. The binding and active sites are formed at the interface between the domains with the active site located at the bottom of the deep cleft. Petsko and colleagues have shown that the enzyme is unable to bind the inhibitor cytidine 2′-monophosphate (2′-CMP) at temperatures lower than −50 °C (Rasmussen et al., 1992). Since no structural differences could be noticed comparing the crystal structures at cryo temperatures and room temperature, they suggested that flexibility is a required property for binding and catalysis (Tilton et al., 1992). Vitagliano and co-workers (2002) could observe in the crystal a significant, reversible, hinge-bending domain motion upon ligand binding which leads to a more compact structure. Upon substrate binding, the hinge region is moved, the angle between the domains decreases, and residues belonging to different domains and involved in substrate binding present a significant reduction of their between distances.

C. *Flexibility of Active Site Residues*

Active site residues play many different roles. They may be involved in catalysis and substrate binding, stabilize the intermediates of the reaction or the structure of the binding cleft. They provide the suitable for the catalysis microenvironments and enable substrates to form enough contact points for strong binding. Some degree of flexibility is inevitable for the active site residues to achieve their functions and accommodate the conformational changes which are necessary during the catalytic cycle. This flexibility facilitates active site rearrangements during the numerous intermediate steps of the reaction. The flexibility is achieved at a cost, usually in form of local strain or instability. A systematic mutational analysis of active site residues of barnase protein was one of the first studies to demonstrate that there is an inverse relationship between stability of active site and activity (Meiering et al., 1992).

The side chains of catalytic residues usually adopt more than one conformation although only some of them may be catalytically competent. In addition, their conformational flexibility has been related to specific steps of the catalytic cycle as, for example, for the proton shuttling. A characteristic example comes from the hydrolytic aldehyde dehydrogenases (ALDHs). Analysis of the crystal structures of several different ALDHs from more than nine different organisms determined in the absence or presence of substrates, cofactors, and products showed that although these proteins do not undergo major conformational changes upon binding of cofactors, they exhibit a high degree of flexibility for the catalytic residues Cys302 and Glu268 (González-Segura et al., 2009; Munoz-Clares et al., 2011; Fig. 2). In particular, Cys302 has been found in two conformations called "resting" and "attacking" conformation, which are far and close to the carbonyl carbon of the bound aldehyde, respectively. The catalytic Glu268 adopts three conformations: (i) the "inside" conformation which activates Cys302 for nucleophilic attack, (ii) the "intermediate," which activates the hydrolytic water molecule, and (iii) the "outside" conformation which releases the proton taken from Cys or water through a proton relay mechanism. Examination of the various residue conformations with respect to the binding cleft architecture in the presence and absence of the cofactors reveals the sterically compatible combinations and helps rationalize the details of the mechanism.

Another recent example comes from the crystal structures of TIM protein complexed with suicide inhibitors (Venkatesan et al., 2011). It is shown that two residues of the active site, the catalytic Glu167 and Glu97, are flexible and can adopt two different conformations when the enzyme exists in the closed liganded state. The differences are mainly associated with changes in the side-chain dihedrals which give rise to two active site geometries. Thus, when the (*S*)-glycidol phosphate ((*S*)-GOP) is bound in the active site, Glu167 adopts its well-known competent conformation and Glu97 is salt bridged. When (*R*)-glycidol phosphate ((*R*)-GOP) is bound

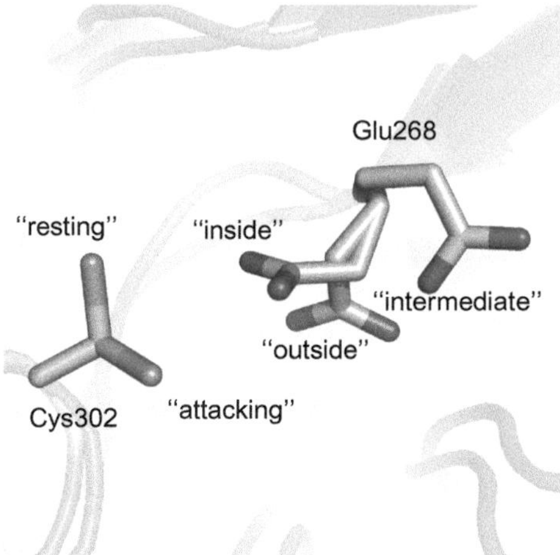

FIG. 2. Local flexibility of the active site residues of aldehyde dehydrogenases is demonstrated by the multiple conformations of Cys302 and Glu268. As it is shown, Cys302 can adopt two conformations called ''resting'' and ''attacking'' while Glu268 adopts three, the ''inside'' ''intermediate'' and ''outside'' conformations. Combinations of these conformations determine different stages of the catalytic cycle (PDB entries 1UZB and 1O02). (For color version of this figure, the reader is referred to the online version of this chapter.)

both residues adopt unusual conformations. The first geometry enables Glu167 to attack the terminal carbon of (S)-GOP in a stereochemically favored (linear) arrangement, but this is not possible for (R)-GOP at the second geometry. These could explain the higher chemical reactivity of (S)-GOP compared with (R)-GOP.

D. *A Specific Example: Flexibility of the* Bc*ZBP Deacetylase*

*Bc*ZBP is a zinc-dependent deacetylase from *Bacillus cereus* whose crystal structure has been reported at the resolution of 1.8 Å (Fadouloglou et al., 2006, 2007). The biological pathway and the function of the protein were unknown until recently when the ortholog from *Bacillus anthracis Ba*BshB (with 97% sequence identity) was identified as a deacetylase involved in the bacillithiol biosynthesis (Newton et al., 2009; Parsonage et al., 2010). *Bc*ZBP is a hexamer and possesses six structurally equivalent active sites which are formed by the association of two monomers. X-ray crystallography, mutagenesis, and molecular dynamics simulations have been used to elucidate aspects related with function and flexibility of the protein.

B-factor analysis of the crystal structure and structural superposition of different states of the enzyme demonstrates a high mobility for the three active site loops (42–51, 129–140, and 180–192) and highlights changes of the active site cavity between states. Structural comparison of *Bc*ZBP with the homologous TT1542, which shares 38% sequence identity and has been used as the search model for the molecular replacement which leaded to the *Bc*ZBP's structure, locates the greater differences between the enzymes in the immediate environment of the active sites. In the case of *Bc*ZBP, the N-terminus of α2-helix has been moved 4 Å closer to the active site together with the preceding loop (42–51) which has been shifted toward the top of the active site. Because *Bc*ZBP represents the structure of apoenzyme and TT1542, the structure without the zinc bound in the active site, we suggest that the observed differences highlight a zinc-induced organization of the active site environment by promoting a more closely packed conformation. On the other hand, the temperature factors of crystal structures demonstrate as the most flexible part of both enzymes the active site loop 180–192, joining helices α5 and α6.

A 50 ns molecular dynamics simulation study (Fadouloglou et al., 2009) supports the X-ray crystallography findings. The analysis resulted in an agreement between the simulation-derived atomic fluctuation and the crystallographically determined atomic temperature factors. Thus, the three loops which frame the active site, have atomic fluctuations which are significantly greater than the average for the rest of the structure demonstrating their high flexibility. The trajectory also reveals a concerted loop motion, which generates the effect of a breathing-like motion around the active site with successive opening and closing events.

The functional significance of the active site loops flexibility was further investigated by mutagenesis of the hinge residues Arg140 and Ala42 (Deli et al., 2010). Arginine 140 is located at the rim of the substrate-binding cleft. In the crystal structure, its side chain adopts two distinct conformations. The one conformer blocks the active site's entry while the other one, which is stabilized by electrostatic interactions, keeps the active site accessible. Arg was replaced by the small, hydrophobic Ala residue and by the oppositely charged Glu residue. In both variants, the ability for distinct conformations at position 140 has been disrupted. Both variants showed a decrease in their catalytic efficiency compared with the wild type. Ala42, on the other hand, which is located in the proximity of the substrate was replaced by a Ser residue. Ser is able to form putative hydrogen bonds either with the substrate or with adjacent amino acids. With either of these ways, the mutation is predicted to increase the rigidity or even to trap the loop to a closed-like conformation. Indeed, the produced variant exhibited a dramatic reduction of efficiency.

IV. Flexibility and Special Aspects of Enzyme Properties

A. *Flexibility and Thermal Enzymatic Adaptation*

Given the environmental variations at the different areas of earth, many organisms have evolved adaptation mechanisms for very low or very high temperatures. As a consequence, there are enzymes isolated from such organisms whose optimal temperature for function is below 10 °C (psychrophilic) or above 45 °C (thermophilic). Catalytic residues are generally conserved in homologous psychrophilic and thermophilic enzymes which imply that cold adaptation resides on other parts of the structure. In the effort to elucidate the molecular basis of enzymatic temperature, adaptation was observed that the cold adaptation is related with a reduced thermal stability (cold adaptation is reviewed by Siddiqui and Cavicchioli, 2006). This led to the hypothesis that optimization of the catalytic activity at low temperatures may be associated with an increased structural flexibility (Feller and Gerday, 1997). Since psychrophilic enzymes must function in low temperatures, the evolutionary pressure to retain structural features responsible for conformational rigidity (e.g., disulfide and salt bridges) is relaxed. In the absence of conformational constraints for stability, they present an increase in flexibility which possibly leads to a reduction of the activation energy (D' Amico et al., 2002). In other words, the increased active site flexibility results in a higher number of conformational states of the enzyme/substrate complex. The energy of activation is used many times as a criterion for the evaluation of psychrophilicity as it is usually lower in the cold-adapted enzymes than to their mesophilic counterparts. Sequence comparison among homologues from thermophilic, mesophilic, and psychrophilic organisms shows that generally in the cold-adapted enzymes (Siddiqui and Cavicchioli, 2006): (i) buried residues tend to be smaller and less hydrophobic, (ii) surface residues tend to be less polar, probably because polar residues confer stability by forming additional intramolecular H-bonds. (iii) Less Ile residues are found, probably due to their good packing properties inside the protein cores, (iv) Lys residues usually replace Arg which facilitate a greater number of electrostatic interactions and H-bonds over Lys, (v) a less number of salt bridges are observed. Radestock and Gohlke (2011) probed by computational means the corresponding states hypothesis which claims that homologues from different thermal-adapted organisms are in corresponding states of similar rigidity and flexibility at their respective optimal temperatures. Comparing a sample of 19 pairs of homologues from meso- and thermophilic enzymes, they showed that adaptive mutations of thermophilic enzymes maintain the balance between overall rigidity important for thermostability and local flexibility important for activity at the respective temperature at which the proteins function.

In several studies, the psychrophilicity has been related to an overall flexibility throughout the protein structure. Studies on the α-amylase from *Pseudoalteromonas haloplanktis* have shown that the cold adaptation strategy for this enzyme leads to a uniformly unstable protein (Feller et al., 1999). Site-directed mutagenesis and comparison with mesophilic and thermostable α-amylases demonstrated a weakening of intramolecular interactions which lead to an overall decrease of the thermostability of the psychrophilic protein. This provides the appropriate plasticity around the active site, necessary to adapt the catalytic efficiency to low temperatures. In the case of Zn-metalloproteases of thermolysin family, an optimization of the overall protein flexibility is achieved via the reduction of the hydrogen bonds stability in the dynamic structure due to a decrease of amino acids which form hydrogen bonds (Xie et al., 2009).

However, the need for a global elevated flexibility was questioned and another cold adaptation model was proposed which implies that the requirement for increased structural flexibility can be limited only to a small, crucial region of the protein structure and, in particular, around and inside the active site (Fields and Somero, 1998). Many studies present evidence which supports this model. Comparing A4-lactase dehydrogenases from Antarctic (optimum temperature of function −2 to 1 °C) and South American (4–10 °C) notothenioid species was found that the active site residues are fully conserved. Combination of kinetic, sequence, and structural data suggested that cold adaptation is based on increasing flexibility in small areas of the molecule, outside the active site, that affect the mobility of adjacent to active site structures (Fields and Somero, 1998). The work of Watanabe and Takada (2004) demonstrates the importance of amino acid substitutions around the active site to the thermal adaptation of isocitrate lyases. Gln207 and Gln217 of the psychrophilic lyase from *Colwellia maris* were replaced by a His and a Lys, respectively, which are the residues occupying the equivalent positions at the lyase from *Escherichia coli* and other organisms and are essential for catalysis. The catalytic activity at low temperatures is decreased or diminished respectively while the enzyme remains active at moderate temperatures. It is believed that each of the two residues affects the thermal adaptation through a different mechanism since the Gln217Lys replacement is associated with structural changes expressed as an increase in the thermostability and changes in the CD spectra (enhancement of α-helical components) while the Gln207His replacement resulted in slight conformational changes. Comparative molecular dynamics studies show that the strategy of members of elastase family for being adapted at cold also concerns increase of localized flexibility (Papaleo et al., 2006; Riccardi and Papaleo, 2010). The most obvious difference between mesophilic and psychrophilic elastases concerns the amino acid composition and flexibility of loops which are clustered around the active site and specificity pocket.

The importance of local flexibility into the active site for cold adaptation has been demonstrated in several cases. The psychrophilic alkaline phosphatase from the Antarctic strain TAB5 has been used as a model system, and the possibility of modifying its psychrophilic properties by introducing—via mutagenesis—predictable flexibility changes to key active site residues (Tsigos et al., 2001) or residues to the direct vicinity of active site (Mavromatis et al., 2002) has explored. Tsigos et al. (2001) have modified the side-chain flexibility of the catalytic residues Trp260 and Ala219 and that of His135 from the Mg^{2+} binding site. Trp260Lys is less active than the wild type at low temperatures, while the double mutant Trp260Lys/Ala219Gln has lost its psychrophilic character, although its activity at elevated temperatures exceeds that of the wild type. Finally, substitution of His135 by Asp resulted in stabilization of the enzyme and in the case of the triple mutant restored a low energy of activation. Thus, the psychrophilic character of an enzyme can be strongly affected by very slight variations of its amino acid sequence which however is expected to drastically change the local flexibility. As it was mentioned above, psychrophilicity has also been related with increased levels of Gly residues which ensure local flexibility. Psychrophilic alkaline phosphatase positions 261 and 262, in the vicinity of active site are occupied by Gly residues. Mavromatis et al. (2002) have investigated their importance in the establishment of the psychrophilic character of the enzyme. To constrain the conformational flexibility of the main chain with the minimum perturbation of the local structure, it was chosen to mutate Gly to Ala. The Gly262Ala mutant is completely inactive while the Gly261Ala has lost the psychrophilic character although it is active at elevated temperatures. Thus, Gly clusters in the vicinity of active sites in combination with their structural environment are frequently essential determinants of the psychrophilic character.

B. Flexibility as an Essential Component of Enzymatic Allostery

Allostery is the process by which conformational changes at one site of a protein (called regulatory or allosteric site) are coupled with changes to a different and usually distant functional site (active site). The most common way for triggering the allosteric response of an enzyme is the binding of a specific allosteric effector molecule which could be an activator or an inhibitor of the enzymatic function. Studies mainly by X-ray crystallography showed that the functional states of an allosteric enzyme are well represented by two, structurally different forms: (i) the "R" form (relaxed), which has an optimal affinity for the substrate and (ii) the "T" form (tensed), which has a minimal affinity. Attempts to explain the nature of the allosteric transition have led to two models, the so-called concerted and sequential models. Both of them have been developed under the condition that allosteric enzymes are symmetric oligomers with identical protomers. Each subunit in the oligomer can adopt one of the "R" or "T" conformations. According to the concerted model (Monod et al., 1965), the conformational changes induced by the binding of an allosteric effector to one subunit is transmitted to all other subunits, thus switching them to the same conformation, corresponding to one of the two possible states, either "T" or "R". On the other hand, the sequential model (Koshland et al., 1966) permits a number of different global states, that is, subunits can change conformations independently and alterations to one of them are not necessarily transmitted to the others. Thus, the oligomer could be found in a number of hybrid "TR" combinations. A relatively new idea for rationalizing the allosteric regulation is based on the concept that allosteric enzymes, as all proteins, exist as a dynamic ensemble of conformational states. According to the ensemble model of allostery, ligand binding to the allosteric site leads to redistribution within the ensemble enabling an altered conformation at the region around the active site (Gunasekaran et al., 2004; Wrabl et al., 2011). The shift in the relative populations of the protein may affect protein function, that is, as the distribution substrate affinities are altered within the conformational ensembles (Goodey and Benkovic, 2008). The dependence of protein function on conformational flexibility favors an expanded view of allostery as an intrinsic property of all dynamic proteins which are potentially allosteric. Based on this idea and in contrast to what was believed in the past, allostery is recognized now as a property of monomeric proteins as well (Gunasekaran et al., 2004). The binding of the effector favors the transition from the relatively rigid inactive T-form to a more flexible active ensemble and the increased flexibility facilitates the conformational transitions during enzyme turnover. Hilser and Thompson (2011) have used the ensemble concept of allostery to discuss the behavior of steroid hormone receptors which are much more dynamic systems than is represented by traditional models. Thus, a dynamic ensemble of structures and the presence of intrinsically disordered segments which are stabilized upon ligand binding can explain the broad variety of ligands which drive remote allosteric responses. The importance of this model to the explanation of cooperativity of the bacterial flagellar switch was recently discussed by Bai et al. (2010). Novinec et al. (2010) showed by intrinsic fluorescence spectroscopy that Cathepsin K, a human cysteine peptidase, has a flexible structure which converts slowly among distinct conformational states. Addition in a protein sample of glucosaminoglycans, which are allosteric regulators of Cathepsin K, causes a change in the enzyme conformation resulting in a rapid binding of the inhibitor stefin A. So, it is proposed that glucosaminoglycans affect the distribution of the preexisting conformational equilibrium.

An intuitively very appealing mechanism for the coupling of the effector binding to one site with the conformational changes to another site is the existence of amino acid networks (Goodey and Benkovic, 2008). According to this model, propagation of the signal through a protein structure is based on networks of physically interconnected and thermodynamically linked residues. Amaro et al. (2007) combining data from crystallography, biochemical kinetic assay, and molecular dynamics simulations have revealed a network of interactions directly correlated to the transmission of the allosteric signal of the imidazole glycerol phosphate synthase. This network is formed by a set of highly conserved amino acids which lead from the allosteric site of the enzyme to its active site, more than 25 Å apart. A correlated motion analysis confirms the involvement of these residues to a stream of coupled motions and their role in the allosteric response.

Two examples which clearly demonstrate that structural flexibility is an essential component of allosteric activation are those of glucosamine-6-phosphate deaminase (Bustos-Jaimes et al., 2002; Rudino-Pinera et al., 2002) and of the human hexameric UDP-glucose dehydrogenase (Kadirvelraj et al., 2011; Sennett et al., 2011). Glucosamine-6-phosphate deaminase catalyzes the reversible isomerization–deamination of glucosamine 6-phosphate and it is allosterically activated by the N-acetyl glucosamine-6-phosphate. Comparing a variety of structures Rudino-Pinera et al. (2002) demonstrated that the main structural differences between the T- and R-states are located at the active site lid. This enzyme possesses a complex active site lid (residues 158–187) formed by a helix–loop segment and a β-strand (Fig. **3**A). The lid is directly connected with both, the active and allosteric sites, that is, Arg172 is an active site residue which participates at the substrate binding while Arg158, Lys160, and Thr161 belong to the allosteric site. Experiments support the correlation of the lid flexibility with the allosteric transition and the substrate-binding properties of the active site. It was shown that the allosteric transition from the "T" to "R" state is not associated with defined geometrical changes of the lid but with pronounced changes to its conformational flexibility. Especially, the atoms in the central segment of the lid show a marked decrease in their crystallographic B-factor from a B average of 80 at the T-state to a B average of 40 $Å^2$ at the R-state. Moreover, when the enzyme is at the R-state, the active site lid has been found in three distinct conformations (Rudino-Pinera et al., 2002). Substrate binding stabilizes the lid in one of the three conformers and produces a general reduction in the atomic vibration of the whole protein.

The connection of the conformational flexibility of the lid and the function of the deaminase was investigated by mutating Phe174 to Ala. This mutation had, as a result, such an increase of the conformational flexibility of the T-form that no electron density was visible for part of the lid (Bustos-Jaimes et al., 2002). The mutant was inactive in the absence of

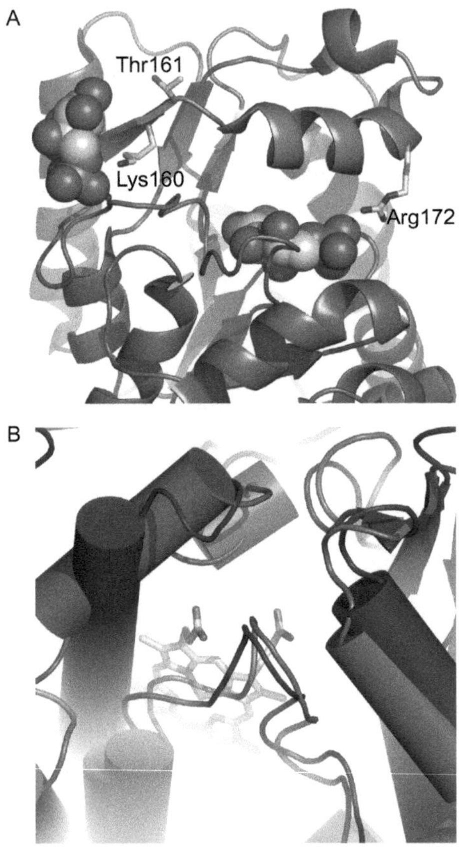

FIG. 3. (A) Cartoon representation of the glucosamine-6-phosphate deaminase. The enzyme possesses a complex active site lid (shown in green) which is directly connected with both, active and allosteric sites. To illustrate the location of these sites, a glucosamine 6-phosphate (space-filling model) is shown bound to each of them (PDB entry 1FS5). (B) Structural comparison of the P450 cytochromes CYP101 (P450cam, PDB entry 2L8M) in gray from *Pseudomonas putida* and CYP108 (P450terp, PDB entry 1CPT) in red, a bacterial enzyme from *Pseudomonas.* The enzymes adopt a conserved fold with a substantial variability around the substrate-binding region. The molecule of heme is shown as stick representation. Small displacements of structural elements which abuts the active/binding toward the heme together with the segmentation of one of the helices (shown on the left of the figure) decreases the size of the active/binding site of CYP101 relative to that of the CYP108. (For interpretation of the references to color in this figure legend, the reader is referred to the Web version of this chapter.)

the allosteric activator, a proof that, conformational flexibility of the active site lid alters the binding properties of the active site.

Except from the active site lid, the 144–154 loop has been shown to be directly associated with the allosteric transition through interactions of Glu148 with residues of the active site lid, which stabilize the R-form or the allosteric site, that is, a salt bridge with Lys 160, which stabilize the T-form. Structural superposition of R-form complexed with either both sites occupied or only the allosteric site shows that there is a correlation between the movement of 144–146 part and the active site lid. When the active site is free, the 144–146 part moves to open the lid and when the active site is occupied moves to close the lid.

The human hexameric UDP-glucose dehydrogenase is an example of an allosteric enzyme whose active site is bifunctional and can bind either substrate or allosteric effector with distinct "induced fit" conformations (Kadirvelraj et al., 2011). In this case, it was shown that packing defects in the protein core in combination with structural flexibility constitute a mechanism for the evolution of allostery. Binding of the UDP-α-D-xylose (UDX) allosteric inhibitor activates a distinct "induced fit" allosteric response: The buried Thr131 loop directly connects the hexamer building interface to the active site and plays the role of an allosteric switch. Upon substrate binding is positioned in a way which supports the formation of a functional binding site. On the other hand, when the allosteric inhibitor binds, this loop moves about 4 Å and rotates about 180°. This movement changes the packing interactions of the protein core and rotates a neighboring helix, resulting in the remarkable repacking of the core and the conversion of the enzyme into an inactive oligomer. Deletion of the buried Val132 traps the enzyme to an open intermediate conformation of the allosteric response (Sennett et al., 2011). Comparison of the closed wild type and the open Δ132 structure identified a hinge-bending axis between two residues of the dimerization domain. A concerted hinge-bending motion between adjacent subunits was proposed to be the basis of the allosteric transition.

C. *Flexibility and Ligand Specificity*

The broad substrate specificity which characterizes some enzymes has been explained under the view of a flexible active site cavity capable of accommodating stereochemically diverse substrates. The different substrates may either induce conformational rearrangements to the broader area of the flexible active site or, if the protein structure is seen as a dynamic ensemble of conformations, each substrate is bound to different conformations with different affinity. The impact of plasticity and flexibility on ligand binding has been illustrated in the literature, among others, with examples from the family of endonucleases V (Feng et al., 2005), glutathione *S*-transferases (Hou et al., 2007), dehydrogenases (Deng et al., 2009), and CYP (Pochapsky et al., 2010). In some cases, the local flexibility around the active site and the possible reposition and repack of only few individual side chains could cause sufficient space and shape alterations inside the active site to switch from one substrate to another without substantial structural changes. However, in other cases, global flexibility seems to be a requirement for promiscuity.

CYPs are a superfamily of enzymes which catalyze the addition of an oxygen to an C—C or C—H bond, that is, they are oxidoreductases catalyzing the oxidation of structurally dissimilar organic substrates. CYPs adopt a similar global fold which is combined with a remarkable adaptivity for substrate recognition. This is possibly due to the modularity of secondary structure features which surround the active site (Fig. 3B). These features vary even between different states of the same enzyme (Pochapsky et al., 2010). Some CYPs show broad substrate and hydroxylation specificity and it has been proposed that their high flexibility and plasticity around the active site is responsible for the wide substrate selectivity and specificity they present. The typical topology of CYPs contains a four alpha helical bundle which also accommodates the heme group which is ligated by a Cys at the beginning of one of the bundle helices. In total, the structure is composed by 14 helices and 5 beta sheets. In an early spectroscopic study of five individual CYPs (Anzenbacher and Hudecek, 2001), it was demonstrated that their active sites exhibit significant differences in their flexibility and stability although in these enzymes these properties cannot be related by a simple relationship since in some cases, a low stability does not necessarily reflect a high flexibility. Among the five CYPs which were tested, one demonstrates a significantly higher promiscuity than the others by being able to accommodate and modify a variety of structurally different substrates. Comparison with the other four cytochromes of higher specificity showed that what differentiates it from them is a more flexible active site together with a less conformationally restricted heme group.

Comparison of the crystal structures of P450 2B4 cytochromes complexed with the ligands 1-CPI (1-(4-chlorophenyl)imidazole) and 4-CPI (4-(4-chlorophenyl)imidazole) highlights a significant shift of the backbone of two helices (F and I) with consequence the reposition of side chains reshaping of the active site and change of its volume almost 80 $Å^2$ (Zhao et al., 2007). Likewise, comparison of the structures of the human P450 2B6 enzyme complexed with the ligands 4-(4-nitrobenzyl)pyridine (4-NBP) and 4-benzylpyridine (4-BP) reveals that small rearrangements of five Phe residues of the active site resulted in an increase of the active site volume sufficient to accommodate the bulkier 4-NBP in comparison to 4-BP (Shah et al., 2011).

The *Plasmodium falciparum* dihydroorotate dehydrogenase (*Pf*DHODH) is an example of enzyme which is able to accommodate a wide range of structurally different classes of inhibitors due to a drastic reorganization which creates different binding pockets (Deng et al., 2009). Structural studies demonstrate that the high flexibility of the enzyme allows for two alternative binding sites close to each other for interactions with different structural classes of ligands. In addition, local flexibility inside the active site pocket contributes to its ability to bind ligands of variable size. The different binding modes result as a consequence of a large conformational change in the position of Phe188 between the two structures which is also accompanied by α-helical shifts.

Studies on glutathione S-transferases suggest that in some cases ligand promiscuity may most easily be achieved by distributing flexibility throughout the structure than limiting it to the active site (Hou et al., 2007). The structurally related isoforms of glutathione, A1-1, and A4-4 represent benchmarks for high catalytic promiscuity and selectivity, respectively. H/D exchange and tryptophan fluorescence showed significantly greater solvent accessibility throughout of the A1-1 sequence compared with A4-4. Moreover, a double mutation to the A1-1 which introduces an aromatic interaction present to the A4-4 and the reverse double mutation to A4-4 were sufficient to change local and global flexibility and partially invert their relative substrate specificities.

Acknowledgment

V. E. F. is supported by a Marie Curie Reintegration grant.

References

Agarwal, P. K. (2006). Enzymes: an integrated view of structure, dynamics and function. *Microb. Cell Fact.* **5**, 2.

Ali, M. H., Imperiali, B. (2005). Protein oligomerization: how and why. *Bioorg. Med. Chem.* **13**, 5013–5020.

Amaro, R. E., Sethi, A., Myers, R. S., Davisson, V. J., Luthey-Schulten, Z. A. (2007). A Network of conserved interactions regulates the allosteric signal in a glutamine amidotransferase. *Biochemistry* **46**, 2156–2173.

Anzenbacher, P., Hudecek, J. (2001). Differences in flexibility of active sites of cytochromes P450 probed by resonance Raman and UV–Vis absorption spectroscopy. *J. Inorg. Biochem.* **85**, 209–213.

Athanasiadis, A., Vlassi, M., Kotsifaki, D., Tucker, P., Wilson, K. S., Kokkinidis, M. (1994). The crystal structure of *Pvu*II endonuclease reveals extensive structural homologies to *Eco*RV. *Nat. Struct. Biol.* **1**, 469–475.

Bai, F., Branch, R. W., Nicolau, D. V., Jr., Pilizota, T., Steel, B. C., Maini, F. K., Berry, R. M. (2010). Conformational spread as a mechanism for cooperativity in the bacterial flagellar switch. *Science* **327**, 685–689.

Biehl, R., Hoffmann, B., Monkenbusch, M., Falus, P., Preost, S., Merkel, R., Richter, D. (2008). Direct observation of correlated interdomain motion in alcohol dehydrogenase. *Phys. Rev. Lett.* **101**, 138102.

Branden, C., Tooze, J. (1999). Introduction to Protein Structure. 2nd edn. pp. 47–50. Garland Publishing, New York, NY.

Bu, Z., Biehl, R., Monkenbusch, M., Richter, D., Callaway, D. J. E. (2005). Coupled protein domain motion in Taq polymerase revealed by neutron spin-echo spectroscopy. *Proc. Natl. Acad. Sci. U.S.A.* **102**, 17646–17651.

Bustos-Jaimes, I., Sosa-Peinado, A., Rudino-Pinera, E., Horjales, E., Calcagno, M. L. (2002). On the role of the conformational flexibility of the active-site lid on the allosteric kinetics of glucosamine-6-phosphate deaminase. *J. Mol. Biol.* **319**, 183–189.

Chapman, H. N., et al. (2009). Femtosecond X-ray protein nanocrystallography. *Nature* **470**, 73–78.

Cheng, X., Balendiran, K., Schildkraut, I., Anderson, J. E. (1994). Structure of PvuII endonuclease with cognate DNA. *EMBO J.* **13**, 3927–3935.

Colonna-Cesari, F., Perahia, D., Karplus, M., Eklund, H., Branden, C. I., Tapia, O. (1986). Interdomain motion in liver alcohol dehydrogenase. *J. Biol. Chem.* **261**, 15273–15280.

D' Amico, S., Claverie, P., Collins, T., Georlette, D., Gratia, E., Hoyoux, A., Meuwis, M.-A., Feller, G., Gerday, C. (2002). Molecular basis of cold adaptation. *Philos. Trans. R. Soc. Lond. B* **357**, 917–925.

Deli, A., Koutsioulis, D., Fadouloglou, V. E., Spiliotopoulou, P., Balomenou, S., Arnaouteli, S., Tzanodaskalaki, M., Mavromatis, K., Kokkinidis, M., Bouriotis, V. (2010). LmbE proteins from *Bacillus cereus* are de-N-acetylases with broad substrate specificity and are highly similar to proteins in *Bacillus anthracis*. *FEBS J.* **277**, 2740–2753.

Deng, X., Gujjar, R., Mazouni, F. E., Kaminsky, W., Malmquist, N. A., Goldsmith, E. J., Rathod, P. K., Phillips, M. A. (2009). Structural plasticity of malaria dihydroorotate dehydrogenase allows selective binding of diverse chemical scaffolds. *J. Biol. Chem.* **284**, 26999–27009.

Dunker, A. K., Oldfield, C. J., Meng, J., Romero, P., Yang, J. Y., Chen, J. W., Vacic, V., Obradovic, Z., Uversky, V. N. (2008). The unfoldomics decade: an update on intrinsically disordered proteins. *BMC Genomics* **9**(Suppl. 2), article S1.

Eklund, H., Samma, J. P., Wallén, L., Brändén, C. I., Akeson, A., Jones, T. A. (1981). Structure of a triclinic ternary complex of horse liver alcohol dehydrogenase at 2.9 Å resolution. *J. Mol. Biol.* **146**, 561–587.

Fadouloglou, V. E., Kotsifaki, D., Gazi, A. D., Fellas, G., Meramveliotaki, C., Deli, A., Psylinakis, E., Bouriotis, V., Kokkinidis, M. (2006). Purification, crystallization and preliminary characterization of a putative LmbE-like deacetylase from *Bacillus cereus*. *Acta Crystallogr.* **F62**, 261–264.

Fadouloglou, V. E., Deli, A., Glykos, N. M., Psylinakis, E., Bouriotis, V., Kokkinidis, M. (2007). Crystal structure of the *Bc*ZBP, a zinc-binding protein from *Bacillus cereus*. *FEBS J.* **274**, 3044–3054.

Fadouloglou, V. E., Stavrakoudis, A., Bouriotis, V., Kokkinidis, M., Glykos, N. M. (2009). Molecular dynamics simulations of *Bc*ZBP, a deacetylase from *Bacillus cereus*: active site loops determine substrate accessibility and specificity. *J. Chem. Theory Comput.* **5**, 3299–3311.

Farago, B., Li, J., Cornilescu, G., Callaway, D. J. E., Bu, J. (2010). Activation of nanoscale allosteric protein domain motion revealed by neutron spin echo spectroscopy. *Biophys. J.* **99**, 3473–3482.

Feller, G., Gerday, C. (1997). Psychrophilic enzymes: molecular basis of cold adaptation. *Cell. Mol. Life Sci.* **53**, 830–841.

Feller, G., D'Amico, D., Gerday, C. (1999). Thermodynamic stability of a cold-active a-amylase from the Antarctic bacterium *Alteromonas haloplanctis*. *Biochemistry* **38**, 4613–4619.

Feng, H., Klutz, A. M., Cao, W. (2005). Active site plasticity of endonuclease V from *Salmonella typhimurium*. *Biochemistry* **44**, 675–683.

Fields, P. A., Somero, G. N. (1998). Hot spots in cold adaptation: localized increases in conformational flexibility in lactate dehydrogenase A4 orthologs of Antarctic notothenioid fishes. *Proc. Natl Acad. Sci. U.S.A.* **95**, 11476–11481.

Fischetti, R. F., Rodi, D. J., Mirza, A., Irving, T. C., Kondrashkina, E., Makowski, L. (2003). High resolution wide angle X-ray scattering of protein solutions. *J. Synch. Res.* **10**, 398–404.

Fisher, E. (1894). Einfluss der configuration auf die wirkung der enzyme. *Ber. Dtsch. Chem. Ges.* **27**, 2985–2993.

Gazi, A. D., Bastaki, M., Charova, S. N., Gkougkoulia, E. A., Kapellios, E. A., Panopoulos, N. J., Kokkinidis, M. (2008). Evidence for a widespread interaction mode of disordered proteins in bacterial type III secretion systems. *J. Biol. Chem.* **283**, 34062–34068.

Gazi, A. D., Charova, S. N., Panopoulos, N. J., Kokkinidis, M. (2009). Coiled-coils in type III secretion systems: structural flexibility, disorder and biological implications. *Cell. Microbiol.* **11**, 719–729.

Glykos, N. M., Cesareni, G., Kokkinidis, M. (1999). Protein plasticity to the extreme: changing the topology of a 4-α-helix bundle with a single amino-acid substitution. *Structure* **7**, 597–603.

González-Segura, L., Rudiño-Piñera, E., Muñoz-Clares, R. A., Horjales, E. (2009). The crystal structure of a ternary complex of betaine aldehyde dehydrogenase from *Pseudomonas aeruginosa* provides new insight into the reaction mechanism and shows a novel binding mode of the 2′-phosphate of $NADP^+$ and a novel cation binding site. *J. Mol. Biol.* **385**, 542–557.

Goodey, N. M., Benkovic, S. J. (2008). Allosteric regulation and catalysis emerge via a common route. *Nat. Chem. Biol.* **4**, 474–482.

Gunasekaran, K., Ma, B., Nussinov, R. (2004). Is allostery an intrinsic property of all dynamic proteins? *Proteins* **57**, 433–443.

Hammes, G. G. (2002). Multiple conformational changes in enzyme catalysis. *Biochemistry* **41**, 8221–8228.

Hammes-Schiffer, S., Benkovic, S. J. (2006). Relating protein motion to catalysis. *Annu. Rev. Biochem.* **75**, 519–541.

Hayward, S. (2004). Identification of specific interactions that drive ligand-induced closure in five enzymes with classic domain movements. *J. Mol. Biol.* **339**, 1001–1021.

Henzler-Wildman, K., Kern, D. (2007). Dynamic personalities of proteins. *Nature* **450**, 964–972.

Hilser, V. J., Thompson, E. B. (2011). Structural dynamics, intrinsic disorder, and allostery in nuclear receptors as transcription factors. *J. Biol. Chem.* **286**, 39675–39682.

Hou, L., Honaker, M. T., Shireman, L. M., Balogh, L. M., Roberts, A. G., Ng, K.-C., Nath, A., Atkins, W. M. (2007). Functional promiscuity correlates with conformational heterogeneity in A-class glutathione *S*-transferases. *J. Biol. Chem.* **282**, 23264–23274.

Kadirvelraj, R., Sennett, N. C., Polizzi, S. J., Weitzel, S., Wood, Z. A. (2011). Role of packing defects in the evolution of allostery and induced fit in human UDP-glucose dehydrogenase. *Biochemistry* **50**, 5780–5789.

Kahra, D., Kovermann, M., Löw, C., Hirschfeld, V., Haupt, C., Balbach, J., Hübner, C. G. (2011). Conformational plasticity and dynamics in the generic protein folding catalyst SlyD unraveled by single-molecule FRET. *J. Mol. Biol.* **411**, 781–790.

Kamerlin, S. C. L., Warshel, A. (2010). At the dawn of the 21st century: is dynamics the missing link for understanding enzyme catalysis? *Proteins* **78**, 1339–1375.

Kember, E. S. (1993). Movable lobes and flexible loops in proteins. Structural deformations that control biochemical activity. *FEBS* **326**, 4–10.

Kendrew, J., Bodo, G., Dintzis, H., Parrish, R., Wyckoff, H., Phillips, D. (1958). A three-dimensional model of the myoglobin molecule obtained by X-ray analysis. *Nature* **181**(4610), 662–666.

Klepeis, J. L., Lindorff-Larsen, K., Dror, R. O., Shaw, D. E. (2009). Long-timescale molecular dynamics simulations of protein structure and function. *Curr. Opin. Struct. Biol.* **19**, 120–127.

Klink, B. U., Scheidig, A. J. (2010). New insights into the dynamic properties and the active site architecture of H-Ras p21 revealed by X-ray crystallography at very high resolution. *BMC Struct. Biol.* **10**, 38.

Klink, B. U., Goody, R. S., Scheidig, A. J. (2006). A newly designed microspectrofluorometer for kinetic studies on protein crystals in combination with X-ray diffraction. *Biophys. J.* **91**, 981–992.

Koshland, D. E. (1958). Application of the theory of enzyme specificity to protein synthesis. *Proc. Natl. Acad. Sci. U.S.A.* **44**, 98–104.

Koshland, D. E., Nemethy, G., Filmer, D. (1966). Comparison of experimental binding data and theoretical models in proteins containing subunits. *Biochemistry* **5**, 365–385.

Kursula, I., Salin, M., Sun, J., Norledge, B. V., Haapalainen, A. M., Sampson, N. S., Wierenga, R. K. (2004). Understanding protein lids: structural analysis of active hinge mutants in triosephosphate isomerase. *Protein Eng. Des. Sel.* **17**, 375–382.

Leonidas, D. D., Shapiro, R., Irons, L. I., Russo, N., Acharya, K. R. (1997). Crystal structures of ribonuclease A complexes with 5′-diphosphoadenosine 3′-phosphate and 5′-diphosphoadenosine 2′-phosphate at 1.7 Å resolution. *Biochemistry* **36**, 5578–5588.

Lesk, A. M., Chothia, C. (1984). Mechanisms of domain closure in proteins. *J. Mol. Biol.* **174**, 175–191.

Linderstrom-Lang, K. U., Schellmann, J. A. (1959). Protein structure and enzyme activity. In: The Enzymes, Boyer, P. D. (Ed.), vol. 1. 2nd edn. pp. 443–510. Academic Press, New York.

Lolis, E., Petsko, G. A. (1990). Crystallographic analysis of the complex between triosephosphate isomerase and 2-phosphoglycolate at 2.5 Å resolution: implications for catalysis. *Biochemistry* **29**, 6619–6625.

Malaban, M. M., Amyes, T. L., Richard, J. P. (2010). A role for flexible loops in enzyme catalysis. *Curr. Opin. Struct. Biol.* **20**, 702–710.

Mavromatis, K., Tsigos, I., Tzanodaskalaki, M., Kokkinidis, M., Bouriotis, V. (2002). Exploring the role of a glycine cluster in cold adaptation of an alkaline phosphatase. *Eur. J. Biochem.* **269**, 2330–2335.

McGeagh, J. D., Ranaghan, K. E., Mulholland, A. J. (2011). Protein dynamics and enzyme catalysis: insights from simulations. *Biochim. Biophys. Acta* **1814**, 1077–1092.

Meiering, E. M., Serrano, L., Fersht, A. R. (1992). Effect of active site residues in barnase on activity and stability. *J. Mol. Biol.* **225**, 585–589.

Mezei, F. (1980). The principles of neutron spin echo. Neutron Spin Echo: Proceedings of a Laue-Langevin Institut Workshop. Springer, Heidelberg, Germany, pp. 3–26.

Monod, J., Wyman, J., Changeux, P. (1965). On the nature of allosteric transitions: a plausible model. *J. Mol. Biol.* **12**, 88–118.

Munoz-Clares, R., Gonzalez-Segura, L., Diaz-Sanchez, A. G. (2011). Crystallographic evidence for active-site dynamics in the hydrolytic aldehyde dehydrogenases. Implications for the deacylation step of the catalyzed reaction. *Chem. Biol. Interact.* **191**, 137–146.

Nashine, V. C., Hammes-Schiffer, S., Benkovic, S. J. (2010). Coupled motions in enzyme catalysis. *Curr. Opin. Chem. Biol.* **14**, 644–651.

Neutze, R., Huldt, G., Hajdu, J., Spoel, D. (2004). Potential impact of an X-ray free electron laser on structural biology. *Rad. Phys. Chem.* **71**, 905–916.

Newton, G. L., Rawat, M., La Clair, J. J., Jothivasan, V. K., Budiarto, T., Hamilton, C. J., Claiborne, A., Helmann, J. D., Fahey, R. C. (2009). Bacillithiol is an antioxidant thiol produced in Bacilli. *Nat. Chem. Biol.* **5**, 625–627.

Novinec, M., Kovacic, L., Lenarcic, B., Baici, A. (2010). Conformational flexibility and allosteric regulation of cathepsin K. *Biochem. J.* **429**, 379–389.

Ohgushi, M., Wada, A. (1983). 'Molten-globule state': a compact form of globular proteins with mobile side-chains. *FEBS* **164**, 21–24.

Okoniewska, M., Tanaka, T., Yada, R. Y. (2000). The pepsin residue glycine-76 contributes to active-site loop flexibility and participates in catalysis. *Biochem. J.* **349**, 169–177.

Oyeyemi, O. A., Sours, K. M., Lee, T., Kohen, A., Resing, K. A., Ahn, N. G., Klinman, J. P. (2011). Comparative hydrogen–deuterium exchange for a mesophilic vs thermophilic dihydrofolate reductase at 25 °C: identification of a single active site region with enhanced flexibility in the mesophilic protein. *Biochemistry* **50**, 8251–8260.

Papaleo, E., Riccardi, L., Villa, C., Fantucci, P., Gioia, L. (2006). Flexibility and enzymatic cold-adaptation: a comparative molecular dynamics investigation of the elastase family. *Biochim. Biophys. Acta* **1764**, 1397–1406.

Parsonage, D., Newton, G. L., Holder, R. C., Wallace, B. D., Paige, C., Hamilton, C. J., Dos Santos, P. C., Redinbo, M. R., Reid, S. D., Claiborne, A. (2010). Characterization of the N-acetyl-α-D-glucosaminyl l-malate synthase and deacetylase functions for bacillithiol biosynthesis in *Bacillus anthracis*. *Biochemistry* **49**, 8398–8414.

Pochapsky, T. C., Kazanis, S., Dang, M. (2010). Conformational plasticity and structure/function relationship in cytochromes P450. *Antioxid. Redox Signal.* **13**, 1273–1290.

Pollet, A., Vandermarliere, E., Lammertyn, J., Strelkov, S. V., Delcour, J. A., Courtin, C. M. (2009). Crystallographic and activity-based evidence for thumb flexibility and its relevance in glycoside hydrolase family 11 xylanases. *Proteins* **77**, 395–403.

Radestock, S., Gohlke, H. (2011). Protein rigidity and thermophilic adaptation. *Proteins* **79**, 1089–1108.

Rasmussen, B. F., Stock, A. M., Ringe, D., Petsko, G. A. (1992). Crystalline ribonuclease A loses function below the dynamical transition at 220 K. *Nature* **357**, 423–424.

Riccardi, L., Papaleo, E. (2010). Unfolding simulations of cold- and warm-adapted elastases. *IIOABJ* **1**, 11–17.

Rudino-Pinera, E., Morales-Arrieta, S., Rojas-Trejo, S. P., Horjales, E. (2002). Structural flexibility, an essential component of the allosteric activation in *Escherichia coli* glucosamine-6-phosphate deaminase. *Acta Crystallogr. D Biol. Crystallogr* **58**, 10–20.

Santoro, J., Gonzalez, C., Bruix, M., Neira, J. L., Nieto, J. L., Rico, M. (1993). High resolution three dimensional structure of ribonuclease A in solution by nuclear magnetic resonance spectroscopy. *J. Mol. Biol.* **229**, 722–734.

Savino, C., Montemiglio, L. C., Sciara, G., Miele, A. E., Kendrew, S. G., Jemth, P., Gianni, S., Vallone, B. (2009). Investigating the structural plasticity of a cytochrome P450. *J. Biol. Chem.* **284**, 29170–29179.

Schlessinger, A., Schaefer, C., Vicedo, E., Schmidberger, M., Punta, M., Rost, B. (2011). Protein disorder—a breakthrough invention of evolution? *Curr. Opin. Struct. Biol.* **21**, 412–418.

Schottea, F., Somanb, J., Olsonb, J. S., Wulff, M., Anfinruda, P. A. (2004). Picosecond time-resolved X-ray crystallography: probing protein function in real time. *J. Struct. Biol.* **147**, 235–246.

Sennett, N. C., Kadirvelraj, R., Wood, Z. A. (2011). Conformational flexibility in the allosteric regulation of human UDP-α-D-glucose 6-dehydrogenase. *Biochemistry* **50**, 9651–9663.

Shah, M. B., Pascual, J., Zhang, Q., Stout, C. D., Halpert, J. R. (2011). Structures of cytochrome P450 2B6 bound to 4-benzylpyridine and 4-(4-nitrobenzyl) pyridine: insight into inhibitor binding and rearrangement of active site side chains. *Mol. Pharmacol.* **80**, 1047–1055.

Siddiqui, K. S., Cavicchioli, R. (2006). Cold-adapted enzymes. *Annu. Rev. Biochem.* **75**, 403–433.

Srajer, V., Royer, W. E., Jr. (2008). Time-resolved X-ray crystallography of heme proteins. *Methods Enzymol.* **437**, 379–395.

Stillman, T. J., Baker, P. J., Britton, K. L., Rice, D. W. (1993). Conformational flexibility in glutamate dehydrogenase. Role of water in substrate recognition and catalysis. *J. Mol. Biol.* **234**, 1131–1139.

Tanaka, T., Yamaguchi, H., Kato, H., Nishioka, T., Katsube, Y., Oda, J. (1993). Flexibility impaired by mutations revealed the multifunctional roles of the loop in glutathione synthetase. *Biochemistry* **32**, 12398–12404.

Teilum, K., Olsen, J. G., Kragelund, B. B. (2011). Protein stability, flexibility and function. *Biochim. Biophys. Acta* **1814**, 969–976.

Thompson, A. A., Jiao, G.-S., Kim, S., Thai, A., Cregar-Hernandez, L., Margosiak, S. A., Johnson, A. T., Han, G. W., O'Malley, S., Stevens, R. C. (2011). Structural characterization of three novel hydroxamate-based zinc chelating inhibitors of the clostridium botulinum serotype A neurotoxin light chain metalloprotease reveals a compact binding site resulting from 60/70 loop flexibility. *Biochemistry* **50**, 4019–4028.

Tilton, R. F. J., Dewan, J. C., Petsko, G. A. (1992). Effects of temperature on protein structure and dynamics: X-ray crystallographic studies of the protein ribonuclease-A at nine different temperatures from 98 to 320 K. *Biochemistry* **31**, 2469–2481.

Todd, A. E., Orengo, C. A., Thornton, J. M. (2002). Plasticity of enzyme active sites. *Trends Biochem. Sci.* **27**, 419–426.

Toiron, C., Gonzalez, C., Bruix, M., Rico, M. (1996). Three-dimensional structure of the complexes of ribonuclease A with 2′,5′-CpA and 3′,5′-d(CpA) in aqueous solution, as obtained by NMR and restrained molecular dynamics. *Protein Sci.* **5**, 1633–1647.

Tsigos, I., Mavromatis, K., Tzanodaskalaki, M., Pozidis, C., Kokkinidis, M., Bouriotis, V. (2001). Engineering the properties of a cold active enzyme through rational redesign of the active site. *Eur. J. Biochem.* **268**, 5074–5080.

Uversky, V. N., Dunker, A. K. (2010). Understanding protein non-folding. *Biochim. Biophys. Acta* **1804**, 1231–1264.

Venkatesan, R., Alahuhta, M., Pihko, P. M., Wierenga, R. K. (2011). High resolution crystal structures of triosephosphate isomerase complexed with its suicide inhibitors: the conformational flexibility of the catalytic glutamate in its closed, liganded active site. *Protein Sci.* **20**, 1387–1397.

Vitagliano, L., Merlino, A., Zagari, A., Mazzarella, L. (2002). Reversible substrate-induced domain motions in ribonuclease A. *Proteins* **46**, 97–104.

Watanabe, S., Takada, Y. (2004). Amino acid residues involved in cold adaptation of isocitrate lyase from a psychrophilic bacterium, *Colwellia maris*. *Microbiology* **150**, 3393–3403.

Westenhoff, S., Nazarenko, E., Malmerberg, E., Davidsson, J., Katona, G., Neutze, R. (2010). Time-resolved structural studies of protein reaction dynamics: a smorgasbord of X-ray approaches. *Acta Crystallogr. A* **A66**, 207–219.

Wlodawer, A., Svensson, L. A., Sjolin, L., Gilliland, G. L. (1988). Structure of phosphate-free ribonuclease A refined at 1.26 Å. *Biochemistry* **27**, 2705–2717.

Wolfenden, R., Snider, M. J. (2001). The depth of chemical time and the power of enzymes as catalysts. *Acc. Chem. Res.* **34**, 938–945.

Woofter, R. T., Maurer, M. C. (2011). Role of calcium in the conformational dynamics of factor XIII activation examined by hydrogen-deuterium exchange coupled with MALDI-TOF MS. *Arch. Biochem. Biophys.* **512**, 87–95.

Wrabl, J. O., Gu, J., Liu, T., Schrank, T. P., Whitten, S. T., Hilser, V. J. (2011). The role of protein conformational fluctuations in allostery, function, and evolution. *Biophys. Chem.* **159**, 129–141.

Xie, B.-B., Bian, F., Chen, X.-L., He, H.-L., Guo, J., Gao, X., Zeng, Y.-X., Chen, B., Zhou, B.-C., Zhang, Y.-Z. (2009). Cold adaptation of Zinc metalloproteases in the thermolysin family from deep sea and arctic sea ice bacteria revealed by catalytic and structural properties and molecular dynamics. *J. Biol. Chem.* **284**, 9257–9269.

Yon, J. M., Perahia, D., Ghelis, C. (1998). Conformational dynamics and enzyme activity. *Biochimie* **80**, 33–42.

Zhao, Y., Sun, L., Muralidhara, B. K., Kumar, S., White, M. A., Stout, C. D., Halpert, J. R. (2007). Structural and thermodynamic consequences of 1-(4-chlorophenyl)imidazole binding to cytochrome P450 2B4. *Biochemistry* **46**, 11559–11567.

MOLECULAR MODELING OF INHIBITORS OF HUMAN DNA METHYLTRANSFERASE WITH A CRYSTAL STRUCTURE: DISCOVERY OF A NOVEL DNMT1 INHIBITOR

By JAKYUNG YOO,* JOO HEE KIM,† KEITH D. ROBERTSON,† AND JOSÉ L. MEDINA-FRANCO*

*Torrey Pines Institute for Molecular Studies, Port St. Lucie, Florida, USA
†Department of Biochemistry and Molecular Biology, Cancer Research Center, CN-2151, Georgia Health Sciences University, Augusta, Georgia, USA

Abstract

DNA methyltransferases (DNMTs) are promising epigenetic targets for the development of novel anticancer drugs and other diseases. Molecular modeling and experimental approaches are being used to identify and develop inhibitors of human DNMTs. Most of the computational efforts conducted so far with DNMT1 employ homology models of the enzyme. Recently, a crystallographic structure of the methyltransferase domain of human DNMT1 bound to unmethylated DNA was published. Following on our previous computational and experimental studies with DNMTs, we herein present molecular dynamics of the crystal structure of human DNMT1. Docking studies of established DNMT1 inhibitors with the crystal structure gave rise to a structure-based pharmacophore model that suggests key interactions of the inhibitors with the catalytic binding site. Results had a good agreement with the docking and pharmacophore models previously developed using a homology model of the catalytic domain of DNMT1. The docking protocol was able to distinguish active DNMT1

DOI: 10.1016/B978-0-12-398312-1.00008-1

Copyright 2012, Elsevier Inc.
All rights reserved.

inhibitors from, for example, experimentally known inactive DNMT1 inhibitors. As part of our efforts to identify novel inhibitors of DNMT1, we conducted the experimental characterization of aurintricarboxylic acid (ATA) that in preliminary docking studies showed promising activity. ATA had a submicromolar inhibition ($IC_{50}=0.68$ μM) against DNMT1. ATA was also evaluated for Dnmt3a inhibition showing an $IC_{50}=1.4$ μM. This chapter illustrates the synergy from integrating molecular modeling and experimental methods to further advance the discovery of novel candidates for epigenetic therapies.

I. Introduction

The genome contains genetic and epigenetic information. While the genome provides the blueprint for the manufacture of all the proteins required to create a living thing, the epigenetic information provides instruction on how, where, and when the genetic information should be used (Robertson, 2001). The major form of epigenetic information within the DNA molecule itself in mammalian cells is DNA methylation, that is, the covalent addition of a methyl group to the 5-position of cytosine, mostly within the CpG dinucleotides in somatic cells (Robertson, 2001). DNA methylation is involved in the control of gene expression, regulation of parental imprinting, and stabilization of X chromosome inactivation as well as maintenance of the genome integrity. It is also implicated in the development of the immune system, cellular reprogramming, and brain function and behavior (Jurkowska et al., 2011). DNA methylation is mediated by a family of DNA methyltransferase enzymes (DNMTs). In mammals, three DNMTs have been identified so far in the human genome, including the two *de novo* methyltransferases (DNMT3A and DNMT3B) and the maintenance methyltransferase (DNMT1), which is generally the most abundant and active of the three (Robertson, 2001; Yokochi and Robertson, 2002; Goll and Bestor, 2005). DNMT1 is responsible for duplicating patterns of DNA methylation during replication and is essential for mammalian development and cancer cell growth (Chen et al., 2007). These enzymes are key regulators of gene transcription, and their roles in carcinogenesis have been the subject of considerable interest over the past decade (Robertson, 2001; Jones and Baylin, 2007). Therefore, specific inhibition of DNA methylation is an attractive and novel approach for cancer therapy (Robertson, 2001; Lyko and Brown, 2005;

Kelly et al., 2010; Portela and Esteller, 2010). DNA methylation inhibitors have also emerged as a promising strategy for the treatment of immunodeficiency and brain disorders (Zawia et al., 2009; Miller et al., 2010).

Human DNMT1 is a protein with 1616 amino acids whose structure can be divided into an N-terminal regulatory domain and a C-terminal catalytic domain (Cheng and Blumenthal, 2008; Lan et al., 2010; Jurkowska et al., 2011). The mechanism of DNA cytosine-C5 methylation is schematically depicted in Fig. 1 (Vilkaitis et al., 2001; Schermelleh et al., 2005; Sippl and Jung, 2009). DNMT forms a complex with DNA and the cytosine which will be methylated flips out from the DNA. The thiol of the catalytic cysteine acts as a nucleophile that attacks the 6-position of the target cytosine to generate a covalent intermediate. The 5-position of the cytosine is activated and conducts a nucleophilic attack on the cofactor *S*-adenosyl-L-methionine (AdoMet/SAM) to form the 5-methyl covalent adduct and

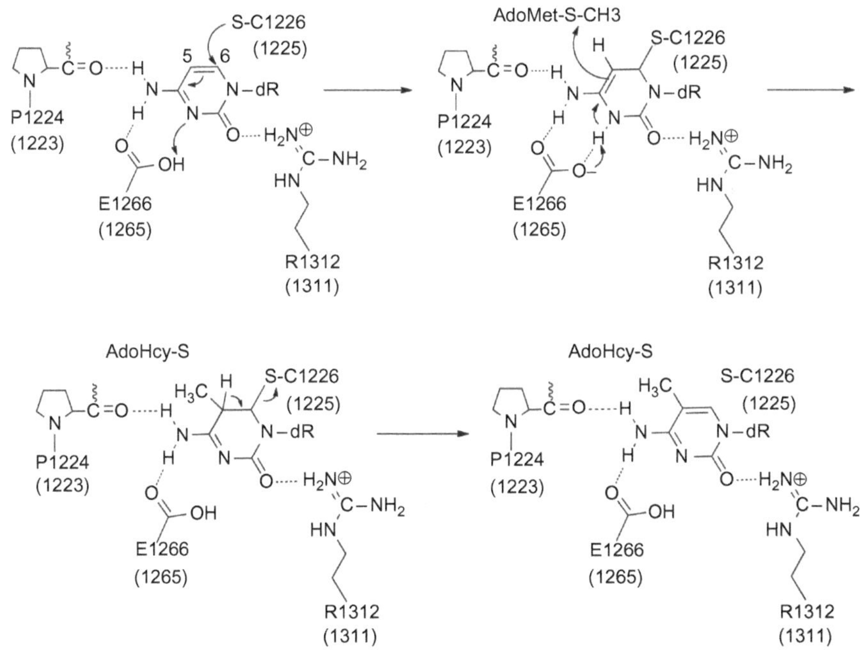

FIG. 1. Mechanism of cytosine DNA methylation. Amino acid residue numbers are based on the crystal structure of hDNMT1. Equivalent residue numbers in parentheses correspond to the homology model.

S-adenosyl-L-homocysteine (AdoHcy/SAH). The attack on the 6-position is assisted by a transient protonation of the cytosine ring at the endocyclic nitrogen atom N3, which is stabilized by a glutamate residue. An arginine residue may participate in the stabilization of the intermediate making a hydrogen bonding interaction with the carbonyl oxygen of cytosine. The covalent complex between the methylated base and the DNA is resolved by deprotonation at the 5-position to generate the methylated cytosine and the free enzyme.

The structure of mammalian DNMTs can be divided into two major parts: a large N-terminal regulatory domain of variable size which has regulatory functions, and a C-terminal catalytic domain which is conserved in eukaryotic and prokaryotic carbon-5 DNMTs (Fig. 2). The N-terminal domain guides the nuclear localization of the enzymes and mediates their interactions with other proteins, DNA, and chromatin. The smaller C-terminal domain harbors the active center of the enzyme and contains 10 amino acid motifs diagnostic for all carbon-5 DNMTs (Jurkowska et al., 2011). The catalytic domain has a core structure shared by all DNMTs. This core structure is termed the "AdoMet-dependent MTase fold" and consists of a six-stranded parallel β sheet with a seventh strand inserted in a parallel fashion between the fifth and sixth strands. Six helices are folded around the central β sheet. Motifs I and X of this domain are involved in the cofactor binding. Motifs IV, VI, and VIII are involved in catalysis. The target recognition domain (TRD) is a nonconserved region between

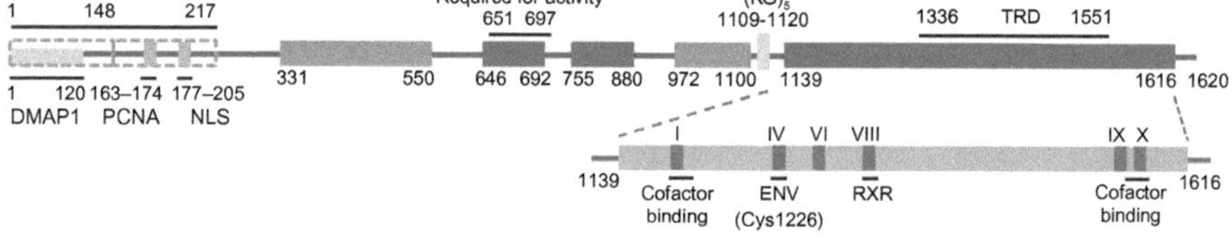

FIG. 2. Schematic representation of human DNMT1. NLS, nuclear localization signal; RFD, replication foci-targeting sequence; BAH, bromo-adjacent homology domain; TRD, target recognition domain. Interaction domains of HDAC1, HDAC2, and the DNMT3s are indicated. The methyltransferase domain comprising six most conserved motifs is enlarged. (For color version of this figure, the reader is referred to the online version of this chapter.)

motifs VIII and IX that is involved in DNA recognition and specificity (Fig. 2) (Jeltsch, 2002; Jurkowska et al., 2011).

Three-dimensional structural information for human DNMT1 is limited. Just recently, a crystal structure of human DNMT1 bound to duplex DNA containing unmethylated cytosine–guanine (CG) sites was published (Song et al., 2011). However, the conformation of the crystal structure is in an ''inactive state'' (the target base from DNA is not flipped out) so that the geometry of the catalytic site does not reflect the catalytic mechanism. Further details of the structure of DNMTs and other available crystal structures of DNMTs are extensively reviewed elsewhere (Cheng and Blumenthal, 2008; Sippl and Jung, 2009; Lan et al., 2010).

DNA methylation inhibitors have been well characterized and tested in clinical trials (Issa and Kantarjian, 2009). To date, only 5-azacytidine and 5-aza-2′-deoxycytidine (decitabine) have been developed clinically. These two drugs are nucleoside analogues, which, after incorporation into DNA, cause covalent trapping and subsequent depletion of DNMTs (Schermelleh et al., 2005; Stresemann and Lyko, 2008). Aza nucleosides are approved by the Food and Drug Administration of the United States for the treatment of myelodysplastic syndrome, where they demonstrate significant, although usually transient, improvement in patient survival and are currently being tested in many solid cancers (Issa et al., 2005; Schrump et al., 2006). However, aza nucleosides have relatively low specificity and are characterized by substantial cellular and clinical toxicity (Stresemann and Lyko, 2008). Their exact mechanism of antitumor action also remains unclear (Issa, 2005; Palii et al., 2008; Fandy et al., 2009). Consequently, there is clear need to identify novel and more specific DNMT inhibitors that do not function via incorporation into DNA.

There are now an increasing number of substances that are reported to inhibit DNMTs (Lyko and Brown, 2005). Compounds such as RG108 and NSC14778 (see below) were identified using virtual screening (Siedlecki et al., 2006; Kuck et al., 2010b). Others are approved drugs for other indications such as hydralazine, procaine, and procainamide. These compounds and several natural products implicated in DNA methylation inhibition are extensively reviewed elsewhere (Hauser and Jung, 2008; Gilbert and Liu, 2010; Li and Tollefsbol, 2010; Medina-Franco and Caulfield, 2011).

Molecular docking, pharmacophore modeling, and molecular dynamics (MD) are some of the computational approaches that our and other research groups are using to explore the ligand-binding interactions of DNMT inhibitors (Evans and Bronowska, 2010; Caulfield and Medina-Franco, 2011; Medina-Franco and Caulfield, 2011; Yoo and Medina-Franco, 2011b,c). Due to the absence of a crystallographic structure of human DNMT1 that was published until recently (Song et al., 2011), molecular modeling studies into the catalytic binding site of DNMTs had been conducted using homology models (Medina-Franco and Caulfield, 2011)

Herein, we describe docking studies of known DNMT1 inhibitors in the catalytic site of the recently published crystal structure of DNMT1 (Song et al., 2011). Prior docking, the conformation of the catalytic site was modeled into an active conformation using MD. To our knowledge, this is the first molecular modeling study conducted with the catalytic binding site of the crystal structure of human DNMT1. Based on the docking results, we developed a structure-based pharmacophore model of DNMT1 inhibitors. In an effort to identify novel nonnucleoside DNMT1 inhibitors, we also report the experimental activity of aurintricarboxylic acid (ATA) that showed submicromolar inhibition against DNMT1. Molecular modeling studies of ATA with the catalytic site of the crystal structure of DNMT1 are also discussed.

II. Molecular Dynamics of the Crystallographic Structure

The recently published X-ray crystal structure of human DNMT1 corresponds to the enzyme bound to DNA-containing unmethylated CpG sites. In this structure, the catalytic loop has an open conformation such that the catalytic cysteine is far away from the binding site. Therefore, the geometry of the catalytic site of the crystal structure does not represent the active conformation of the enzyme that corresponds to the catalytic mechanism of DNA methylation. In contrast, in the X-ray structure of bacterial M.*Hha*I (Klimasauskas et al., 1994; Kumar et al., 1997) and in our previously developed homology model (Yoo and Medina-Franco, 2011b), the DNA is embedded in the catalytic core and the catalytic loops are also close to the inserted target cytosine in the active site (Fig. 3). We therefore used MD to model the crystal structure of the human DNMT1–DNA into an active state. To this end, the crystal structure of the DNMT1–DNA complex with bound AdoHcy was obtained from the Protein Data Bank (PDB code: 3PTA). The unmethylated DNA and other domains, except the methyltransferase domain, were removed. The DNA double helix reconstructed from the structure of M.*Hha*I (PDB: 1MHT) (Klimasauskas et al.,

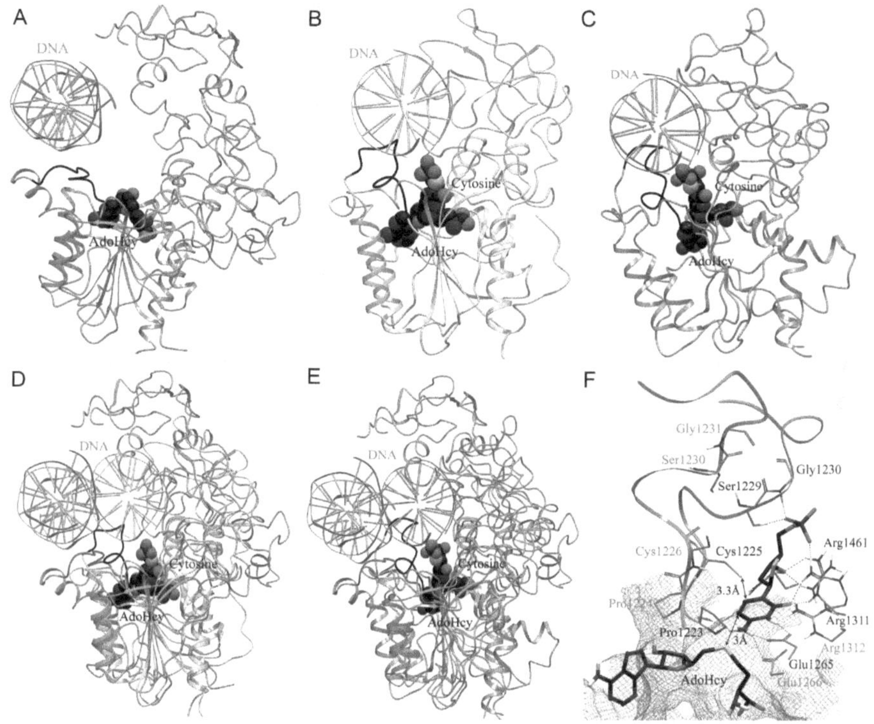

Fig. 3. (A) Methyltransferase domain of human DNMT1 (PDB: 3PTA) (pink). (B) Published structure of the M.*Hha*I–DNA complex (PDB: 1MHT) (yellow). (C) Previously developed homology model of DNMT1 (green). In (A)–(C), the catalytic loops are in black. Superposition of the crystal structure of human DNMT1 (1135–1600)–DNA complex with (D) the M.*Hha*I–DNA complex and (E) the homology model. The catalytic loop of the crystal structure is in red. AdoHcy and the flipped cytosine are shown in space-filling view. (F) Binding model of deoxycytidine (carbon atoms in black) with key amino acid residues of the crystal structure (carbon atoms in pink) and the homology model (carbon atoms in green). Hydrogen bonding interactions are depicted with dashes. (For interpretation of the references to color in this figure legend, the reader is referred to the online version of this chapter.)

1994), in the catalytic core, and target cytosine inserted into the active site. The missing residues (1480–1483) were modeled using Prime (version 2.2, Schrödinger LLC, New York, NY, 2010). The refined model was treated according to the Protein Preparation Wizard implemented in Maestro, which optimizes H-bond networks and flip orientations/tautomeric states of Gln, Asn, and His residues, and performed a geometry optimization to a maximum root mean square deviation (RMSD) of 0.3 Å with OPLS2005 force field. We have recently employed a similar procedure to prepare the structure of other proteins (Hernández-Campos et al., 2010). The stability of the refined model was probed by an MD simulation using MacroModel (version 9.8, Schrödinger LLC, New York, NY, 2010). A stochastic dynamics (SD) method was used with the OPLS force field and implicit water solvation under the following conditions: (a) equilibration time 1.0 ps at 300 K, (b) 100 ps of simulation time at 300 K, (c) time steps = 1.5 fs, and (d) 100 conformations stored. To make a direct comparison of the energy of the conformers obtained by SD, the collection of SD conformations was submitted to a multiminimization step with the Polak-Ribiere Conjugate Gradient (PRCG). Convergence in the SD simulation was judged complete by the RMSD of the 100 stored structures and the interaction between key amino acid residues and target cytosine.

Figure 4 shows the conformation of the modified X-ray structure with the catalytic loop in an active conformation. Although the RMSD of the Cα atoms of the initial and modified X-ray structure is 1.92 Å, the catalytic loop of the modified structure adopts a different conformation with respect to the X-ray structure. In the X-ray structure, the distance of superimposed target cytosine C6 to the sulfur atom of Cys1226 in the catalytic loop is 9.5 Å (Fig. 4D). In contrast, the catalytic loop of the modified structure is located above the cytosine as an "active site lid." The target cytosine lies between the nucleophile Cys1226 and sulfur atom of AdoHcy. The distance of cytosine C6 to the sulfur atom of Cys1226 is 3.4 Å, and cytosine C5 atom is 3.5 Å away from the sulfur atom of AdoHcy. In addition, the N3 protonated form of cytosine can make hydrogen bond with the acidic side chain of Glu1266, Pro1224, Arg1310, and Arg1312 (Fig. 4E). The α-phosphate backbone of deoxycytidine makes a hydrogen bond network with Arg1312, Ser1230, Gly1231, and Lys1535. In addition, the 3′-OH of the sugar moiety forms hydrogen bonds with Thr1528 and Gly1577. Interestingly, the key interactions with Glu1266 in the ENV motif, Arg1312 in the RXR motif, Ser1230 and Gly1231 in the catalytic loop are observed in the homology model (Yoo and Medina-Franco, 2011b). The catalytic cores of the modified crystal structure and homology model of DNMT1 (Yoo and Medina-Franco, 2011b) have similar conformation of the catalytic loop. These insights suggest that the modified structure and homology model are reliable to explore the binding mode of inhibitors of human DNMT1.

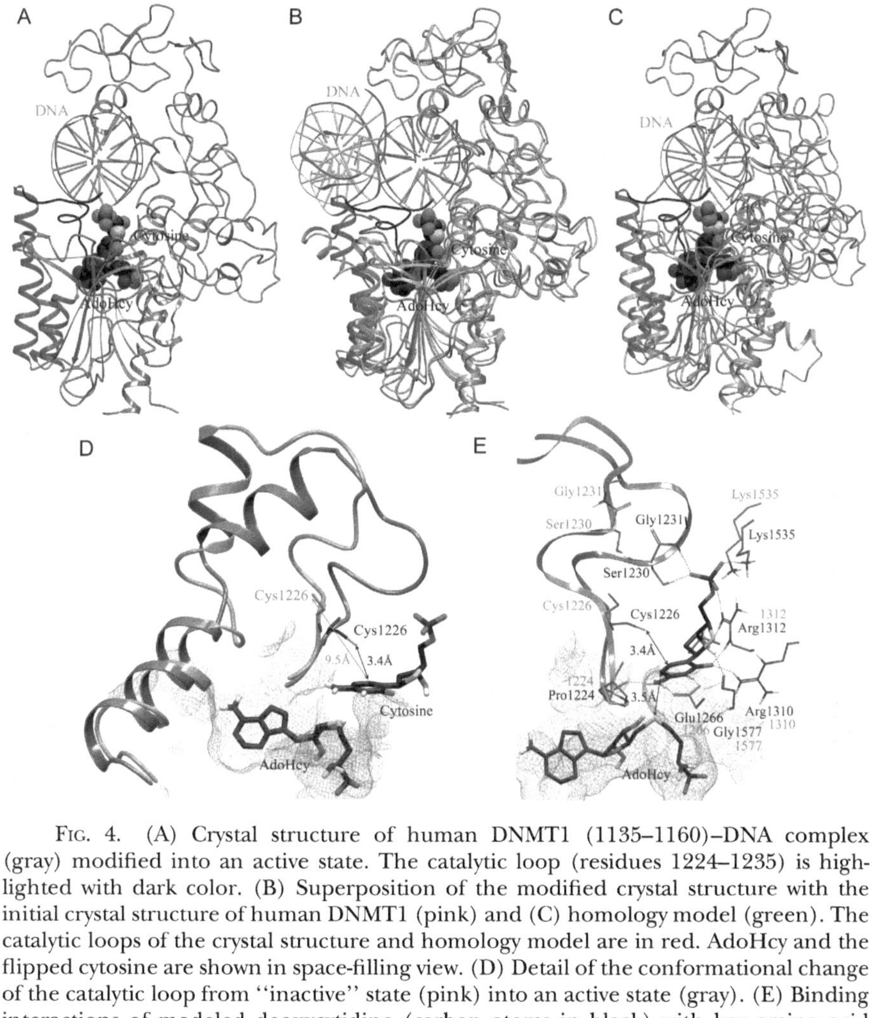

FIG. 4. (A) Crystal structure of human DNMT1 (1135–1160)–DNA complex (gray) modified into an active state. The catalytic loop (residues 1224–1235) is highlighted with dark color. (B) Superposition of the modified crystal structure with the initial crystal structure of human DNMT1 (pink) and (C) homology model (green). The catalytic loops of the crystal structure and homology model are in red. AdoHcy and the flipped cytosine are shown in space-filling view. (D) Detail of the conformational change of the catalytic loop from "inactive" state (pink) into an active state (gray). (E) Binding interactions of modeled deoxycytidine (carbon atoms in black) with key amino acid residues in the modified crystal structure (carbon atoms in gray) and crystal structure (carbon atoms in pink). Hydrogen bonding interactions are depicted with dashes. (For interpretation of the references to color in this figure legend, the reader is referred to the online version of this chapter.)

III. Docking

The docking of known active DNMT inhibitors into a homology model of DNMT1 has been published (Yoo and Medina-Franco, 2011b). In that study, the binding mode of 14 inhibitors (except ATA and NSC97317) was compared for the first time using the same molecular modeling protocol. Despite the fact that the inhibitors have different structural scaffolds, all compounds in that study showed common interactions with key amino acid residues such as Glu1265, Arg1311, Arg1461, Ser1229, and Gly1230. In addition, the insights of the molecular docking supported the proposed mechanism of inhibition of compounds such as curcumin and parthenolide that are putative covalent blockers of the catalytic site (Yoo and Medina-Franco, 2011b).

Building on our previous work, we performed the docking of 16 established inhibitors into the catalytic site of the X-ray structure of hDNMT1 modified to an active conformation (see previous section). For reference (as negative control), we also docked 19 compounds that previously have shown very weak or no enzymatic inhibitory activity or that were used as decoys (NSC4092, NSC21970, NSC57278, NSC19555, NSC27292, and NSC303530; Siedlecki et al., 2006; Kuck et al., 2010b). The chemical structures of both groups of compounds are shown in Fig. 5. It is worth noting that all 35 compounds in Fig. 5 have not been screened under the same experimental conditions; therefore, a quantitative comparison between the docking scores and enzymatic inhibition is not feasible. However, it is possible to assess whether molecular docking is able to separate actives from inactives/decoys. Also, and more importantly in this case, we seek to explore the putative binding site of established inhibitors into the catalytic binding site using the crystallographic structure of DNMT1.

To conduct the docking, the chemical structures of the 35 compounds were built in Maestro 9.1 (Schrödinger LLC, New York, NY, 2010) and minimized with the OPLS2005 force field and gradient termination at 0.001 kJ/mol-Å (MacroModel, 2010). Ligands were prepared with conformational search in MacroModel and LigPrep (version 2.4, Schrödinger LLC, New York, NY, 2010) to generate all the possible configurations (LigPrep, 2010). The conformational analysis was carried out with Monte Carlo Multiple Minimum and Low-Mode conformational search method, employing the OPLS force field using GB/SA water solvation model. The lowest energy conformation of each ligand was docked with

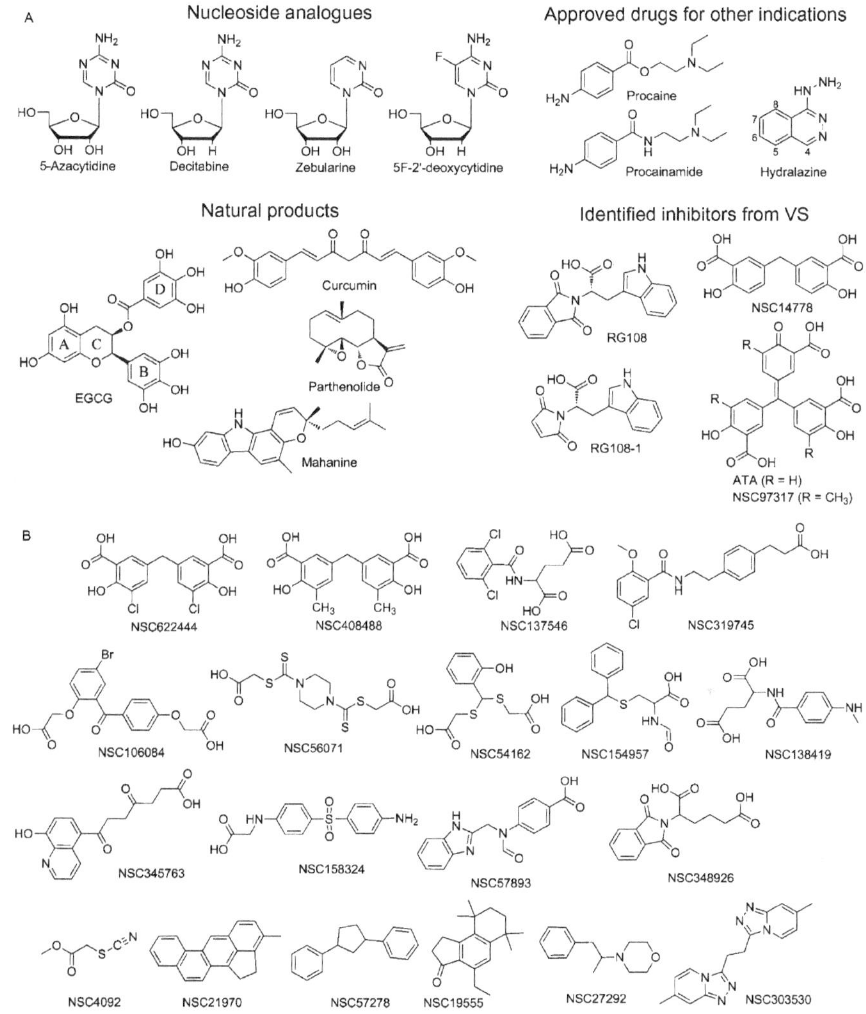

FIG. 5. Compounds considered in this study that have been previously tested for inhibition of DNMT1: (A) active and (B) inactive/decoys.

the X-ray structure modified to an active conformation (see above) using the program Glide (version 5.6, Schrödinger LLC, New York, NY, 2010). Our group has successfully used Glide to develop binding models of inhibitors of DNMTs (Kuck et al., 2010a,b; Caulfield and Medina-Franco,

2011; Medina-Franco et al., 2011; Yoo and Medina-Franco, 2011b,c). Three flexible docking methods, including high-throughput virtual screening (HTVS), standard precision (SP), and extra precision (XP), implemented in Glide were used for comparison. For each method, a receptor grid box of 23 Å on each side, with a default inner box ($10 \times 10 \times 10$ Å^3), was centered on the target cytosine. Input partial charges were used and no constrains were applied during docking. In the case of the HTVS and SP docking methods, one pose per ligand was intended for large database screening applications, and 10 poses per ligand for XP docking method was set to optimize more precise poses of the top-scoring ligands by HTVS and SP docking. The postdocking minimization was applied to investigate their effect on the accuracy in the three docking protocols: five docking poses per ligand were minimized for HTVS and SP, and 10 poses per ligand for XP. The best docking poses of XP docking were selected to generate the pharmacophore features (see below).

Table I summarizes the results of docking into the catalytic site of the crystal structure using Glide HTVS, SP, and XP. For comparison, the docking scores obtained using the previously reported homology model of DNMT1 (Yoo and Medina-Franco, 2011b) are shown. The docking results were able to predict energetically favorable binding modes with accurate scoring of docking conformations. Docking results with Glide HTVS and SP with the modified crystal structure and homology model showed significant impact on ranking score with remarkable consistency as shown in graph of Table I. In the docking results of HTVS and SP, the ranking of active and that of inactive have similar variations. Almost all active compounds have more stable energy values than inactives. Of note, all known inhibitors except procaine and procainamide ranked higher than decoys. Nucleoside inhibitors including 5-azacytidine and 5-aza-2′-deoxycytidine (decitabine) showed the best Glide XP docking scores using the modified crystallographic and homology models. It is clear from these studies that optimal docking and scoring combination will decrease the number of false positive and false negatives in virtual screening.

Figure 6 shows the corresponding binding mode of selected DNMT inhibitors generated with Glide XP into the catalytic binding site of the modified crystal structure. The binding modes of nucleoside inhibitors closely match with the target cytidine. The α-phosphate of 5-azacytidine and zebularine makes a hydrogen bond network with Ser1230 and Gly1231 in the catalytic loop, with Lys1535 in the TRD, and Arg1312.

TABLE I
Summary of Docking Results

	Homology model			Crystal structure		
Compound	HTVS	SP	XP	HTVS	SP	XP
Azacytidine	−6.72	−6.53	−9.98	−7.58	−7.71	−11.53
Decitabine	−5.00	−6.03	−8.72	−7.56	−7.82	−11.17
EGCG	−4.67	−4.87	−9.37	−4.83	−5.92	−11.16
Zebularine	−5.77	−6.62	−10.12	−6.75	−7.17	−11.14
5F-2′-dC	−6.61	−6.52	−8.84	−7.21	−7.65	−10.97
Cytosine	−5.13	−6.74	−10.03	−7.59	−7.82	−10.63
ATA	−3.85	−4.38	−9.52	−4.92	−6.52	−10.19
NSC97317	−3.98	−4.21	−8.42	−5.21	−6.39	−8.89
Curcumin	−3.99	−3.86	−9.00	−4.68	−5.41	−8.56
NSC14778	−4.25	−3.89	−7.45	−6.62	−7.31	−7.96
RG108-1	−4.38	−5.47	−7.01	−5.04	−6.20	−6.90
RG108	−4.08	−5.57	−6.67	−4.50	−5.30	−6.70
Hydralazine	−4.54	−4.65	−5.05	−4.63	−5.04	−5.51
NSC408488	−3.16	−3.61	−6.22	−3.27	−3.57	−5.49
NSC622444	−3.44	−3.10	−7.13	−2.99	−3.37	−5.25
NSC54162	−3.02	−2.01	−5.90	−2.72	−1.60	−5.25
NSC138419	−3.53	−4.09	−5.56	−1.80	−2.92	−5.19
Mahanine	−3.94	−4.58	−5.46	−3.93	−3.97	−5.08
Procaine	−3.25	−2.92	−4.83	−3.40	−3.26	−5.08
NSC57893	−4.38	−3.91	−4.74	−3.83	−3.76	−5.05
NSC137546	−1.98	−1.86	−4.34	−2.18	−3.25	−4.88
Parthenolide	−3.66	−3.67	−3.52	−3.99	−4.23	−4.77
Procainamide	−2.84	−3.05	−4.72	−3.04	−3.79	−4.70
NSC56071	−3.95	−3.37	−4.98	−3.10	−3.20	−4.63
NSC319745	−2.96	−1.62	−3.85	−1.52	−2.50	−4.57
NSC106084	−3.37	−2.55	−1.90	−2.29	−2.92	−4.45
NSC348926	−4.34	−3.40	−4.65	−2.57	−3.53	−4.42
NSC19555	−2.66	−2.96	−3.46	−2.96	−3.60	−4.38
NSC158324	−4.27	−3.46	−3.46	−2.46	−2.90	−4.32
NSC303530	−2.41	−2.86	−3.26	−2.85	−2.90	−4.15
NSC21970	−3.10	−3.10	−3.71	−3.31	−3.17	−3.78
NSC154957	−2.80	−2.69	−1.81	−2.18	−2.72	−3.69
NSC27292	−1.90	−1.89	−3.17	−1.91	−1.97	−3.59
NSC345763	−1.93	−2.34	−3.72	−2.37	−2.64	−3.49
NSC4092	−1.51	−1.46	−2.77	−1.49	−1.56	−2.91

(*Continued*)

Table I
Summary of Docking Results (*Continued*)

Compound	Homology model			Crystal structure		
	HTVS	SP	XP	HTVS	SP	XP

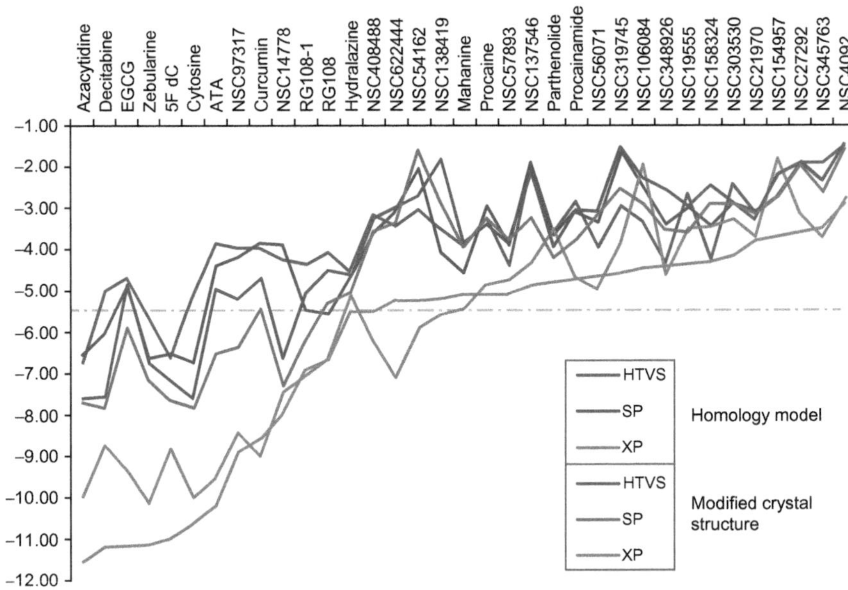

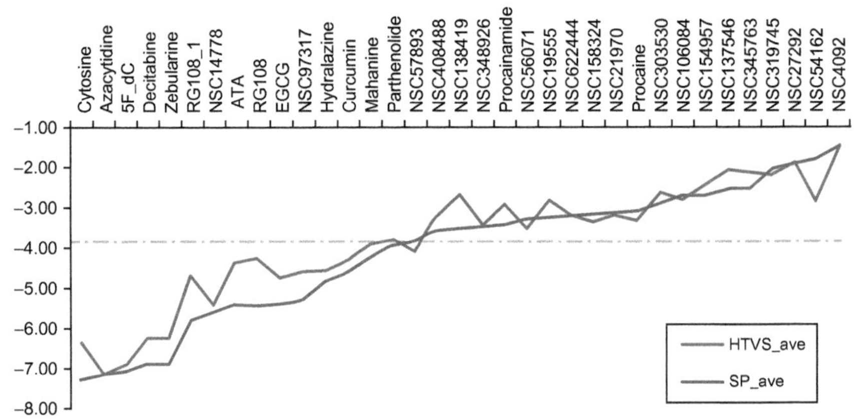

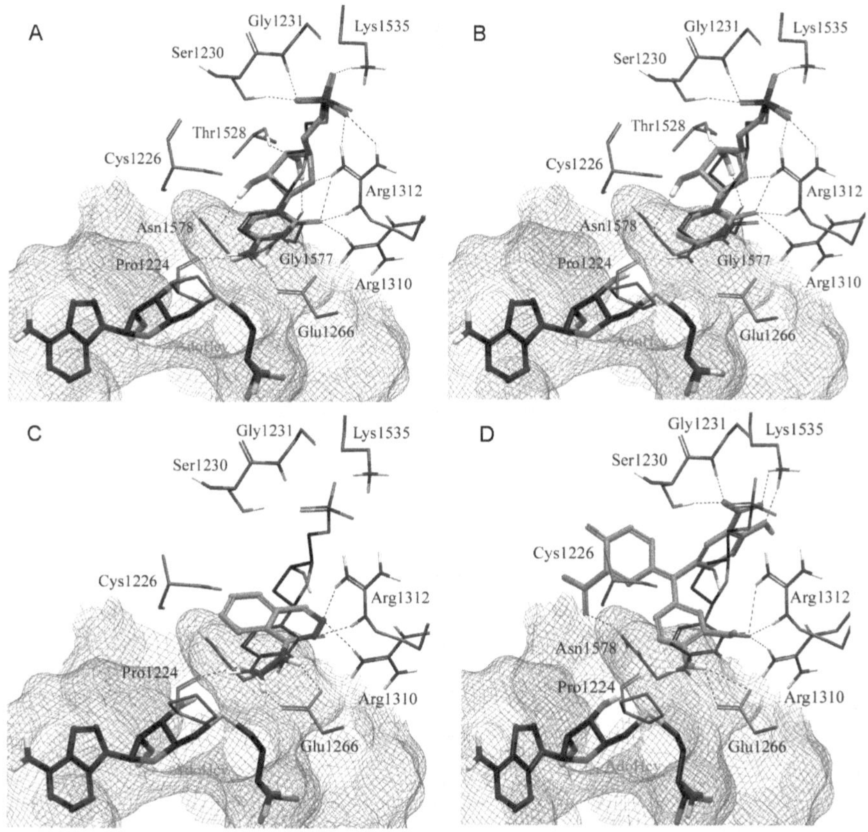

FIG. 6. Binding mode of representative inhibitors of DNMT1 (carbon atoms in green) into the catalytic site of the modified X-ray structure of DNMT1. The binding mode of deoxycytidine (carbon atoms in dark gray) is shown for reference: (A) 5-azacytidine, (B) zebularine, (C) hydralazine, (D) ATA. (For interpretation of the references to color in this figure legend, the reader is referred to the online version of this chapter.)

The ribose groups form additional hydrogen bonds with Thr1528, Gly1577, and Asn1578. The target base of most nucleoside inhibitors makes key interactions with the conserved glutamate residue in the ENV motif (motif IV: E/Glu1266, N/Asn1267, V/Val1268) and arginine residues in the RXR motif (motif VIII: R/Arg1310, X/1311, R/Arg1312), and Pro1224. Although zebularine does not form hydrogen bonds with

Glu1266 and Pro1224 (due to the fact that zebularine lacks the amino group at the C4 position), the distance of 3.7 Å between Cys1226 and C6 is sufficient to start the catalytic mechanism. In addition, the N3 of zebularine and $O^{\varepsilon 1}$ of Glu1266 are in an available hydrogen bonding distance for N3 protonation (Fig. 6). These observations are in agreement with the docking poses in the catalytic site of homology model of hDNMT1 (Yoo and Medina-Franco, 2011b). In the homology model, nucleoside inhibitors make hydrogen bond networks with common key amino acid residues, Pro1223, Glu1265 Arg1311, Ser1229, and Gly1230. The RMSD values of the top-scored poses of nucleoside inhibitors between modified crystal structure and homology are less than 1.0 Å. These comparisons show that similar binding modes were obtained in the crystal structure modeled to an active state and the previously generated homology model of hDNMT1. Therefore, these results suggest that both models of DNMT1 are useful for scoring and pose selection for virtual screening. In the next few paragraphs, we describe in more detail the binding mode of representative inhibitors of DNMT1 into the catalytic site of the crystal structure of human DNMT1 modeled into an active conformation.

Hydralazine is an antihypertensive drug and has antitumor effect when used in combination with valproic acid (a histone deacetylase inhibitor) (Arce et al., 2006; Dueñas-González et al., 2008). Figure 6C shows the comparison of the binding modes of hydralazine with target cytidine. The amino group of hydralazine and cytidine matched well, forming a hydrogen bond with Glu1266 and Pro1224. The nitrogen of the phthalazine ring overlapped with the carbonyl oxygen and formed hydrogen bonds with Arg1310 and Arg1312 in the RXR motif. Similar binding mode of hydralazine with low RMSD (0.52 Å) was obtained using a homology model of DNMT1 with same docking protocol (Yoo and Medina-Franco, 2011b). According to the binding pose, the small structure of hydralazine could not occupy the site of the sugar ring and phosphate backbone of nucleoside analogues. We therefore can design analogues which are substituted at the C4 position of hydralazine with enhanced enzymatic affinity (Yoo and Medina-Franco, 2011b).

Natural products, such as epigallocatechin gallate (EGCG), curcumin, parthenolide, and mahanine; inhibitors identified from virtual screening, such as RG108 and its analogue RG108-1; and approved drugs for other indications, such as procaine and procainamide, were docked in the catalytic binding site of modified crystal structure and compared with the binding mode in the homology model. Table II summarizes the common amino acid residues participating in hydrogen bond interactions.

The sesquiterpene lactone parthenolide, which is identified in the anti-inflammatory medicinal herb feverfew, was suggested to be a covalent blocker. Previous docking studies with the homology model supported this hypothesis (Yoo and Medina-Franco, 2011b). The binding pose in the modified crystal structure overlaps with previous docking model at low RMSD (0.17 Å) and makes hydrogen bond with common arginine residue. The docking poses of procaine, procainamide, and mahanine in the two catalytic binding sites show similar interactions with key amino acid residues within 1.5 Å RMSD. Although the binding poses of EGCG, curcumin, RG108, and RG108-1 were different, they all form hydrogen bonds with common amino acid residues which are important to the catalytic mechanism.

IV. Pharmacophore Modeling

For 16 established DNMT inhibitors in Fig. 5A, we developed an energy-optimized pharmacophore (e-pharmacophore) hypothesis using the best-scoring pose for each compound and descriptors calculated with Glide XP. We employed the structure-based pharmacophore approach that we previously used to model DNMT inhibitors using a homology model of the catalytic site (Yoo and Medina-Franco, 2011b). This approach is based on mapping the energetic terms from the Glide XP scoring function onto atom centers and has the advantage of combining pharmacophore perception with protein–ligand energetic terms to rank the importance of the pharmacophore features (Salam et al., 2009). The docking models of the inhibitors were refined using Glide XP, the Glide XP scoring terms were computed, and the energies were mapped into atoms. Then, the pharmacophoric sites were automatically generated with Phase (version 3.2, Schrödinger, LLC, New York, NY, 2010) using the default set of six chemical features, namely, hydrogen bond acceptor (A), hydrogen bond donor (D), hydrophobic (H), negative ionizable (N), positive ionizable (P), and aromatic ring (R) site. The Glide XP energies from the atoms that comprise each pharmacophoric site were summed. The pharmacophoric sites were then ranked based on the summed energies, and the most favorable sites were selected for the pharmacophore hypothesis (Salam

TABLE II

Summary of the Interactions of the Docked Compounds in the Homology Model and the Modified Crystal Structure of Human DNMT1

	Homology model	3PTA		Common amino acid residues participating in H-bonding				
Compound	XP	XP	RMSD (Å)	Arg1312 (1311)	Glu1266 (1265)	Pro1224 (1223)	Ser1230 (1229)	Gly1231 (1230)
Azacytidine	−9.98	−11.34	0.76–1.64	HE, NH1, NH2	OE2	CO	OH	NH
5F-dC	−8.84	−11.22	0.52–0.88	HE, NH1, NH2	OE2	CO	OH	NH
EGCG	−9.37	−11.16	3.75–5.62	HE, NH1	OE1			
Decitabine	−8.72	−11.00	0.40–0.63	HE, NH1, NH2	OE2	CO	OH	NH
Zebularine	−10.12	−10.97	0.81–1.47	HE, NH1, NH2	OE2		OH	NH
Cytosine	−10.03	−10.73	0.35–0.72	HE, NH1, NH2	OE1, OE2	CO	OH	NH
ATA	−9.52	−10.19	0.57–4.13	NH1	OE1		OH	
NSC97317	−8.42	−8.89	1.48–5.02	HE, NH1	OE1		OH	
Curcumin	−9.00	−8.56	2.90–3.57	HE, NH1	OE1			
NSC14778	−7.45	−7.96	1.47–2.44	NH1			OH	NH
RG108-1	−7.01	−6.90	2.82–4.37	NH1, NH2			OH	NH
RG108	−6.67	−6.70	2.96–4.55	NH1, NH2			OH	NH
Hydralazine	−5.05	−5.51	0.52–0.92	HE, NH1	OE1, OE2	CO		
Mahanine	−5.46	−5.08	1.24–1.79	HE	OE1			
Procaine	−4.83	−5.08	1.15–1.89		OE2	CO		
Procainamide	−4.72	−4.70	1.30–1.78		OE2	CO		
Parthenolide	−3.14	−3.75	0.17–0.22	HE, NH1				

et al., 2009). These features were used to develop a common pharmacophore model that was evaluated on its ability to reproduce known inhibitors using Phase. The distance matching tolerance was set to 2.0 Å. To account for protein flexibility and lessen the effects of minor steric clashes, excluded volume spheres were created for all receptor atoms within 5 Å around each ligand. Each sphere has a radius corresponding to 50% of the van der Waals radius of the receptor atom. Receptor atoms less than 1.5 Å from the ligand were ignored. The combination of best-scoring features that matched a minimum of two sites in each known inhibitor was chosen and regenerated (Salam et al., 2009).

Figure 7A shows the five-feature pharmacophore model for the 16 DNMT1 inhibitors. The energetic value and amino acid residues participating in the interaction are shown in the figure. The best-scoring feature

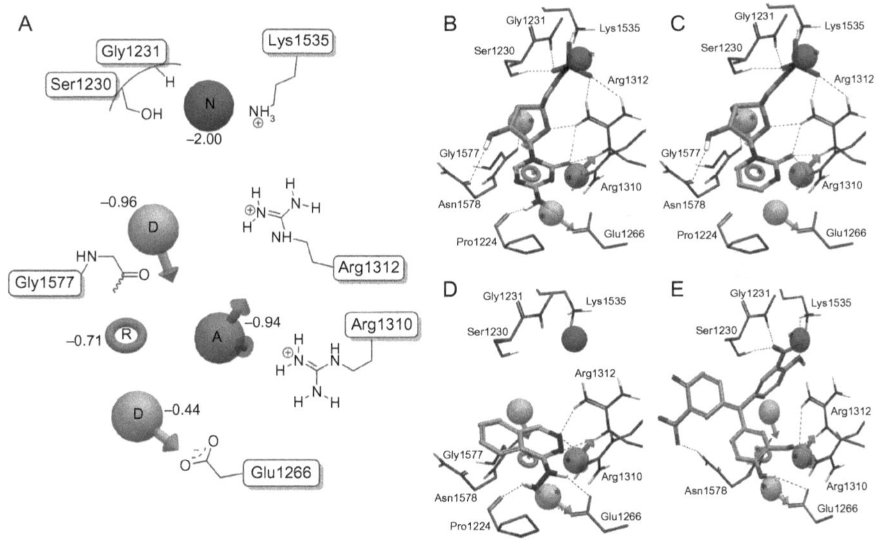

FIG. 7. (A) Structure-based pharmacophore model using binding modes of known inhibitors in the catalytic binding site of the modified crystal structure. *Red sphere* negative ionizable (N), *pink sphere* hydrogen bond acceptor (A), *blue sphere* hydrogen bond donors (D), and *orange ring* aromatic ring (R). Selected amino acid residues in the catalytic site are schematically depicted for reference. Comparison between the binding mode and pharmacophore hypothesis for representative DNMT inhibitors: (B) 5-azacytidine, (C) zebularine, (D) hydralazine, (E) ATA. (For interpretation of the references to color in this figure legend, the reader is referred to the online version of this chapter.)

is a negative charge (N) which is close to the side chains of Ser1230, Gly1231, Lys1535, and Arg1312. The second and third favorable features are a donor site (D) that is near the side chain of Gly1577 and an acceptor site (A) that is in close proximity to the side chain of Arg1310 and Arg1312 in the RXR motif. The fourth ranked feature is an aromatic ring (R) that stabilizes the binding conformations of ligands between AdoHcy and Cys1226. The fifth ranked feature is a donor site (D) that is near the side chain of Glu1266, which is a residue implicated in the mechanism of methylation (Fig. 7A). These pharmacophore features represent the most important interactions of the 16 DNMT1 inhibitors with the catalytic site. Notably, glutamate in the ENV motif and the arginines in the RXR motif are highly conserved residues in catalytic domains of DNMTs (Jurkowska et al., 2011). Interestingly, the five-feature pharmacophore model derived with the modified crystal structure of DNMT1 (Fig. 7A) is consistent with the pharmacophore hypothesis previously reported with the homology model of DNMT1, although the ranking of the features and nearby amino acids are slightly different. Most of the inhibitors matched several pharmacophore features considering a distance tolerance of 2 Å. Representative examples are shown in Fig. 7B–E. All nucleoside analogues except zebularine matched with all five pharmacophore features (Fig. 7B and C). Zebularine matches four features, but it does not satisfy the hydrogen bonding donor feature interacting with Glu1266 because zebularine does not have the amino group on the base ring. However, the donor feature (D) can match the N3 protonated form of zebularine.

Despite the fact that hydralazine has a small structure, it matches the aromatic ring (R), donor (D), and acceptor (A) features that are close to Glu1266, Arg1310, and Arg1312 (Fig. 7D). This is in agreement with our previous findings of the modeling of hydralazine with homology models of DNMT1 (Singh et al., 2009; Yoo and Medina-Franco, 2011b). Based on these results, it is expected that hydralazine analogues that match the top-ranked negative charge feature (N) will have improved potency, as we have suggested previously (Yoo and Medina-Franco, 2011a).

Interestingly, RG108 and RG108-1, both with a carboxylate functional group, matched the negative feature (N). The indole ring of both structures satisfied the ring feature (R). EGCG also matched with four pharmacophoric features; the hydroxyl group of ring B matched with the ring feature (R), acceptor (A) and donor (D) features close to glutamate and arginine residues; the ketone of the gallate moiety matched with negative feature (N). Curcumin and mahanine matched with the aromatic ring (R) and donor feature (D) nearby Arg1310 and Arg1312. Although parthenolide matched only the acceptor feature (A), it is possible that the γ-methylene lactone alkylates the catalytic Cys1226, as previously suggested (Liu et al., 2009). These results are similar to the insights previously reported with a homology model of the catalytic domain of DNMT1.

In contrast, the 19 compounds, inactive or previously used as decoys (Fig. 5B), did not match with any of the pharmacophore features or satisfied only one feature. Even if most of these molecules have a carboxylate group like known inhibitors, they did not match the negative feature (N) and the donor feature (D) interacting with Glu1266. Taken together, results of the structure-based pharmacophore modeling with the modified crystal structure were in good agreement with the docking studies described above and previously reported docking and pharmacophore modeling with a homology model of DNMT1.

V. Experimental and Molecular Modeling Studies of a Novel Inhibitor of Human DNMT1

It is well recognized that integration of computational and experimental approaches boosts the design of novel enzyme and other molecular target inhibitors and further advances drug candidates (López-Vallejo et al., 2011). This has been demonstrated in the area of inhibitors of DNMT1 (Medina-Franco and Caulfield, 2011). For example, virtual screening of large compound databases followed by experimental characterization leads to the identification of novel DNMT1 inhibitors such as RG108 and 5,5-methylenedisalicylic acid (NSC14478; Fig. 5A) (Siedlecki et al., 2006; Kuck et al., 2010b).

There is now an increased interest in the biological activity of 5,5-methylenedisalicylic acid and related compounds. For example, 5,5-methylenedisalicylic acid also emerged as an experimentally validated hit of a virtual screening with the viral NS5 RNA methyltransferase, a promising drug target against flaviviruses which are the causative agents of severe diseases such as dengue or yellow fever (Podvinec et al., 2010). In addition, in a separate virtual screening, a structural analogue of 5,5-methylenedisalicylic acid, ATA (Fig. 5A) was identified as a potent inhibitor of the methyltransferase activities on flaviviral methyltransferases (Milani et al., 2009).

Also, a recent study shows that ATA inhibits two AdoMet-dependent RNA methyltransferases of the severe acute respiratory syndrome (SARS) coronavirus (Bouvet et al., 2010). A recent study showed that NSC97317, a structural analogue of ATA (Fig. 5A), is an inhibitor of DNMT1 (Yoo and Medina-Franco, 2011c). Preliminary docking studies of ATA with a homology model of DNMT1 strongly suggested that this compound would show enzymatic inhibitory activity. To experimentally test this hypothesis and, as part of an ongoing effort to identify novel inhibitors of DNMT1, we conducted the experimental characterization of ATA as an enzymatic inhibitor of DNMT1 using *in vitro* assays.

A. *Enzymatic Inhibition of ATA*

In order to experimentally test the hypothesis that ATA is an inhibitor of DNMT1, the enzymatic inhibitory activity of this compound was measured using a nonradioactive *in vitro* enzymatic assay. For this assay, full-length human DNMT1 was expressed by baculovirus infection of Sf9 insect cells for 48 h. DNMT1 in this system (FastBac-HT system, Invitrogen) is hexahistidine tagged and purified on nickel NTA resin as we have described previously (Yokochi and Robertson, 2002) and dialyzed to remove imidazole. DNMT1 was used with the colorimetric 96-well-plate-based DNMT activity kit from Epigentek (an ELISA-like assay using anti-5-methylcytosine antibody to detect methylation). After a 1-h incubation of DNMT1 with substrate plus or minus inhibitor, samples were read in a plate reader at 450 nm to determine activity. Figure 8A illustrates the dose–response plot over a range of DNMT1 concentrations for this assay. This figure (Fig. 8B–D) also shows inhibition plots for selected inhibitors that were used as positive controls, namely, SAH, the natural product EGCG, and the L-tryptophan derivative RG108. SAH and EGCG showed IC_{50} values of 8.6 and 0.7 μM, respectively. Under these assay conditions, RG108 showed only 11% inhibition at 100 μM. Figure 9A shows the inhibition plot for ATA, revealing an $IC_{50}=0.68$ μM against DNMT1. Of note, the enzymatic inhibitory activity of ATA against DNMT1 was greater than the activity of SAH and RG108 and it was comparable with the activity of EGCG under the current assay conditions. We also measured the Dnmt3a inhibition by ATA that showed an $IC_{50}=1.4$ μM (Fig. 9B), indicating that ATA is somewhat selective toward DNMT1 (approximately two-fold).

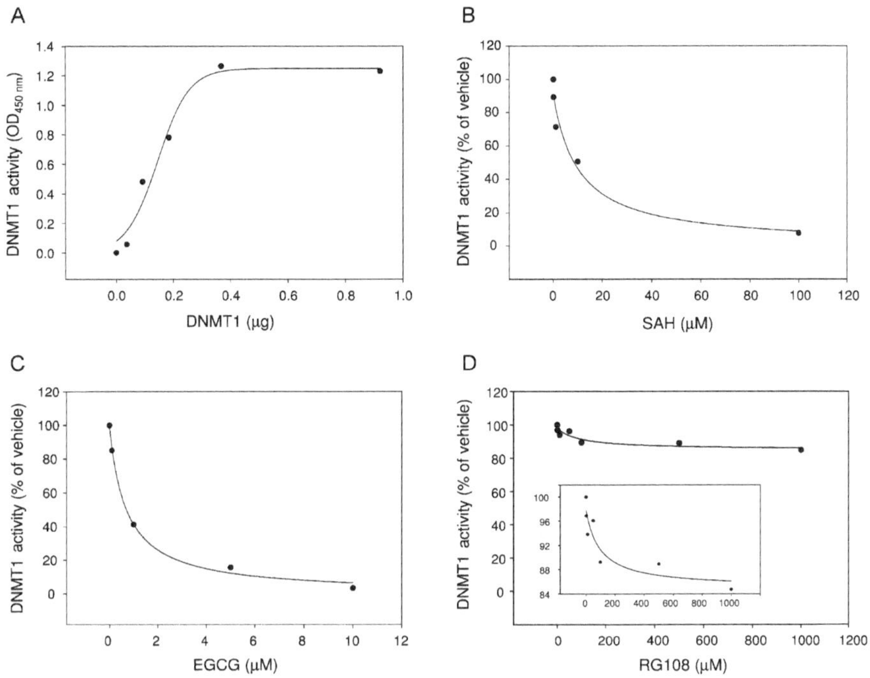

FIG. 8. (A) Titration of recombinant DNMT1 enzymatic activity using a colorimetric assay. DNMT1 activity was measured using EpiQuik DNA methyltransferase activity/inhibition assay kit (Epigentek). (B–D) Dose–response plots for each compound against DNMT1. Data are presented in terms of percentage activity versus the results using vehicle only treated control, which was assigned a value of 100%. The IC_{50} concentrations of compounds were determined by enzyme assay under identical conditions (0.3 μg/1.6 pmol of DNMT1, incubated for 1 h).

B. Docking and Pharmacophore Modeling of ATA

In order to explore the binding mode of ATA with DNMT1 at the molecular level, we conducted molecular modeling of ATA. Docking of ATA with the crystal structure of DNTM1 modified into an active conformation suggested two binding poses. The first docking pose of ATA is shown in Fig. 6D as a deprotonated form (expected in aqueous solution). According to this binding model, ATA forms a strong hydrogen bond network with the same amino acid residues that interact with the nucleoside inhibitors. The hydroxybenzoic acid moieties of ATA form hydrogen

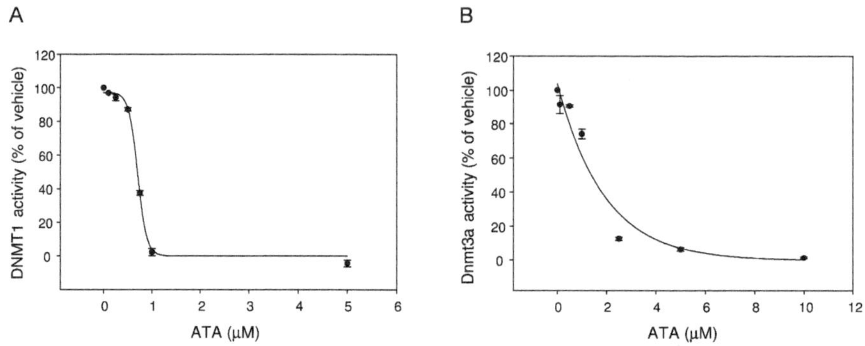

FIG. 9. (A) Dose–response plots for ATA inhibition against human DNMT1. Data are presented in terms of percent of vehicle only and the IC_{50} concentration of ATA was determined by enzyme assay with 0.3 μg of DNMT1 (1.6 pmol). (B) Dose–response plots for ATA inhibition against murine Dnmt3a. Data are presented in terms of percent of vehicle only control and the IC_{50} concentration of ATA was determined by enzyme assay with 4.0 μg of DNMT3A (33.3 pmol).

bonds with Glu1266, Arg1310, Arg1312, Ser1230, Gly1231, and Lys1535. The other carboxylate group makes interactions with Asn1578. The structurally related 5,5-methylenedisalicylic acid (NSC14778) showed weak inhibitory activity against DNMT1 in a biochemical assay (Kuck et al., 2010b). Although the binding pose of NSC14778 partially overlaps ATA showing same interactions with Ser1230, Gly1231, and Lys1535, NSC14778 does not make key interactions with Glu1266 and Arg1312 (not shown). A similar binding model of NSC14778 was also determined in our previous study using the homology model (Yoo and Medina-Franco, 2011b). The second pose of ATA was consistent with the recently reported binding mode in the catalytic site of homology model (Yoo and Medina-Franco, 2011c). Of note, ATA shows a good docking score comparable to those of nucleoside inhibitors (Table I).

We also compared the binding model of ATA obtained with Glide XP with the pharmacophore model developed with the modified crystal structure (Fig. 7E). The binding position of ATA matches with the best-scoring feature, that is, negative charge (N), hydrogen bond acceptor (A) (interaction with Glu1266), donor (D) (interaction with Arg1310 and Arg1312), and aromatic ring (R) (Fig. 7E). In contrast, NSC14778 satisfies only one, the negative feature (N). Therefore, this model suggests that introduction of a substituent that could match the donor or acceptor feature associated with the interaction of glutamate or arginine residues would enhance the affinity of the analogue NSC14778. The binding position of NSC97317 matches with three pharmacophoric features, that is, negative charge (N), hydrogen bond acceptor (A), and aromatic ring (R). These findings are further supported by the experimental observation and analysis of docking energy.

VI. Conclusions and Perspectives

DNMTs are involved in epigenetic regulation of the genome and are promising targets for therapeutic intervention in cancer and other diseases. Computational approaches are increasingly being used to identify promising compounds as potential therapeutic agents or as research tools to study epigenetic mechanisms involving DNMTs. Also, computational methods are employed to better understand the mechanism of established inhibitors of DNA methylation. Herein, we describe molecular modeling studies with a recently published crystallographic structure of human DNMT1 that contains unmethylated DNA and the enzyme in an "inactive" conformation. The conformation of the crystallographic structure was modeled in an active state using molecular dynamics. Experimentally known DNMT1 inhibitors were docked in parallel into the modified crystal structure with Glide HTVS, SP, and XP. The binding modes were characterized by common interactions with amino acid residues that participate in the proposed mechanism of DNA C5-methylation and other key amino acid residues including Ser1230, Gly1231, Glu1266, Arg1310, Arg1312, and Lys1535. The docking scores were also able to distinguish active inhibitors from known inactive and previously used decoy compounds. Indeed, most of the established inhibitors of DNMT1 had high ranking scores. Based on the docking results, a structure-based pharmacophore model was developed that was characterized by five features, including a negative charge, an aromatic ring, a hydrogen bond acceptor, and two donor sites. The pharmacophore model represents the key protein–ligand interactions of the DNMT1 inhibitors obtained in docking. Results of the docking and pharmacophore modeling had a good agreement with the insights of the modeling studies published previously using a homology model of the catalytic domain of DNMT1. As part of an ongoing interest of identifying novel inhibitors, we experimentally tested ATA in an enzymatic inhibition assay. Preliminary docking studies of ATA with DNMT1 had strongly suggested that this structural analogue of 5,5-methylenedisalicylic acid would show enzymatic inhibitory activity. Indeed, ATA showed an $IC_{50} = 0.68$ μM against human DNMT1. We also measured the DNMT3A inhibition by ATA, which showed an $IC_{50} = 1.4$ μM, indicating that ATA is somewhat selective toward DNMT1 (approximately twofold).

A major follow-up study of this work is to perform virtual screening of large compound data sets using our validated docking protocol and pharmacophore model with the crystal structure of human DNMT1 modeled into an active conformation to identify potential novel DNMT1 inhibitors. In order to further explore and perhaps someday enhance the enzymatic selectivity toward DNMT1 over murine Dnmt3a, we are currently employing molecular docking and other modeling approaches of ATA with the published crystal structure of DNMT3A (Jia et al., 2007). Results of these insights from the molecular modeling will be published in a separate communication in due course.

Acknowledgments

This work was supported by the Menopause & Women's Health Research Center, the State of Florida, Executive Office of the Governor's Office of Tourism, Trade, and Economic Development, and R01 CA116028 (K. D. R.).

References

Arce, C., Segura-Pacheco, B., Perez-Cardenas, E., Taja-Chayeb, L., Candelaria, M., Dueñnas-Gonzalez, A. (2006). Hydralazine target: from blood vessels to the epigenome. *J. Transl. Med.* **4**, 10.

Bouvet, M., Debarnot, C., Imbert, I., Selisko, B., Snijder, E. J., Canard, B., et al. (2010). In vitro reconstitution of SARS-Coronavirus mRNA cap methylation. *PLoS Pathog.* **6**, e1000863.

Caulfield, T., Medina-Franco, J. L. (2011). Molecular dynamics simulations of human DNA methyltransferase 3B with selective inhibitor nanaomycin A. *J. Struct. Biol.* **176**, 185–191.

Chen, T. P., Hevi, S., Gay, F., Tsujimoto, N., He, T., Zhang, B. L., et al. (2007). Complete inactivation of DNMT1 leads to mitotic catastrophe in human cancer cells. *Nat. Genet.* **39**, 391–396.

Cheng, X. D., Blumenthal, R. M. (2008). Mammalian DNA methyltransferases: a structural perspective. *Structure* **16**, 341–350.

Dueñas-González, A., García-López, P., Herrera, L. A., Medina-Franco, J. L., González-Fierro, A., Candelaria, M. (2008). The prince and the pauper. A tale of anticancer targeted agents. *Mol. Cancer* **7**, 82.

Evans, D. A., Bronowska, A. K. (2010). Implications of fast-time scale dynamics of human DNA/RNA cytosine methyltransferases (DNMTs) for protein function. *Theor. Chem. Acc.* **125**, 407–418.

Fandy, T. E., Herman, J. G., Kerns, P., Jiemjit, A., Sugar, E. A., Choi, S. H., et al. (2009). Early epigenetic changes and DNA damage do not predict clinical response in an overlapping schedule of 5-azacytidine and entinostat in patients with myeloid malignancies. *Blood* **114**, 2764–2773.

Gilbert, E. R., Liu, D. (2010). Flavonoids influence epigenetic-modifying enzyme activity: structure-function relationships and the therapeutic potential for cancer. *Curr. Med. Chem.* **17**, 1756–1768.

Goll, M. G., Bestor, T. H. (2005). Eukaryotic cytosine methyltransferases. *Annu. Rev. Biochem.* **74**, 481–514.

Hauser, A. T., Jung, M. (2008). Targeting epigenetic mechanisms: potential of natural products in cancer chemoprevention. *Planta Med.* **74**, 1593–1601.

Hernández-Campos, A., Velázquez-Martínez, I., Castillo, R., López-Vallejo, F., Jia, P., Yu, Y., et al. (2010). Docking of protein kinase B inhibitors: implications in the structure-based optimization of a novel scaffold. *Chem. Biol. Drug Des.* **76**, 269–276.

Issa, J.-P. (2005). Optimizing therapy with methylation inhibitors in myelodysplastic syndromes: dose, duration, and patient selection. *Nat. Clin. Pract. Oncol.* **2**(Suppl 1), S24–S29.

Issa, J.-P. J., Kantarjian, H. M. (2009). Targeting DNA methylation. *Clin. Cancer Res.* **15**, 3938–3946.

Issa, J. P. J., Kantarjian, H. M., Kirkpatrick, P. (2005). Azacitidine. *Nat. Rev. Drug Discov.* **4**, 275–276.

Jeltsch, A. (2002). Beyond Watson and Crick: DNA methylation and molecular enzymology of DNA methyltransferases. *ChemBioChem* **3**, 274–293.

Jia, D., Jurkowska, R. Z., Zhang, X., Jeltsch, A., Cheng, X. (2007). Structure of Dnmt3a bound to Dnmt3L suggests a model for de novo DNA methylation. *Nature* **449**, 248–251.

Jones, P. A., Baylin, S. B. (2007). The epigenomics of cancer. *Cell* **128**, 683–692.

Jurkowska, R. Z., Jurkowski, T. P., Jeltsch, A. (2011). Structure and function of mammalian DNA methyltransferases. *ChemBioChem* **12**, 206–222.

Kelly, T. K., De Carvalho, D. D., Jones, P. A. (2010). Epigenetic modifications as therapeutic targets. *Nat. Biotechnol.* **28**, 1069–1078.

Klimasauskas, S., Kumar, S., Roberts, R. J., Cheng, X. D. (1994). Hhal methyltransferase flips its target base out of the DNA helix. *Cell* **76**, 357–369.

Kuck, D., Caulfield, T., Lyko, F., Medina-Franco, J. L. (2010a). Nanaomycin A selectively inhibits DNMT3B and reactivates silenced tumor suppressor genes in human cancer cells. *Mol. Cancer Ther.* **9**, 3015–3023.

Kuck, D., Singh, N., Lyko, F., Medina-Franco, J. L. (2010b). Novel and selective DNA methyltransferase inhibitors: docking-based virtual screening and experimental evaluation. *Bioorg. Med. Chem.* **18**, 822–829.

Kumar, S., Horton, J. R., Jones, G. D., Walker, R. T., Roberts, R. J., Cheng, X. (1997). DNA containing 4′-thio-2′-deoxycytidine inhibits methylation by HhaI methyltransferase. *Nucleic Acids Res.* **25**, 2773–2783.

Lan, J., Hua, S., He, X. N., Zhang, Y. (2010). DNA methyltransferases and methyl-binding proteins of mammals. *Acta Biochim. Biophys. Sin.* **42**, 243–252.

Li, Y., Tollefsbol, T. O. (2010). Impact on DNA methylation in cancer prevention and therapy by bioactive dietary components. *Curr. Med. Chem.* **17**, 2141–2151.

LigPrep., version 2.2. Schrödinger, LLC, New York, NY.

Liu, Z. F., Liu, S. J., Xie, Z. L., Pavlovicz, R. E., Wu, J., Chen, P., et al. (2009). Modulation of DNA methylation by a sesquiterpene lactone parthenolide. *J. Pharmacol. Exp. Ther.* **329**, 505–514.

López-Vallejo, F., Caulfield, T., Martínez-Mayorga, K., Giulianotti, M. A., Nefzi, A., Houghten, R. A., et al. (2011). Integrating virtual screening and combinatorial chemistry for accelerated drug discovery. *Comb. Chem. High Throughput Screen.* **14**, 475–487.

Lyko, F., Brown, R. (2005). DNA methyltransferase inhibitors and the development of epigenetic cancer therapies. *J. Natl. Cancer Inst.* **97**, 1498–1506.

MacroModel, version 9.8. Schrödinger, LLC, New York, NY.

Medina-Franco, J. L., Caulfield, T. (2011). Advances in the computational development of DNA methyltransferase inhibitors. *Drug Discovery Today* **16**, 418–425.

Medina-Franco, J., López-Vallejo, F., Kuck, D., Lyko, F. (2011). Natural products as DNA methyltransferase inhibitors: a computer-aided discovery approach. *Mol. Diversity* **15**, 293–304.

Milani, M., Mastrangelo, E., Bollati, M., Selisko, B., Decroly, E., Bouvet, M., et al. (2009). Flaviviral methyltransferase/RNA interaction: structural basis for enzyme inhibition. *Antiviral Res.* **83**, 28–34.

Miller, C. A., Gavin, C. F., White, J. A., Parrish, R. R., Honasoge, A., Yancey, C. R., et al. (2010). Cortical DNA methylation maintains remote memory. *Nat. Neurosci.* **13**, 664–666.

Palii, S. S., Van Emburgh, B. O., Sankpal, U. T., Brown, K. D., Robertson, K. D. (2008). DNA methylation inhibitor 5-aza-2′-deoxycytidine induces reversible genome-wide DNA damage that is distinctly influenced by DNA methyltransferases 1 and 3B. *Mol. Cell. Biol.* **28**, 752–771.

Podvinec, M., Lim, S. P., Schmidt, T., Scarsi, M., Wen, D., Sonntag, L.-S., et al. (2010). Novel inhibitors of dengue virus methyltransferase: discovery by in vitro-driven virtual screening on a desktop computer grid. *J. Med. Chem.* **53**, 1483–1495.

Portela, A., Esteller, M. (2010). Epigenetic modifications and human disease. *Nat. Biotechnol.* **28**, 1057–1068.

Robertson, K. D. (2001). DNA methylation, methyltransferases, and cancer. *Oncogene* **20**, 3139–3155.

Salam, N. K., Nuti, R., Sherman, W. (2009). Novel method for generating structure-based pharmacophores using energetic analysis. *J. Chem. Inf. Model.* **49**, 2356–2368.

Schermelleh, L., Spada, F., Easwaran, H. P., Zolghadr, K., Margot, J. B., Cardoso, M. C., et al. (2005). Trapped in action: direct visualization of DNA methyltransferase activity in living cells. *Nat. Methods* **2**, 751–756.

Schrump, D. S., Fischette, M. R., Nguyen, D. M., Zhao, M., Li, X. M., Kunst, T. F., et al. (2006). Phase I study of decitabine-mediated gene expression in patients with cancers involving the lungs, esophagus, or pleura. *Clin. Cancer Res.* **12**, 5777–5785.

Siedlecki, P., Boy, R. G., Musch, T., Brueckner, B., Suhai, S., Lyko, F., et al. (2006). Discovery of two novel, small-molecule inhibitors of DNA methylation. *J. Med. Chem.* **49**, 678–683.

Singh, N., Dueñas-González, A., Lyko, F., Medina-Franco, J. L. (2009). Molecular modeling and dynamics studies of hydralazine with human DNA methyltransferase 1. *ChemMedChem* **4**, 792–799.

Sippl, W., Jung, M. (2009). DNA methyltransferase inhibitors. In: Epigenetic Targets in Drug Discovery, Sippl, W. and Jung, M. (Eds.), pp. 163–183. Wiley-VCH, Weinheim.

Song, J., Rechkoblit, O., Bestor, T. H., Patel, D. J. (2011). Structure of DNMT1-DNA complex reveals a role for autoinhibition in maintenance DNA methylation. *Science* **331**, 1036–1040.

Stresemann, C., Lyko, F. (2008). Modes of action of the DNA methyltransferase inhibitors azacytidine and decitabine. *Int. J. Cancer* **123**, 8–13.

Vilkaitis, G., Merkiene, E., Serva, S., Weinhold, E., Klimasauskas, S. (2001). The mechanism of DNA cytosine-5 methylation: kinetic and mutational dissection of HhaI methyltransferase. *J. Biol. Chem.* **276**, 20924–20934.

Yokochi, T., Robertson, K. D. (2002). Preferential methylation of unmethylated DNA by mammalian de novo DNA methyltransferase Dnmt3a. *J. Biol. Chem.* **277**, 11735–11745.

Yoo, J., Medina-Franco, J. L. (2011a). Discovery and optimization of inhibitors of DNA methyltransferase as novel drugs for cancer therapy. In: Drug Development - A Case Study Based Insight into Modern Strategies, Dr., Rundfeldt, Chris. (Ed.). 978-953-307-257-9,. InTech, Croatia Available at: http://www.intechopen.com/books/drug-development-a-case-study-based-insight-into-modern-strategies/discovery-and-optimization-of-inhibitors-of-dna-methyltransferase-as-novel-drugs-for-cancer-therapy.

Yoo, J., Medina-Franco, J. L. (2011b). Homology modeling, docking, and structure-based pharmacophore of inhibitors of DNA methyltransferase. *J. Comput.-Aided Mol. Des.* **25**, 555–567.

Yoo, J., Medina-Franco, J. L. (2011c). Trimethylaurintricarboxylic acid inhibits human DNA methyltransferase 1: insights from enzymatic and molecular modeling studies. *J. Mol. Model* **18**, 1583–1589.

Zawia, N. H., Lahiri, D. K., Cardozo-Pelaez, F. (2009). Epigenetics, oxidative stress, and Alzheimer disease. *Free Radic. Biol. Med.* **46**, 1241–1249.

IN SILICO STRATEGIES TOWARD ENZYME FUNCTION AND DYNAMICS

By SÍLVIA G. ESTÁCIO

Research Institute for Medicines and Pharmaceutical Sciences (iMed.UL), Faculty of Pharmacy, University of Lisbon. Av. Prof. Gama Pinto, Lisbon, Portugal
Condensed Matter Physics Centre (CFMC), University of Lisbon, Av. Prof. Gama Pinto 2, 1649-003 Lisbon, Portugal

Abstract

Enzymes are outstanding biocatalysts involved in a plethora of chemical reactions occurring in the cell. Despite their incommensurable importance, a comprehensive understanding of enzyme catalysis is still missing. This task becomes more laborious given the unavoidability of including the inherent dynamic nature of enzymes into that description. As such, it is essential to ascertain the nature and contribution of enzyme conformational changes to catalysis and to evaluate the adequacy of the proposal associating protein internal motions to the rate enhancement achieved.

Dynamic events in enzymes span a wide range of time- and length-scales which have led to a surge in multiscale methodologies targeting enzyme function and dynamics. Computational strategies assume a preponderant role in such studies by allowing the atomic detail investigation of the fundamental mechanisms of enzyme catalysis thus surpassing what is achievable through experiments. While high-accuracy quantum mechanical methods are indicated to uncover the details of the chemical reaction occurring at the active site, molecular mechanical force fields and molecular dynamics approaches provide powerful means to access the conformational energy landscape accessible to enzymes. This review outlines

DOI: 10.1016/B978-0-12-398312-1.00009-3
Copyright 2012, Elsevier Inc. All rights reserved.

some of the most important *in silico* methodologies in this area, highlighting examples of problems tackled and the insights obtained.

I. Introduction

Enzymes are key players in most biological processes, catalyzing nearly all important chemical reactions in the body as well as being involved in the control of the transcription of genetic information, signal transduction, and cell regulation (Fersht, 1998; Kamerlin and Warshel, 2010a). These highly pro-efficient catalysts specifically enhance the rates of chemical reactions in the cell by 10^8–10^{20}-fold (Radzicka and Wolfenden, 1995; Wolfenden and Snider, 2001), thus making feasible processes that otherwise would not occur on biologically relevant timescales. Despite more than 50 years of research, an in-depth, universal theory for the precise molecular origins of the rate enhancement enabled by enzymes is still missing (Neet, 1998; Agarwal, 2006; Klinman, 2009). A theory of enzyme catalysis would definitely impact the first-principles design of new, fully functional enzymatic catalysts as well as the rational underlying current structure-based drug design methodologies. The wealth of studies covering topics which range from the three-dimensional structures of enzymes (X-ray and NMR studies) to several aspects of their mechanistic pathways, including the rates of various steps, have not been able to elucidate which physicochemical properties of enzymes cause the acceleration of chemical reactions (Agarwal, 2006).

The catalytic activity of enzymes has frequently been associated with their structures through the proposal of descriptions relying on the structural interactions between enzymes and substrates. This structure–function relation constitutes the main dogma of structural biology. The fundamental mechanisms of enzyme catalysis have been the subject of intensive research since the early proposal by Fischer (1894) of a "lock and key" model. The induced-fit hypothesis introduced in 1958 by Koshland (Koshland, 1958) was built upon the "lock and key" model but further extends it to incorporate flexibility, thus yielding a new perspective into substrate recognition and enzyme catalysis. In this model, the active site side chains are molded into their favorable, productive alignments by interaction with the substrate (McGeagh et al., 2011). Examples of an induced-fit behavior are diverse (de Oliveira et al., 2003; Sullivan and Holyoak, 2008; Estácio et al., 2011). Currently, both this model and the conformational selection model constitute two of the major theories used to rationalize the mechanism by which an enzyme adopts its catalytically productive/functional state (Sullivan and Holyoak, 2008). In contrast to the induced-fit model, for which the enzyme acquires its catalytically productive conformation through changes in protein structure that are induced by substrate binding, the conformational selection model captures the multiple conformational states adopted by the enzyme, with the substrate having a higher affinity for the active state (Boehr et al., 2009; Pingali et al., 2011).

Such models, however, fail to describe allostericity (change of a protein conformation resulting in change of function as, for instance, noncompetitive inhibition; Lee and Craik, 2009) and cooperative effects (Koshland and Hamadani, 2002; Qian and Shi, 2009). Contrary to the generally accepted idea, which assumes that each enzymatic active site catalyzes a single specific biochemical reaction, the archaeal fructose-1,6-bisphosphate aldolase/phosphatase (FBPA/P) enzyme, which consists of a single catalytic domain, catalyzes two chemically distinct reactions of gluconeogenesis. It was recently shown that this dual role is due to a dramatic conformational change at the active site (Fushinobu et al., 2011). The structure–function paradigm of structural biology has, thus, been progressively reassessed and extended to include dynamic effects as well as some structural disorder (Agarwal, 2006; Borbulevych et al., 2009; Vendruscolo, 2010; He et al., 2011; Lu et al., 2011; Schramm, 2011b). Currently, it is widely accepted that proteins in the intracellular environment permanently undergo conformational fluctuations which span a plethora of timescales, ranging from

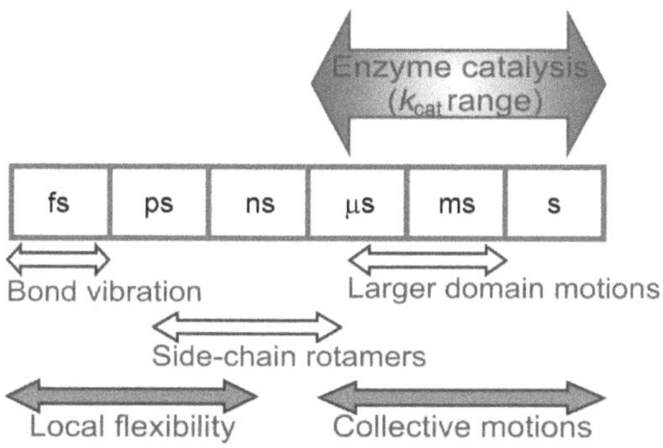

FIG. 1. Range of timescales spanned by proteins' dynamic events. Scheme inspired in Henzler-Wildman and Kern (2007) and Schramm (2011a). (For color version of this figure, the reader is referred to the online version of this chapter.)

femto- to milliseconds (and beyond), and amplitudes (see Fig. 1). The conformational diversity displayed by proteins arises from local fast-timescale side-chain fluctuations and movements of active site loops (Takuya et al., 2002; Amaro et al., 2009; Estácio et al., 2011) and slow-timescale global structural rearrangements including secondary structure reorganizations (Skinner et al., 2008), interconversions of protein fold (Tuinstra et al., 2008; Yadid et al., 2010), and closed-to-open conformational transitions (de Oliveira et al., 2011; Masterson et al., 2011) phenomena. The overall dynamic nature of proteins is embossed into a complex conformational landscape with a variety of conformations of similar energy and small energy barriers separating the corresponding minima. The dynamic perspective thus adds ruggedness (Solomatin et al., 2010) into the, otherwise smooth, energy landscape depicted in the "thermodynamic hypothesis" of protein folding proposed by Anfinsen (1973), which assumes that amino acid sequences evolved to specifically favor the native conformation, the thermodynamically most stable structure in the intracellular medium. The labile nature of an enzyme, which translates into the ability to fluctuate among different conformers, has been regarded an outcome of evolutionary adaptability (Tokuriki and Tawfik, 2009; McGeagh et al., 2011). In fact, the functions of these inherently plastic entities are often related with excursions between multiple substates (Bouvignies et al., 2011; He et al., 2011; Masterson et al., 2011; Ramanathan et al., 2011).

Intrinsically disordered proteins, which perform biological functions despite lacking well-defined tertiary structures under physiological conditions, further revolutionized the classical structure–function paradigm disseminated among structural biologists (Dyson and Wright, 2005; Tompa and Fuxreiter, 2008; Tompa, 2011). These proteins constitute potentially attractive systems for investigating dynamical contributions to function. Astonishingly, enzymes, whose dynamic character had previously been strictly restricted to sets of tightly folded native conformations, have been observed in disordered states which can be as catalytically efficient as their native counterparts (Pervushin et al., 2007).

In recent years, a myriad of studies have grossly contributed to the notion that protein internal dynamics and enzyme catalysis are irrevocably interconnected given the similarities between the timescales spanned. Cyclophilin A (CypA), an enzyme which catalyzes the reversible *cis–trans* isomerization of prolyl peptide bonds, has been the research subject of a number of biophysical (Eisenmesser et al., 2002, 2005) and theoretical

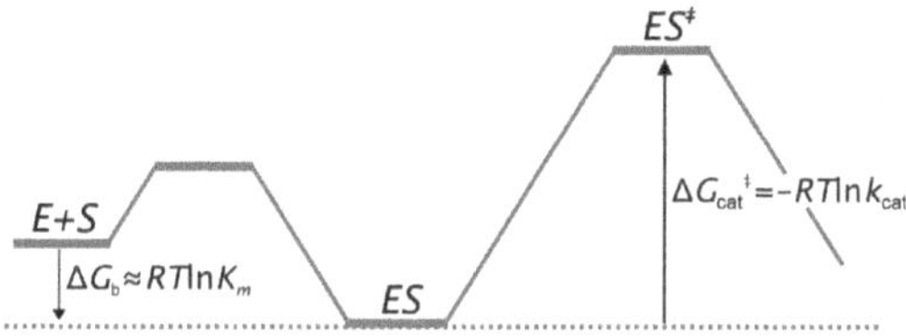

FIG. 2. Schematic representation of the energetic of enzymatic reactions, adapted from Warshel (1998). *E*, *S*, *ES*, and $ES^{\ddagger}$ represent the enzyme, substrate, enzyme–substrate complex (Michaelis complex), and transition state, respectively. (For color version of this figure, the reader is referred to the online version of this chapter.)

studies (Agarwal, 2004, 2005) due to a perceived link between protein dynamics and enzymatic catalysis. NMR dispersion experiments undertaken by Kern and coworkers on CypA (Eisenmesser et al., 2002, 2005) have revealed the existence of a coincidence between a global conformational change process and the peptidyl-prolyl isomerization of the substrate. These authors additionally detected characteristic collective motions during catalysis which are common to the free CypA, with frequencies similar to the turnover numbers, k_{cat} (refer to Fig. 2). The theoretical investigation of the catalytic activity of CypA (Agarwal, 2004, 2005), which relied on a combination of molecular dynamics (MD) and umbrella sampling (US) (Torrie and Valleau, 1977; Kästner, 2011) techniques, allowed the identification of fluctuations in certain regions of the enzyme backbone and the discovery of a network of protein vibrations which play a role in promoting the isomerization reaction. Flomenbom et al. (2005) single-molecule experiments on *Candida antarctica* lipase B have provided evidence of multiple active conformers that slowly interconvert on the same timescale of catalytic activity. Kern and coworkers (Henzler-Wildman et al., 2007a,b) have observed that an allosteric change in the enzyme adenylate kinase (Adk) structure, apprehended while investigating conformational changes around its equilibrium basin, impacts the overall enzyme turnover rate. The combined use of X-ray crystallography, NMR, single-molecule FRET, normal mode analysis, and MD simulations enabled the identification of partially closed conformational states along the pathway toward the catalytically active fully closed state in the free Adk. Though sampled in a pico- to nanosecond timescale, it was shown that these local, high-frequency fluctuations promote the large amplitude, micro- to millisecond timescale, slower lid motions that produce the fully closed catalytically competent state (Henzler-Wildman et al., 2007a). Finally, Gáspári et al. (2010) recognized the significant role of fast pico- to nanosecond timescale dynamics on the molecular recognition of canonical inhibitors by serine proteases, thus reconciling the ''lock and key'' model conventionally associated to the efficiency of those inhibitors with a dynamic insight grounded on the enzymes' loop flexibility.

The impact of solvent dynamics cannot be dismissed in this discussion (Fenimore et al., 2004; Frauenfelder et al., 2009). Recently, Grossman et al. (2011) were able to prove the existence of a gradient of fast-to-slow coupled water–enzyme motions highly correlated with rearrangements occurring at the active site of a zinc-dependent human metalloproteinase by using a combination of terahertz spectroscopy, X-ray absorption, and MD techniques. It is presumed that those motions may assist enzyme–substrate interactions through water polarization brought about by the catalytic zinc and the enzyme active site.

From the above, it is undeniable that the relative importance of small-amplitude fast vibrations or large-amplitude slow conformational motions in enzymes depends on the type of enzymatic reaction. Large conformational changes on the microsecond to second timescales are, for instance, typically associated with substrate/cofactor binding and product release (Fersht, 1998; Agarwal, 2006; McGeagh and Mulholland, 2010) or the attainment of a specific chemical environment at the active site (Jiménez et al., 2008). However, the suggestion that an enzyme's internal motions may contribute to enhance the reaction rates of chemical steps in catalysis remains elusive and has raised important discussions between both advocates and assailants of such proposal (Pisliakov et al., 2009; Kamerlin and Warshel, 2010a,b; Karplus, 2010). This controversy has been amplified by the theoretical and experimental difficulties in ascertaining the connection between protein dynamics and catalytic reaction rates. In fact, a valid quantification of an enzyme's catalytic efficiency must necessarily rely on a (differential) comparison between the enzymatic reaction and the equivalent, uncatalyzed reference reaction in solution (Warshel, 2003; Kamerlin and Warshel, 2010a).

One of the major guiding lines underlying the theoretical investigation of the catalytic rate enhancement provided by enzymes is concerned with the identification and characterization of the transition state (TS). The access to information regarding the TS endeavors the design of TS analogs to be used as inhibitors, the analysis of dynamic motions that promote TS formation, and the separation between ground state (GS) and TS effects

(Schwartz and Schramm, 2009; Schramm, 2011a). Particularly important is the definition of its location and the quantification of the respective free energy barrier (activation free energy) which endeavors the calculation of rates of enzymatic reaction through the generalized transition-state theory (TST)-inspired (Horiuti, 1938; Wigner, 1938) Eyring equation (Evans and Polanyi, 1935; Eyring, 1935):

$$k(T) = k(T)\frac{k_B T}{h}\exp\left[-\frac{\Delta G^{\ddagger}(T)}{RT}\right] \quad (1)$$

where $\Delta G^{\ddagger}(T)$ is the Gibbs free energy difference between the reactants and the transition state and $k(T)$ is the temperature-dependent generalized transmission coefficient (Garcia-Viloca et al., 2004; Pu et al., 2006; Mata, 2010). This coefficient corrects some of the assumptions underlying the original, classical, pseudo-equilibrium transition-state theory (CTST) (in CTST, $k(T) = 1$) by introducing dynamic, nonequilibrium, and quantum-mechanical tunneling effects (Garcia-Viloca et al., 2004; Kamerlin and Warshel, 2010a). It includes a contribution from dynamical recrossing which accounts for the behavior of a reactive trajectory that recrosses the TS region going back to the reactants' side. Another contribution relates to the inclusion of the effects of quantum mechanical tunneling. Quantum tunneling is considerably important in reactions involving proton transfer: protons can tunnel across the potential barrier into the products side thus producing a large reaction rate enhancement (Sutcliffe and Scrutton, 2006; Sutcliffe et al., 2006). The third contribution accounts for deviations from the Boltzmann equilibrium distribution (Garcia-Viloca et al., 2004). TST is still widely used as a first approximation to understand enzymatic catalysis (Olsson et al., 2006) despite recent opposition (Schwartz and Schramm, 2009).

The concept of TS stabilization, which basically presumes that the enzyme binds the TS stronger than the GS, was first introduced by Haldane (1930) and Pauling (1946, 1948) and recently formalized by Radzicka and Wolfenden (1995). Though still widely accepted, in the sense that most current theories put forward to explain catalytic efficiency assume some degree of complementarity to the TS, it is considered insufficient (Neet, 1998). Currently, there is an active debate on whether the catalytic power of enzymes involves TS stabilization or GS destabilization (Warshel, 2003; Giraldo et al., 2006).

It is well known that many enzymes evolved by optimizing k_{cat}/K_m (Warshel, 1998; Wolfenden and Snider, 2001), where $K = k_1/k_{-1}$ and

$K_m = (k_{-1} + k_{cat})/k_1 \approx k_{-1}/k_1$. The rate enhancement provided by enzymes can, therefore, involve a destabilization of the GS (by reducing K_m; Zhang and Schramm, 2011) or a favorable differential stabilization of the TS (by increasing k_{cat}; Ranaghan et al., 2004; Szefczyk et al., 2004) (see Fig. 2). The key question lies on the clarification of the factors responsible for the reduction of the activation free energy barrier in the chemical step. Some of the proposals put forward to explain the rate enhancement achieved by enzymes have incorporated some aspects of entropy as, for instance, the binding entropy and orbital steering theories (Mesecar et al., 1997), among others. The contributions of low-barrier hydrogen bonds (LBHB; Jencks, 1986) and nuclear tunneling (Klinman, 2009) have also been considered. Warshel and collaborators (Warshel, 1998; Warshel et al., 2006) firmly sustain that TS formation is driven exclusively by electrostatics, with no meaningful contribution from the enzyme dynamic motions. The proposal by Schramm and coworkers (Schwartz and Schramm, 2009; Schramm, 2011a) merges both fast- and slow-timescale enzymatic motions in the description of the catalytic rate enhancement and passage across the TS barrier. According to this perspective, the TS lifetime lies in the same femtosecond range of bond vibrations—the crossing of the TS barrier is thus fast in comparison with the rate of the enzymatic reaction (see Fig. 1). It is further assumed that slow conformational changes occurring in the same timescale of enzymatic turnover (k_{cat}) do not interfere with the chemistry of the catalysis, contributing exclusively to the attainment of a particular geometry of the TS; access to the transient, short-lived TS is enabled by rarely occurring vibrations at the catalytic active site.

Modern computational methodologies represent a precious contributor to the study of enzyme reactivity, enabling the identification and characterization of the TS and transiently formed, scarcely populated conformers which are inaccessible to most biophysical methods (Bouvignies et al., 2011; Ramanathan et al., 2011). It is thus possible to attain a detailed description of the transition pathways. Computational simulations are insuperable in their capability to provide in-depth atomistic insights into protein motions along a wide range of timescales thus allowing to assess in which degree do those motions impact enzyme catalysis.

In the next section, two of the most consensually used *in silico* methods that target enzyme function and dynamics are revised, namely, quantum-mechanical/molecular-mechanical (QM/MM) and MD methodologies.

II. *In Silico* Methods: From *Ab Initio* Quantum Mechanical Calculations to Classical Molecular Mechanical Force Fields

Enzymes are large and complex molecules comprising thousands of atoms which, as such, represent a huge challenge to the computational modeling techniques available. This inherent complexity is further enhanced by the inclusion of the enzyme's environment, whether a solvent, a lipid membrane, cofactors, or DNA which may be bound to the enzyme.

The major aims underlying the computational study of an enzymatic reaction are the clarification of its chemical mechanism (e.g., roles of the amino acids located at the active site, identification of short-lived intermediate/transition state(s), etc.), the identification of specific interactions stabilizing the intermediate/transition state(s) (with immediate application in structure-based drug design methodologies of new ligands devoted to pharmaceutical applications), and comprehension of the effects produced by an enzyme's conformational fluctuations, particularly their influence over the reaction rates of enzymatic reactions (Mulholland and Grant, 2007; Lonsdale et al., 2010) (Fig. 3). Computational modeling and simulation studies rely upon the preexistence of X-ray crystallography and/or NMR structures of the particular system of interest. In fact, the

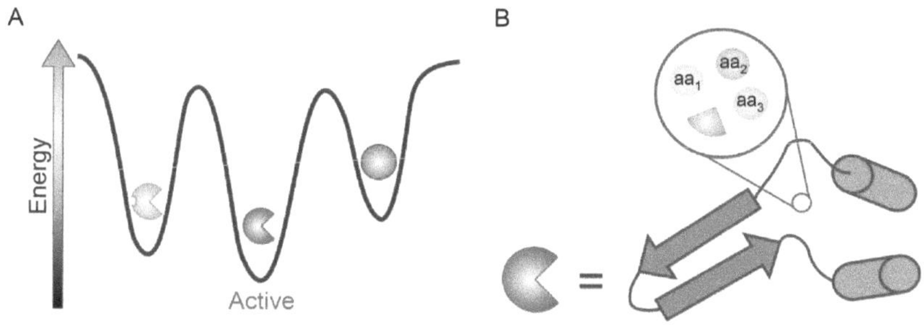

Fig. 3. Multiscale aims underlying the use of computational enzymology approaches. (A) Investigation of the dynamic, high-dimensional, free energy landscape of an enzyme. Each minimum (well) along the energy surface is a conformational substate, while a maximum between wells represents a transition state. (B) Characterization of an enzyme catalytic mechanism which involves, among others, the clarification of the roles performed by catalytic amino acids and the identification of the short-lived transition or intermediate states and stabilizing factors/interactions. Scheme adapted from Lee and Craik (2009). (For color version of this figure, the reader is referred to the online version of this chapter.)

appearance of high-resolution X-ray crystallography structures was the main trigger surrounding the rise in studies focusing on the atomistic details of protein function (Henzler-Wildman and Kern, 2007).

Though enzymology-targeted computational methodologies attained an unprecedented level of accuracy in recent years, as a consequence of the rapid increase in the available computational power, the application of high-level quantum mechanical calculations to the totality of atoms in a macromolecule is still strictly prohibitive. The development of multiscale strategies for modeling biological systems assumes, thus, a preponderant role. Some such strategies might include a high-accuracy quantum mechanical representation of a small part of the system (the chemically reactive site, where the phenomenon of chemical interest occurs) which is embedded in an external layer (that can, or not, include a portion of the environment in which the macromolecule is inserted) treated by a more coarse-grained approximation (Kamerlin et al., 2011) as, for instance, a force field. These hybrid, multilayer methods, which include QM/MM (Lin and Truhlar, 2007; Senn and Thiel, 2009; Mata, 2010) and cluster approaches (Himo and Siegbahn, 2003; Siegbahn and Himo, 2009, 2011; Liao et al., 2011), have made feasible to perform calculations on a wide variety of environments, such as aqueous solutions or membranes. The explicit inclusion of the region around the active site offers significant advantages over the use of continuum models including the possibility to include, and investigate, the effects of conformational changes beyond the quantum mechanical region.

As extensively debated in Section I, it is now consensually accepted that the mechanisms underlying many of the functions of biological macromolecules are often characterized by large-scale structural changes. In that regard, enzymes' catalytic functions often depend on excursions between conformers in the energy landscape which are only transiently formed and scarcely populated. Understanding the mechanisms of conformational transition among conformers is of the utmost importance to drug-design methodologies aimed at modulating protein function (Yang et al., 2009).

The characterization of the conformational transitions among the various stable and metastable states and their associated dynamic behavior is currently out of reach to the majority of biophysical tools available due to the transient nature of some of these conformers. This task becomes even more daunting and prohibitively expensive when exploring the multidimensional energy landscape of biomolecules with weights of the order of 10^6 Da. MD simulations, however, provide a valid alternative despite the restricted timescales spanned by conventional simulation techniques which are unable to tackle biologically relevant conformational changes occurring on the micro- to millisecond timescale range (Hampton et al., 2010) (Fig. 1). High-accuracy first-principles MD simulations are typically restricted to shorter timescales and small molecules or small portions of a macromolecule. The use of coarse-grained models, that is, simplified representations of the interaction potential, molecular structure, or both, reduces the complexity of the problem and enables simulations which span longer timescales (Clementi, 2008; de Marco and Varnai, 2009; Potestio et al., 2009; Wohlert and Berglund, 2011). Network- (Chng and Yang, 2008; Yang et al., 2009) and structure-based Gō models (Taketomi et al., 1975; Faísca et al., 2005; Whitford et al., 2009; Scheraga et al., 2011) are common coarse-grained potential energy functions used to simplify force fields.

A. *Hybrid Quantum Mechanical/Molecular Mechanical (QM/MM) Methods*

The use of quantum mechanical (QM) methods, which directly take into account the electronic structure of a molecule by solving the Schrödinger equation, is unavoidable when investigating chemical reactions (i.e., bond formation/breaking events) and other processes involving movement of electrons such as charge transfer (CT; Mikołajczyk et al., 2011; Ufimtsev et al., 2011; Woiczikowski et al., 2011) or electronic excitation (Tozer and Handy, 2000; Steindal et al., 2011; Tachikawa et al., 2011). Despite the huge computational power currently available, the use of high-accuracy electronic structure methods is still strictly restricted to a few hundred atoms which limit their applicability to the chemically active regions (i.e., catalytic cores) of biomolecules. It is, therefore, essential to resort to the use of force-field-based molecular mechanics (MM) methods to ensure the correct treatment of the remainder thousands of atoms in those macromolecules. The combined (or hybrid) QM/MM methodology, first introduced by Warshel and Levitt (1976), has become the method of choice for the current state-of-the-art modeling of reactivity in biomolecular systems at an affordable computational cost. Over the past decade, numerous review papers have put forward overviews of the innovations in the field of QM/MM methodologies while exposing their applicability to a plethora of issues in biomolecular systems (Lin and Truhlar, 2007; Acevedo and Jorgensen, 2009; Senn and Thiel, 2009; Ranaghan and Mulholland, 2010; Kamerlin et al., 2011).

QM/MM methods work by splitting the system under study into two main layers: (i) a smaller inner region, corresponding to the chemically relevant part of the macromolecule, that is treated quantum mechanically (QM region) and (ii) an extended outer region, mostly nonoverlapping to the former, described at a lower level of accuracy through the use of MM-derived methods (e.g., a force field) (MM region). The explicit representation of an enzyme beyond its active site offers significant advantages over the use of continuum models (e.g., cluster approach of Siegbahn and coworkers (Himo and Siegbahn, 2003; Siegbahn and Himo, 2009; Siegbahn and Himo, 2011)) such as the possibility to capture the contribution of the enzyme structure and dynamics to catalysis.

1. Choice of QM and MM Methodologies

The QM/MM formalism is amenable to variety of combinations of QM and MM methods. Concerning the option for a particular QM method, post-Hartree–Fock *ab initio* electron correlation methods, such as those based on Møller–Plesset perturbation theory (MPn) or coupled-cluster theory (e.g., CCSD with single and double excitations or CCSD(T)), constitute the top-notch QM methods of choice (Shavitt and Bartlett, 2009). Though the applicability of such high-accuracy methods involves a high demand in computational resources, it is now possible to study up to several tens of atoms with the development of linear-scaling local electron correlation methods (e.g., LMP2, LCCSD; Saebo and Pulay, 1993; Senn and Thiel, 2009). Despite the recent advances which have endeavored the application of high-accuracy *ab initio* QM/MM methods to the study of enzymatic reactions (Claeyssens et al., 2006), density functional theory (DFT; Claeyssens et al., 2011; Tian and Eriksson, 2011) or semiempirical methods (Geerke et al., 2008; Acevedo and Jorgensen, 2009) remain the most popular choices to represent the QM region given the inherently favorable ratio between computational effort and accuracy. The use of such methods may, however, introduce additional sources of error. The DFT B3LYP functional (Becke, 1988, 1993a,b), "the main working-horse of computational chemistry" (Zhang et al., 2010) whose use is widely spread in a variety of QM/MM studies applied to enzymatic reactions, has been shown to provide accurate predictions of thermochemistry of small covalent systems. However, its incomplete treatment of dispersion brings about a number of problems (Mata, 2010; Zhang et al., 2010) which include the incorrect description of van der Waals (vdW) interactions (Zhao and Truhlar, 2005), the underestimation of reaction barrier heights (Zhao and Truhlar, 2008; Zhang et al., 2009), and the incorrect estimation of CT excitations (Dreuw et al., 2003). Semiempirical methods, based on the neglect of diatomic differential overlap (NDDO), are about 10^3-fold faster than their *ab initio* counterparts which deem them particularly attractive to the investigation of biomolecular problems. MNDO (Dewar and Thiel, 1977), MNDO/d (Thiel and Voityuk, 1996), AM1 (Dewar et al., 1985), PM3 (Stewart, 1989), PDDG/PM3 (Repasky et al., 2002), PDDG/MNDO (Repasky et al., 2002), and SCC-DFTB (Elstner, 2006) are among some of the most used semiempirical (SQM) formalisms. Though allowing for less expensive calculations, their use is particularly flawed if the system being studied is too different from the molecules in the database used to parameterize the method.

Though not an electronic structure method *per se*, the empirical valence bond (EVB) method of Warshel and coworkers (Warshel and Weiss, 1980; Rosta et al., 2006; Kamerlin et al., 2011) is usually included in the classification of QM/MM methods. In EVB, the chemical reaction is described through a valence bond approach in which the system wavefunction is given by a linear combination of ionic–covalent resonance structures which define the Hamiltonian for the isolated molecule and the potential energy is obtained by solving the corresponding secular equation. The energy of each resonance form is given by an empirical force field and the Hamiltonian is calibrated to reproduce available experimental (or, exceptionally, *ab initio* QM) data for a given solution reaction as, for instance, barrier heights (Warshel and Weiss, 1980). This method has been successfully applied to the modeling of solvent and protein environment contributions on enzymatic reactions, whether as a component of a QM/MM methodology or as a reference potential for the sampling of the conformational space (Rosta et al., 2006; Kamerlin et al., 2011).

Constrained or frozen density functional theories (CFDT/FDFT; Wesolowski and Warshel, 1993) split the system into two regions both treated using DFT. However, in the external region the atomic electron densities are either frozen or constrained.

Regarding the choice of MM method, there are a number of biomolecular force fields available whose validity has been amply investigated and tested. AMBER (Cornell et al., 1995), GROMOS (Scott et al., 1999), CHARMM (MacKerell et al., 1998b), and OPLS-AA (Jorgensen et al., 1996) force fields are among the most popular. Ideally, force fields should be derived from *ab initio* QM calculations. In practice, however, force field parameters are fitted to bulk properties of proteins and lipids in solution (for additional details see Section II.B.1).

Some of the most commonly used combinations of methodologies QM and MM are schematized in Fig. 4.

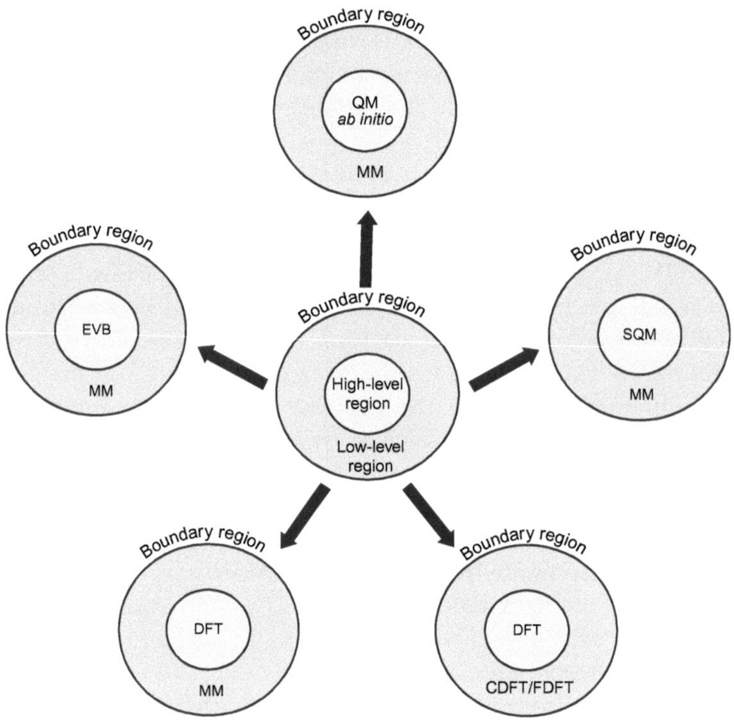

FIG. 4. Schematic representation of distinct quantum mechanical/molecular mechanical (QM/MM) approaches, adapted from Kamerlin et al. (2011). (For color version of this figure, the reader is referred to the online version of this chapter.)

2. *Additive and Subtractive QM/MM Schemes*

The total QM/MM energy can be obtained through two schemes of calculation: an additive and a subtractive approach.

In the additive QM/MM scheme, the total system (T) energy is given by (Senn and Thiel, 2009):

$$E_{\text{additive}}(T) = E_{\text{QM}}(I) + E_{\text{MM}}(E) + E_{\text{QM/MM}}(I/E) \quad (2)$$

The QM/MM energy of the entire system is thus obtained through the sum of three contributions: $E_{\text{QM}}(I)$, the QM energy of the inner region; $E_{\text{MM}}(E)$, the MM energy of the external region, without the QM core; and $E_{\text{QM/MM}}(I/E)$, an explicit coupling term which accounts for the interactions across the boundary region between the two system partitions. This coupling term includes bonded, nonbonded vdW and electrostatic contributions—the former two are treated at the MM level while electrostatic interactions are modeled through electrostatic or polarization embedding schemes. In the electrostatic embedding model, the MM force field point charges are included in the QM Hamiltonian. Polarized embedding additionally covers the polarization of the MM region by the QM one.

In the subtractive procedure, the energy of the entire system is obtained according to the following expression (Senn and Thiel, 2009):

$$E_{\text{subtractive}}(T) = E_{\text{QM}}(I) + E_{\text{MM}}(T) - E_{\text{MM}}(I) \quad (3)$$

This requires a QM calculation on the inner region, $E_{\text{QM}}(I)$, an MM calculation on the entire system, $E_{\text{MM}}(T)$, and again an MM calculation on the internal region, $E_{\text{MM}}(I)$, which avoids double counting of the energy of the inner core. In this interpolating scheme, the coupling between both regions (i.e., QM/MM interactions) is handled entirely at the MM level which means that the influence of the MM region on the QM properties of the inner core is completely neglected. This constitutes the so-called mechanical embedding model of QM/MM interactions.

Despite being easy to implement, the subtractive formalism possesses a few limitations. These are substantiated in an incorrect representation of the electrostatic QM/MM interactions as a consequence of the general MM treatment of both regions which, additionally, raises the problem of finding realistic MM force field parameters for the inner QM region (Banáš et al., 2009; Senn and Thiel, 2009; Shaik et al., 2010). It remains, however, an attractive QM/MM scheme given the easy implementation of a disparity of low-accuracy theories other than an MM force field and the possibility to generalize it to a multilayer representation of which the most well known is the ONIOM (Own N-layer Integrated molecular Orbital molecular Mechanics) formalism of Morokuma and collaborators (Svensson et al., 1996; Vreven et al., 2006; Chung et al., 2011). ONIOM combines any number of molecular orbital methods as well as MM methods.

The progresses made so far have not yet solved all the challenges associated with the use of hybrid methods as, for instance, the choice of the covalent bonds through which the system is to be split into two regions or the treatment of the underlying boundary region. In particular, the use of link atoms (Singh and Kollman, 1986), special boundary atoms, or localized orbitals (in the generalized hybrid orbital – GHO (Gao et al., 1998) or local self-consistent field – LSCF (Théry et al., 1994) methods is extensively discussed in the literature (Vreven et al., 2006; Lin and Truhlar, 2007; Acevedo and Jorgensen, 2009; Senn and Thiel, 2009; Ranaghan and Mulholland, 2010; Shaik et al., 2010).

Likewise, the option for a mechanical, electrostatic, or polarized embedding scheme to treat the electrostatic coupling between both QM and MM regions should be carefully assessed (Senn and Thiel, 2009).

Another critical issue inherent to the implementation of a QM/MM methodology is the inclusion of long-range electrostatic effects which may be due to the bulk solvent or the outer region of the macromolecule. Though the inclusion of all electrostatic interactions is strictly forbidden, the use of cutoffs may introduce incoherencies given the long-range nature of those interactions. For a retrospect on some of the methods used to tackle long-range electrostatic effects in QM/MM studies of enzymatic reactions, it is advisable to consult references (Laino et al., 2005; Schaefer et al., 2005; Senn and Thiel, 2009; Liu, 2010). Benighaus and Thiel (2009, 2011) have, recently, proposed the application of the solvated macromolecule boundary potential (SMBP) which enables the evaluation of the electrostatic contributions of both the outer macromolecule region and the bulk solvent. Steindal et al. (2011) opted, instead, for the use of the solvation polarizable continuum model (PCM) combined with a fully polarizable QM/MM scheme in which the reciprocal polarization of both PCM and MM parts is attained through an iterative scheme involving the update of MM-induced dipoles and PCM surface charges until convergence. This scheme was shown to be effective in including both short-range effects of the first solvation shell and long-range electrostatics due to the bulk solvent.

3. QM/MM Applications

The main applications of hybrid QM/MM methods include the clarification of the structures and energetics of stationary points (local minima and saddle points) on potential energy surfaces and the prediction of free energy differences (e.g., solvation, reaction, and activation free energy barriers) of organic or enzymatic reactions.

QM/MM methodologies, which yield the energy and gradients for a given structure accounting for a single stationary point on a potential energy surface, are amenable to on-the-fly combination with appropriate optimization, MD or Monte Carlo (MC) simulation protocols given that the latter use gradients/forces information to update structures.

QM/MM optimization algorithms take advantage of the partition of the system into two regions for which the evaluation of energies and forces involve distinct requirements of computational resources. One of the most popular, the microiterative scheme, involves alternating optimizations (i.e., geometry relaxation) of both QM and MM regions until mutual convergence is achieved. Currently, there are numerous variations of this algorithm in use (Vreven et al., 2003; Prat-Resina et al., 2004; Kästner et al., 2007).

The outputs of QM/MM geometry optimizations of single stationary points along an enzymatic reaction path should be considered with care given the inherent size and complexity of the corresponding conformational space which are especially important at the finite temperatures of the calculations. It is, thus, essential to perform an extensive sampling of such complex landscape, with special focus on the active site. Warshel and coworkers have previously discussed the importance of having several representative TSs and their corresponding minima while performing QM/MM optimization studies (Kamerlin et al., 2009, 2011). An easy, albeit incomplete, take on this importance sampling problem is the use of a set of snapshots extracted from a classical MD trajectory.

The analysis of free energy differences underlies any theoretical investigation on enzymatic reactivity. QM/MM MD (or MC) simulation schemes provide an exhaustive sampling of the conformational space which enables the calculation of free energy differences by means of US (Wang et al., 2007; Ke et al., 2009a; Kästner, 2011), thermodynamic integration (TI; Senn et al., 2005), free energy perturbation (FEP; Kästner et al., 2006), or Jarzynski's equality-derived (Jarzynski, 1997) nonequilibrium treatments (e.g., steered molecular dynamics simulations (SMD; Park et al., 2003; Xiong et al., 2006; Bu et al., 2011)).

QM/MM MD simulations, which include both Born–Oppenheimer (BOMD) and Car–Parrinello MD (CPMD) formulations, have been widely applied to the study of biomolecular systems and enzymatic catalysis (Colombo et al., 2002; Wang et al., 2007; Ke et al., 2009b). More details on BOMD and CPMD simulations are available in Section II.B.2. First-principles QM/MM MD (MC) simulations are extremely resource demanding and slow converging, so much so that the use of semiempirical QM representations and/or implicit or continuum solvation models (e.g., PCM (Tomasi et al., 2005), generalized Born model, GB (Onufriev et al., 2002), Smx models of Cramer and Truhlar (Cramer and Truhlar, 2008; Marenich et al., 2009), or the more recent AGBNP2 (Gallicchio et al., 2009)) is the common practice when investigating biomolecules (Cui, 2002; Jaramillo and Wodak, 2005; Mennucci, 2010). Recently, Acevedo and Jorgensen (2009) introduced a QM/MM MC methodology which includes a semiempirical PDDG/PM3 treatment of the QM region along with a MC sampling algorithm. In this scheme, the solvent is included as part of the MM region and the solute–solvent interactions consist of Coulomb and Lennard-Jones terms. The former make use of the QM and MM charges and the latter use standard MM force field parameters. The consensus between the results, obtained for numerous organic reactions and for several enzyme-catalyzed processes, and the available experimental data was considered extremely promising regarding the use of semiempirical QM representations.

Warshel and coworkers (Rosta et al., 2006; Kamerlin et al., 2011), on the other hand, popularized the concept of reference classical potential for the sampling of the phase space to efficiently derive activation free energy barriers in enzymes. The most challenging issue in this EVB mapping formulation is the evaluation of the free energy of transfer from this reference potential to the *ab initio* one. This free energy can be estimated either by means of an FEP approach or, preferentially, through a linear response approach (LRA). Warshel's implementations use EVB as a powerful and computationally inexpensive reference potential for sampling.

Regarding the evaluation of free energies, it is important to additionally mention the transition-path sampling (TPS) technique (Bolhuis et al., 2002). TPS, an MC walk through trajectory space between the reactant and product basins, enables access to the femtosecond-lived TSs and to the fast (femto- to picosecond) dynamic motions promoting TS barrier crossing. It is usually combined with a QM/MM methodology which allows taking into account the rest of the enzyme and/or its environment while sampling an ensemble of transition-path trajectories (Crehuet and Field, 2007; Schwartz and Schramm, 2009). Information relative to the reaction mechanism, TSs, and free energies are extracted from this ensemble (Dellago et al., 2002). The use of QM/MM-based TPS was crucial to elucidate the mechanism of the hydride and proton transfer reaction catalyzed by the *Bacillus stearothermophilus* lactate dehydrogenase: Quaytman and Schwartz (2008) were able to demonstrate that hydride transfer happens simultaneously to the moment of maximal compression of the active site which is promoted by a specific vibration.

B. Molecular Dynamics Simulations

MD simulations provide access to the thermodynamic properties and time-dependent kinetic phenomena underlying the evolution in time of a system. These features underline the successful use of MD simulations on the investigation of dynamic aspects surrounding molecular recognition and enzyme catalysis (Karplus and Kuriyan, 2005). Though an MD trajectory *per se* provides limited information on the atomic positions, velocities, and single-point energies, such microscopic aspects are readily translated into meaningful kinetic (e.g., reaction rates) and thermodynamic (e.g., free energy differences) macroscopic properties through application of statistical mechanics formalisms.

1. Implementation: Integration Algorithms, Force Fields, Solvation, and Electrostatics

MD simulations involve the iterative calculation of the forces acting on the system particles and the posterior update of their positions. The particles move in response to their interactions following equations of motion which, in classical MD, are defined in the framework of Newtonian mechanics. There are a number of algorithms available to integrate the equations of motions of which the most widely used are the Verlet, leapfrog, velocity Verlet, Beeman, and predictor–corrector algorithms (Verlet, 1967; Gear, 1971; Beeman, 1976; Swope et al., 1982). In classical MD simulations, the adoption of the Born–Oppenheimer approximation (Born and Oppenheimer, 1927) legitimizes the neglect of quantum dynamical effects—the concomitant restriction of the calculations to the nuclear coordinates enables the efficient application of this methodology to extensive biomolecular systems.

One of the major challenges surrounding the implementation of an MD simulation scheme, which greatly influences its accuracy and stability, is the definition of the potential energy function describing the inter- and intramolecular interactions in the system. The efficient integration of the equations of motion relies on the immediate availability of first and second derivatives of the potential energy function with respect to the particle positions (forces and force gradients, respectively).

The typical functional form of a classical MM potential energy function comprises bonded (bond stretching, angle bending, and torsions/dihedrals) and nonbonded pairwise additive terms describing long-range electrostatics and vdW interactions (Wang et al., 2001):

$$\begin{aligned} V(r) = & \sum_{\text{bonds}} \frac{1}{2} k_b (b - b_0)^2 + \sum_{\text{angles}} \frac{1}{2} k_\theta (\theta - \theta_0)^2 \\ & + \sum_{\text{dihedrals}} \frac{1}{2} V_n [1 + \cos(n\varphi - \delta)] + \sum_{\text{electrostatics}}^{i>j} \frac{q_i q_j}{4\pi\varepsilon_0 r_{ij}} + \sum_{\text{vdW}}^{i<j} \left(\frac{A_{ij}}{r_{ij}^{12}} - \frac{B_{ij}}{r_{ij}^{6}} \right) \end{aligned} \quad (4)$$

vdW terms are usually computed through a Lennard-Jones-type potential, while electrostatic contributions are given by Coulombic interactions between rigid point charges. These functions usually consist of a large number of parameterized terms, fitted to experimental and/or *ab initio* electronic structure calculations on small molecules or fragments, which are assumed to be transferable to the larger molecules of interest. The set of energy functions and associate parameters constitute the so-called force field.

The AMBER (Cornell et al., 1995), GROMOS (Scott et al., 1999), CHARMM (MacKerell et al., 1998b), and OPLS-AA (Jorgensen et al., 1996) biomolecular force fields previously mentioned follow, more or less, the energy expression of Eq. (4). GROMOS is a united-atom force field, while CHARMM force fields include both all-atom and united-atom variants. The AMBER ff99SB (Hornak et al., 2006; Wickstrom et al., 2009; Cerutti et al., 2010), CHARMM22 (MacKerell et al., 1998a), and CHARMM22 with CMAP corrections (CHARMM22/CMAP; Mackerell et al., 2004) are among the most popular force fields for proteins. However, it has been recently invoked that the level of accuracy in energy calculations delivered by the currently used force fields is far from being satisfactory (Faver et al., 2011).

Though allowing the relatively inexpensive study of extensive biomolecular systems, MM force fields present some important limitations regarding condensed phase simulations (Mackerell, 2004). The representation of a molecule's charge distribution through effective partial fixed charges, assigned to the atoms irrespective of the environment, does not account for any adaptation to changes in the local electric field. The importance of including polarization effects was first disclosed in the 1976 work by Warshel and Levitt (1976) which focused on the lysozyme catalytic reaction—it was demonstrated that the reaction intermediate is energetically unstable being stabilized only by the formation of induced dipoles in the enzyme–substrate complex. As a response to this major limitation, a new generation of polarizable force fields is being brought forward. Currently, the most widely used methodologies to incorporate explicit electronic polarization (Cieplak et al., 2009; Luque et al., 2011) include inducible multipoles (Dang, 1998; Wang et al., 2011a) (i.e., inclusion of both partial atomic charges and inducible multipoles) and fluctuating charge models (Patel and Brooks, 2006; Lipparini and Barone, 2011). Fluctuating charge models tackle both polarization and CT effects. The AMOEBA polarizable force field developed by Ponder and coworkers (Ren and Ponder, 2002; Ponder et al., 2010) is a popular representative of this generation of force fields implemented in the TINKER software (Ponder, 2011). Its major difference from traditional biomolecular force fields resides in the replacement of the fixed partial charges distribution with polarizable atomic multipoles. AMOEBA represents an improvement over nonpolarizable force fields and is considered an excellent candidate for hybrid QM/MM schemes (Woodcock et al., 2011) given the attainment of accurate structural and thermodynamic properties for small proteins as well as reliable transport properties (Ponder et al., 2010).

The inclusion of polarizability adds increased complexity to the energy and force evaluation in each MD time step. The use of methods in which the polarizable degrees of freedom are included as dynamic variables as, for instance, the extended Lagrangian method, is essential to insure a lower computational cost than the one associated with the use of a self-consistent method (Wang and Skeel, 2005; Lopes et al., 2009).

MD simulations are usually preceded by an energy minimization procedure aimed at refining the molecule structure. The most used minimization algorithms include those that use first-order (e.g., steepest descent and conjugate gradient) and second-order (e.g., Newton–Raphson) derivatives (Höltje et al., 2008).

The minimization procedure is usually followed by the inclusion of the environment (i.e., solvent, membrane, etc.) around the molecule of interest so as to properly characterize it through MD. In the specific case of biological macromolecules, it is essential to include some implicit or explicit representation of water. There are a number of force fields available to model explicit water, for example, SPC (Berendsen et al., 1981), SPC/E (Berendsen et al., 1987), TIP3P (Jorgensen et al., 1983), TIP4P (Jorgensen and Madura, 1985), TIP5P (Mahoney and Jorgensen, 2000) and the polarizable SPC/P (Ahlström et al., 1989), TIP4P/P (Chen et al., 2000), TIP4P/FQ (Rick, 2001), SWM4-DP (Lamoureux et al., 2003), SWM4-NDP (Lamoureux et al., 2006), and AMOEBA (Ren and Ponder, 2004). Though the explicit inclusion of water endeavors a correct description of electrostatic effects, it occasionally represents an intolerable request for computational resources, especially when dealing with large proteins. Implicit, also designated continuum, solvent models (see Section II.A.3 for a brief list) represent instantaneous dielectric responses of the medium and, as such, their use involves lower computational requirements. These models also exhibit several advantages over explicit models as, for instance, a much more straightforward estimation of the solute's free energy of solvation and an improved conformational sampling (Onufriev, 2008).

The computationally most costly part of an MD simulation is concerned with the treatment of nonbonded interactions, particularly long-range electrostatic interactions. In biomolecular simulations, it is essential to correctly represent electrostatic interactions given their key roles in determining the structure, dynamics, and function of biologically relevant molecules (Hünenberger et al., 2001).

While the number of bonded interactions is a linear function of the number of atoms, the number of Coulombic interactions is directly proportional to the square of the number of charges/atoms in the system. Given their magnitude and long-range nature, the choice of an efficient treatment of long-range electrostatics is crucial so as to avoid the need to evaluate a prohibitively large number of interactions. Early MD applications used the spherical truncation of interactions with cutoff criteria (Loncharich and Brooks, 1989). This formalism was implemented in, for example, shifting and switching function methods (Norberg and Nilsson, 2000). More sophisticated, efficient, long-range electrostatics evaluation methods include Ewald sum and particle-mesh Ewald (PME;

York et al., 1993; Abraham and Gready, 2011), fast multipole (FM; Deng and Cai, 2007), and reaction field (Xu et al., 2009) methods.

2. *First-Principles Molecular Dynamics*

Despite having been advantageously applied to the investigation of the conformational changes and dynamics involved in enzymatic reactions, the application of empirical MM force fields to the modeling of chemical reactions is strictly interdicted (McGeagh et al., 2011). Beyond the drawback related with the noninclusion of polarization effects, MM force fields establish a preconnectivity between atoms which makes them unable to tackle phenomena involving qualitative changes of electronic structure along the MD trajectories, namely bond breaking and formation events. First-principles MD simulations emerged as an effective response to these issues by allowing a first-principles description of the chemical reaction occurring at the enzyme active site while simultaneously including the dynamical effects from the surrounding protein environment. It is thus possible to explore the enzyme's free energy surface while focusing on a high-level description of the enzymatic reaction. In these methodologies, the forces acting over the "point-like" atomic nuclei are obtained from electronic structure calculations performed concurrently to the generation of the classical phase-space MD trajectory. The option for a specific force field is, in this case, replaced by the option for the theoretical approach used to solve Schrödinger's equation. DFT has been widely used as the electronic structure method "by excellence" in a number of first-principles MD applications (Dal Peraro et al., 2007). The explicit treatment of the electronic structure allows including polarization effects and bond breaking/formation events which are described within the accuracy inherent to the representation of the electronic part. In CPMD, the classical MD trajectory for the atomic nuclei is obtained on-the-fly with the electronic structure calculation so that the QM motions of electrons are taken into consideration in the forces calculation. The computationally more demanding Born–Oppenheimer implementation requires the electronic structure problem to be solved self-consistently at each MD time step. CPMD circumvents the requirement of an expensive self-consistent iterative minimization at each time step by using a fictitious dynamics of the electronic degrees of freedom that keeps the electrons in the GS, thus thriving to maintain the validity of the adiabatic separation between the motions of nuclei and electrons inherent to the Born–Oppenheimer approximation. The main disadvantage of the CPMD scheme resides in the fact that a very small time step must be used in order to generate stable MD trajectories. The groups of Rothlisberger and Carloni have been especially active in the development and application of CPMD methodologies to the study of complex chemical and biochemical phenomena (Colombo et al., 2002; Dal Peraro et al., 2007; Brunk et al., 2011). In the recent literature, it is also possible to find a plethora of examples of the application of BOMD simulations to the characterization of enzymatic reaction mechanisms (Ke et al., 2009a; Wu et al., 2009; Zhou et al., 2010; Panek et al., 2011).

QM/MM MD implementations (i.e., CPMD/MM or BOMD/MM schemes) grant an effective way to overcome the limitations associated with the application of first-principles calculations to the study of extensive biologically meaningful systems. Wu et al. (2009), for instance, have put in evidence the high-accuracy features associated with the application of QM/MM BOMD simulations to the investigation of the dynamics, and concomitant coordination changes, of the catalytic zinc ion in thermolysin (TLN) and histone deacetylase 8 (HDAC8). In this particular scheme, the zinc and its coordinating ligands were treated with the B3LYP DFT functional, while all other atoms in the enzyme molecule were described by the AMBER ff99SB MM force field. The use of QM/MM first-principles simulations appears particularly suited to the evaluation of the catalytic role played by the calalytic ion in metalloenzymes as the recent work by Morokuma and coworkers (Abdel-Azeim et al., 2011) on the cobalamin-dependent methionine synthase (MetH) has further demonstrated.

Despite the obvious benefits associated with the use of first-principles MD simulations, it is essential to bear in mind that these methods are restricted to ~20–40 Å length-scales and ~10–100 ps timescales which render them unable to capture slower timescale, large-scale protein motions.

3. *Advanced Sampling Techniques*

In computational enzymology, the comprehension of an enzyme reaction mechanism and evaluation of the corresponding reaction and activation free energy barriers (and reaction rates through TST) require a thorough characterization of the free energy landscape, stable basins, and the concomitant reaction paths. These requirements are partially hindered in current conventional MD methodologies which cannot access biologically relevant protein dynamics and conformational changes on the micro- to millisecond timescale (Henzler-Wildman and Kern, 2007). In the case of large biological molecules, this timescale limitation translates into an inability of the system to overcome the traps of local minima. It may also hamper the adequate sampling of regions of the free energy landscape separated from the initial basin by high energy barriers (infrequent or rare events).

In order to circumvent this issue, numerous alternatives which "accelerate" MD by enhancing the sampling of phase space, in particular regions of high free energy (e.g., TSs) difficult to access, have been proposed. There are three main classes of enhanced sampling methodologies (Schlick, 2009). In one of these classes, sampling of the conformational space is impacted through a biasing potential. In the second, non-Boltzmann sampling techniques, which increase the probability of occurrence of high-energy states, are used. The third class includes methodologies which, contrary to the previous ones, do not alter the potential energy function or the sampling method but, instead, the MD algorithm acting by, for example, freezing the fastest degrees of freedom (which allows the use of larger time steps). The first group includes US (Torrie and Valleau, 1977; Kästner, 2011), accelerated MD (Hamelberg et al., 2004; Perez et al., 2009), and metadynamics (Laio and Parrinello, 2002; Ensing et al., 2006; Barducci et al., 2011) techniques. The second encloses steered MD (Park et al., 2003; Xiong et al., 2006), simulated annealing (SA; Kirkpatrick et al., 1983), and replica exchange (REMD; Sugita and Okamoto, 1999) (based on parallel tempering MC (Hansmann, 1997)) techniques which use high temperatures to overcome energetic barriers. Among the methods that alter the MD algorithm, one finds the popular SHAKE algorithm (van Gunsteren and Berendsen, 1977) and multiple time-step methods such as the reversible reference system propagation algorithm (r-RESPA; Tuckerman et al., 1992). Some of these methods will be briefly discussed in this section. For further information, see references (Adcock and McCammon, 2006; Schlick, 2009, 2010; Zuckerman, 2011).

US is one of the most widely known integrants of the first class of biasing potential methods. In this method, bias potentials (e.g., harmonic potentials) applied along a, one- or multidimensional, reaction coordinate (usually geometric parameters such as distances or root-mean-square deviations (RMSD)) accelerate the system dynamics. The bias potential drives the reaction coordinate to a target value. US is typically performed in a series of windows, each of which comprises an MD simulation at a specific, restrained value of the reaction coordinate. The windows thus sampled are finally combined by using the weighted histogram analysis (WHAM; Kumar et al., 1995; Roux, 1995) or umbrella integration (Kästner, 2009) methods which allow to obtain the potential of mean force (PMF) along the specified reaction coordinate.

In the accelerated MD (aMD) method of Hamelberg et al. (2004), the rationale promoting an enhanced conformational space sampling involves a reduction of the energy barriers separating the different substates accessible to a system. The energy landscape is thus modified by adding a bias potential, $\Delta V(r)$, to the true potential such that the minima below a given threshold energy level, E, are raised, while all other basins (energy wells) above that level remain unchanged (Hamelberg et al., 2004; Wang et al., 2011b):

$$\Delta V(r) = \begin{cases} 0, & V(r) \geq E \\ \dfrac{(E - V(r))^2}{\alpha + (E - V(r))}, & V(r) < E \end{cases} \tag{5}$$

The parameter α modulates the depth of the potential energy basins. All the statistics collected with the biasing potential are later corrected to remove the effect of the bias. This scheme enables easier transitions between adjacent energy basins and allows sampling infrequent events typically inaccessible to conventional MD. aMD simulations have been successfully applied to the identification of large-scale conformational changes of *Trypanosoma cruzi* proline racemase enzyme (TcPR), a therapeutic target for Chagas' disease (de Oliveira et al., 2011). The characterization of the TcPR closed-to-open conformational interconversion mechanism is an undoubtedly important contribution for future structure-based drug design methodologies targeting Chagas'. Likewise, aMD simulations have confirmed the existence of a scarcely populated, "hidden," semiclosed state accessible to the ligand-free maltose binding protein (MBP), a prototype for periplasmic binding proteins (PBPs) which play a key role in nutrient uptake in Gram-negative bacteria. The rationale used has further provided a detailed picture of the transition between the open and partially closed states and a more complete understanding of the ligand recognition mechanism involved (Bucher et al., 2011).

Metadynamics is a relatively recent MD methodology introduced by Laio and Parrinello (2002). It works as a typical biasing potential method, through the addition of an external bias potential to the system's Hamiltonian. This potential acts along a selected number of degrees of freedom (order parameters), the so-called collective variables (CVs), and is constructed as a sum of Gaussians centered along the system's trajectory in the CVs space. Its attractiveness resides in the ability to be proficiently used both for accelerating rare events and reconstruct the underlying free energy surface (Ensing et al., 2006; Barducci et al., 2011). Metadynamics is usually used in combination with classical (Leone et al., 2009; Parker et al., 2011) or first-principles MD methods, particularly CPMD (Alfonso-Prieto et al., 2009; Ardèvol and Rovira, 2011; Biarnés et al., 2011). In a recent investigation by Ardèvol and Rovira (2011), hybrid QM/MM Car–Parrinello metadynamics simulations were instrumental in identifying a very short-lived oxocarbenium ion-pair intermediate in the, mostly planar, free energy profile for the glycosylation reaction catalyzed by the trehalose-6-phosphate synthase (OtsA), a glycosyltransferase that intervenes in oligosaccharide synthesis. This observation represented a considerate breakthrough regarding the attainment of a more complete description of the type of mechanism involved in the formation of trehalose-6-phosphate from uridyldiphosphate glucose (UDP-Clc) and glucose-6-phosphate (Glc-6P) catalyzed by OtsA. While first-principles metadynamics simulations are particularly useful in uncovering details of reaction steps, force field-based metadynamics simulations provide a more qualitative insight into the conformational changes that characterize an enzymatic mechanism.

In the second class of methods, of which REMD is one of the most popular representatives, the system is forced to sample the remainder conformational space by having the amount of time spent in local energy minima reduced. In REMD (Sugita and Okamoto, 1999), various noninteracting replicas of the system of interest are simultaneously simulated, at several different temperatures. Periodically, at fixed intervals of time, one attempts to swap the conformations in neighboring replicas (i.e., temperatures) according to a Metropolis-type criterion (Metropolis et al., 1953) that "weights" the probability of sampling each conformation at the other temperature. The transition probability is formulated so as to maintain each replica's equilibrium distribution. The rationale behind REMD intends to promote an acceleration of the crossing of energy barriers thus allowing a broader exploration of the conformational space. REMD advantageously includes in the sampling conformations belonging to high-energy transition regions which would otherwise be rarely reached at lower temperatures thus significantly increasing the chances of observing the full ensemble of transition pathways (Deng et al., 2011). Since the computational cost associated with the use of REMD simulations greatly increases with the size of the system (due to a proportional increase in the number of replicas and number of accessible conformations), various treatments have been put forward to improve the slow convergence of REMD (Hritz and Oostenbrink, 2007; Okur et al., 2007; Roitberg et al., 2007). There are several examples in the recent literature of the application of REMD to the unveiling of important conformational changes which underline enzymatic activity (Campbell et al., 2010; Vijayan and Biggin, 2010; Deng et al., 2011). Particularly, Deng et al. (2011) used REMD in combination with a transition-path theory (TPT) analysis (Noé et al., 2009; Weinan and Vanden-Eijnden, 2010) to characterize the conformational dynamics of the flaps of HIV-1 protease, a major drug target in the fight against AIDS whose flaps play a key role in the mechanism of substrate binding. The methodology used allowed the acquisition of an ensemble of transition paths proving that, contrary to the idea transmitted by conventional MD studies, conformational changes can occur through a variety of paths which are temperature dependent.

Though the advances in the implementation of enhanced sampling techniques have been extraordinary, the development of new, more sophisticated methodologies is still very much a work in progress as some recent examples demonstrate (Dickson et al., 2011; Kunz and van Gunsteren, 2011).

III. Conclusions

Computational methodologies, as the ones highlighted in this review, can provide new insights into enzyme function and dynamics which complement and extend the evidence obtained through experimental work. The information obtained ranges from atomistic-level details on the catalytic mechanism to the evaluation of the coupling between chemical and conformational changes. This review further acknowledges the importance of using multiscale approximations given the breadth of timescales and amplitudes relevant to enzyme catalysis and dynamics.

Ultimately, computational studies on enzyme catalysis provide invaluable clues for structure-based drug design methodologies and the development of synthetic catalysts.

Acknowledgments

The author gratefully acknowledges financial support from Fundação para a Ciência e a Tecnologia (FCT, Portugal) with a postdoctoral fellowship (SFRH/BPD/46313/2008).

References

Abdel-Azeim, S., Li, X., Chung, L. W., Morokuma, K. (2011). Zinc–homocysteine binding in cobalamin-dependent methionine synthase and its role in the substrate activation: DFT, ONIOM, and QM/MM molecular dynamics studies. *J. Comput. Chem.* **32**, 3154–3167.

Abraham, M. J., Gready, J. E. (2011). Optimization of parameters for molecular dynamics simulation using smooth particle-mesh Ewald in GROMACS 4.5. *J. Comput. Chem.* **32**, 2031–2040.

Acevedo, O., Jorgensen, W. L. (2009). Advances in quantum and molecular mechanical (QM/MM) simulations for organic and enzymatic reactions. *Acc. Chem. Res.* **43**, 142–151.

Adcock, S. A., McCammon, J. A. (2006). Molecular dynamics: survey of methods for simulating the activity of proteins. *Chem. Rev.* **106**, 1589–1615.

Agarwal, P. K. (2004). Cis/trans isomerization in HIV-1 capsid protein catalyzed by cyclophilin A: insights from computational and theoretical studies. *Proteins* **56**, 449–463.

Agarwal, P. K. (2005). Role of protein dynamics in reaction rate enhancement by enzymes. *J. Am. Chem. Soc.* **127**, 15248–15256.

Agarwal, P. N. (2006). Enzymes: an integrated view of structure, dynamics and function. *Microb. Cell Fact.* **5**, 2.

Ahlström, P., Wallqvist, A., Engström, S., Jönsson, B. (1989). A molecular dynamics study of polarizable water. *Mol. Phys.* **68**, 563–581.

Alfonso-Prieto, M., Biarnés, X., Vidossich, P., Rovira, C. (2009). The molecular mechanism of the catalase reaction. *J. Am. Chem. Soc.* **131**, 11751–11761.

Amaro, R. E., Cheng, X., Ivanov, I., Xu, D., McCammon, J. A. (2009). Characterizing loop dynamics and ligand recognition in human and avian-type influenza neuraminidases via generalized Born molecular dynamics and end-point free energy calculations. *J. Am. Chem. Soc.* **131**, 4702–4709.

Anfinsen, C. B. (1973). Principles that govern the folding of protein chains. *Science* **181**, 223–230.

Ardèvol, A., Rovira, C. (2011). The molecular mechanism of enzymatic glycosyl transfer with retention of configuration: evidence for a short-lived oxocarbenium-like species. *Angew. Chem. Int. Ed.* **50**, 10897–10901.

Banáš, P., Jurečka, P., Walter, N. G., Šponer, J., Otyepka, M. (2009). Theoretical studies of RNA catalysis: hybrid QM/MM methods and their comparison with MD and QM. *Methods* **49**, 202–216.

Barducci, A., Bonomi, M., Parrinello, M. (2011). Metadynamics. *WIREs Comput. Mol. Sci.* **1**, 826–843.

Becke, A. D. (1988). Density-functional exchange-energy approximation with correct asymptotic behavior. *Phys. Rev. A* **38**, 3098–3100.

Becke, A. D. (1993a). A new mixing of Hartree-Fock and local density-functional theories. *J. Chem. Phys.* **98**, 1372–1377.

Becke, A. D. (1993b). Density-functional thermochemistry. III. The role of exact exchange. *J. Phys. Chem.* **98**, 5648–5652.

Beeman, D. (1976). Some multistep methods for use in molecular dynamics calculations. *J. Comput. Phys.* **20**, 130–139.

Benighaus, T., Thiel, W. (2009). A general boundary potential for hybrid QM/MM simulations of solvated biomolecular systems. *J. Chem. Theory Comput.* **5**, 3114–3128.

Benighaus, T., Thiel, W. (2011). Long-range electrostatic effects in QM/MM studies of enzymatic reactions: application of the solvated macromolecule boundary potential. *J. Chem. Theory Comput.* **7**, 238–249.

Berendsen, H. J. C., Postma, H. J. C., Gunsteren, W. F. V., Hermans, J. (1981). Interaction models for water in relation to protein hydration. In: Intermolecular Forces, Pullman, B. (Ed.), pp. 331–338. D. Reidel Publishing Company, Dordrecht, The Netherlands.

Berendsen, H. J. C., Grigera, J. R., Straatsma, T. P. (1987). The missing term in effective pair potentials. *J. Phys. Chem.* **91**, 6269–6271.

Biarnés, X., Ardevol, A., Iglesias-Fernández, J., Planas, A., Rovira, C. (2011). Catalytic itinerary in 1,3-1,4-β-glucanase unravelled by QM/MM metadynamics. Charge is not yet fully developed at the oxocarbenium ion-like transition state. *J. Am. Chem. Soc.* **133**, 20301–20309.

Boehr, D. D., Nussinov, R., Wright, P. E. (2009). The role of dynamic conformational ensembles in biomolecular recognition. *Nat. Chem. Biol.* **5**, 789–796.

Bolhuis, P. G., Chandler, D., Dellago, C., Geissler, P. L. (2002). Transition path sampling: throwing ropes over rough mountain passes, in the dark. *Annu. Rev. Phys. Chem.* **53**, 291–318.

Borbulevych, O. Y., Piepenbrink, K. H., Gloor, B. E., Scott, D. R., Sommese, R. F., Cole, D. K., et al. (2009). T cell receptor cross-reactivity directed by antigen-dependent tuning of peptide-MHC molecular flexibility. *Immunity* **31**, 885–896.

Born, M., Oppenheimer, R. (1927). Zur quantentheorie der molekeln. *Ann. Phys.* **389**, 457–484.

Bouvignies, G., Vallurupalli, P., Hansen, D. F., Correia, B. E., Lange, O., Bah, A., et al. (2011). Solution structure of a minor and transiently formed state of a T4 lysozyme mutant. *Nature* **477**, 111–114.

Brunk, E., Ashari, N., Athri, P., Campomanes, P., de Carvalho, F. F., Curchod, B. F. E., et al. (2011). Pushing the frontiers of first-principles based computer simulations of chemical and biological systems. *Chimia* **65**, 667–671.

Bu, L., Beckham, G. T., Shirts, M. R., Nimlos, M. R., Adney, W. S., Himmel, M. E., et al. (2011). Probing carbohydrate product expulsion from a processive cellulase with multiple absolute binding free energy methods. *J. Biol. Chem.* **286**, 18161–18169.

Bucher, D., Grant, B. J., Markwick, P. R., McCammon, J. A. (2011). Accessing a hidden conformation of the maltose binding protein using accelerated molecular dynamics. *PLoS Comput. Biol.* **7**, e1002034.

Campbell, Z. T., Baldwin, T. O., Miyashita, O. (2010). Analysis of the bacterial luciferase mobile loop by replica-exchange molecular dynamics. *Biophys. J.* **99**, 4012–4019.

Cerutti, D. S., Freddolino, P. L., Duke, R. E., Case, D. A. (2010). Simulations of a protein crystal with a high resolution X-ray structure: evaluation of force fields and water models. *J. Phys. Chem. B* **114**, 12811–12824.

Chen, B., Xing, J., Siepmann, J. I. (2000). Development of polarizable water force fields for phase equilibrium calculations. *J. Phys. Chem. B* **104**, 2391–2401.

Chng, C.-P., Yang, L.-W. (2008). Coarse-grained models reveal functional dynamics–II. Molecular dynamics simulation at the coarse-grained level—theories and biological applications. *Bioinform. Biol. Insights* **2**, 171–185.

Chung, L. W., Hirao, H., Li, X., Morokuma, K. (2012). The ONIOM method: its foundation and applications to metalloenzymes and photobiology. *WIREs Comput. Mol. Sci.* **2**, 327–350.

Cieplak, P., Dupradeau, F.-Y., Duan, Y., Wang, J. (2009). Polarization effects in molecular mechanical force fields. *J. Phys. Condens. Matter* **21**, 333102.

Claeyssens, F., Harvey, J. N., Manby, F. R., Mata, R. A., Mulholland, A. J., Ranaghan, K. E., et al. (2006). High-accuracy computation of reaction barriers in enzymes. *Angew. Chem. Int. Ed.* **45**, 6856–6859.

Claeyssens, F., Ranaghan, K. E., Lawan, N., Macrae, S. J., Manby, F. R., Harvey, J. N., et al. (2011). Analysis of chorismate mutase catalysis by QM/MM modelling of enzyme-catalysed and uncatalysed reactions. *Org. Biomol. Chem.* **9**, 1578–1590.

Clementi, C. (2008). Coarse-grained models of protein folding: toy models or predictive tools? *Curr. Opin. Struct. Biol.* **18**, 10–15.

Colombo, M. C., Guidoni, L., Laio, A., Magistrato, A., Maurer, P., Piana, S., et al. (2002). Hybrid QM/MM Car—Parrinello simulations of catalytic and enzymatic reactions. *ChemInform* **56**, 13–19.

Cornell, W. D., Cieplak, P., Bayly, C. I., Gould, I. R., Merz, K. M., Ferguson, D. M., et al. (1995). A second generation force field for the simulation of proteins, nucleic acids, and organic molecules. *J. Am. Chem. Soc.* **117**, 5179–5197.

Cramer, C. J., Truhlar, D. G. (2008). A universal approach to solvation modeling. *Acc. Chem. Res.* **41**, 760–768.

Crehuet, R., Field, M. J. (2007). A transition path sampling study of the reaction catalyzed by the enzyme chorismate mutase. *J. Phys. Chem. B* **111**, 5708–5718.

Cui, Q. (2002). Combining implicit solvation models with hybrid quantum mechanical/molecular mechanical methods: a critical test with glycine. *J. Chem. Phys.* **117**, 4720–4728.

Dal Peraro, M., Ruggerone, P., Raugei, S., Gervasio, F. L., Carloni, P. (2007). Investigating biological systems using first principles Car–Parrinello molecular dynamics simulations. *Curr. Opin. Struct. Biol.* **17**, 149–156.

Dang, L. X. (1998). Importance of polarization effects in modeling the hydrogen bond in water using classical molecular dynamics techniques. *J. Phys. Chem. B* **102**, 620–624.

de Marco, G., Varnai, P. (2009). Molecular simulation of conformational transitions in biomolecules using a combination of structure-based potential and empirical valence bond theory. *Phys. Chem. Chem. Phys.* **11**, 10694–10700.

de Oliveira, C. A. F., Guimarães, C. R. W., Barreiro, G., de Alencastro, R. B. (2003). Investigation of the induced-fit mechanism and catalytic activity of the human cytomegalovirus protease homodimer via molecular dynamics simulations. *Proteins: Struct. Funct. Bioinform.* **52**, 484–491.

de Oliveira, C. A. F., Grant, B. J., Zhou, M., McCammon, J. A. (2011). Large-scale conformational changes of Trypanosoma cruzi proline racemase predicted by accelerated molecular dynamics simulation. *PLoS Comput. Biol.* **7**, e1002178.

Dellago, C., Bolhuis, P. G., Geissler, P. L. (2002). Transition path sampling. *Adv. Chem. Phys.* **123**, 1–84.

Deng, S., Cai, W. (2007). Extending the fast multipole method for charges inside a dielectric sphere in an ionic solvent: high-order image approximations for reaction fields. *J. Comput. Phys.* **227**, 1246–1266.

Deng, N. J., Zheng, W., Gallicchio, E., Levy, R. M. (2011). Insights into the dynamics of HIV-1 protease: a kinetic network model constructed from atomistic simulations. *J. Am. Chem. Soc.* **133**, 9387–9394.

Dewar, M. J. S., Thiel, W. (1977). Ground states of molecules. 38. The MNDO method. Approximations and parameters. *J. Am. Chem. Soc.* **99**, 4899–4907.

Dewar, M. J. S., Zoebisch, E. G., Healy, E. F., Stewart, J. J. P. (1985). Development and use of quantum mechanical molecular models. 76. AM1: a new general purpose quantum mechanical molecular model. *J. Am. Chem. Soc.* **107**, 3902–3909.

Dickson, A., Maienschein-Cline, M., Tovo-Dwyer, A., Hammond, J. R., Dinner, A. R. (2011). Flow-dependent unfolding and refolding of an RNA by nonequilibrium umbrella sampling. *J. Chem. Theory Comput.* **7**, 2710–2720.

Dreuw, A., Weisman, J. L., Head-Gordon, M. (2003). Long-range charge-transfer excited states in time-dependent density functional theory require non-local exchange. *J. Chem. Phys.* **119**, 2943–2946.

Dyson, H. J., Wright, P. E. (2005). Intrinsically unstructured proteins and their functions. *Nat. Rev. Mol. Cell Biol.* **6**, 197–208.

Eisenmesser, E. Z., Bosco, D. A., Akke, M., Kern, D. (2002). Enzyme dynamics during catalysis. *Science* **295**, 1520–1523.

Eisenmesser, E. Z., Millet, O., Labeikovsky, W., Korzhnev, D. M., Wolf-Watz, M., Bosco, D. A., et al. (2005). Intrinsic dynamics of an enzyme underlies catalysis. *Nature* **438**, 117–121.

Elstner, M. (2006). The SCC-DFTB method and its application to biological systems. *Theor. Chem. Acc.* **116**, 316–325.

Ensing, B., De Vivo, M., Liu, Z., Moore, P., Klein, M. L. (2006). Metadynamics as a tool for exploring free energy landscapes of chemical reactions. *Acc. Chem. Res.* **39**, 73–81.

Estácio, S. G., Moreira, R., Guedes, R. C. (2011). Characterizing the dynamics and ligand-specific interactions in the human leukocyte elastase through molecular dynamics simulations. *J. Chem. Inf. Model.* **51**, 1690–1702.

Evans, M. G., Polanyi, M. (1935). Some applications of the transition state method to the calculation of reaction velocities, especially in solution. *Trans. Faraday Soc.* **31**, 875–894.

Eyring, H. (1935). The activated complex in chemical reactions. *J. Chem. Phys.* **3**, 107.

Faísca, P. F. N., Telo da Gama, M. M., Nunes, A. (2005). The Gō model revisited: native structure and the geometric coupling between local and long-range contacts. *Proteins* **60**, 712–722.

Faver, J. C., Benson, M. L., He, X., Roberts, B. P., Wang, B., Marshall, M. S., et al. (2011). The energy computation paradox and ab initio protein folding. *PLoS One* **6**, e18868.

Fenimore, P. W., Frauenfelder, H., McMahon, B. H., Young, R. D. (2004). Bulk-solvent and hydration-shell fluctuations, similar to α- and β-fluctuations in glasses, control protein motions and functions. *Proc. Natl. Acad. Sci. USA* **101**, 14408–14413.

Fersht, A. R. (1998). Structure and Mechanism in Protein Science: A Guide to Enzyme Catalysis and Protein Folding. W. H. Freeman, New York.

Fischer, E. (1894). Einfluss der configuration auf die wirkung der enzyme. *Ber. Dtsch. Chem. Ges.* **27**, 2984–2993.

Flomenbom, O., Velonia, K., Loos, D., Masuo, S., Cotlet, M., Engelborghs, Y., et al. (2005). Stretched exponential decay and correlations in the catalytic activity of fluctuating single lipase molecules. *Proc. Natl. Acad. Sci. USA* **102**, 2368–2372.

Frauenfelder, H., Chen, G., Berendzen, J., Fenimore, P. W., Jansson, H., McMahon, B. H., et al. (2009). A unified model of protein dynamics. *Proc. Natl. Acad. Sci. USA* **106**, 5129–5134.

Fushinobu, S., Nishimasu, H., Hattori, D., Song, H.-J., Wakagi, T. (2011). Structural basis for the bifunctionality of fructose-1,6-bisphosphate aldolase/phosphatase. *Nature* **478**, 538–541.

Gallicchio, E., Paris, K., Levy, R. M. (2009). The AGBNP2 implicit solvation model. *J. Chem. Theory Comput.* **5**, 2544–2564.

Gao, J., Amara, P., Alhambra, C., Field, M. J. (1998). A generalized hybrid orbital (GHO) method for the treatment of boundary atoms in combined QM/MM calculations. *J. Phys. Chem. A* **102**, 4714–4721.

Garcia-Viloca, M., Gao, J., Karplus, M., Truhlar, D. G. (2004). How enzymes work: analysis by modern rate theory and computer simulations. *Science* **303**, 186–195.

Gáspári, Z., Várnai, P., Szappanos, B., Perczel, A. (2010). Reconciling the lock-and-key and dynamic views of canonical serine protease inhibitor action. *FEBS Lett.* **584**, 203–206.

Gear, C. W. (1971). Numerical Initial Value Problems in Ordinary Differential Equations. Prentice Hall PTR, Upper Saddle River, NJ.

Geerke, D. P., Thiel, S., Thiel, W., van Gunsteren, W. F. (2008). QM-MM interactions in simulations of liquid water using combined semi-empirical/classical Hamiltonians. *Phys. Chem. Chem. Phys.* **10**, 297–302.

Giraldo, J., Roche, D., Rovira, X., Serra, J. (2006). The catalytic power of enzymes: conformational selection or transition state stabilization? *FEBS Lett.* **580**, 2170–2177.

Grossman, M., Born, B., Heyden, M., Tworowski, D., Fields, G. B., Sagi, I., et al. (2011). Correlated structural kinetics and retarded solvent dynamics at the metalloprotease active site. *Nat. Struct. Mol. Biol.* **18**, 1102–1108.

Haldane, J. B. S. (1930). Enzymes. Longmans Green, New York.

Hamelberg, D., Mongan, J., McCammon, J. A. (2004). Accelerated molecular dynamics: a promising and efficient simulation method for biomolecules. *J. Chem. Phys.* **120**, 11919–11929.

Hampton, S., Agarwal, P. K., Alam, S. R., Crozier, P. S. (2010). Towards microsecond biological molecular dynamics simulations on hybrid processors. 2010 International Conference on High Performance Computing and Simulation (HPCS), Caen, France, pp. 98–107.

Hansmann, U. H. E. (1997). Parallel tempering algorithm for conformational studies of biological molecules. *Chem. Phys. Lett.* **281**, 140–150.

He, Y., Li, Y., Mukherjee, S., Wu, Y., Yan, H., Lu, H. P. (2011). Probing single-molecule enzyme active-site conformational state intermittent coherence. *J. Am. Chem. Soc.* **133**, 14389–14395.

Henzler-Wildman, K., Kern, D. (2007). Dynamic personalities of proteins. *Nature* **450**, 964–972.

Henzler-Wildman, K. A., Lei, M., Thai, V., Kerns, S. J., Karplus, M., Kern, D. (2007a). A hierarchy of timescales in protein dynamics is linked to enzyme catalysis. *Nature* **450**, 913–916.

Henzler-Wildman, K. A., Thai, V., Lei, M., Ott, M., Wolf-Watz, M., Fenn, T., et al. (2007b). Intrinsic motions along an enzymatic reaction trajectory. *Nature* **450**, 838–844.

Himo, F., Siegbahn, P. E. M. (2003). Quantum chemical studies of radical-containing enzymes. *Chem. Rev.* **103**, 2421–2456.

Höltje, H.-D., Sippl, W., Rognan, D., Folkers, G. (2008). Molecular Modeling: Basic Principles and Applications. 3rd edn. Wiley-VCH, Weinheim.

Horiuti, J. (1938). On the statistical mechanical treatment of the absolute rate of chemical reaction. *Bull. Chem. Soc. Jpn.* **13**, 210–216.

Hornak, V., Abel, R., Okur, A., Strockbine, B., Roitberg, A., Simmerling, C. (2006). Comparison of multiple Amber force fields and development of improved protein backbone parameters. *Proteins* **65**, 712–725.

Hritz, J., Oostenbrink, C. (2007). Optimization of replica exchange molecular dynamics by fast mimicking. *J. Chem. Phys.* **127**, 204104.

Hünenberger, P. H., Börjesson, U., Lins, R. D. (2001). Electrostatic interactions in biomolecular systems. *Chimia* **55**, 861–866.

Jaramillo, A., Wodak, S. J. (2005). Computational protein design is a challenge for implicit solvation models. *Biophys. J.* **88**, 156–171.

Jarzynski, C. (1997). Equilibrium free-energy differences from nonequilibrium measurements: a master-equation approach. *Phys. Rev. E* **56**, 5018–5035.
Jencks, W. P. (1986). Catalysis in Chemistry and Enzymology. Dover Publications, New York.
Jiménez, A., Clapés, P., Crehuet, R. (2008). A dynamic view of enzyme catalysis. *J. Mol. Model.* **14**, 735–746.
Jorgensen, W. L., Madura, J. D. (1985). Temperature and size dependence for Monte Carlo simulations of TIP4P water. *Mol. Phys.* **56**, 1381–1392.
Jorgensen, W. L., Chandrasekhar, J., Madura, J. D., Impey, R. W., Klein, M. L. (1983). Comparison of simple potential functions for simulating liquid water. *J. Chem. Phys.* **79**, 926–935.
Jorgensen, W. L., Maxwell, D. S., Tirado-Rives, J. (1996). Development and testing of the OPLS all-atom force field on conformational energetics and properties of organic liquids. *J. Am. Chem. Soc.* **118**, 11225–11236.
Kamerlin, S. C. L., Warshel, A. (2010a). At the dawn of the 21st century: is dynamics the missing link for understanding enzyme catalysis? *Proteins: Struct., Funct., Bioinf.* **78**, 1339–1375.
Kamerlin, S. C. L., Warshel, A. (2010b). Reply to karplus: conformational dynamics have no role in the chemical step. *Proc. Natl. Acad. Sci. USA* **107**, E72.
Kamerlin, S. C. L., Haranczyk, M., Warshel, A. (2009). Progress in ab initio QM/MM free-energy simulations of electrostatic energies in proteins: accelerated QM/MM studies of pKa, redox reactions and solvation free energies. *J. Phys. Chem. B* **113**, 1253–1272.
Kamerlin, S. C. L., Vicatos, S., Dryga, A., Warshel, A. (2011). Coarse-grained (multiscale) simulations in studies of biophysical and chemical systems. *Annu. Rev. Phys. Chem.* **62**, 41–64.
Karplus, M. (2010). Role of conformation transitions in adenylate kinase. *Proc. Natl. Acad. Sci. USA* **107**, E71.
Karplus, M., Kuriyan, J. (2005). Molecular dynamics and protein function. *Proc. Natl. Acad. Sci. USA* **102**, 6679–6685.
Kästner, J. (2009). Umbrella integration in two or more reaction coordinates. *J. Chem. Phys.* **131**, 034109.
Kästner, J. (2011). Umbrella sampling. *WIREs Comput. Mol. Sci.* **1**, 932–942.
Kästner, J., Senn, H. M., Thiel, S., Otte, N., Thiel, W. (2006). QM/MM free-energy perturbation compared to thermodynamic integration and umbrella sampling: application to an enzymatic reaction. *J. Chem. Theory Comput.* **2**, 452–461.
Kästner, J., Thiel, S., Senn, H. M., Sherwood, P., Thiel, W. (2007). Exploiting QM/MM capabilities in geometry optimization: a microiterative approach using electrostatic embedding. *J. Chem. Theory Comput.* **3**, 1064–1072.
Ke, Z., Wang, S., Xie, D., Zhang, Y. (2009a). Born–Oppenheimer ab initio QM/MM molecular dynamics simulations of the hydrolysis reaction catalyzed by protein arginine deiminase 4. *J. Phys. Chem. B* **113**, 16705–16710.
Ke, Z., Zhou, Y., Hu, P., Wang, S., Xie, D., Zhang, Y. (2009b). Active site cysteine is protonated in the PAD4 Michaelis complex: evidence from Born–Oppenheimer ab initio QM/MM molecular dynamics simulations. *J. Phys. Chem. B* **113**, 12750–12758.

Kirkpatrick, S., Gelatt, C. D. J., Vecchi, M. P. (1983). Optimization by simulated annealing. *Science* **220**, 671–680.

Klinman, J. P. (2009). An integrated model for enzyme catalysis emerges from studies of hydrogen tunneling. *Chem. Phys. Lett.* **471**, 179–193.

Koshland, D. E. (1958). Application of a theory of enzyme specificity to protein synthesis. *Proc. Natl. Acad. Sci. USA* **44**, 98–104.

Koshland, D. E., Hamadani, K. (2002). Proteomics and models for enzyme cooperativity. *J. Biol. Chem.* **277**, 46841–46844.

Kumar, S., Rosenberg, J. M., Bouzida, D., Swendsen, R. H., Kollman, P. A. (1995). Multidimensional free-energy calculations using the weighted histogram analysis method. *J. Comput. Chem.* **16**, 1339–1350.

Kunz, A.-P. E., van Gunsteren, W. F. (2011). A method for conformational sampling of loops in proteins based on adiabatic decoupling and temperature or force scaling. *ChemPhysChem* **12**, 2609–2614.

Laino, T., Mohamed, F., Laio, A., Parrinello, M. (2005). An efficient real space multigrid QM/MM electrostatic coupling. *J. Chem. Theory Comput.* **1**, 1176–1184.

Laio, A., Parrinello, M. (2002). Escaping free-energy minima. *Proc. Natl. Acad. Sci. USA* **99**, 12562–12566.

Lamoureux, G., MacKerell, A. D., Roux, B. (2003). A simple polarizable model of water based on classical Drude oscillators. *J. Chem. Phys.* **119**, 5185–5197.

Lamoureux, G., Harder, E., Vorobyov, I. V., Roux, B., MacKerell, A. D., Jr. (2006). A polarizable model of water for molecular dynamics simulations of biomolecules. *Chem. Phys. Lett.* **418**, 245–249.

Lee, G. M., Craik, C. S. (2009). Trapping moving targets with small molecules. *Science* **324**, 213–215.

Leone, V., Lattanzi, G., Molteni, C., Carloni, P. (2009). Mechanism of action of cyclophilin a explored by metadynamics simulations. *PLoS Comput. Biol.* **5**, e1000309.

Liao, R.-Z., Yu, J.-G., Himo, F. (2011). Quantum chemical modeling of enzymatic reactions: the case of decarboxylation. *J. Chem. Theory Comput.* **7**, 1494–1501.

Lin, H., Truhlar, D. G. (2007). QM/MM: what have we learned, where are we, and where do we go from here? *Theor. Chem. Acc.* **117**, 185–199.

Lipparini, F., Barone, V. (2011). Polarizable force fields and polarizable continuum model: a fluctuating charges/PCM approach. 1. Theory and implementation. *J. Chem. Theory Comput.* **7**, 3711–3724.

Liu, H. (2010). QM/MM energy functions, configuration optimizations, and free energy simulations of enzyme catalysis. In: Kinetics and Dynamics: From Nano- to Bio-scale, Challenges and Advances in Computational Chemistry and Physics, Paneth, P. and Dybala-Defratyka, A. (Eds.), pp. 331–353. Springer, Netherlands.

Loncharich, R. J., Brooks, B. R. (1989). The effects of truncating long-range forces on protein dynamics. *Proteins* **6**, 32–45.

Lonsdale, R., Ranaghan, K. E., Mulholland, A. J. (2010). Computational enzymology. *Chem. Commun.* **46**, 2354–2372.

Lopes, P. E. M., Harder, E., Roux, B., Mackerell, A. (2009). Formalisms for the explicit inclusion of electronic polarizability in molecular modeling and dynamics studies. In Multi-scale Quantum Models for Biocatalysis: Modern Techniques and Applications, Challenges and Advances in Computational Chemistry and Physics, York, D. M. and Lee, T.-S. (Eds.) pp. 219–257. Springer, Netherlands.

Lu, J., Kadakkuzha, B. M., Zhao, L., Fan, M., Qi, X., Xia, T. (2011). Dynamic ensemble view of the conformational landscape of HIV-1 TAR RNA and allosteric recognition. *Biochemistry* **50**, 5042–5057.

Luque, F. J., Dehez, F., Chipot, C., Orozco, M. (2011). Polarization effects in molecular interactions. *WIREs Comput. Mol. Sci.* **1**, 844–854.

Mackerell, A. D. (2004). Empirical force fields for biological macromolecules: overview and issues. *J. Comput. Chem.* **25**, 1584–1604.

MacKerell, A. D., Bashford, D., Bellott, M., Dunbrack, R. L., Evanseck, J. D., Field, M. J., et al. (1998a). All-atom empirical potential for molecular modeling and dynamics studies of proteins. *J. Phys. Chem. B* **102**, 3586–3616.

MacKerell, A., Brooks, C., Nilsson, L., Roux, B., Won, Y., Karplus, M. (1998b). CHARMM: the energy function and its parameterization with an overview of the program. In The Encyclopedia of Computational Chemistry, Schleyer, P. v. R. (Ed.) pp. 271–277. John Wiley & Sons, Chichester.

Mackerell, A. D., Feig, M., Brooks, C. L. (2004). Extending the treatment of backbone energetics in protein force fields: limitations of gas-phase quantum mechanics in reproducing protein conformational distributions in molecular dynamics simulations. *J. Comput. Chem.* **25**, 1400–1415.

Mahoney, M. W., Jorgensen, W. L. (2000). A five-site model for liquid water and the reproduction of the density anomaly by rigid, nonpolarizable potential functions. *J. Chem. Phys.* **112**, 8910–8922.

Marenich, A. V., Cramer, C. J., Truhlar, D. G. (2009). Universal solvation model based on the Generalized Born approximation with asymmetric descreening. *J. Chem. Theory Comput.* **5**, 2447–2464.

Masterson, L. R., Shi, L., Metcalfe, E., Gao, J., Taylor, S. S., Veglia, G. (2011). Dynamically committed, uncommitted, and quenched states encoded in protein kinase A revealed by NMR spectroscopy. *Proc. Natl. Acad. Sci. USA* **108**, 6969–6974.

Mata, R. A. (2010). Application of high level wavefunction methods in quantum mechanics/molecular mechanics hybrid schemes. *Phys. Chem. Chem. Phys.* **12**, 5041–5052.

McGeagh, J. D., Mulholland, A. J. (2010). Enzyme dynamics and catalysis: insights from simulations. In: Kinetics and Dynamics: From Nano- to Bio-scale, Challenges and Advances in Computational Chemistry and Physics, Paneth, P. and Dybala-Defratyka, A. (Eds.), pp. 375–395. Springer, Netherlands.

McGeagh, J. D., Ranaghan, K. E., Mulholland, A. J. (2011). Protein dynamics and enzyme catalysis: insights from simulations. *Biochimica et Biophysica Acta, Proteins Proteomics* **1814**, 1077–1092.

Mennucci, B. (2010). Continuum solvation models: what else can we learn from them? *J. Phys. Chem. Lett.* **1**, 1666–1674.

Mesecar, A. D., Stoddard, B. L., Koshland, D. E. (1997). Orbital steering in the catalytic power of enzymes: small structural changes with large catalytic consequences. *Science* **277**, 202–206.

Metropolis, N., Rosenbluth, A. W., Teller, A. H., Teller, E. (1953). Equation of state calculations by fast computing machines. *J. Chem. Phys.* **21**, 1087–1092.

Mikołajczyk, M. M., Zaleśny, R., Czyżnikowska, Z., Toman, P., Leszczynski, J., Bartkowiak, W. (2011). Long-range corrected DFT calculations of charge-transfer integrals in model metal-free phthalocyanine complexes. *J. Mol. Model.* **17**, 2143–2149.

Mulholland, A. J., Grant, I. M. (2007). Computational enzymology: insights into enzyme mechanism and catalysis from modelling. In: Molecular Materials with Specific Interactions-Modeling and Design, Sokalski, W. A. (Ed.) pp. 275–304. Springer, Netherlands.

Neet, K. E. (1998). Enzyme catalytic power minireview series. *J. Biol. Chem.* **273**, 25527–25528.

Noé, F., Schütte, C., Vanden-Eijnden, E., Reich, L., Weikl, T. R. (2009). Constructing the equilibrium ensemble of folding pathways from short off-equilibrium simulations. *Proc. Natl. Acad. Sci. USA* **106**, 19011–19016.

Norberg, J., Nilsson, L. (2000). On the truncation of long-range electrostatic interactions in DNA. *Biophys. J.* **79**, 1537–1553.

Okur, A., Roe, D. R., Cui, G., Hornak, V., Simmerling, C. (2007). Improving convergence of replica-exchange simulations through coupling to a high-temperature structure reservoir. *J. Chem. Theory Comput.* **3**, 557–568.

Olsson, M. H., Mavri, J., Warshel, A. (2006). Transition state theory can be used in studies of enzyme catalysis: lessons from simulations of tunnelling and dynamical effects in lipoxygenase and other systems. *Philos. Trans. R. Soc. Lond. B Biol. Sci.* **361**, 1417–1432.

Onufriev, A. (2008). Implicit solvent models in molecular dynamics simulations: a brief overview. In: Annual Reports in Computational Chemistry, Wheeler, R. A. and Spellmeyer, D. C. (Eds.) pp. 125–137. Elsevier, Amsterdam.

Onufriev, A., Case, D. A., Bashford, D. (2002). Effective Born radii in the generalized Born approximation: the importance of being perfect. *J. Comput. Chem.* **23**, 1297–1304.

Panek, J. J., Mazzarello, R., Novič, M., Jezierska-Mazzarello, A. (2011). Impact of Mercury(II) on proteinase K catalytic center: investigations via classical and Born-Oppenheimer molecular dynamics. *Mol. Divers.* **15**, 215–226.

Park, S., Khalili-Araghi, F., Tajkhorshid, E., Schulten, K. (2003). Free energy calculation from steered molecular dynamics simulations using Jarzynski's equality. *J. Chem. Phys.* **119**, 3559–3566.

Parker, L. J., Italiano, L. C., Morton, C. J., Hancock, N. C., Ascher, D. B., Aitken, J. B., et al. (2011). Studies of glutathione transferase P1-1 bound to a platinum(IV)-based anticancer compound reveal the molecular basis of its activation. *Chem. Eur. J.* **17**, 7806–7816.

Patel, S., Brooks, C. L. (2006). Fluctuating charge force fields: recent developments and applications from small molecules to macromolecular biological systems. *Mol. Simul.* **32**, 231–249.

Pauling, L. (1946). Molecular architecture and biological reactions. *Chem. Eng. News* **24**, 1375–1377.

Pauling, L. (1948). Nature of forces between large molecules of biological interest. *Nature* **161**, 707–709.

Perez, D., Uberuaga, B. P., Shim, Y., Amar, J. G., Voter, A. F. (2009). Accelerated molecular dynamics methods: introduction and recent developments. In: Annual Reports in Computational Chemistry, Wheeler, R. A. and Spellmeyer, D. C. (Eds.) pp. 79–98. Elsevier, Amsterdam.

Pervushin, K., Vamvaca, K., Vogeli, B., Hilvert, D. (2007). Structure and dynamics of a molten globular enzyme. *Nat. Struct. Mol. Biol.* **14**, 1202–1206.

Pingali, S. V., O'Neill, H. M., McGaughey, J., Urban, V. S., Rempe, C. S., Petridis, L., et al. (2011). Small angle neutron scattering reveals pH-dependent conformational changes in Trichoderma reesei Cellobiohydrolase I. *J. Biol. Chem.* **286**, 32801–32809.

Pisliakov, A. V., Cao, J., Kamerlin, S. C. L., Warshel, A. (2009). Enzyme millisecond conformational dynamics do not catalyze the chemical step. *Proc. Natl. Acad. Sci. USA* **106**, 17359–17364.

Ponder, J. W. (2011). TINKER-Software Tools for Molecular Design. Washington University School of Medicine, Saint Louis.

Ponder, J. W., Wu, C., Ren, P., Pande, V. S., Chodera, J. D., Schnieders, M. J., et al. (2010). Current status of the AMOEBA polarizable force field. *J. Phys. Chem. B* **114**, 2549–2564.

Potestio, R., Pontiggia, F., Micheletti, C. (2009). Coarse-grained description of protein internal dynamics: an optimal strategy for decomposing proteins in rigid subunits. *Biophys. J.* **96**, 4993–5002.

Prat-Resina, X., Bofill, J. M., González-Lafont, À., Lluch, J. M. (2004). Geometry optimization and transition state search in enzymes: different options in the microiterative method. *Int. J. Quantum Chem.* **98**, 367–377.

Pu, J., Gao, J., Truhlar, D. G. (2006). Multidimensional tunneling, recrossing, and the transmission coefficient for enzymatic reactions. *Chem. Rev.* **106**, 3140–3169.

Qian, H., Shi, P.-Z. (2009). Fluctuating enzyme and its biological functions: positive cooperativity without multiple states. *J. Phys. Chem. B* **113**, 2225–2230.

Quaytman, S. L., Schwartz, S. D. (2008). Comparison studies of the human heart and Bacillus stearothermophilus lactate dehydrogreanse by transition path sampling. *J. Phys. Chem. A* **113**, 1892–1897.

Radzicka, A., Wolfenden, R. (1995). A proficient enzyme. *Science* **267**, 90–93.

Ramanathan, A., Savol, A. J., Langmead, C. J., Agarwal, P. K., Chennubhotla, C. S. (2011). Discovering conformational sub-states relevant to protein function. *PLoS One* **6**, e15827.

Ranaghan, K. E., Mulholland, A. J. (2010). Investigations of enzyme-catalysed reactions with combined quantum mechanics/molecular mechanics (QM/MM) methods. *Inter. Rev. Phys. Chem.* **29**, 65–133.

Ranaghan, K. E., Ridder, L., Szefczyk, B., Sokalski, W. A., Hermann, J. C., Mulholland, A. J. (2004). Transition state stabilization and substrate strain in enzyme catalysis: ab initio QM/MM modelling of the chorismate mutase reaction. *Org. Biomol. Chem.* **2**, 968–980.

Ren, P., Ponder, J. W. (2002). Consistent treatment of inter- and intramolecular polarization in molecular mechanics calculations. *J. Comput. Chem.* **23**, 1497–1506.

Ren, P., Ponder, J. W. (2004). Temperature and pressure dependence of the AMOEBA water model. *J. Phys. Chem. B* **108**, 13427–13437.

Repasky, M. P., Chandrasekhar, J., Jorgensen, W. L. (2002). PDDG/PM3 and PDDG/MNDO: improved semiempirical methods. *J. Comput. Chem.* **23**, 1601–1622.

Rick, S. W. (2001). Simulations of ice and liquid water over a range of temperatures using the fluctuating charge model. *J. Chem. Phys.* **114**, 2276–2283.

Roitberg, A. E., Okur, A., Simmerling, C. (2007). Coupling of replica exchange simulations to a non-Boltzmann structure reservoir. *J. Phys. Chem. B* **111**, 2415–2418.

Rosta, E., Klähn, M., Warshel, A. (2006). Towards accurate ab initio QM/MM calculations of free-energy profiles of enzymatic reactions. *J. Phys. Chem. B* **110**, 2934–2941.

Roux, B. (1995). The calculation of the potential of mean force using computer simulations. *Comput. Phys. Commun.* **91**, 275–282.

Saebo, S., Pulay, P. (1993). Local treatment of electron correlation. *Annu. Rev. Phys. Chem.* **44**, 213–236.

Schaefer, P., Riccardi, D., Cui, Q. (2005). Reliable treatment of electrostatics in combined QM/MM simulation of macromolecules. *J. Chem. Phys.* **123**, 1–14.

Scheraga, H. A., Khalili, M., Liwo, A. (2011). Protein-folding dynamics: overview of molecular simulation techniques. *Annu. Rev. Phys. Chem.* **58**, 57–83.

Schlick, T. (2009). Molecular dynamics-based approaches for enhanced sampling of long-time, large-scale conformational changes in biomolecules. *F1000 Biol. Rep.* **1**, 1–9.

Schlick, T. (2010). Molecular Modeling and Simulation: An Interdisciplinary Guide. 2nd edn. Springer-Verlag New York, New York.

Schramm, V. L. (2011a). Enzymatic transition states, transition-state analogs, dynamics, thermodynamics, and lifetimes. *Annu. Rev. Biochem.* **80**, 703–732.

Schramm, V. L. (2011b). Introduction: principles of enzymatic catalysis. *Chem. Rev.* **106**, 3029–3030.

Schwartz, S. D., Schramm, V. L. (2009). Enzymatic transition states and dynamic motion in barrier crossing. *Nat. Chem. Biol.* **5**, 551–558.

Scott, W. R. P., Hünenberger, P. H., Tironi, I. G., Mark, A. E., Billeter, S. R., Fennen, J., et al. (1999). The GROMOS biomolecular simulation program package. *J. Phys. Chem. A* **103**, 3596–3607.

Senn, H. M., Thiel, W. (2009). QM/MM methods for biomolecular systems. *Angew. Chem. Int. Ed.* **48**, 1198–1229.

Senn, H. M., Thiel, S., Thiel, W. (2005). Enzymatic hydroxylation in p-hydroxybenzoate hydroxylase: a case study for QM/MM molecular dynamics. *J. Chem. Theory Comput.* **1**, 494–505.

Shaik, S., Cohen, S., Wang, Y., Chen, H., Kumar, D., Thiel, W. (2010). P450 enzymes: their structure, reactivity, and selectivity-modeled by QM/MM calculations. *Chem. Rev.* **110**, 949–1017.

Shavitt, I., Bartlett, R. J. (2009). Many-Body Methods in Chemistry and Physics: MBPT and Coupled-Cluster Theory (Cambridge Molecular Science). Cambridge University Press, Cambridge.

Siegbahn, P. E. M., Himo, F. (2009). Recent developments of the quantum chemical cluster approach for modeling enzyme reactions. *J. Biol. Inorg. Chem.* **14**, 643–651.

Siegbahn, P. E. M., Himo, F. (2011). The quantum chemical cluster approach for modeling enzyme reactions. *WIREs Comput. Mol. Sci.* **1**, 323–336.

Singh, U. C., Kollman, P. A. (1986). A combined ab initio quantum mechanical and molecular mechanical method for carrying out simulations on complex molecular systems: applications to the CH3Cl + Cl− exchange reaction and gas phase protonation of polyethers. *J. Comput. Chem.* **7**, 718–730.

Skinner, J. J., Wood, S., Shorter, J., Englander, S. W., Black, B. E. (2008). The Mad2 partial unfolding model: regulating mitosis through Mad2 conformational switching. *J. Cell Biol.* **183**, 761–768.

Solomatin, S. V., Greenfeld, M., Chu, S., Herschlag, D. (2010). Multiple native states reveal persistent ruggedness of an RNA folding landscape. *Nature* **463**, 681–684.

Steindal, A. H., Ruud, K., Frediani, L., Aidas, K., Kongsted, J. (2011). Excitation energies in solution: the fully polarizable QM/MM/PCM method. *J. Phys. Chem. B* **115**, 3027–3037.

Stewart, J. J. P. (1989). Optimization of parameters for semi-empirical methods I. Method. *J. Comput. Chem.* **10**, 209–220.

Sugita, Y., Okamoto, Y. (1999). Replica-exchange molecular dynamics method for protein folding. *Chem. Phys. Lett.* **314**, 141–151.

Sullivan, S. M., Holyoak, T. (2008). Enzymes with lid-gated active sites must operate by an induced fit mechanism instead of conformational selection. *Proc. Natl. Acad. Sci. USA* **105**, 13829–13834.

Sutcliffe, M. J., Scrutton, N. S. (2006). Computational studies of enzyme mechanism: linking theory with experiment in the analysis of enzymic H-tunnelling. *Phys. Chem. Chem. Phys.* **8**, 4510–4516.

Sutcliffe, M. J., Masgrau, L., Roujeinikova, A., Johannissen, L. O., Hothi, P., Basran, J., et al. (2006). Hydrogen tunnelling in enzyme-catalysed H-transfer reactions: flavoprotein and quinoprotein systems. *Philos. Trans. R. Soc. Lond. B Biol. Sci.* **361**, 1375–1386.

Svensson, M., Humbel, S., Froese, R. D. J., Matsubara, T., Sieber, S., Morokuma, K. (1996). ONIOM: a multilayered integrated MO+MM method for geometry optimizations and single point energy predictions. A test for Diels–Alder reactions and Pt(P(t-Bu)3)2 + H2 oxidative addition. *J. Phys. Chem.* **100**, 19357–19363.

Swope, W. C., Andersen, H. C., Berens, P. H., Wilson, K. R. (1982). A computer simulation method for the calculation of equilibrium constants for the formation of physical clusters of molecules: application to small water clusters. *J. Chem. Phys.* **76**, 637–649.

Szefczyk, B., Mulholland, A. J., Ranaghan, K. E., Sokalski, W. A. (2004). Differential transition-state stabilization in enzyme catalysis: quantum chemical analysis of interactions in the chorismate mutase reaction and prediction of the optimal catalytic field. *J. Am. Chem. Soc.* **126**, 16148–16159.

Tachikawa, H., Yabushita, A., Kawasaki, M. (2011). Ab initio theoretical calculations of the electronic excitation energies of small water clusters. *Phys. Chem. Chem. Phys.* **13**, 20745–20749.

Taketomi, H., Ueda, Y., Gō, N. (1975). Studies on protein folding, unfolding and fluctuations by computer simulation. I. The effect of specific amino acid sequence represented by specific inter-unit interactions. *Int. J. Pept. Protein Res.* **7**, 445–459.

Takuya, O., Hakoshima, T., Itakura, M., Yamamori, S., Takahashi, M., Hashimoto, Y., et al. (2002). Role of loop structures of neuropsin in the activity of serine protease and regulated secretion. *J. Biol. Chem.* **277**, 14724–14730.

Théry, V., Rinaldi, D., Rivail, J.-L., Maigret, B., Ferenczy, G. G. (1994). Quantum mechanical computations on very large molecular systems: the local self-consistent field method. *J. Comput. Chem.* **15**, 269–282.

Thiel, W., Voityuk, A. A. (1996). Extension of MNDO to d orbitals: parameters and results for the second-row elements and for the zinc group. *J. Phys. Chem.* **100**, 616–626.

Tian, B.-X., Eriksson, L. A. (2011). Catalytic mechanism and roles of Arg197 and Thr183 in the staphylococcus aureus sortase A enzyme. *J. Phys. Chem. B* **115**(44), 13003–13011.

Tokuriki, N., Tawfik, D. S. (2009). Protein dynamism and evolvability. *Science* **324**, 203–207.

Tomasi, J., Mennucci, B., Cammi, R. (2005). Quantum mechanical continuum solvation models. *Chem. Rev.* **105**, 2999–3094.

Tompa, P. (2011). Unstructural biology coming of age. *Curr. Opin. Struct. Biol.* **21**, 419–425.

Tompa, P., Fuxreiter, M. (2008). Fuzzy complexes: polymorphism and structural disorder in protein–protein interactions. *Trends Biochem. Sci.* **33**, 2–8.

Torrie, G. M., Valleau, J. P. (1977). Nonphysical sampling distributions in Monte Carlo free-energy estimation: umbrella sampling. *J. Comput. Phys.* **23**, 187–199.

Tozer, D. J., Handy, N. C. (2000). On the determination of excitation energies using density functional theory. *Phys. Chem. Chem. Phys.* **2**, 2117–2121.

Tuckerman, M., Berne, B. J., Martyna, G. J. (1992). Reversible multiple time scale molecular dynamics. *J. Chem. Phys.* **97**, 1990–2001.

Tuinstra, R. L., Peterson, F. C., Kutlesa, S., Elgin, E. S., Kron, M. A., Volkman, B. F. (2008). Interconversion between two unrelated protein folds in the lymphotactin native state. *Proc. Natl. Acad. Sci. USA* **105**, 5057–5062.

Ufimtsev, I. S., Luehr, N., Martinez, T. J. (2011). Charge transfer and polarization in solvated proteins from ab initio molecular dynamics. *J. Phys. Chem. Lett.* **2**, 1789–1793.

van Gunsteren, W., Berendsen, H. (1977). Algorithms for macromolecular dynamics and constraint dynamics. *Mol. Phys.* **34**, 1311–1327.

Vendruscolo, M. (2010). Enzymatic activity in disordered states of proteins. *Curr. Opin. Chem. Biol.* **14**, 671–675.

Verlet, L. (1967). Computer "experiments" on classical fluids. I. Thermodynamical properties of Lennard-Jones molecules. *Phys. Rev.* **159**, 98–103.

Vijayan, R., Biggin, P. C. (2010). Conformational preferences of a 14-residue fibrillogenic peptide from acetylcholinesterase. *Biochemistry* **49**, 3678–3684.

Vreven, T., Morokuma, K., Farkas, O., Schlegel, H. B., Frisch, M. J. (2003). Geometry optimization with QM/MM, ONIOM, and other combined methods. I. Microiterations and Constraints. *J. Comput. Chem.* **24**, 760–769.

Vreven, T., Byun, K. S., Komáromi, I., Dapprich, S., Montgomery, J. A., Morokuma, K., et al. (2006). Combining quantum mechanics methods with molecular mechanics methods in ONIOM. *J. Chem. Theory Comput.* **2**, 815–826.

Wang, W., Skeel, R. D. (2005). Fast evaluation of polarizable forces. *J. Chem. Phys.* **123**, 164107.

Wang, W., Donini, O., Reyes, C. M., Kollman, P. A. (2001). Biomolecular simulations: recent developments in force fields, simulations of enzyme catalysis, protein-ligand, protein-protein, and protein-nucleic acid noncovalent interactions. *Annu. Rev. Biophys. Biomol. Struct.* **30**, 211–243.

Wang, S., Hu, P., Zhang, Y. (2007). Ab initio quantum mechanical/molecular mechanical molecular dynamics simulation of enzyme catalysis: the case of histone lysine methyltransferase SET7/9. *J. Phys. Chem. B* **111**, 3758–3764.

Wang, J., Cieplak, P., Li, J., Wang, J., Cai, Q., Hsieh, M., et al. (2011a). Development of polarizable models for molecular mechanical calculations II: induced dipole models significantly improve accuracy of intermolecular interaction energies. *J. Phys. Chem. B* **115**, 3100–3111.

Wang, Y., Harrison, C. B., Schulten, K. (2011b). Implementation of accelerated molecular dynamics in NAMD. *Comput. Sci. Discov.* **4**, 015002.

Warshel, A. (1998). Electrostatic origin of the catalytic power of enzymes and the role of preorganized active sites. *J. Biol. Chem.* **273**, 27035–27038.

Warshel, A. (2003). Computer simulations of enzyme catalysis: methods, progress, and insights. *Annu. Rev. Biophys. Biomol. Struct.* **32**, 425–443.

Warshel, A., Levitt, M. (1976). Theoretical studies of enzymic reactions: dielectric, electrostatic and steric stabilization of the carbonium ion in the reaction of lysozyme. *J. Mol. Biol.* **103**, 227–249.

Warshel, A., Weiss, R. M. (1980). An empirical valence bond approach for comparing reactions in solutions and in enzymes. *J. Am. Chem. Soc.* **102**, 6218–6226.

Warshel, A., Sharma, P. K., Kato, M., Xiang, Y., Liu, H., Olsson, M. H. M. (2006). Electrostatic basis for enzyme catalysis. *Chem. Rev.* **106**, 3210–3235.

Weinan, E., Vanden-Eijnden, E. (2010). Transition-path theory and path-finding algorithms for the study of rare events. *Annu. Rev. Phys. Chem.* **61**, 391–420.

Wesolowski, T. A., Warshel, A. (1993). Frozen density functional approach for ab initio calculations of solvated molecules. *J. Phys. Chem.* **97**, 8050–8053.

Whitford, P. C., Noel, J. K., Gosavi, S., Schug, A., Sanbonmatsu, K. Y., Onuchic, J. N. (2009). An all-atom structure-based potential for proteins: bridging minimal models with all-atom empirical forcefields. *Proteins* **75**, 430–441.

Wickstrom, L., Okur, A., Simmerling, C. (2009). Evaluating the performance of the ff99SB force field based on NMR scalar coupling data. *Biophys. J.* **97**, 853–856.

Wigner, E. (1938). The transition state method. *Trans. Faraday Soc.* **34**, 29–41.

Wohlert, J., Berglund, L. A. (2011). A coarse-grained model for molecular dynamics simulations of native cellulose. *J. Chem. Theory Comput.* **7**, 753–760.

Woiczikowski, P. B., Steinbrecher, T., Kubař, T., Elstner, M. (2011). Nonadiabatic QM/MM simulations of fast charge transfer in Escherichia coli DNA photolyase. *J. Phys. Chem. B* **115**, 9846–9863.

Wolfenden, R., Snider, M. J. (2001). The depth of chemical time and the power of enzymes as catalysts. *Acc. Chem. Res.* **34**, 938–945.

Woodcock, H. L., Miller, B. T., Hodoscek, M., Okur, A., Larkin, J. D., Ponder, J. W., et al. (2011). MSCALE: a general utility for multiscale modeling. *J. Chem. Theory Comput.* **7**, 1208–1219.

Wu, R., Hu, P., Wang, S., Cao, Z., Zhang, Y. (2009). Flexibility of catalytic zinc coordination in thermolysin and HDAC8: a Born–Oppenheimer ab initio QM/MM molecular dynamics study. *J. Chem. Theory Comput.* **6**, 337–343.

Xiong, H., Crespo, A., Marti, M., Estrin, D., Roitberg, A. E. (2006). Free energy calculations with non-equilibrium methods: application of the Jarzynski relationship. *Theor. Chem. Acc.* **116**, 338–346.

Xu, Z., Deng, S., Cai, W. (2009). Image charge approximations of reaction fields in solvents with arbitrary ionic strength. *J. Comput. Phys.* **228**, 2092–2099.

Yadid, I., Kirshenbaum, N., Sharon, M., Dym, O., Tawfik, D. S. (2010). Metamorphic proteins mediate evolutionary transitions of structure. *Proc. Natl. Acad. Sci. USA* **107**, 7287–7292.

Yang, Z., Májek, P., Bahar, I. (2009). Allosteric transitions of supramolecular systems explored by network models: application to chaperonin GroEL. *PLoS Comput. Biol.* **5**, e1000360.

York, D., Darden, T., Pedersen, L. (1993). The effect of long-range electrostatic interactions in simulations of macromolecular crystals: a comparison of the ewald and truncated list methods. *J. Chem. Phys.* **99**, 8345–8348.

Zhang, Y., Schramm, V. L. (2011). Ground-state destabilization in orotate phosphoribosyltransferases by binding isotope effects. *Biochemistry* **50**, 4813–4818.

Zhang, Y., Xu, X., Goddard, W. A. (2009). Doubly hybrid density functional for accurate descriptions of nonbond interactions, thermochemistry, and thermochemical kinetics. *Proc. Natl. Acad. Sci. USA* **106**, 4963–4968.

Zhang, I. Y., Wu, J., Xu, X. (2010). Extending the reliability and applicability of B3LYP. *Chem. Commun.* **46**, 3057–3070.

Zhao, Y., Truhlar, D. G. (2005). Design of density functionals that are broadly accurate for thermochemistry, thermochemical kinetics, and nonbonded interactions. *J. Phys. Chem. A* **109**, 5656–5667.

Zhao, Y., Truhlar, D. G. (2008). Density functionals with broad applicability in chemistry. *Acc. Chem. Res.* **120**, 215–241.

Zhou, Y., Wang, S., Zhang, Y. (2010). Catalytic reaction mechanism of acetylcholinesterase determined by Born–Oppenheimer ab initio QM/MM molecular dynamics simulations. *J. Phys. Chem. B* **114**, 8817–8825.

Zuckerman, D. M. (2011). Equilibrium sampling in biomolecular simulations. *Annu. Rev. Biophys.* **40**, 41–62.

DYNAMIC SIMULATIONS AS A COMPLEMENT TO EXPERIMENTAL STUDIES OF ENZYME MECHANISMS

By ANDREW CHRISTOFFERSON,* LIFENG ZHAO,† AND QIANG PEI†

*School of Applied Sciences, Royal Melbourne Institute of Technology, Melbourne, Australia
†National Institute of Biological Sciences, Beijing, PR China

Abstract

Dynamic simulations are a useful means of studying certain aspects of enzyme reaction mechanisms that cannot be understood easily through current experimental methods. With increases in computing power and innovations with computer hardware, more and more accurate techniques for modeling biological systems at the atomic level are under development. Here, we discuss two enzyme systems that illustrate the utility of dynamic simulations in conjunction with experimental studies. The first involves the use of X-ray crystal structures, kinetic studies, and molecular dynamics simulations in the study of the reduction of the prodrug CB1954 (5-(aziridin-1-yl)-2,4-dinitrobenzamide) by the flavoenzyme nitroreductase from *Escherichia coli*. The second involves the study of the mechanism of the *Salmonella serovars* phosphothreonine lyase SpvC using X-ray crystal structures, biochemical assays, molecular dynamics simulations, and potential mean force calculations. It is only through a combination of experimental and computational techniques that a thorough atomic-level understanding of these enzyme reaction mechanisms is achieved.

ADVANCES IN PROTEIN CHEMISTRY AND STRUCTURAL BIOLOGY, Vol. 87
DOI: 10.1016/B978-0-12-398312-1.00010-X

Copyright 2012, Elsevier Inc.
All rights reserved.

I. Introduction

Computers are playing an increasing role in both drug discovery and the study of enzymes (Leach, 2001; Karplus and McCammon, 2002; Karplus and Kuriyan, 2005; Brent and Bruck, 2006; Dodson et al., 2008). While it is impossible to come to definitive conclusions based on computational studies alone, due to the many approximations utilized by the currently available software, these techniques are nevertheless important tools for aiding the work of experimental researchers.

One such technique is molecular dynamics (MD). MD is a molecular modeling method aimed at studying the behavior of a system over time. It has the advantage of taking an atomic-level look at the effects of solvent, temperature, and pressure on the system (Rapaport, 1995; Deng et al., 2001). It is particularly useful for looking at properties such as protein conformation and flexibility, the interactions between a protein and a ligand, hydrogen bonding, and solvent behavior.

As MD simulations are the evolutions of consecutive, discrete, configurational changes over time, a chief application of MD is to investigate dynamic motions and behaviors of biomolecules. Generally, MD simulations are used to model kinetic processes at a time scale from nanoseconds to hundreds of nanoseconds depending on the availability of computing resources (Van Voorhis et al., 2010). This time scale is mainly related to ligand binding in proteins or DNA, motions of residue loops, and domain vibrations, which are the general problems to which MD simulations are applied (Dodson et al., 2008). In recent years, with the increasing development of computational techniques, especially techniques utilizing graphics processing units (Anderson et al., 2008), large-scale, parallel, molecular simulations are increasingly applied to biomolecular systems. Trial simulations are even carried out to the millisecond time scale (Shaw et al., 2008; Voelz et al., 2010) with as many as 10^9 atoms in the system (Wilson et al., 2002). However, these mammoth-level simulations are often for hardware testing and seldom affordable for scientific purpose by general biophysical researchers.

Another more general and broadly used application is making use of the sampling power of MD simulations, as long-time simulations can be deemed time-independent Markov chains (Bacallado et al., 2009). Although the simulation time scale of MD simulations is limited to less than a microsecond in most simulations, it is more effective to sample a narrow configurational space in biomolecular systems using MD simulations for parallel-computing efficiency than other methods like Monte Carlo. For longer time scale motions, sampling schemes were designed to enhance sampling at a series of discrete points along the procedure path, such as umbrella sampling (Torrie and Valleau, 1974, 1977). With these sampling methods, MD simulations are not only simple procedures for evolution analysis but can be powerful for statistical investigations of thermodynamic properties. Specifically, a number of algorithms (Gilson et al., 1997; Chipot and Pohorille, 2007) such as free energy perturbation, thermodynamic integration, and potential of mean force (PMF), to name a few, were developed and broadly applied to biomolecular systems to evaluate thermodynamic properties such as ΔG, ΔH, ΔU, and ΔS. As a continuous procedure is broken into consecutive discrete points, more complicated, time consuming, and even unphysical processes like relative binding energies between ligand–protein and protein–protein systems can be investigated using MD simulations.

In recent years, the number of protein sequences uploaded to freely available databases has increased dramatically. Additionally, there is a large amount of structural information from X-ray and NMR data for proteins with unknown function. In fact, function determination is now the key limiting step in developing a detailed understanding of biological systems. MD simulations can help address this issue. Simulations have been shown to be in good agreement with experimental work (Offman et al., 2011), and may be used to guide experimental work (in the case of drug design), or simply to gain insight into poorly understood enzyme reaction mechanisms.

In this work, we will look at the use of MD simulations in the investigation of the reaction mechanism of the reduction of the prodrug CB1954 by *Escherichia coli* nitroreductase (NTR), and the study of the mechanism of the *Salmonella serovars* phosphothreonine lyase SpvC. By illustrating the use of MD simulations as a complement to experimental studies in these two systems, the value of MD simulation methods may be clearly seen.

II. The Reduction of CB1954 by *E. coli* NTR

Virus-Directed Enzyme Prodrug Therapy (VDEPT) is a gene therapy approach to cancer treatment aimed at overcoming the obstacle of dose-limiting toxicity. In VDEPT, a virus is genetically modified to carry the gene for a specific enzyme not found within normal human cells. When the virus is injected into a tumor, the tumor cells are directed to synthesize the enzyme, causing the expression of the enzyme within the cancer cells. The patient is then given an inactive, nontoxic, prodrug, which reacts with the enzyme within the cancer cells to produce a cytotoxic compound that kills the cells. It has been found that the prodrug CB1954 (5-(aziridin-1-yl)-2,4-dinitrobenzamide) and the flavoenzyme NTR from *E. coli* (see Fig. 1) are a good potential combination for VDEPT (Grove et al., 1999; Weedon et al., 2000; Palmer et al., 2004). However, the mechanism for the reduction of CB1954 by NTR is poorly understood.

Wild-type *E. coli* NTR, encoded by the *nfs*B gene, was initially identified as a target for antibiotics such as nitrofurazone and nitrofurantoin (McCalla et al., 1978; Whiteway et al., 1998; Race et al., 2005). Although its actual function within the bacteria has yet to be discovered, it is known to reduce a wide range of nitroaromatics and quinones (Anlezark et al., 1992; Zenno et al., 1996). Expression of NTR in the bacteria is induced by the MarA protein during oxidative stress (Barbosa and Levy, 2002), so it may function as a general two-electron reductant to limit free radical formation.

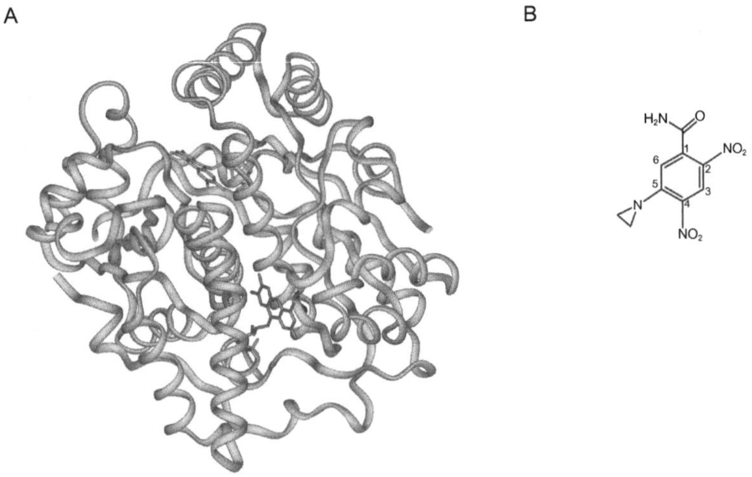

FIG. 1. (A) The prodrug 5-(aziridin-1-yl)-2,4-dinitrobenzamide (CB1954). (B) Structure of *E. coli* nitroreductase NfsB (NTR) from PDB file 1YKI.

The enzyme exists as a dimer, with each of the identical subunits containing 217 amino acids and a flavin mononucleotide (FMN) cofactor. NTR has two active sites, each formed at the dimer interface by residues from both subunits (Lovering et al., 2001). The protein itself has a relatively rigid structure and primarily acts as a framework for the reaction between FMN and the substrate.

NTR is referred to as an ''oxygen-insensitive'' enzyme, because the reduced form does not easily give up one electron to molecular oxygen to form the semiquinone radical (Bryant and DeLuca, 1991; Koder and Miller, 1998). Thus two-electron oxidation and reduction is observed, but one-electron oxidation or reduction is not. This may be due to the nature of the amino acids surrounding FMN and their tendency to not stabilize the semiquinone state of the bound flavin (Haynes et al., 2002).

It is known that the CB1954 prodrug is activated by an enzyme-initiated reduction of a nitro group to a hydroxylamine. The 4-nitro hydroxylamine derivative then reacts with acetylthioesters naturally found in the cell to form a toxic *N*-acetoxy derivative (Knox et al., 1988; Grove et al., 1999). The initial conversion from a strongly electron-withdrawing nitro group to an electron-donating hydroxylamine effectively acts as a ''switch'' to convert the relatively harmless prodrug into a strongly cytotoxic compound. This activated compound causes DNA–DNA interstrand cross-links to form within the cancer cells. These cross-links are poorly repaired, and cause cell death in both dividing and nondividing cells. NTR may reduce either of the 2-nitro groups on CB1954, but not both groups of the same molecule. With wild-type NTR, both hydroxylamine species are formed in equal proportions at the same rate (Anlezark et al., 1992). While the 4-nitro derivative acts as a bifunctional alkylating agent and is vastly more cytotoxic, the 2-nitro derivative has been observed to have a greater bystander effect *in vivo* (Helsby et al., 2004).

The primary limitation of the NTR/CB1954 combination is a low affinity of NTR for the prodrug. CB1954 is not a natural substrate of NTR, and so does not naturally bind well to the enzyme. As a result, a low yield of activated prodrug is observed (Anlezark et al., 1992). Site directed mutagenesis has been used to improve the NTR/CB1954 combination for VDEPT. To do this rationally, it is necessary to know the mechanism of the reaction.

It has been determined that the nitro group of CB1954 is reduced to a hydroxylamine via two successive two-electron transfers (Anlezark et al., 1992). Although the reduction of the nitro group to the unstable nitroso intermediate must be performed by the enzyme, it is not necessary for

NTR to be involved in the reduction of the nitroso intermediate to hydroxylamine, as the nitroso may be directly reduced by NAD(P)H in solution (Race et al., 2005). Additionally, kinetic studies have shown that the reduction of the nitro group by the enzyme to the nitroso intermediate is the rate-limiting step (Race et al., 2005). Therefore, the focus of the work examining the mechanism has been on the enzyme-catalyzed reduction of the 4-nitro group of CB1954 to a nitroso group.

The overall reduction from nitro to nitroso requires the addition of two protons and two electrons. While the electrons must come from the FMN cofactor of NTR and at least one of the protons must come from solution, the source of the other proton may be either the FMN or the solution. Hydride transfer is considered to be the concerted transfer of one proton and two electrons from the same source—in this case the FMN. The addition of the proton from solution (for the total of two protons and two electrons) may occur either before or after this hydride transfer. Hydride transfer requires that one of the oxygen atoms of the nitro group comes within van der Waals contact of HN5 of FMN (see Fig. 2). In contrast, the other possible mechanism, with solely electron transfer from the FMN (where both electrons come from FMN but both protons come from solution) only requires that some part of CB1954 is within van der Waals contact of some part of the π-system of FMN, while the nitro group is exposed to solution. Electron transfer over a wide range of conformations is possible as both the HOMO of FMN and LUMO of CB1954 are delocalized across much of their respective molecules (see Fig. 2).

There is empirical evidence that argues both for hydride transfer and for electron transfer. FMN has been observed to undergo both one-electron and two-electron transfers as a cofactor in other enzymes (Kyte, 1995). The initial reduction of FMN by NAD(P)H must occur via hydride transfer to give $NAD(P)^+$. Although it is difficult to prove true hydride transfer experimentally, if a two-electron transfer is followed immediately by a proton transfer from the FMN, this is essentially the same as a hydride transfer, termed "net hydride transfer" hereafter.

For the electron transfer mechanism to be viable, FMN must donate two electrons to CB1954, then lose the N5 proton to solution before it can be reduced again by NAD(P)H. The p*Ka* of the oxidized, protonated FMN is 0.2 (Kyte, 1995), which allows for this possibility. Gas-phase quantum mechanics calculations show that either reaction mechanism (net hydride transfer or electron transfer) is kinetically viable (Christofferson and Wilkie, 2009).

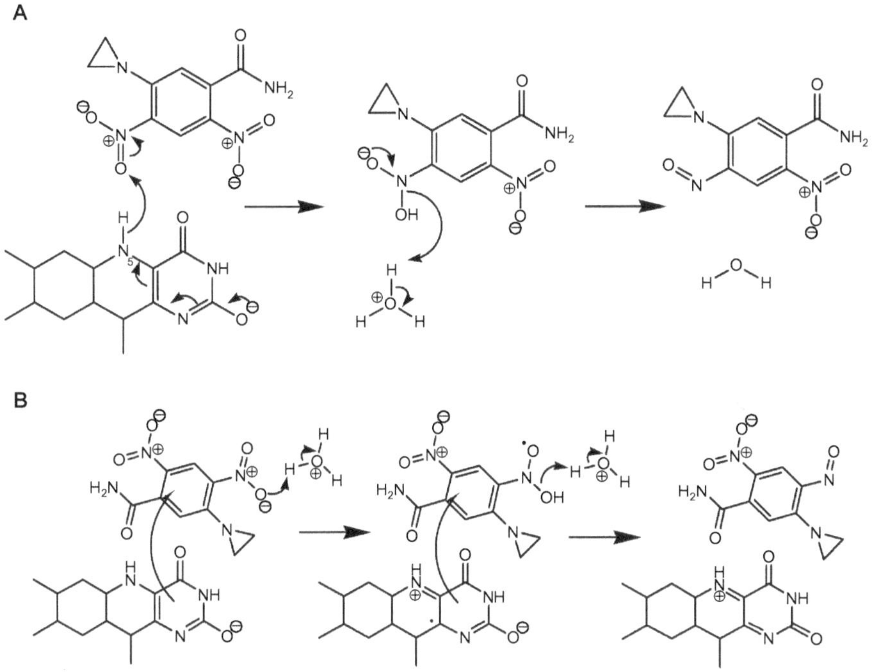

FIG. 2. (A) Proposed hydride transfer mechanisms for the reduction of a nitro group to a nitroso by FMN. (B) Proposed electron transfer mechanism for the reduction of the 4-nitro group of CB1954 by FMN.

Knowledge of the exact reaction mechanism is critical, as it will help determine what amino acids should be mutated in order to increase activity. For a hydride transfer mechanism, the enzyme should be mutated in such a way as to maximize the binding of CB1954 in an orientation that favors the reduction of the 4-nitro group. For an electron transfer mechanism, it may be more beneficial to mutate the enzyme in a way that modifies the redox potential of FMN.

A. *Experimental Work on the CB1954/NTR System*

Steady-state kinetics studies have shown that the enzyme reacts via a ping-pong bi-bi mechanism (Race et al., 2005). First, NAD(P)H binds to the enzyme and donates two electrons to the FMN cofactor. $NAD(P)^+$ is then released, allowing the opportunity for the substrate to bind to the active site of the reduced enzyme and thereby be reduced itself. The active site cannot accommodate NAD(P)H and a second substrate simultaneously (Lovering et al., 2001).

The primary feature of aromatic substrate binding appears to be hydrophobic ring stacking, with the substrate sandwiched between the FMN ring system and the Phe124 residue, at a distance of approximately 3.5 Å from each (Lovering et al., 2001). An example of this can be seen in the 1ICR crystal structure (Lovering et al., 2001), which shows nicotinic acid bound to NTR, and provides a good example of aromatic substrate binding. The lack of many other interactions between the substrate and protein may explain both NTR's broad substrate range, and its affinity for aromatic substrates. Additionally, the carboxyl group of nicotinic acid has a hydrogen bond to the backbone of the Thr41 residue (Lovering et al., 2001). This hydrogen bond is also seen with other ligands, such as nitrofurazone and acetate, in the oxidized state of the enzyme. Acetate can only bind in the active site in the oxidized state of the enzyme, implying that carboxylic acid groups may bind preferentially to the oxidized state of the enzyme (Race et al., 2005).

There is a large change in electron distribution in FMN upon reduction (Wouters et al., 1997). In free flavin, the pK_a of N5 changes from 8.5 in the oxidized state (where all three rings are aromatic) to 6.8 in the reduced state (where only the outer two rings are aromatic). This change in electron distribution may explain differences in binding between the oxidized and reduced enzyme. The flavoenzyme pentaerythritol tetranitrate reductase has also been reported to have different binding properties for the oxidized and reduced states of the enzyme (Barna et al., 2001).

Following the determination of the protein structure, site-directed mutagenesis was done on NTR to try to improve its activity for CB1954. Initially, nine residues in the active site of NTR surrounding FMN that could have an effect on catalysis or substrate binding were identified for single mutation: Ser40, Thr41, Tyr68, Phe70, Asn71, Gly120, Phe124, Glu165, and Gly166 (Grove et al., 2003). For each position, all possible amino acid substitutions were generated and cloned into the *E. coli* bacteria. The clones were replica plated on agar plates, and a different concentration of CB1954 was added to each plate to determine the effect of the amino acid substitution on the reaction with the prodrug, as a decrease in bacterial growth correlates to prodrug reduction. Of the nine potential residues, only Gly166 was found to be essential for prodrug reduction. Additionally, no significant improvements were found with any mutation of Glu165 or Gly120 (Grove et al., 2003).

The most active single mutants were found at Phe124, with 15 amino acid substitutions yielding increased sensitivity to CB1954. The large hydrophobic side-chain of phenylalanine likely imposes restrictions in binding in the wild-type enzyme, and various studies have shown that mutations at this residue may be key in modulating substrate specificity (Zenno et al., 1998; Grove et al., 2003). Additional improvements were found by mutating Ser40 to residues with small side-chains such as glycine and alanine. Thr41 had improvements with residues with long hydrophobic groups, polar groups, or glycine. Tyr68 had improvements with glycine or asparagine. Phe70 had several residue substitutions with improvement, possibly due to its location near the opening of the active site, where mutations could allow greater access to the active site, as well as reduce the difficulty of product release. Finally, Asn71 had improvements primarily with polar residues such as serine (Grove et al., 2003).

Kinetic parameters for wild-type NTR and the best mutants with both CB1954 and nitrofurazone were determined in order to give a more quantitative comparison of the mutants' efficacy (Race et al., 2005). Although the global turnover rate (k_{cat}) and global binding constant (K_m) for wild-type NTR and CB1954 could not be determined due to the limited solubility of the prodrug, the specificity constant (k_{cat}/K_m) could be measured and is independent of NADPH concentration for a bi-bi substituted enzyme mechanism (Cleland, 1989). The overall reaction rate is determined by k_{cat}/K_m at low concentrations of CB1954, and the concentration of CB1954 that can be used for gene therapy applications is much lower than the solubility limit (Chung-Faye et al., 2001).

All of the single mutants except for the Asn71 to serine (N71S) showed both an improvement over wild type and an improvement in selectivity for CB1954 over nitrofurazone. This selectivity is important, as there are many potential competitors and inhibitors *in vivo*. The N71S mutant did show some improvement over wild type with CB1954 (Race et al., 2007). Crystal structures of this mutant show that while the direct hydrogen bond from the asparagine side-chain to FMN is lost, due to the shorter side-chain of the serine, this is replaced by a hydrogen bond from a conserved water molecule bridging the serine side-chain and the FMN. It is believed that this mutation affects the redox potential of FMN, rather than substrate binding, as the improvements in the specificity constant for CB1954 and nitrofurazone are similar (Race et al., 2007).

For the Thr41 to lysine (T41L) and Tyr68 to glycine (Y68G) mutants, the improvement in the specificity constant came from an increase in apparent k_{cat}, while the apparent K_m values (at 100 μM NADH) remained largely the same as wild type. For all of the other mutants, the effect was reversed—improvements in k_{cat}/K_m arose from a decrease in apparent K_m, while there was little change in apparent k_{cat}. Additionally, the T41L mutant showed a preference for the 4-nitro reduction over the 2-nitro by a 3:1 ratio (Race et al., 2007). Because, none of these mutations have an appreciable effect on nitrofurazone, it is believed that improvements are due to interactions with the substrate, rather than a modulation of the redox potential of FMN (Race et al., 2007).

All four of the mutants at Phe124 exhibited different k_{cat}/K_m values, suggesting that each mutant had a different affect on the substrate. Of the four, asparagine (F124N) had the best specificity constant and the greatest selectivity for CB1954 over nitrofurazone.

Several crystal structures of single mutants have been determined, both for the free holoenzyme and with enzyme bound to nicotinate. Regardless of the presence or absence of nicotinate, all mutants were found to be structurally similar to wild-type NTR. Although there was naturally some variation in side-chain orientation, there was very little difference in backbone conformation, demonstrating that single mutations have little effect on protein structure (Race et al., 2007).

Following this, double mutants were made by combining selected single mutants (Jaberipour et al., 2010). The best double mutant was found to be a combination of the Thr41 to leucine and Asn71 to serine single mutants (T41L/N71S). A novel genetic selection system based on lambda phage for selecting the most active triple mutants was developed (Guise et al., 2007). The best triple mutant was a combination of Thr41 to glutamine, Asn71 to serine, and Phe124 to threonine (T41Q/N71S/F124T) (see Table I). Further kinetic work was done on the best single, double, and triple mutants, using both CB1954 and nitrofurazone as substrates (Jarrom et al., 2009).

This data shows that there is very little improvement in global turnover rate with the mutations for either CB1954 or nitrofurazone. The fact that k_{cat} values for the two substrates are rather similar also implies that the rate-limiting step may be independent of the redox reaction and may in

TABLE I
Kinetic Parameters for the Reduction of CB1954 and Nitrofurazone by Wild-Type NTR and the Best Single, Double, and Triple Mutants

Protein (with CB1954)	k_{cat} (s^{-1})	$K_{mCB1954}$ (μM)	$k_{cat}/K_{mCB1954}$ ($\mu M^{-1} s^{-1}$)
Wild type	140±32	17200±4800	0.007
F124N	95±7	3081±465	0.031±0.006
T41L/N71S	153±8	216±33	0.71±0.08
T41Q/N71S/ F124T	181±7	569±45	0.318±0.016

Protein (with nitrofurazone)	k_{cat} (s^{-1})	$K_{mNitrofurazone}$ (μM)	$k_{cat}/K_{mNitrofurazone}$ ($\mu M^{-1} s^{-1}$)
Wild type	225±34	1850±400	0.15±0.02
F124N	110±12	972±208	0.11±0.01
T41L/N71S	103±8	237±39	0.38 ±0.03
T41Q/N71S/ F124T	142±21	1039±261	0.137±0.02

Reprinted (adapted) with permission from (Jarrom et al., 2009). Copyright (2009) American Chemical Society.

fact be product release. Additionally, the greatest improvements in specificity constants come from improvements in K_m, which implies that binding may be key to enzyme efficiency. It was also found that the T41L/N71S mutant reduces the 4-nitro group of CB1945 exclusively (Jarrom et al., 2009), while the triple mutant has no selectivity for either nitro group.

While mutation of selected amino acids has resulted in a 100-fold improvement in specificity for CB1954, the sheer number of possible combinations of residue mutations makes it difficult to produce further improvements without knowing more about the reaction mechanism of NTR and in particular, how the substrate binds.

Like nicotinic acid and acetate, nitrofurazone has a hydrogen bond from a carbonyl oxygen to the Thr41 backbone (see Fig. 3). There is an additional hydrogen bond from the nitrofurazone amide group to the Glu165 side-chain. This may be an amide binding pocket, which would allow compounds with amide groups to bind within van der Waals contact of FMN, allowing the redox reaction to take place. In this structure,

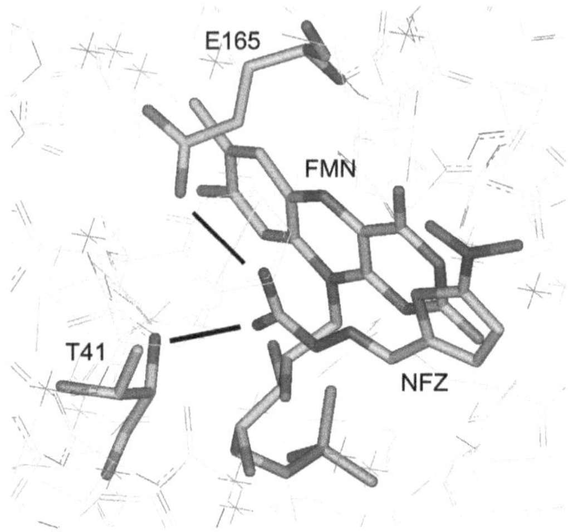

FIG. 3. Nitrofurazone bound to oxidized NTR from PDB file 1YKI.

nitrofurazone is bound with the amide group over FMN and the nitro group exposed to solution—ideal for electron transfer and prohibitive for hydride transfer—but both NTR and nitrofurazone are present in their oxidized form, so this is a dead end complex.

The 1IDT crystal structure purports to show CB1954 bound to oxidized NTR in two different orientations: Active Site "A" (see Fig. 4A) with the 2-nitro group close to N5 of FMN and Active Site "B" (see Fig. 4B) with the 4-nitro group close to N5 of FMN (Johansson et al., 2003). Unlike Active Site A, there are several close contacts between CB1954 and NTR in Active Site B. Both the amide and 2-nitro groups of CB1954 have hydrogen bonds with the Lys14 side-chain and one oxygen of the 4-nitro group has a hydrogen bond with the Thr41 backbone. Additionally, the aziridine ring of CB1954 has hydrophobic contacts with both Phe124 and Phe70.

While this structure with two different sites would explain why NTR produces 2-nitro and 4-nitro reduction products in equal proportions, it seems unlikely that this structure occurs in solution, considering the fact that the two active sites are identical. It is also important to note that the crystallization was performed at pH 4.6, and CB1954 is very unstable below pH 6. Additionally, it is very difficult to form a complex with any ligand when acetate is used as the buffer solution, as acetate tends to bind

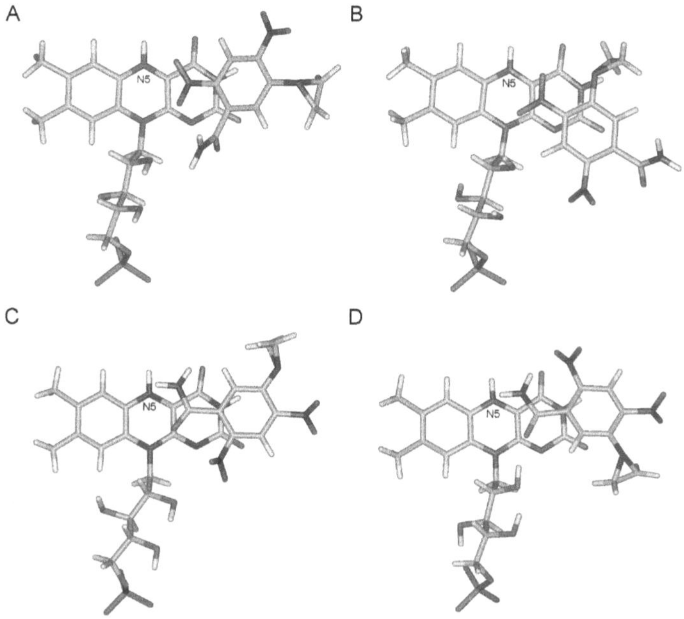

FIG. 4. CB1954 orientations with respect to FMN. (A) 2-Nitro hydride transfer orientation. (B) 4-Nitro hydride transfer orientation. (C) Electron transfer amide orientation 1. (D) Electron transfer amide orientation 2.

preferentially. Nevertheless, this crystal structure does provide *possible* binding orientations for CB1954 bound to NTR in a manner suitable for hydride transfer, and no other crystal structures with CB1954 and NTR exist to date.

B. MD Simulations on CB1954 in Wild-Type and Mutant NTR

In order to gain more insight into the possible reaction mechanisms for the reduction of CB1954 by NTR, MD simulations were performed with CB1954 in the active site of the reduced form of wild-type NTR, as well as with the best single, double, and triple mutants experimentally determined (Jarrom et al., 2009), in binding orientations suitable for net hydride transfer, or for electron transfer.

The model structure for the hydride transfer mechanism orientation was obtained from the crystallographic coordinates of the wild-type NTR with one CB1954 bound with the 2-nitro group in a position to receive a hydride from FMN in one active site of NTR, and a CB1954 bound in the other active site in a position for the 4-nitro group to receive a hydride from FMN (PDB accession code: 1IDT; Johansson et al., 2003). The model structure for the electron transfer mechanism orientation was obtained from the crystallographic coordinates of wild-type NTR with nitrofurazone bound in the active site (PDB accession code: 1YKI; Race et al., 2005). An overlay of the CB1954 amide group with the amide group of nitrofurazone in the crystal structure was performed, using the visualization program InsightII (Accelrys Inc., 2000), and the nitrofurazone was subsequently removed. There are two possible CB1954 orientations that have the amide group bound in the active site in a suitable position for electron transfer, defined as orientation 1 and orientation 2 in this work (see Fig. 4). The electron transfer simulation model contained one CB1954 in each orientation.

For both the hydride transfer and the electron transfer models, mutants were created by residue replacement in InsightII to give models for the single mutant F124N, the double mutant T41L/N71S, and triple mutant T41Q/N71S/F124T. Hydrogen atoms were added to the reduced FMN cofactor and CB1954 in InsightII, and parameter files for these molecules were generated with the Antechamber and Parmchk programs of the AMBER 8 package (Case et al., 2004), using the AM1-BCC charge method (Jakalian et al., 2000). Hydrogen atoms were added to the amino acids of the proteins in the Xleap program of AMBER 8, according to physiological pH and a visual inspection of hydrogen bonding patterns. The ff03 force field (Ponder and Case, 2003) was applied to determine amino acid parameters, and the general AMBER force field (GAFF) was used for the FMN cofactor and CB1954 ligand. In each case, the system was solvated with a TIP3P water box (Jorgensen et al., 1983) with a minimum distance from protein surface to box edge of 10 Å.

Prior to the MD simulation, a molecular mechanics minimization was performed on each structure, employing the steepest descent method for 10 cycles, followed by conjugate gradient until the default cut-off energy values were reached. Each simulation run consisted of 200 ps of equilibration, followed by ten 1 ns runs, for a total of 10 ns of data gathering. Simulations were performed at 300 K, with a 2-fs time step, periodic boundary conditions, and a nonbonded cut-off of 8 Å. Coordinates and

TABLE II
Minimum, Mean, and Maximum Alpha-Carbon RMSD Values

	Min	Mean	Max
Hydride transfer			
Wild Type	0.88	1.41	1.69
F124N	1.46	1.76	1.97
T41L/N71S	1.35	1.55	1.78
T41Q/N71S/F124T	1.19	1.70	2.07
Electron transfer			
Wild Type	1.12	1.53	1.87
F124N	1.43	1.07	1.79
T41L/N71S	1.09	1.72	2.10
T41Q/N71S/F124T	1.40	1.81	2.12

velocities were recorded every 50 ps. Root-mean-square deviation (RMSD) of protein alpha-carbons and hydrogen bonding analysis was performed by the Ptraj program of AMBER 8.

Alpha-carbon RMSD values show that in every case, the protein structure remains relatively rigid and stable throughout the simulation (see Table II). For the double and triple mutants, both containing the N71S mutation, the hydrogen bond between Asn71 and FMN is replaced by a hydrogen bond between Ser71, a water molecule, and FMN, as for the N71S single mutant (Race et al., 2007).

In order for CB1954 to be reduced by NTR via net hydride transfer, there must be a stable binding orientation for the prodrug in the active site of the enzyme, with one of the oxygen of the nitro groups within ~4 Å of N5 of FMN. Additionally, as experimental data shows that the wild-type enzyme reduces the 2-nitro and 4-nitro groups in equal proportions, there must be binding orientations for both nitro groups in the active site, of approximately equal energy, differing by less than 1 kcal/mol in the experimental system.

For the hydride transfer model in every simulation, the CB1954 moved out of the active site during the equilibration calculations. In no case was there an oxygen of a nitro group with an average distance of less than 6 Å from the N5 of FMN. This suggests that direct hydride transfer from the FMN to CB1954 is not likely with the substrate in the initial orientation in the binding site.

Interestingly, in one simulation of the wild-type protein and one of the T41Q/N71S/F124T mutant, the 4-nitro CB1954 came out of the active site and spontaneously reoriented to the orientation 1 amide binding (electron transfer) orientation (see Fig. 5). In the case of the wild-type protein, this occurred in the first nanosecond, while for the triple mutant, it occurred in the third nanosecond of the simulation. For the wild-type protein, the distances for the key interactions are in good agreement with those of the simulation of the CB1954 that began in that orientation, but for the triple mutant the mean distances are larger than those of the CB1954 that began in the analogous orientation, possibly indicating that the CB1954 is less tightly bound in the triple mutant.

For the electron transfer orientation, there is little difference in CB1954 binding between the wild-type NTR and mutants. In each case, there is a consistent hydrogen bond between the amide oxygen of CB1954 and the backbone nitrogen of residue 41 (Thr in the wild type or single mutant, Leu in the double mutant, and glutamine in the triple mutant), as well as a second hydrogen bond from this amide oxygen to an hydroxyl group in the ribityl tail of FMN. Additionally, there is a hydrogen bond from the nitrogen of the amide group of CB1954 to the Glu165 side-chain oxygen (see Fig. 6).

There are, however, some slight differences between the mutants. As Table III shows, for the wild type, single, and triple mutants, the hydrogen bond from the CB1954 amide oxygen to the Thr or Gln41 backbone nitrogen is longer in orientation 2 than in orientation 1, while in the double mutant the reverse is true (although only slightly so). In all cases, the hydrogen bond from the CB1954 amide hydrogen to the Ser40 side-chain oxygen is shorter in orientation 2.

In orientation 2 of the double mutant (T41L/N71S), the hydrogen bond from the amide oxygen of CB1954 to the ribityl tail of FMN is somewhat longer than in any other simulation, while the distance to the Leu41 backbone nitrogen is slightly shorter. In both orientations of the triple mutant (T41Q/N71S/F124T), the hydrogen bond from the CB1954 amide oxygen to the Gln41 backbone nitrogen is somewhat longer than in any other simulation—slightly more so for orientation 2—while the hydrogen bonds from the CB1954 amide oxygen are similar to the wild type and other mutants.

Hydrogen bonding patterns indicate that, for both orientation 1 and orientation 2 of the amide binding (electron transfer) mode, the 4-nitro

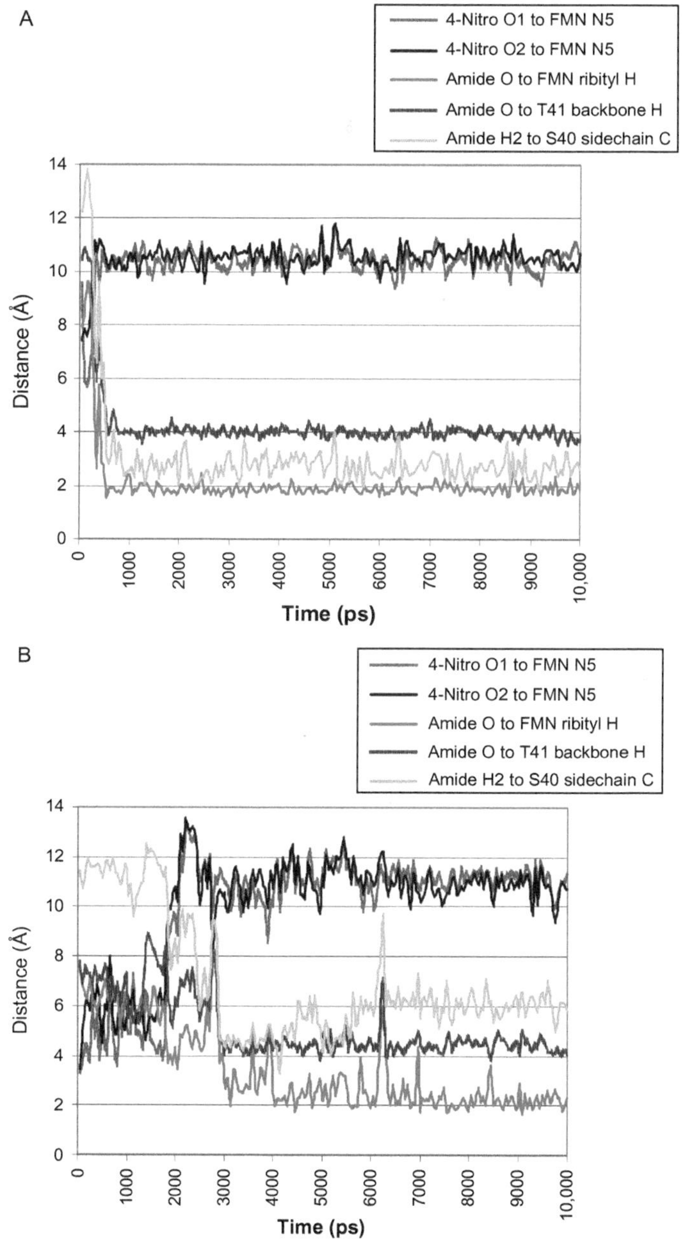

FIG. 5. Distances between CB1954 and enzyme for (A) wild-type 4-nitro hydride transfer orientation and (B) triple mutant 4-nitro hydride transfer orientation.

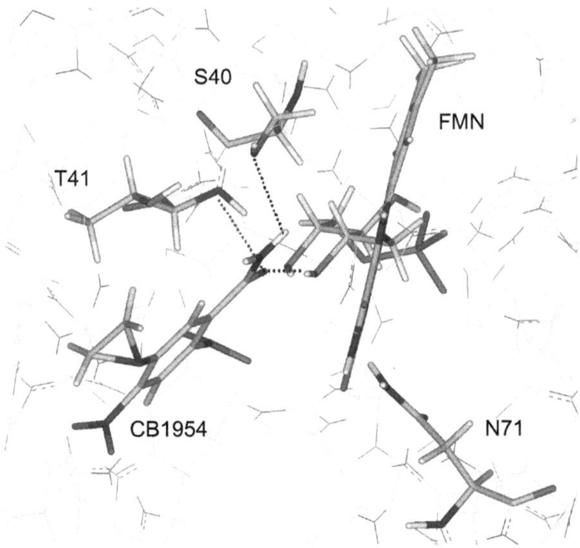

Fig. 6. Key interactions of the amide binding (electron transfer) orientation 1.

group of CB1954 is involved in a greater number of hydrogen bonds than the 2-nitro group (see Table IV), but this is more pronounced in orientation 2. For the triple mutant, in orientation 2 there are no hydrogen bonds with water during the simulation when a 3-Å cut-off is used, and only 1% occupancy when a 3.5-Å cut-off is employed. This suggests that orientation 2 may not be a viable orientation for the triple mutant. In all other cases, water molecule exchange is extensive, indicating that there is plenty of water moving within van der Waals contact of the nitro groups. Space-filling models show that for both orientations, both nitro groups are sufficiently exposed to solution to gain protons in the wild-type active site (see Fig. 7).

C. *Utilization of the MD Simulation Results*

For the electron transfer orientations, in both the wild type and mutants there is significantly less hydrogen bonding to solvent molecules for the 2-nitro group of CB1954 in orientation 2. Experimentally, it has been found that the wild-type enzyme, and the single and triple mutants reduce both 2-nitro and 4-nitro groups in equal proportions, while the double mutant

Table III

Mean Distances in Angstroms (Å) for Selected Atoms

	CB1954 nitro O1—FMN N5	CB1954 nitro O2—FMN N5	CB1954 amide O—FMN ribityl H	CB1954 amide O—Res[a]41 backbone N	CB1954 amide H—Ser40 side-chain O
Hydride transfer					
Wild-type 2-nitro	11.13	11.51			
Wild-type 4-nitro[b]	9.52	8.70	2.11	4.21	3.10
F124N 2-nitro	6.84	8.00			
F124N 4-nitro	7.74	7.65			
T41L/N71S 2-nitro	8.45	7.85			
T41L/N71S 4-nitro	10.35	9.71			
T41Q/N71S/F124T 2-ntiro	12.28	11.14			
T41Q/N71S/F124T 4-nitro[b]	9.48	9.64	3.72	5.29	7.39
Electron transfer					
Wild-type orientation 1			2.11	3.89	2.69
Wild-type orientation 2			2.10	4.03	2.20
F124N orientation 1			1.97	3.87	2.46
F124N orientation 2			2.14	4.00	2.26
T41L/N71S orientation 1			2.00	3.95	2.26
T41L/N71S orientation 2			2.65	3.83	2.13
T41Q/N71S/F124T orientation 1			2.13	4.34	2.64
T41Q/N71S/F124T orientation 2			2.13	5.00	2.13

[a]Thr in wild type and single mutant, Leu in double mutant, and Gln in triple mutant.
[b]Spontaneously reorients to electron transfer orientation 1.

TABLE IV
Percent Occupancy of Nitro Group Oxygens Having Hydrogen Bond with Solvent

	3.0 Å cut-off				3.5 Å cut-off			
	2-Nitro		4-Nitro		2-Nitro		4-Nitro	
	O1	O2	O1	O2	O1	O2	O1	O2
Wild-type orientation 1	3	16	8	14	19	53	31	58
Wild-type orientation 2	0	10	13	15	5	34	42	53
F124N orientation 1	2	10	25	14	15	30	68	39
F124N orientation 2	0	6	20	8	2	20	44	31
T41L/N71S orientation 1	24	12	19	15	56	47	58	66
T41L/N71S orientation 2	0	13	20	14	3	29	60	50
T41Q/N71S/F124T orientation 1	3	6	13	16	11	27	47	55
T41Q/N71S/F124T orientation 2	0	0	8	16	0	1	23	47

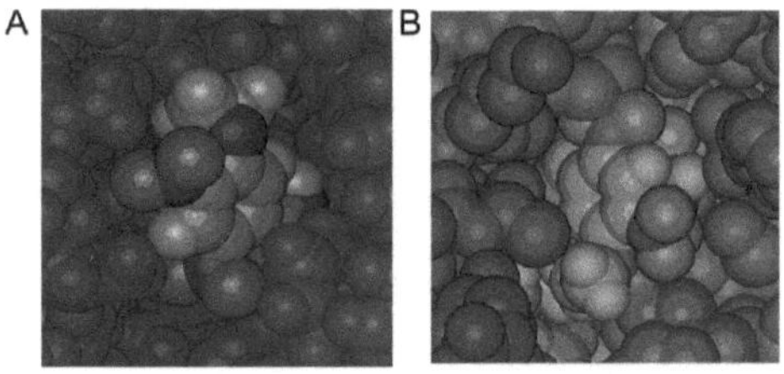

FIG. 7. (A) Space-filling model of CB1954 in the active site of NTR, orientation 1. (B) Space-filling model of CB1954 in the active site of NTR, orientation 2. NTR has been colored purple, CB1954 in atomic colors to show groups accessible to solvent. (For color version of this figure, the reader is referred to the online version of this chapter.)

reduces the 4-nitro group almost exclusively (Jarrom et al., 2009). It is possible that CB1954 binds to the wild-type enzyme, single and triple mutants in orientation 1 and in orientation 2 to the T41L/N71S mutant. This change in binding orientation could be explained by a potentially favorable interaction between the aziridine group of CB1954 and the Leu41 residue—found only in orientation 2 of the double mutant, although this has not yet been studied.

These calculations suggest that the reduction of CB1954 by NTR is by the electron transfer mechanism and have shown its binding orientations. This knowledge has the potential to aid the development of new and better NTR mutants and thereby increase the efficacy of the CB1954/NTR combination in cancer chemotherapy. In order to have a more efficient therapy, it is necessary to improve K_m for CB1954, and it may be useful to be able to modulate the proportions of 4-nitro (which is more cytotoxic) and 2-nitro (which has a better bystander effect) reductions. As a result of the work described here, it may be possible to create mutants with a greater affinity for amide groups, favor orientation 1 or orientation 2, or to shield one or the other of the nitro groups from solution, as desired. However, care must be taken not to detrimentally affect the first step of the reaction—the reduction of NTR by NAD(P)H.

Visual inspection of the binding site of NTR in the vicinity of the bound CB1954 highlights that many of the residues have already been subject to mutation. For example, mutating Thr41 to a larger hydrophobic side-chain may help stabilize the orientation 2 conformation but we have already seen that the T41L mutant improves binding (Grove et al., 2003). In this case, the experimental data is in good agreement with the computational results but the computational results are not able to provide any new insights for mutants. Alternatively, it may be more useful to consider modifications to CB1954 to aid binding, possibly in conjunction with mutations to the enzyme. For example, the binding orientation shown in Fig. 7B reveals a significant cavity close to the aziridine ring, thus, modification to this group may allow further interactions with amino acids lining this pocket, for example, Gly120, Ala116, or Lys14.

III. Investigation of the Enzymatic Mechanisms of Phosphothreonine Lyase

Mitogen-activated protein kinases (MAPKs) play an important role in modulating cellular activities, such as immune functions, by reacting to extracellular stimuli (Tena et al., 2001; Ausubel, 2005; Ashwell, 2006; Shizuo et al., 2006). MAPKs operate via a protein kinase cascade, where a trio of MAP kinases activates in sequence. When a MAP3K is activated by an external stimulus, it in turn phosphorylates a MAP2K on its serine and threonine residues, which in turn causes the MAP2K to phosphorylate the threonine and tyrosine residues of the third MAP kinase.

Pathogenic bacteria can repress a host's innate immunity by inactivating the MAPK signaling pathway through the use of specific enzymes (Dong et al., 2002; Shan et al., 2007). Phosphothreonine lyase is a newly discovered enzyme family from Gram-negative bacterial pathogenic type III Secretion System effectors and includes enzymes such as SpvC from nontyphoid *Salmonella,* HopAI1 from the plant pathogen *Pseudomonas syringae,* and OspF from *Shigella* spp. (Arbibe et al., 2007; Kramer et al., 2007; Shan et al., 2007; Zhang et al., 2007; Zhu et al., 2007). These enzymes can permanently inactivate the host MAPKs by cleaving the C_{β}—O_{γ} bond from the phosphothreonine (pT) of a MAPK activation loop, which results in a β-elimination methyldehydroalanine (Mdha) product (Li et al., 2007; Zhang et al., 2007; Zhu et al., 2007; Chen et al., 2008). Such catalytic activity in the absence of cofactors or metal ions is unique in currently known enzymes with respect to their function of carrying out β-elimination of peptide substrates containing phosphothreonine or phosphoserine (pS). Due to their specific action in suppressing the host immune system, phosphothreonine lyases may hold promise as drug targets to inhibit these pathogenic bacteria and understanding the catalyzed reactions involved will facilitate the discovery of novel antibiotics.

A. *Crystallographic Studies of SpvC*

Conformational transitions are always coupled with enzyme activity. However, the mechanisms of the coupling between protein conformation and bio-functions for each individual enzyme are usually different and largely remain unknown (Gao et al., 2004, 2006; Priyakumar and MacKerell, 2006). X-ray crystal structures of SpvC, mutant complexes of SpvC with sulfate groups (SpvC–sulfate), and the inactive mutants complexed with substrate peptides (SpvC–Erk2) certainly reflect the conformational changes upon substrates binding (Chen et al., 2008). Conformational changes in SpvC–Erk2 complex mainly occur close to the substrate binding site, which is a shallow groove formed by loops L1 (residues 90–103), L2 (residues 142–153), and L3 (residues 214–227) and the twisted five-stranded β-sheet (β3, β5, β6, β4, and β7). Based on the structural characteristics for the three conformations of SpvC and its complexes characterized by X-ray crystallography, we are able to propose a fundamental functional cycle of SpvC catalyzing MAPKs' dephosphorylation based on available crystal structures as well as our previous understandings of characteristics of SpvC enzymatic activities. As shown in Scheme 1, the first step is substrate recognition and binding, which leads to a major conformational change in the entire cycle. Loop L1 is

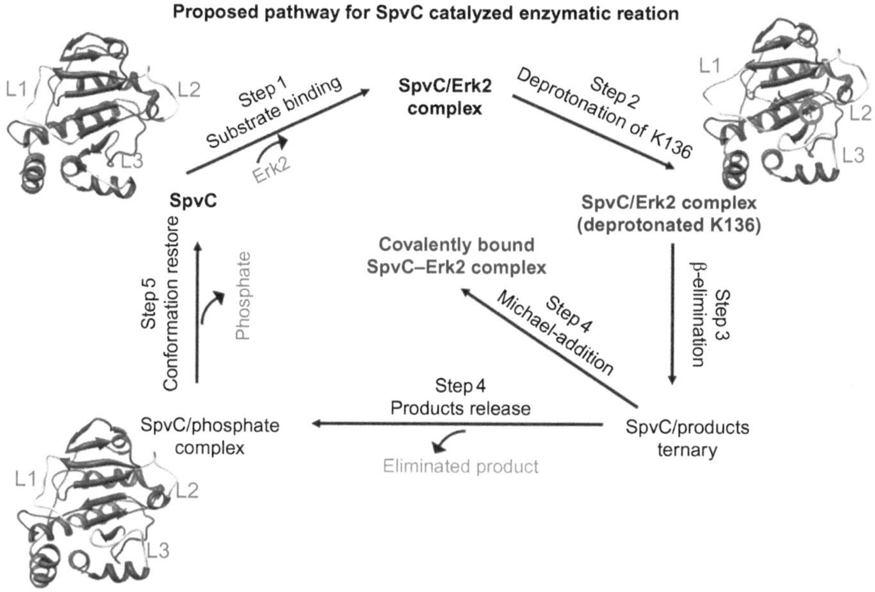

SCHEME 1. Schematic diagram of the complete SpvC-catalyzed reactions. The primary reaction is a β-elimination reaction via E2 mechanism; the secondary reaction is a Michael-addition initialized from the product of the primary reaction that suicidally breaks the enzymatic cycle. Both reactions use deprotonated K136 as catalytic residue; however, K136 acts as catalytic base in the primary elimination reaction but a nucleophile in the secondary addition reaction.

stretched into an open shape upon the binding of phosphotyrosine (pY158). R148 and V149 in loop L2 are components of the phosphothreonine (pT183) binding site, with the formation of the pT183 binding site leading to the unfolding of the α-helix (α5) at the terminal of L2, while loop L3 undergoes a significant movement toward pT183. The maximum movement of SpvC conformations is up to 10 Å (R220) and an electronic interaction is formed between the positively charged R220 and the negatively charged substrates. The second step is the catalytic reaction, the mechanism of which is discussed below. In the final step, the produced Mdha-containing peptide and phosphate group are released from the binding site one after another and the conformation of SpvC returns to its initial state.

Strong electrostatic interactions between positively charged binding site residues (i.e., K104, H106, R148, R213, and R220) and the substrate phosphate group pT183, as well the hydrophobic interactions between residues F86 and V149 and the methyl group of pT183, contribute to forming a fully desolvated substrate binding pocket (see Fig. 8) (Zhu et al., 2007; Chen et al., 2008). This structural evidence, combined with biochemical assays, strongly supports a β-elimination reaction mechanism, in which the highly conserved residue K136 acts as a catalytic base to abstract the H_α from pT183 and H106 acts as catalytic acid to donate a proton to the phosphate group, which results in the cleavage of the $C_\beta—O_\gamma$ bond and the formation of a $C_\alpha{=}C_\beta$ double bond (Zhu et al., 2007; Chen et al., 2008).

However, SpvC was also reported to react with peptides containing dehydroalanine (Dha) but it is insensitive to free Mdha. It has been proposed that K136 acts as a nucleophile to attack the unsaturated C_β atom of Dha and form a covalently bound complex via a nucleophilic addition mechanism (Zhu et al., 2007). Similarly, we detected a covalently bound SpvC–peptide complex in addition to the primary β-elimination product; a product with a molecular weight of approximately 130 kD that was observed when the phosphorylated Erk2 (a phosphothreonine-containing MAPK peptide) was used as substrate. Experiments showed

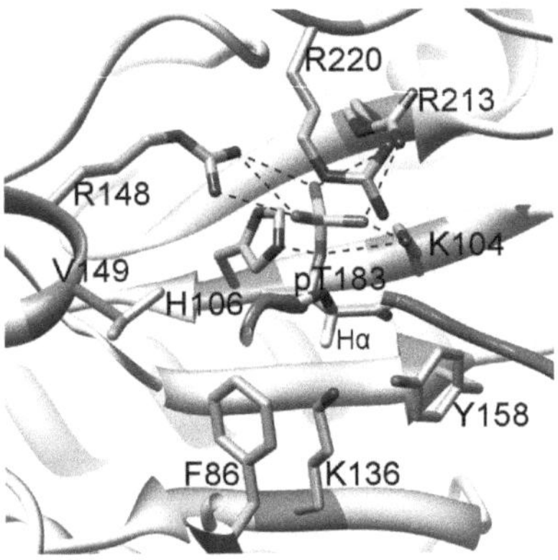

FIG. 8. Specific recognition of phosphothreonine by SpvC. K136 is modeled according to the crystal structure of apo SpvC (PDB code: 1Z8M). Hydrogen bonds between phosphate moiety of pT183 and binding site residues are represented by dashed lines. All H atoms except Hα of pT183 are omitted for clarity. Reprinted with permission from Pei et al. (2011). Copyright (2011) Elsevier.

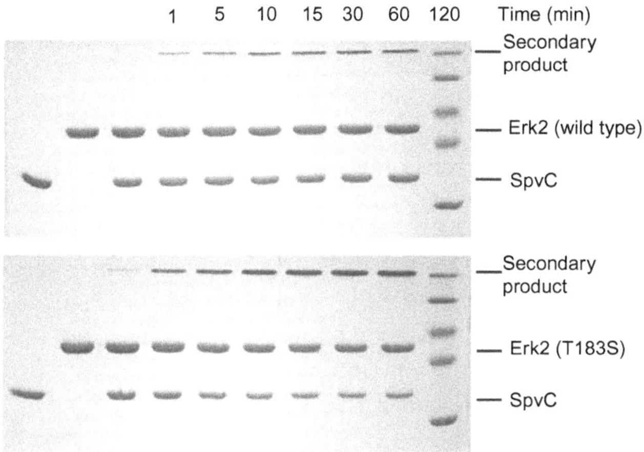

FIG. 9. Coomassie blue-stained gel showing the reaction product of SpvC with Erk2. The wild-type Erk2 (top panel) and its mutant T183S (bottom panel) were phosphorylated by coexpression with MEK1 in bacteria and used as the substrates of SpvC. Reprinted with permission from Pei et al. (2011). Copyright (2011) Elsevier.

the product is a covalently bound complex of SpvC–Erk2 through a secondary reaction mechanism, and its amount progressively increased over time whereas both Erk2 and SpvC decreased. The top panel of Fig. 9 shows the coomassie blue-stained gel of the covalently bound product of SpvC with wild-type Erk2. As comparison, mutant T183S in Erk2 (bottom panel) was phosphorylated by coexpression with MEK1 in bacteria and used as the substrate for SpvC. The amount of secondary product from the Erk2 mutant was significantly larger compared to that from the wild-type Erk2. Apparently, SpvC suicidally terminates the enzymatic cycle, which may provide additional information for developing novel antibiotics in inhibiting the pathogenic bacteria.

Previous biochemical studies have demonstrated that substitution of phosphothreonine with phosphoserine compromises both peptide substrate recognition and catalytic efficiency (Zhu et al., 2007). Based on the crystallographic structures, it is evident that important interactions occur between binding site residue V149 and the methyl group of pT183, which supports the binding preference of phosphothreonine over phosphoserine in this enzyme family (Zhu et al., 2007; Chen et al., 2008).

B. *MD Simulations and PMF Calculations on SpvC Complexes*

The crystal structure of the inactive SpvC K136A mutant complexed with a phosphothreonine-containing Erk2 peptide (2Z8P) (Chen et al., 2008) was used to model the wild-type enzyme–substrate complex (referred to as SpvC–Erk2 hereafter) according to the coordinates of K136 in SpvC ligand-free structure (2Z8M) (Chen et al., 2008). The Erk2 T183S mutant complex (SpvC–T183S) was modeled by replacing the methyl group of pT183 with a hydrogen atom. A total of four reactant complexes were constructed: SpvC–Erk2 and SpvC–T183S with K136 protonated, and SpvC–Erk2 and SpvC–T183S with K136 deprotonated. All treatments are essentially the same for the set up and simulation of the SpvC–Erk2 and SpvC–T183S systems and thereafter only SpvC–Erk2 system will be described unless otherwise noted. The protonation states of other titratable residues were carefully checked based on their local environment (e.g., the putative catalytic acid H106 was protonated). SpvC-product complexes (referred to as SpvC–Mdha and SpvC–Dha hereafter) were also constructed by simply replacing the substrate with the elimination product. The force field parameters for unnatural amino acids Mdha and Dha were generated by assigning closely related atom types from the CHARMM General force field (Vanommeslaeghe et al., 2009).

MD simulations were performed using the CHARMM program (version c34b1) (Brooks et al., 1983) with the all-hydrogen protein force field (MacKerell et al., 1998) including the dihedral cross-term corrections (CMAP) (MacKerell et al., 2004), the TIP3P water model (Jorgensen et al., 1983), the phosphothreonine and phosphoserine parameters (Feng et al., 1996), and sodium parameters (Beglov and Roux, 1997). MD simulations applied 2 fs integration time step, SHAKE of covalent bonds involving hydrogen (Ryckaert et al., 1977), Periodic boundary condition (PBC), and Particle mesh Ewald (PME) method (Essmann et al., 1995). All calculations used an atom-based truncation scheme updated heuristically with a list cut-off of 16 Å, a nonbond cut-off of 12 Å and with the Lennard-Jones (LJ) smoothing function initiated at 10 Å.

Hydrogen atoms were added using the HBUILD module in CHARMM. The complex system was overlaid with a 70-Å cubic water box, oriented with respect to the center of mass of protein complex, with ions added to obtain electrical neutrality. The equilibration of the solvated system consisted of an energy minimization and a 20-ps NVT MD simulation with heavy atoms of solute restrained using a force constant of 2 kcal/mol/Å^2. Following this, restraints were removed and the system was fully equilibrated by energy minimization, a 40-ps NVT MD simulation and a 20-ps NPT MD simulation using the Nosé-Hoover temperature coupling scheme. Finally, a production run of 10 ns MD simulation was performed in constant-NPT ensemble.

One of the outstanding questions for understanding the action of SpvC is how the positively charged lysine K136 is deprotonated in order to act as a catalytic base. We performed 10 ns MD simulations of SpvC–peptide complexes with K136 in both deprotonated and protonated states, and undertook extensive structural and energetic analyses with emphasis on the atomic events. Note that the analysis is only presented for SpvC–Erk2 if similar behaviors are observed for the T183S mutant.

We performed 10 ns MD simulations of SpvC–peptide complexes with K136 in both deprotonated and protonated states. Figure 10A shows the number of hydrogen bonds formed between head groups of K136 in protonated and deprotonated states, respectively. It is easy to demonstrate that the protonated K136 is exposed in water solvent and the deprotonated K136 is favorable in a relatively water free environment. Structure analysis shows that the deprotonated K136 is greatly stabilized inside the fully desolvated binding pocket surrounded by conserved hydrophobic residues F86 and V149 as well as the substrate peptide. Deprotonated K136 forms a stable hydrogen bond with the substrate peptide (carbonyl O of A182 adjacent to pT183) and rarely interacts with solvent waters (see Fig. 10B).

The motivation for the utilization of MD simulations for SpvC-products complexes was to observe the conformational change in the pT183 binding site after the primary β-elimination. For RMSD measured for binding site residues including P86, K104, H106, K136, R148, V149, Y158, R213, R220, and the β-elimination products, the phosphate group and β-Mdha have larger fluctuations compared to the SpvC–Erk2 wild-type complex. However, similar to the SpvC–Erk2 wild-type complex, the hydrogen bond analysis (see Fig. 11) shows that the deprotonated K136 is stabilized inside the desolvated active site and positioned toward the C_β atom.

Examination of the RMSD of the binding site residues and substrate peptides with respect to the crystal structure indicates that SpvC–peptide complex structures are stable in all four simulations (see Fig. 12A);

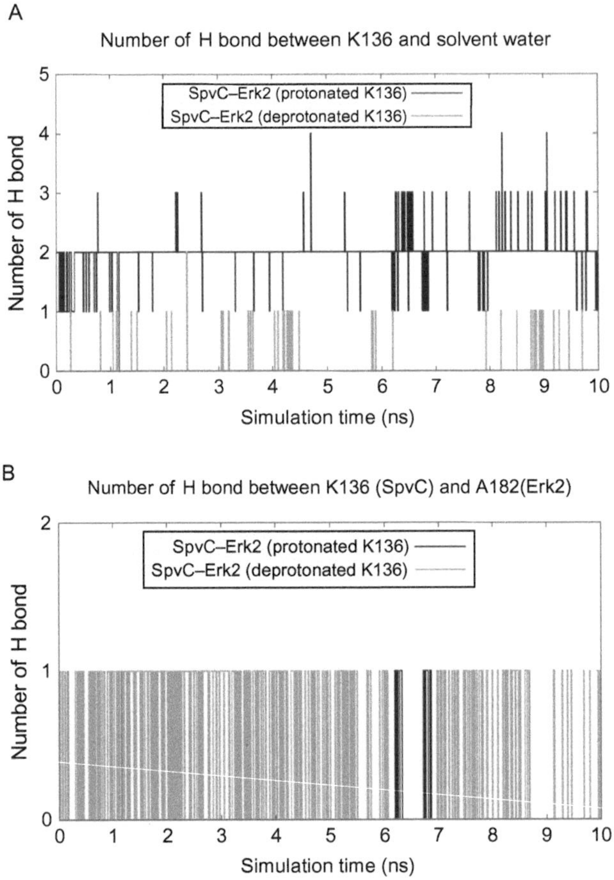

FIG. 10. (A) Number of hydrogen bond between head group of K136 and water solvent in two MD simulations of SpvC–Erk2 with K136 in both protonated (black line) and deprotonated (gray line) states. (B) Number of hydrogen bond form between H_ζ of K136 and carboxyl O of A182 (O) in Erk2 in two MD simulations of SpvC–Erk2 with K136 in both protonated (black line) and deprotonated (gray line) states. Reprinted with permission from Pei et al. (2011). Copyright (2011) Elsevier.

however, critical structural perturbations are observed. The most notable event occurring in the deprotonated K136 system is that the deprotonated K136 is greatly stabilized inside the fully desolvated binding pocket surrounded by conserved hydrophobic residues F86 and V149 as well as the

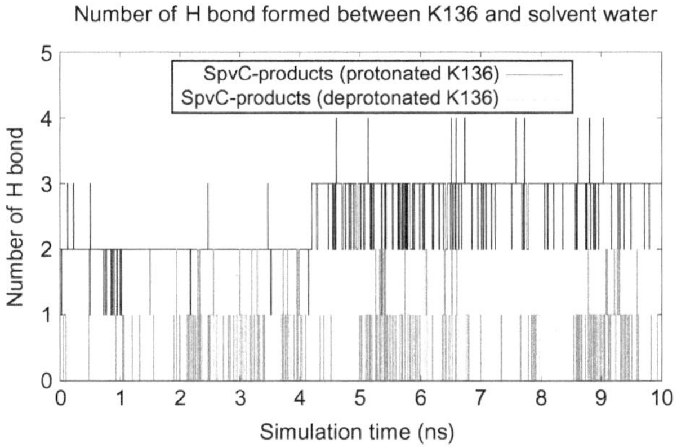

FIG. 11. Number of hydrogen bond form between head group of K136 and water solvent in two MD simulations of SpvC-products with K136 in both protonated (black line) and deprotonated (gray line) states. Reprinted with permission from Pei et al. (2011). Copyright (2011) Elsevier.

substrate peptide, forming a stable hydrogen bond with substrate peptide (carbonyl O of A182 adjacent to pT183) and rarely interacting with solvent waters (see Fig. 10). As shown in Fig. 12, these interactions help to stabilize the lone pair of electrons of K136 poised toward the H_α of pT183 ($r_{H_\alpha - N_\zeta} = 2.8 \pm 0.3$ Å). In contrast, the charged ammonium group of protonated K136 is exposed to the solvent, forming an average of two hydrogen bonds with water molecules and rarely interacting with substrate or other binding site residues, with a significantly longer $r_{H_\alpha - N_\zeta}$ distance (4.6 ± 0.5 Å) (see Fig. 12C). The F86D and V149D mutants were shown to significantly decrease enzymatic activity (Chen et al., 2008). It seems that these two hydrophobic residues not only contribute to substrate recognition but are also involved in catalysis by stabilizing K136 in its deprotonated form. The present results suggest the deprotonation of K136 is coupled with the reorganization of the substrate binding pocket, and the deprotonated K136 is stabilized in the fully desolvated binding pocket, poised at the ideal position to abstract the alpha hydrogen atom of pT183. A similar scenario was observed for a catalytic lysine in acetoacetate decarboxylase (Ho et al., 2009).

Consistent with experimental observations, our MD simulations on SpvC–Erk2 and its mutant T183S reveal the atomic details in depth (Fig. 13A and B). In SpvC–Erk2, the presence of the methyl group of

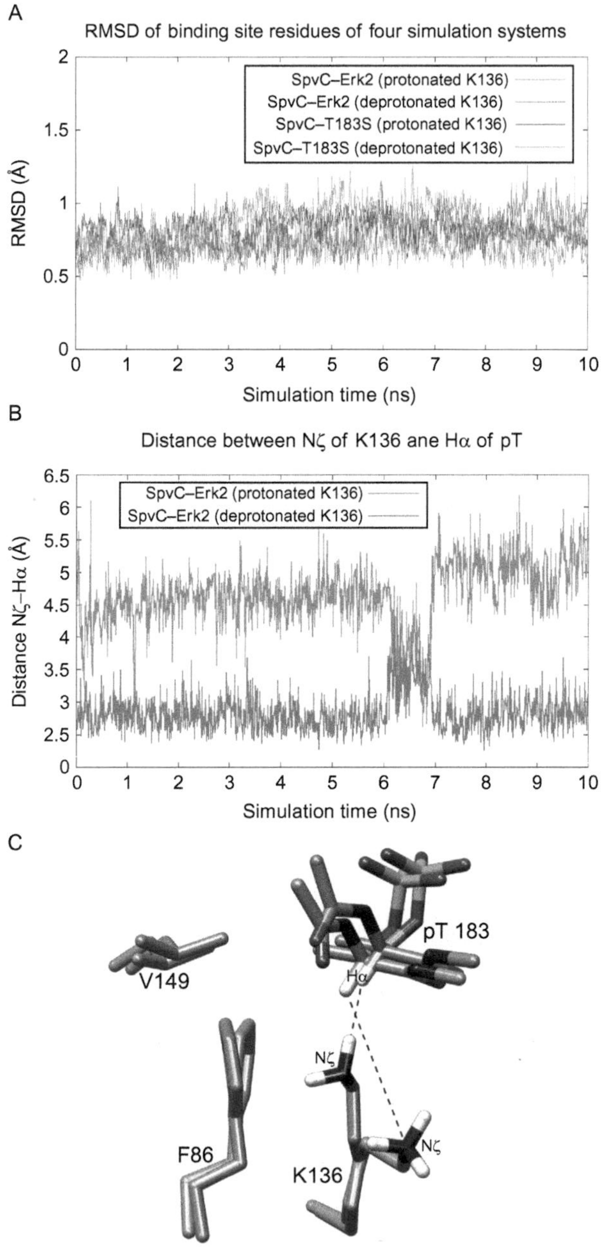

FIG. 12. (A) RMSD of binding site residues and substrate in SpvC–Erk2 and SpvC–T183S with K136 in both protonated and deprotonated states. Binding site residues

pT183 facilitates the longer lifetime of the strong interaction between the H_ε of the putative catalytic acid H106 and the O_γ of pT183 (ca. 1.89 Å) compared to the analogous interaction in SpvC–T183S (ca. 2.40 Å). Such strengthening of the interaction might enhance the donation of the H_ε atom of H106 to O_γ of pT183, thus promoting the β-elimination reaction. Those coupled variations suggest that the hydrophobic interaction between the methyl group of pT183 and the binding site residue V149 not only contribute to substrate binding, but also actively participate in the primary β-elimination reaction.

To fully understand the entire functional cycle, there are still questions that need to be answered. The catalytic residue K136 must be deprotonated in order to catalyze the β-elimination reaction, which is stabilized by a solvent-free environment. However, as we can observe from the crystal structure and MD simulation study of SpvC–Erk2 with K136 in different protonation states, K136 is located on the substrate peptide on the opposite side from loop L3 and has no direct contact with the phosphate moiety pT183 (Pei et al., 2011). Therefore, a question of critical functional importance is: what properties of SpvC ensure a tight coupling between substrate binding induced conformational change and activation catalytic activity, specifically the deprotonation of catalytic residue K136? Moreover, product release usually plays the essential role to restore conformation of enzymes to initialize the enzymatic cycle, which is also under consideration in our study.

A key quantity in studying conformational changes induced enzymatic reactions is the free energy difference, $\Delta\Delta G^{\ddagger}(T)$ (Jorgensen, 1989; Gao et al., 2006; Yu et al., 2007). One of the most powerful methods, which are generally used is the PMF. By applying an external restraint to force domain movement, this method can sample a longer time scale movement of bio-molecules, for example, loop flipping (Priyakumar and MacKerell,

include F86, K104, H106, K136, R148, V149, Y158, R213, R220, and pT183. (B) Distance between Nζ of K136 and Hα of pT183 in SpvC–Erk2 with K136 in protonated (orange line) and deprotonated (turquoise line) states. (C) Superimposed structures of SpvC–Erk2 with K136 in both protonated and deprotonated states. Carbon atoms are colored in orange in protonated K136 system and colored in turquoise in deprotonated K136; O, N, H, and P atoms are colored in red, blue, white, and cyan, respectively. Reprinted with permission from Pei et al. (2011). Copyright (2011) Elsevier. (For interpretation of the references to color in this figure legend, the reader is referred to the online version of this chapter.)

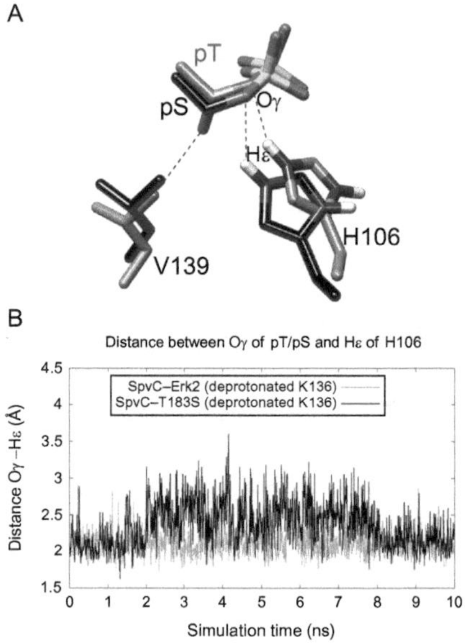

FIG. 13. (A) Superimposed structures of SpvC–Erk2 and its mutant SpvC–T183S. Hydrophobic interaction between V149 and methyl group of pT stabilizes the Oγ atom poising toward the catalytic acid H106. (B) Distance between Oγ and Hε of H106 in simulations of SpvC–Erk2 (gray line) and SpvC–T183S (black line) with K136 in the deprotonated state. Reprinted with permission from Pei et al. (2011). Copyright (2011) Elsevier.

2006; Yu et al., 2007). Simulation systems were set up to represent the loop movement during the processes of substrate binding and product release, representing different state of conformational changes. All the simulation systems are summarized in Table V. The structures of the two end points in each simulation system were generated based on crystal structures. The choice of an appropriate order parameter to represent enzyme conformational changes is important for the success of PMF calculation. This parameter should well separate the two end point states and provide minimal restraint on the nature of conformations adopted along the path between these states (Banavali and Roux, 2005; Arora and Brooks, 2007). In our study, the difference between the RMSD (dRMSD) of flexible loops of two end point structures were chosen as the order parameter, which contains information about numerous individual degrees of freedom that contribute conformational transitions

TABLE V
Summary of the PMF Simulation Systems

Process of study	PMF simulation system	Reference 1 (PDB ID)	Reference 2 (PDB ID)	Ligand
Substrate binding	SpvC–Erk2 (protonated K136)	2Z8M	2Z8P	Erk2 peptide
	SpvC–Erk2 (deprotonated K136)	2Z8M	2Z8P	Erk2 peptide
	SpvC (protonated K136)	2Z8M	2Z8P	None
Product release	SpvC–phosphate	2Z8P	2Z8N	Phosphate
	SpvC-products (Phosphate and product peptide)	2Z8P	2Z8N	Phosphate and product peptide
Phosphate release	SpvC–phosphate	2Z8N	2Z8M	Phosphate

PDB codes of the X-ray crystal structures of the reference conformations of the two end point structures of SpvC used to set up umbrella sampling are shown.

(Priyakumar and MacKerell, 2006; Yu et al., 2007). The NEB module (Chu et al., 2003) in CHARMM program was used to generate 51 intermediate conformations including the two end points. The simulation systems were energetically minimized and then soaked in a 45-Å water ball. A series of pre-equilibration runs of the water ball and simulation system were performed with a restraint on the order parameter dRMSD from 1000 to 75 kcal/mol/Å^2. An equilibration run of 1 ns with a restrain on dRMSD of 75 kcal/mol/Å^2 was then performed and followed by a production run for 3 ns with the same restraint. The weighted histogram analysis was used to obtain the potential mean force along the order parameter from the saved time series variable reaction coordinate every time step.

The convergence analysis performed on each 1 ns of production run shows a good convergency of each simulation. For the substrate binding process, the profiles of the potential mean force in each simulation system are shown in Fig. 14. For SpvC–Erk2 (protonated K136), the open state is ~1.0 kcal/mol more favorable than the closed state. However, for SpvC–

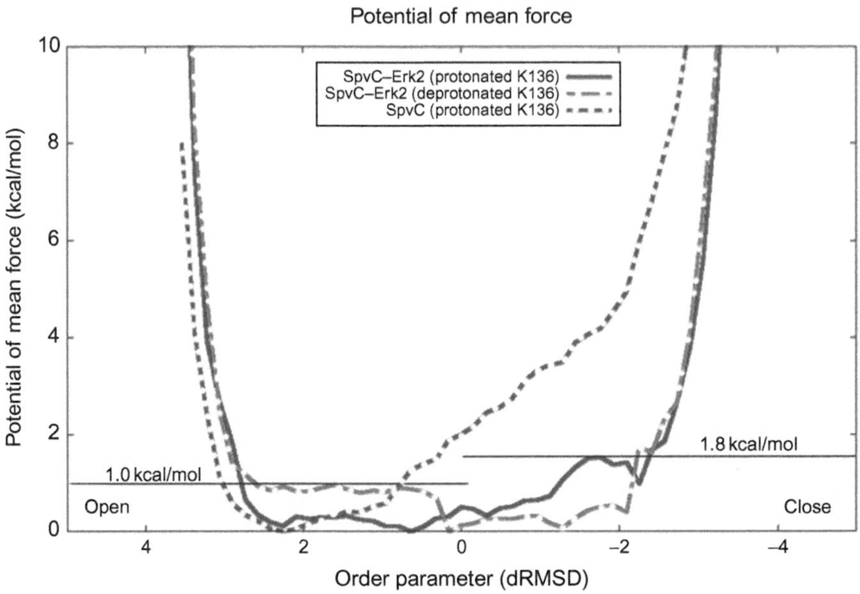

FIG. 14. Profiles of PMF to study the conformational change induced by Erk2 peptide binding on SpvC with the key catalytic residue K136 in both protonated and deprotonated states (SpvC–Erk2 (protonated K136) and SpvC–Erk2 (deprotonated K136)). A simulation system without a substrate (SpvC (protonated K136)) was also set up for comparison.

Erk2 (deprotonated K136), the closed state is ~1.5 kcal/mol more favorable than the open state. For the apo structure (without substrate binding), the K136 is most likely to be found in the protonated state, which is also confirmed with experimental results and our MD simulation results. The comparison of PMF profiles for SpvC holo structures indicates that the substrate binding is initialed by substrate recognition, followed by a conformational changes of flexible loops (L1, L2, L3), which form a solvent free binding site for and stabilized by deprotonation of K136. The profile of PMF calculations for product, the release process is shown in Fig. 15. The setup system is used to discover the sequence of product release. The calculation results indicate that with or without product peptide present, an energy barrier of ~3.5 kcal/mol needs to be crossed for the conformation of SpvC to transfer from a closed state to an open

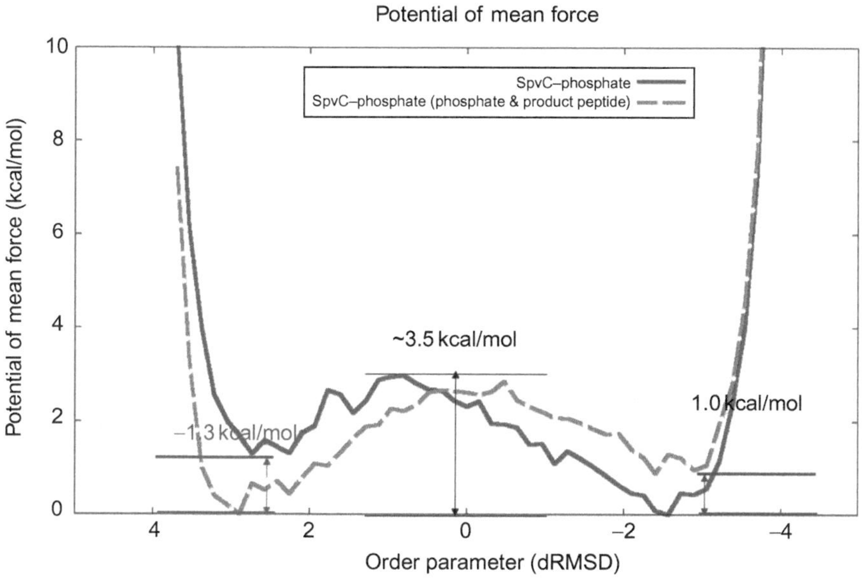

FIG. 15. Profiles of PMF to study the order of products release and the correlated conformational change. The two simulation systems are SpvC-products (product peptide and phosphate) and SpvC–phosphate, respectively.

state. However, by comparing two PMF profiles, it is observed that the product peptide is easier to release from the binding site when SpvC adapts a closed conformation. In the following step, a relatively high energy barrier (~12 kcal/mol) needs to be crossed for the conformation of SpvC to transfer from the closed state to the open state, while the product phosphate is released from the binding site (see Fig. 16). These result indicates that phosphate release might be a rate-determine step in the enzymatic cycle.

C. *Implications of the Computational Analysis of SpvC*

By combining theoretical calculations and experimental observations, a complete enzymatic pathway for SpvC-catalyzed reactions may be obtained (Scheme 1). Calculated results are in good agreement with the experimental measurements, and reveal the correlation between conformation and activity of SpvC, the critical stages of SpvC-catalyzed reactions and

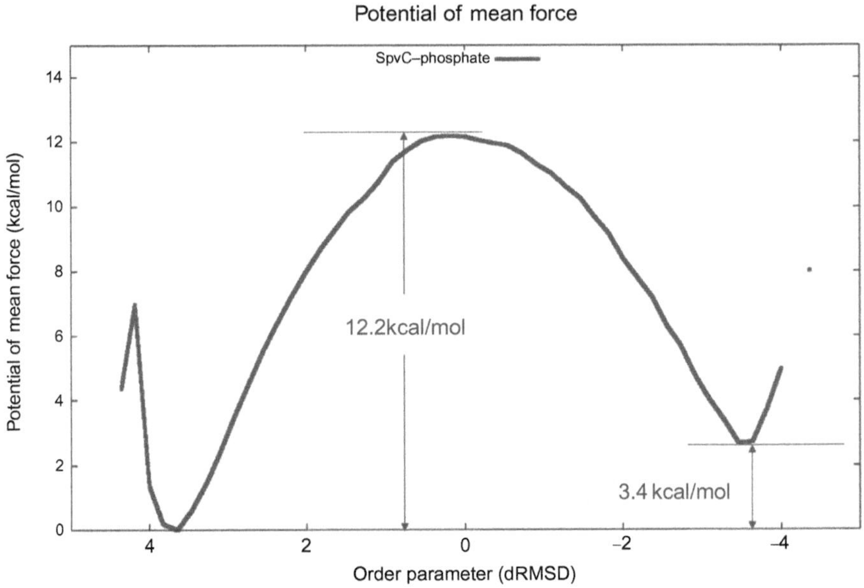

FIG. 16. Profiles of PMF to study the phosphate release and the conformational restoration of SpvC to the open state.

reconcile the roles of several important binding site residues in catalytic reactions, providing a description of the dynamic behavior of the enzymatic pathway of SpvC. Specifically, MD simulation results indicate that the deprotonation of the catalytic residue K136 is facilitated by the formation of the fully desolvated active site upon substrate binding, thus the deprotonated K136 is stabilized and precisely located to act as a catalytic base. The PMF calculation results indicate that the deprotonation of catalytic residue K136 is initiated by the formation of the closed binding site due to the conformational change. The hydrophobic binding site residues F86 and V149 are not only involved in substrate recognition but also play an important role in stabilizing the deprotonated K136. In addition to dynamic simulations, a truncated active site model was applied to explore the elimination reaction pathway using quantum mechanics methods. The 2D minimal potential energy surface, resulted from thoroughly scanning the reaction coordinates of alpha hydrogen abstraction by catalytic base K136 and proton donation by catalytic acid H106, suggests that the β-elimination occurs via an E2-like mechanism. The activation energies for the wild-type substrate and its T183S mutant were estimated to be 18.22 and 20.36 kcal/mol, respectively (Pei et al., 2011). In this β-elimination reaction, binding site residue K104 was identified to play critical role in determining the structural and energetics of transition state. Finally, the subsequent deprotonation of K136 happens to be a natural sequel of the primary elimination reaction, restores its nucleophile capacity to attack the unsaturated C_β atom of elimination product, and leads to a covalently bound complex via a Michael-addition mechanism. The activation energies in this step for corresponding wild-type substrate and its T183S mutant were calculated to be 16.0 and 12.92 kcal/mol, respectively (Pei et al., 2011). In contrast to its favorable role in β-elimination reaction, the methyl group in wild-type substrate introduces steric hindrance for nucleophilic addition reaction to occur. Two products produced from this reaction, which are the phosphate and product peptide, are released from the SpvC binding site one after the other. It is easier for product peptide to release when SpvC adapts a closed conformation. A conclusion that the following release of product phosphate might be the rate-determine step on the enzymatic cycle can be proposed based on our PMF calculations. Since such SpvC-catalyzed enzymatic reactions are novel, our present studies may provide additional structural and mechanistic information to facilitate the development of effective antibiotics in inhibiting the pathogenic bacteria.

IV. Future Perspectives

Currently, dynamics simulations alone cannot definitively determine enzyme reaction mechanisms. Likewise, current experimental methods alone cannot give an atomic-level understanding of many biological processes and it is only through a combination of these two methods that an in-depth understanding of many enzyme reaction methods may be achieved. As computer storage and processing power increases, new and more detailed computational algorithms will be developed that can provide a more thorough and detailed understanding of enzyme reaction mechanisms and other biomolecular reactions—in conjunction with experimental approaches.

References

Anderson, J. A., Lorenz, C. D., Travesset, A. (2008). General purpose molecular dynamics simulations fully implemented on graphics processing units. *J. Comput. Phys.* **227**, 5342–5359.

Anlezark, G. M., Knox, R. J., Friedlos, F., Sherwood, R. F., Melton, R. G. (1992). The bioactivation of 5-(aziridin-1-yl)-2,4-dinitrobenzamide (CB1954)-II. A comparison of an Escherichia coli nitroreductase and Walker DT diaphorase. *Biochem. Pharmacol.* **44**, 2297–2301.

Arbibe, L., Kim, D. W., Batsche, E., Pedron, T., Mateescu, B., Muchardt, C., et al. (2007). An injected bacterial effector targets chromatin access for transcription factor NF-kappaB to alter transcription of host genes involved in immune responses. *Nat. Immunol.* **8**, 47–56.

Arora, K., Brooks, C. L. (2007). Large-scale allosteric conformational transitions of adenylate kinase appear to involve a population-shift mechanism. *Proc. Nat. Acad. Sci. USA* **104**, 18496–18501.

Ashwell, J. D. (2006). The many paths to p38 mitogen-activated protein kinase activation in the immune system. *Nat. Rev. Immunol.* **6**, 532–540.

Ausubel, F. M. (2005). Are innate immune signaling pathways in plants and animals conserved? *Nat. Immunol.* **6**, 973–979.

Bacallado, S., Chodera, J. D., Pande, V. (2009). Bayesian comparison of Markov models of molecular dynamics with detailed balance constraint. *J. Chem. Phys.* **131**, 045106–045110.

Banavali, N. K., Roux, B. (2005). Free energy landscape of A-DNA to B-DNA conversion in aqueous solution. *J. Am. Chem. Soc.* **127**, 6866–6876.

Barbosa, T. M., Levy, S. B. (2002). Activation of the Escherichia coli nfnB gene by MarA through a highly divergent marbox in a class II promoter. *Mol. Microbiol.* **45**, 191–202.

Barna, T. M., Khan, H., Bruce, N. C., Barsukov, I., Scrutton, N. S., Moody, P. C. E. (2001). Crystal structure of pentaerythritol tetranitrate reductase: "Flipped" binding geometries for steroid substrates in different redox states of the enzyme. *J. Mol. Biol.* **310**, 433–447.

Beglov, D., Roux, B. (1997). An integral equation to describe the solvation of polar molecules in liquid water. *J. Phys. Chem. B* **101**, 7821–7826.

Brent, R., Bruck, J. (2006). 2020 Computing: can computers help to explain biology? *Nature* **440**, 416–417.

Brooks, B. R., Bruccoleri, R. E., Olafson, B. D., States, D. J., Swaminathan, S., Karplus, M. (1983). CHARMM: a program for macromolecular energy, minimization, and dynamics calculations. *J. Comput. Chem.* **4**, 187–217.

Bryant, C., DeLuca, M. (1991). Purification and characterization of an oxygen-insensitive NAD(P)H nitroreductase from Enterobacter cloacae. *J. Biol. Chem.* **266**, 4119–4125.

Case, D. A., Darden, T. A., Cheatham, T. E., Simmerling, C. L., Wang, J., Duke, R. E., et al. (2004). AMBER 8 (U. o. California, Ed.). University of California, San Francisco.

Chen, L., Wang, H., Zhang, J., Gu, L., Huang, N., Zhou, J.-M., et al. (2008). Structural basis for the catalytic mechanism of phosphothreonine lyase. *Nat. Struct. Mol. Biol.* **15**, 101–102.

Chipot, C., Pohorille, A. (2007). Free Energy Calculations Theory and Applications in Chemistry and Biology. Springer, Berlin Heidelberg.

Christofferson, A., Wilkie, J. (2009). Mechanism of CB1954 reduction by Escherichia coli nitroreductase. *Biochem. Soc. Trans.* **37**, 413–418.

Chu, J.-W., Trout, B. L., Brooks, B. R. (2003). A super-linear minimization scheme for the nudged elastic band method. *J. Chem. Phys.* **119**, 12708–12717.

Chung-Faye, G., Palmer, D., Anderson, D., Clark, J., Downes, M., Baddeley, J., et al. (2001). Virus-directed, enzyme prodrug therapy with nitroimidazole reductase: a phase I and pharmacokinetic study of its prodrug, CB1954. *Clin. Cancer Res.* **7**, 2662–2668.

Cleland, W. W. (1989). The kinetics of enzyme-catalyzed reactions with two or more substrates or products. I. Nomenclature and rate equations. *Biochim. Biophys. Acta* **1000**, 213–2201963.

Deng, H., Zhadin, N., Callender, R. (2001). Dynamics of protein ligand binding on multiple time scales: NADH binding to lactate dehydrogenase. *Biochemistry* **40**, 3767–3773.

Dodson, G. G., Lane, D. P., Verma, C. S. (2008). Molecular simulations of protein dynamics: new windows on mechanisms in biology. *EMBO Rep.* **9**, 144–150.

Dong, C., Davis, R. J., Flavell, R. A. (2002). Map kinase in the immune response. *Annu. Rev. Immunol.* **20**, 55–72.

Essmann, U., Perera, L., Berkowitz, M., Darden, T., Lee, H., Pedersen, L. (1995). A smooth particle mesh Ewald method. *J. Chem. Phys.* **103**, 8577–8593.

Feng, M.-H., Philippopoulos, M., MacKerell, A. D., Lim, C. (1996). Structural characterization of the phosphotyrosine binding region of a high-affinity SH2 domain-phosphopeptide complex by molecular dynamics simulation and chemical shift calculations. *J. Am. Chem. Soc.* **118**, 11265–11277.

Gao, J., Byun, K. L., Kluger, R. (2004). Catalysis by enzyme conformational change. *Top. Curr. Chem.* **238**, 113–136.

Gao, J., Ma, S., Major, D. T., Nam, K., Pu, J., Truhlar, D. G. (2006). Mechanisms and free energies of enzymatic reactions. *Chem. Rev.* **106**, 3188–3209.

Gilson, M. K., Given, J. A., Bush, B. L., McCammon, J. A. (1997). The statistical-thermodynamic basis for computation of binding affinities: a critical review. *Biophys. J.* **72**, 1047–1069.

Grove, J. I., Searle, P. F., Weedon, S. J., Green, N. K., McNeish, I. A., Kerr, D. J. (1999). Virus-directed enzyme prodrug therapy using CB1954. *Anticancer Drug Des.* **14**, 461–472.

Grove, J. I., Lovering, A. L., Guise, C., Race, P. R., Wrighton, C. J., White, S. A., et al. (2003). Generation of Escherichia coli nitroreductase mutants conferring improved cell sensitization to the prodrug CB1954. *Cancer Res.* **63**, 5532–5537.

Guise, C. P., Grove, J. I., Hyde, E. I., Searle, P. F. (2007). Direct positive selection for improved nitroreductase variants using SOS triggering of bacteriophage lambda lytic cycle. *Gene Ther.* **14**, 690–698.

Haynes, C. A., Koder, R. L., Miller, A.-F., Rodgers, D. W. (2002). Structures of Nitroreductase in Three States. *J. Biol. Chem.* **277**, 11513–11520.

Helsby, N. A., Ferry, D. M., Patterson, A. V., Pullen, S. M., Wilson, W. R. (2004). 2-Amino metabolites are key mediators of CB 1954 and SN 23862 bystander effects in nitroreductase GDEPT. *Br. J. Cancer* **90**, 1084–1092.

Ho, M.-C., Menetret, J.-F., Tsuruta, H., Allen, K. N. (2009). The origin of the electrostatic perturbation in acetoacetate decarboxylase. *Nature* **459**, 393–397.

Jaberipour, M., Vass, S. O., Guise, C. P., Grove, J. I., Knox, R. J., Hu, L., et al. (2010). Testing double mutants of the enzyme nitroreductase for enhanced cell sensitisation to prodrugs: effects of combining beneficial single mutations. *Biochem. Pharmacol.* **79**, 102–111.

Jakalian, A., Bush, B. L., Jack, D. B., Bayly, C. I. (2000). Fast, efficient generation of high-quality atomic charges. AM1-BCC model: I. Method. *J. Comput. Chem.* **21**, 132–146.

Jarrom, D., Jaberipour, M., Guise, C., Daff, S., Hicks, M. R., Burns, G., et al. (2009). Steady-state and stopped-flow kinetic studies of Escherichia coli NfsB mutants with enhanced activity for the prodrug CB1954. *Biochemistry* **48**, 7665–7672.

Johansson, E., Parkinson, G. N., Denny, W. A., Neidle, S. (2003). Studies on the nitroreductase prodrug-activating system. Crystal structures of complexes with the inhibitor dicoumarol and dinitrobenzamide prodrugs and of the enzyme active form. *J. Med. Chem.* **46**, 4009–4020.

Jorgensen, W. L. (1989). Free energy calculations: a breakthrough for modeling organic chemistry in solution. *Acc. Chem. Res.* **22**, 184–189.

Jorgensen, W. L., Chandrasekhar, J., Madura, J. D., Impey, R. W., Klein, M. L. (1983). Comparison of simple potential functions for simulating liquid water. *J. Chem. Phys.* **79**, 926–935.

Karplus, M., Kuriyan, J. (2005). Molecular dynamics and protein function. *Proc. Nat. Acad. Sci. USA* **102**, 6679–6685.

Karplus, M., McCammon, J. A. (2002). Molecular dynamics simulations of biomolecules. *Nat. Struct. Biol.* **9**, 646–652.

Knox, R. J., Friedlos, F., Jarman, M., Roberts, J. J. (1988). A new cyto-toxic, DNA interstrand crosslinking agent, 5-(aziridin-1-Yl)-4-hydroxylamino-2-nitrobenzamide, is formed from 5-(aziridin-1-Yl)-2,4-dinitrobenzamide (Cb-1954) by a nitroreductase enzyme in Walker carcinoma-cells. *Biochem. Pharmacol.* **37**, 4661–4669.

Koder, R. L., Miller, A.-F. (1998). Steady-state kinetic mechanism, stereospecificity, substrate and inhibitor specificity of Enterobacter cloacae nitroreductase. *Biochim. Biophys. Acta* **1387**, 395–405.

Kramer, R. W., Slagowski, N. L., Eze, N. A., Giddings, K. S., Morrison, M. F., Siggers, K. A., et al. (2007). Yeast functional genomic screens lead to identification of a role for a bacterial effector in innate immunity regulation. *PLoS Pathog.* **3**, e21.

Kyte, J. (1995). Mechanism in Protein Chemistry. Garland Publishing, Inc., New York.

Leach, A. R. (2001). Molecular Modelling: Principles and Applications. Pearson Education, Harlow.

Li, H., Xu, H., Zhou, Y., Zhang, J., Long, C., Li, S., et al. (2007). The phosphothreonine lyase activity of a bacterial type III effector family. *Science* **315**, 1000–1003.

Lovering, A. L., Hyde, E. I., Searle, P. F., Scott, A. W. (2001). The structure of Escherichia coli nitroreductase complexed with nicotinic acid: three crystal forms at 1.7 (A)over-circle, 1.8 (A)over-circle and 2.4 (A)over-circle resolution. *J. Mol. Biol.* **309**, 203–213.

MacKerell, A. D., Jr., Brooks, B., Brooks, C. L., III, Nilsson, L., Roux, B., Won, Y., et al. (1998). CHARMM: the energy function and its parameterization with an overview of the program. *In* The encyclopedia of Computational Chemistry Wiley, New York.

MacKerell, A. D., Jr., Feig, M., Brooks, C. L., III (2004). Extending the treatment of backbone energetics in protein force fields: limitations of gas-phase quantum mechanics in reproducing protein conformational distributions in molecular dynamics simulations. *J. Comput. Chem.* **25**, 1400–1415.

McCalla, D. R., Kaiser, C., Green, M. H. L. (1978). Genetics of nitrofurazone resistance in Escherichia coli. *J. Bacteriol.* **133**, 10–16.

Offman, M. N., Krol, M., Rost, B., Silman, I., Sussman, J. L., Futerman, A. H. (2011). Comparison of a molecular dynamics model with the X-ray structure of the N370S acid-beta-glucosidase mutant that causes Gaucher disease. *Protein Eng. Des. Sel.* **24**, 773–775.

Palmer, D. H., Mautner, V., Mirza, D., Oliff, S., Gerritsen, W., van der Sijp, J. R., et al. (2004). Virus-directed enzyme prodrug therapy: intratumoral administration of a replication-deficient adenovirus encoding nitroreductase to patients with resectable liver cancer. *J. Clin. Oncol.* **22**, 1546–1552.

Pei, Q., Christofferson, A., Zhang, H., Chai, J., Huang, N. (2011). Computational investigation of the enzymatic mechanisms of phosphothreonine lyase. *Biophys. Chem.* **157**, 16–23.

Ponder, J. W., Case, D. A. (2003). Force fields for protein simulations. *Adv. Protein Chem.* **66**, 27–85.

Priyakumar, U. D., MacKerell, A. D., Jr. (2006). Computational approaches for investigating base flipping in oligonucleotides. *Chem. Rev.* **106**, 489–505.

Race, P. R., Lovering, A. L., Green, R. M., Ossor, A., White, S. A., Searle, P. F., et al. (2005). Structural and mechanistic studies of Escherichia coli nitroreductase with the antibiotic nitrofurazone. *J. Biol. Chem.* **280**, 13256–13264.

Race, P. R., Lovering, A. L., White, S. A., Grove, J. I., Searle, P. F., Wrighton, C. W., et al. (2007). Kinetic and structural characterisation of Escherichia coli nitroreductase mutants showing improved efficacy for the prodrug substrate CB1954. *J. Mol. Biol.* **368**, 481–492.

Rapaport, D. C. (1995). The Art of Molecular Dynamics Simulation. Cambridge University Press, Cambridge, UK.

Ryckaert, J.-P., Ciccotti, G., Berendsen, H. J. C. (1977). Numerical integration of the cartesian equations of motion of a system with constraints: molecular dynamics of n-alkanes. *J. Comput. Phys.* **23**, 327–341.

Shan, L., He, P., Sheen, J. (2007). Intercepting host MAPK signaling cascades by bacterial type III effectors. *Cell Host Microbe* **1**, 167–174.

Shaw, D. E., Deneroff, M. M., Dror, R. O., Kuskin, J. S., Larson, R. H., Salmon, J. K., et al. (2008). Anton, a special-purpose machine for molecular dynamics simulation. *Commun. ACM* **51**, 91–97.

Shizuo, A., Satoshi, U., Osamu, T. (2006). Pathogen recognition and innate immunity. *Cell* **124**, 783–801.

Tena, G., Asai, T., Chiu, W.-L., Sheen, J. (2001). Plant mitogen-activated protein kinase signaling cascades. *Curr. Opin. Plant Biol.* **4**, 392–400.

Torrie, G. M., Valleau, J. P. (1974). Monte Carlo free energy estimates using non-Boltzmann sampling: application to the sub-critical Lennard-Jones fluid. *Chem. Phys. Lett.* **28**, 578–581.

Torrie, G. M., Valleau, J. P. (1977). Nonphysical sampling distributions in Monte Carlo free-energy estimation: umbrella sampling. *J. Comput. Phys.* **23**, 187–199.

Van Voorhis, T., Kowalczyk, T., Kaduk, B., Wang, L.-P., Cheng, C.-L., Wu, Q. (2010). The diabatic picture of electron transfer, reaction barriers, and molecular dynamics. *Annu. Rev. Phys. Chem.* **61**, 149–170.

Vanommeslaeghe, K., Hatcher, E., Acharya, C., Kundu, S., Zhong, S., Shim, J., et al. (2009). CHARMM general force field: a force field for drug-like molecules compatible with the CHARMM all-atom additive biological force fields. *J. Comput. Chem.* **31**, 671–690.

Voelz, V. A., Bowman, G. R., Beauchamp, K., Pande, V. S. (2010). Molecular simulation of *ab initio* protein folding for a millisecond folder NTL9(1−39). *J. Am. Chem. Soc.* **132**, 1526–1528.

Weedon, S. J., Green, N. K., McNeish, I. A., Gilligan, M. G., Mautner, V., Wrighton, C. J., et al. (2000). Sensitisation of human carcinoma cells to the prodrug CB1954 by adenovirus vector-mediated expression of E-coli nitroreductase. *Int. J. Cancer* **86**, 848–854.

Whiteway, J., Koziarz, P., Veall, J., Sandhu, N., Kumar, P., Hoecher, B., et al. (1998). Oxygen-insensitive nitroreductases: analysis of the roles of nfsA and nfsB in development of resistance to 5-nitrofuran derivatives in Escherichia coli. *J. Bacteriol.* **180**, 5529–5539.

Wilson, W. R., Pullen, S. M., Hogg, A., Helsby, N. A., Hicks, K. O., Denny, W. A. (2002). Quantitation of bystander effects in nitroreductase suicide gene therapy using three-dimensional cell cultures. *Cancer Res.* **62**, 1425–1432.

Wouters, J., Durant, F., Champagne, B., André, J.-M. (1997). Electronic properties of flavins: implications on the reactivity and absorption properties of flavoproteins. *Int. J. Quantum Chem.* **64**, 721–733.

Yu, H., Ma, L., Yang, Y., Cui, Q. (2007). Mechanochemical coupling in the myosin motor domain. I. Insights from equilibrium active-site simulations. *PLoS Comput. Biol.* **3**, 199–213.

Zenno, S., Koike, H., Tanokura, M., Saigo, K. (1996). Gene cloning, purification, and characterization of NfsB, a minor oxygen-insensitive nitroreductase from Escherichia coli, similar in biochemical properties to FRase I, the major flavin reductase in Vibrio fischeri. *J. Biochem.* **120**, 736–744.

Zenno, S., Kobori, T., Tanokura, M., Saigo, K. (1998). Conversion of NfsA, the major Escherichia coli nitroreductase, to a flavin reductase with an activity similar to that of Frp, a flavin reductase in Vibrio harveyi, by a single amino acid substitution. *J. Bacteriol.* **180**, 422–425.

Zhang, J., Shao, F., Li, Y., Cui, H., Chen, L., Li, H., et al. (2007). A pseudomonas syringae effector inactivates MAPKs to suppress PAMP-induced immunity in plants. *Cell Host Microbe* **1**, 175–185.

Zhu, Y., Li, H., Long, C., Hu, L., Xu, H., Liu, L., et al. (2007). Structural insights into the enzymatic mechanism of the pathogenic MAPK phosphothreonine lyase. *Mol. Cell* **28**, 899–913.

THE INCREASING ROLE OF QM/MM IN DRUG DISCOVERY

By ALESSIO LODOLA* AND MARCO DE VIVO†

*Dipartimento Farmaceutico, Università degli Studi di Parma, Parma, Italy
†Drug Discovery and Development, Italian Institute of Technology, Genova, Italy

Abstract

Since its first appearance in 1976, the quantum mechanics/molecular mechanics (QM/MM) approach has mostly been used to study the chemical reactions of enzymes, which are frequently the target of drug discovery programs. In principle, a detailed understanding of the enzymatic mechanism should help researchers to design a potent enzyme inhibitor or new drug. However, QM/MM has not yet had a widespread impact on structure-based drug design. This is mostly due to its high computational cost. We expect this to change with the recent and extraordinary increases in computational power, and with the availability of more efficient algorithms for QM/MM calculations. Here, we report on some representative examples of QM/MM studies, including our own research, of pharmaceutically relevant enzymes, such as ribonuclease H and fatty acid amide hydrolase (FAAH). We aim to show how QM/MM has traditionally been used to study enzymatic catalysis. In this regard, we discuss its potential to become a routinely used drug design tool. To support this, we also discuss selected computational studies where QM/MM insights have been helpful in improving the potency of covalent inhibitors of FAAH.

ADVANCES IN PROTEIN CHEMISTRY AND STRUCTURAL BIOLOGY, Vol. 87
DOI: 10.1016/B978-0-12-398312-1.00011-1

Copyright 2012, Elsevier Inc.
All rights reserved.

I. Introduction

Many drug discovery programs target enzymes (Bachovchin and Cravatt, 2011). These are proteins that catalyze a specific chemical reaction of a biological pathway. The drug binds to the catalytic pocket of the pharmacologically relevant enzyme(s). This prevents the enzyme from exerting its natural function, leading to the desired pharmacological effects. Understanding enzymatic catalysis is thus potentially crucial in designing more effective enzyme inhibitor drugs (Mulholland, 2005; Singh et al., 2011).

In the past decade, there has been a significant increase in the availability of high-resolution protein structures, mainly enzymes that have been deposited in the protein data bank. However, a single structure does not contain all the information required to properly describe a catalytic reaction pathway. A chemical reaction can involve intermediates and transition states (TSs). For many of those states, no detailed structural information is available. Even when several significant stationary points are structurally characterized, it is not easy to determine the energies and connections of those configurations (Mulholland, 2008).

Computational methods offer one way of understanding how enzymes work. They can provide detailed information about the enzymatic reaction mechanism, which is not available from the readily accessible experimental data (Warshel, 2003; Garcia-Viloca et al., 2004; Gao et al., 2006). In particular, hybrid quantum mechanics/molecular mechanics (QM/MM) computations combine the accuracy of quantum mechanics (QM) methods in describing the reactive chemistry, with the ability of molecular mechanics (MM) force fields to provide a good description of conformational energetics and nonbonded interactions in large systems (Masgrau et al., 2006; Senn and Thiel, 2009). This is why, since its first appearance in 1976 (Warshel and Levitt, 1976), the QM/MM approach has mainly been used to study enzymatic catalysis.

Here, we discuss some representative QM/MM studies of pharmaceutically interesting targets. We believe these once-prohibitive computations will increasingly affect structure-based drug design (SBDD; Peters et al., 2006; Raha et al., 2007; Zhou et al., 2010; De Vivo, 2011; Lodola et al., 2011b). For example, QM/MM can be used to describe polarization effects and charge transfer and, ultimately, to more accurately estimate the interaction energy between the ligand (drug) and its receptor (Cavalli et al., 2006; Menikarachchi and Gascon, 2010). We will focus on how QM/MM is used to study bond-forming/breaking reactions, such as those catalyzed by enzymes. The TS theory proposes TS analogs as potent inhibitors. It exemplifies how drug design is connected with the elucidation of enzymatic reaction mechanisms. In other words, a detailed understanding of biocatalysis can be translated into the rational design of TS-based drugs (Schramm, 2005).

This chapter makes no attempt to exhaustively review computational enzymatic catalysis, which is already well documented in the literature (Carloni et al., 2002; Warshel, 2003; Garcia-Viloca et al., 2004; Marti et al., 2004; Friesner and Guallar, 2005; Mulholland, 2005; Warshel et al., 2006; Acevedo and Jorgensen, 2010; Lonsdale et al., 2010). Rather, we focus on examples from our own research. In particular, we discuss QM/MM simulations that have been used to characterize the enzymatic phosphoryl transfer reaction. This includes proteins that are important SBDD targets such as ribonuclease H (RNase H) for antiviral drugs (Shaw-Reid et al., 2005). We also discuss the use of QM/MM computations to address the mechanism of covalent inhibitors of the fatty acid amide hydrolase (FAAH). FAAH is a promising enzymatic target for treating pain and other diseases (Petrosino and Di Marzo, 2010).

Ultimately, these examples show how QM/MM can be used to understand enzymatic catalysis and initiate SBDD studies, demonstrating the growing role of QM/MM in drug discovery.

II. The QM/MM Method

QM and MM combine to form the QM/MM method, which is often used when dealing with macromolecules (Leach, 1996; Jensen, 1999; Cramer, 2004). In QM/MM studies of enzymatic mechanisms, the reactive region of the active site is described at a suitable QM level. An MM force field is used to classically describe the majority of the system (most of the enzyme and the solvent), which is not directly involved in the process, but which affects the reaction by means of nonbonded interactions (Field et al., 1990; Fig. 1).

The interaction of both subsystems must be taken into account correctly, so that the total energy may be simply expressed as (Ferrer et al., 2011):

$$E_{\text{total}} = E_{\text{QM}} + E_{\text{MM}} + E_{\text{QM/MM}}$$

E_{QM} and E_{MM} are the energies of the QM and MM regions, respectively, calculated in a standard way. $E_{\text{QM/MM}}$ describes the interaction between

FIG. 1. Scheme of a system partitioned into QM and MM regions. (For color version of this figure, the reader is referred to the online version of this chapter.)

the QM and MM regions, and can be treated in various ways, as described below.

In principle, the quantum region can be described by any QM method. For instance, Hartree–Fock-based semiempirical (SE) methods such as AM1 and PM3 can be used, with the advantage of allowing large QM systems (hundreds of atoms). However, these SE are often inaccurate and, in some cases, difficult to use (e.g., for many transition metals; Lodola et al., 2008b). More reliable, but expensive, calculations can be carried out using the density functional theory (DFT) or electron correlation approaches. These include the MP2 perturbation method and coupled-cluster theory (Claeyssens et al., 2006). The self-consistent charge density functional tight binding (SCC-DFTB) method is a relatively recent alternative to expensive *ab initio* methods and rather inaccurate SE methods (de et al., 2007). In several cases, SCC-DFTB has given results that are in good agreement with QM/MM calculations performed at higher levels of theory (i.e., B3LYP) and a greater computational cost.

Standard protein force fields such as CHARMM (Brooks et al., 2009) or AMBER (Hornak et al., 2006) are normally used to treat atoms outside the QM region. These force fields provide a remarkably good description of the conformational energetics of macromolecules (including proteins and nucleic acids) and nonbonded interactions in large systems (Karplus and McCammon, 2002). For these reasons, they are currently being used in standard QM/MM implementations.

As noted above, the available QM/MM hybrid schemes must combine the quantum and classical parts to obtain a meaningful $E_{QM/MM}$ contribution. Several QM/MM schemes have been proposed (Senn and Thiel, 2009). They basically differ on the form of the $E_{QM/MM}$ term. Thus, QM/MM methods are usually grouped into three distinct formulations: mechanical, electrostatic, and polarization coupling. In mechanical coupling, the QM Hamiltonian is not influenced by the point charges of the MM system. Instead, a force field is used to account for the interactions between the QM and MM regions. Classical atom types are assigned to the QM atoms, and the QM/MM interaction energy is calculated by directly summing the electrostatic energy and the Van der Waals contribution. These, in turn, are calculated by applying the Coulomb law and the Lennard–Jones potential, respectively (Senn and Thiel, 2009).

In electrostatic coupling, the electrostatic interaction between the MM and the QM part is achieved by inserting the MM charges directly in the QM Hamiltonian, regardless of whether the QM is SE, HF, DFT, or a correlated method. The interaction of QM electrons with MM point charges moves from the $E_{QM/MM}$ term (where it is described with a force field) to the quantum energy $E_{QM.}$ It is described as an interaction of a charge density with point charges. This kind of electrostatic coupling is the most popular embedding scheme used today, at least for biomolecular applications.

A further step in the complexity of the QM/MM method is the polarization coupling scheme, where classical charges become flexible. The QM and MM regions are subjected to a mutual polarization process, which improves the description of the electrostatic forces but dramatically increases the computational cost.

Traditionally, QM calculations are used to characterize the potential energy surface (PES) of a process via geometry optimization, followed by computation of the second derivatives. Researchers can thus identify the stationary points along a minimum energy path, including TSs. This approach implies direct calculation, storage, and manipulation of the Hessian matrix, which contains the topological information of the PES of interest. It becomes extremely difficult with large protein systems with many degrees of freedom.

To overcome this, the "adiabatic mapping" approach offers one basic way to potentially model reaction paths (Lodola et al., 2008b). The energy of the system is calculated by minimizing the energy at a series of fixed (or restrained, e.g., by harmonic forces) values of a reaction coordinate, for example, the distance between two atoms. This approach has been successful for reactions involving negligible structural changes for a small number of groups (Ranaghan and Mulholland, 2010). In this specific condition, the minimization of the potential energy along the specified reaction coordinate provides a reasonable approximation of the "enthalpic component" of the free energy profile for the considered reaction.

However, approaches based only on energy minimization are intrinsically inaccurate. This is because they account for neither structural fluctuations in the protein active site, nor entropy effects (Ranaghan and Mulholland, 2010). Multiple starting structures offer a partial solution to the lack of conformational fluctuations in the adiabatic mapping approach (Klahn et al., 2005). For a given path, multiple PESs can be built. A correlation between the available configurational space of the enzyme and energy barrier can thus be found (Lodola et al., 2010).

According to TS theory, the rate constant of a chemical transformation is related to the free energy barrier and not to the potential energy barrier. Techniques that sample configurations along a reaction coordinate give a more sophisticated and extensive description. They do this by taking account of multiple conformations and estimating entropic effects. This can be essential for modeling some types of enzyme reactions, particularly those involving significant structural rearrangement of the protein or solvent (Warshel et al., 2006; Lonsdale et al., 2010). Simulations of this type can provide estimates of the free energy profile along a specific reaction coordinate (Hu and Yang, 2008).

The unrestrained molecular dynamics method can in principle be used to calculate free energy profiles. However, it does not provide an exhaustive sampling of low-populated regions such as TSs. In order to sample regions of higher energy, a biasing potential is typically used, as in the case of umbrella sampling (Gao et al., 2006). The introduced bias can subsequently be removed using appropriate statistical methods to obtain an unbiased free energy profile.

In this regard, Car and Parrinello proposed a new method of performing DFT-based molecular dynamics (the Car–Parrinello molecular dynamics (CPMD) method) in 1985. This method simulates (biological) systems at their actual temperature, thus allowing the inclusion of entropic effects (Car and Parrinello, 1985). CPMD is an *ab initio* method that differs from traditional classical MD because it does not rely on empirically determined potentials. Instead, internuclear forces are determined from "on the fly" electronic structure calculations employing the Kohn–Sham formulation of DFT (Hohenberg and Kohn, 1964; Kohn and Sham, 1965; Car, 2002). The core electrons are treated through the use of atomic pseudopotentials, while the wave function for the valence electrons is expressed in a plane-wave basis set. CPMD, in its QM/MM implementation (Laio et al., 2002), can be used to characterize the free energy surface of enzymatic reactions. It has been successfully applied to a number of pharmaceutically relevant targets, as reported in these reviews (Carloni et al., 2002; Dal Peraro et al., 2007). Some examples of CPMD studies are reported below.

III. QM/MM Studies of Pharmaceutically Relevant Targets

Here, we focus on some representative QM/MM studies of therapeutically relevant enzymes, including from our own research. These studies will show how QM-based methods can be used to investigate enzymatic mechanisms, including the dynamic features of an enzyme's catalytic site. The following examples are just a few of the many available in the literature. The interested reader is referred to the several informative reviews that concern computational studies of enzymatic reaction mechanisms (Carloni et al., 2002; Warshel, 2003; Garcia-Viloca et al., 2004; Marti et al., 2004; Friesner and Guallar, 2005; Mulholland, 2005; Warshel et al., 2006; Acevedo and Jorgensen, 2010; Lonsdale et al., 2010).

First, it is important to stress the connection between the atomic level understanding of the enzymatic mechanism and the rational design of ligands/drugs, which relies primarily on TS theory, as mentioned in Section I. TS theory defines TS analogs as potent inhibitors. This is because TS theory holds that the catalytic power of enzymes derives from their ability to lower the TS energy along the enzymatic reaction pathway. The lowering of the enzymatic reaction barrier is a direct consequence of the tight (i.e., favorable) interaction between the substrate and the receptor in the TS geometry. Therefore, a ligand resembling the geometry and physicochemical properties of the enzymatic TS is a potent inhibitor (i.e., a TS analog). Given the fleeting nature of the enzymatic TS, computations can be used to determine which TS characteristics to use as template for designing TS analogs. These potent TS inhibitors are optimal starting points for rational drug design (Schramm, 2005).

A. *Enzymatic Catalysis Studied Via Car–Parrinello QM/MM Simulations*

Herein, we report on computational studies of two metalloenzymes that catalyze the transfer of the phosphate group. Hydrolysis and subsequent transfer of phosphates is the central chemical reaction of many metabolic pathways, including kinase cascade activation, membrane transport, and gene transcription (Morrison and Heyde, 1972; Knowles, 1980; Jackson and Denu, 2001; Johnson and Lewis, 2001; Kennelly, 2001; Dzeja and Terzic, 2003). The nature of the enzymatic TS of the phosphoryl transfer mechanism can be dissociative or associative, according to the metaphosphate-like or phosphorane-like nature of the TS geometry (Branduardi et al., 2011). Importantly, it is still unclear whether a stable penta-coordinated phosphate intermediate exists along the phosphoryl transfer. Only one crystal structure, the β-phosphoglucomutase crystal (Lahiri et al., 2003), has shown such an intermediate. Additionally and intriguingly, many enzymes catalyze phosphoryl transfers using positively charged metal ions (e.g., Mg^{2+}, Zn^{2+}, Mn^{2+}, Ca^{2+}), which help stabilize the overall negative charge of the scissile phosphate in the catalytic site. The catalytic role of the metal ions in increasing the efficiency of the enzymatic reaction is a subject of great interest (Dupureur, 2008; Ho et al., 2010). Researchers are still debating how the metal ions help during the two chemical steps of phosphoryl transfer reactions, that is, the protonation of the leaving group and deprotonation of the attacking nucleophile. Moreover, these reactions often involve water-bridge (WB) proton shuttles, likely catalyzed by electrostatic stabilization due to the presence of the metal ion(s).

First, we present some work on the phosphoryl transfer reaction catalyzed by the RNase H, which is a target for developing antiviral drugs. This enzyme requires two Mg^{2+} ions contained in the catalytic site to catalyze a nucleotidyl transfer (NT) reaction. The second example is the soluble epoxide hydrolase (sEH), which carries out the phosphatase activity in its N-terminal lobe, with a single Mg^{2+} ion in the catalytic site. The sEH is a promising target for hypertension and acute respiratory syndrome treatment. It is interesting to compare the catalytic mechanism of different enzymes (i.e., RNase H vs. sEH), which use two versus one Mg^{2+} ion in catalyzing the same enzymatic reaction.

1. QM/MM Studies of Ribonuclease H

RNase H, which belongs to the NT superfamily, catalyzes the metal ion-dependent phosphodiester hydrolysis of the P—O3′ bond of either RNA or DNA strands (Fig. 2), degrading RNA·DNA hybrids (Broccoli et al., 2004; Yang et al., 2006). These enzymes are ubiquitous in nature, being crucial in nucleic acid synthesis, recombination, processing, and degradation. RNase

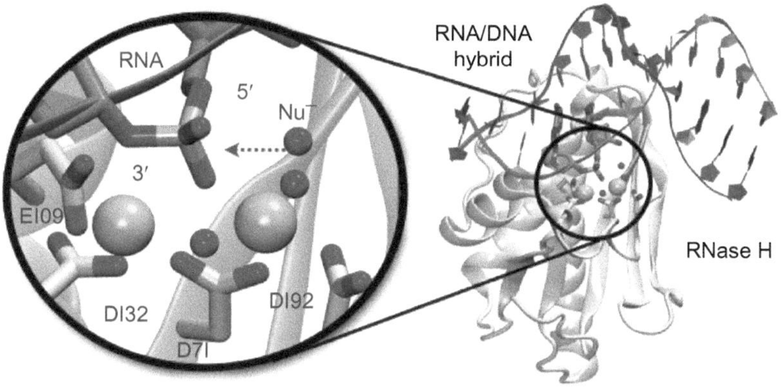

FIG. 2. Right: The crystallographic structure of the complex RNase H/RNA·DNA hybrid. Left: A close view of the catalytic site shows two Mg^{2+} ions surrounded by carboxylic groups, the scissile phosphate of the substrate RNA strand and the nucleophilic oxygen atom.

H activity in HIV reverse transcriptase represents a promising target for anti-HIV drug design (Klumpp et al., 2003; Shaw-Reid et al., 2005). Therefore, understanding the role of metal ions in endonuclease activity is of paramount importance for discovering new structural probes or drugs acting on these vital enzymes, with the ultimate goal of modulating their activity.

Given the large negative charge on the backbone of RNA and DNA strands, positively charged metal ions are often found in the catalytic pocket of these proteins. Indeed, most metalloendonucleases contain Mg^{2+} ions in the catalytic site, though other divalent metals such as Mn^{2+} or Zn^{2+} have been identified. Nevertheless, some metals (e.g., Mg^{2+} and Mn^{2+}) can promote the enzymatic function when in the correct concentration, while others such as Ca^{2+} inhibit the RNase H endoribonuclease activity. This evidence comes from very informative studies, which have also provided several high-resolution crystal structures revealing the arrangement of the complex formed by RNase H from *Bacillus halodurans* and its substrate, the RNA·DNA hybrid (Nowotny et al., 2005, 2007, 2008; Nowotny and Yang, 2006).

QM/MM CPMD has recently been used to clarify the bimetal-aided NT reaction in RNase H (De Vivo et al., 2008). These QM/MM simulations use the DFT/BLYP level of theory for the QM part. Two reagent states have been considered in studying this enzymatic reaction (Fig. 2). In one, the nucleophilic species is a water molecule (P_{WAT}). In the other, a hydroxide ion (P_{OH^-}) is the reactive nucleophile. Starting from these different reactants, we sought to clarify the following important aspects of the catalytic reaction: (1) the nature of the enzymatic mechanism (concerted one-step or stepwise, with formation of a stable phosphorane intermediate); (2) the energetics and formation mechanism of the nucleophilic hydroxide ion; (3) the role of the pro-R_p oxygen during the catalysis; and finally (4) the role of the two metal cofactors in aiding the catalysis.

Both P_{OH^-} and P_{WAT} show an in-line SN_2-like nucleophilic attack on the scissile phosphorus (Fig. 3), leading to an associative mechanism with phosphorane-like TSs. P_{OH^-} shows the lowest free energy barrier—~ 10.5 kcal mol^{-1}—due to the high nucleophilicity of the OH^- group. Nevertheless, P_{WAT} shows a competitive mechanism, which includes a metastable pentavalent phosphorane intermediate, as observed so far only in the debated β-phosphoglucomutase crystal. This mechanism shows a free energy barrier of ~16 kcal mol^{-1} if dehydration energy is also taken into account (~3 kcal mol^{-1}). These free energy values are qualitatively in agreement with the experimental data for substrate analogs for HIV-1 RNase H activity (Shaw-Reid et al., 2005). Interestingly, P_{WAT} also includes a metastable pentavalent phosphorane intermediate, observed so far only in the debated β-phosphoglucomutase crystal. Elsässer et al. have applied high level QM/MM calculations to the study of RNase A. They have also suggested the presence of such an intermediate (Fig. 4).

A key aspect of the RNase H reaction mechanism is how the two Mg^{2+} ions act in a cooperative fashion to promote and facilitate both nucleophile formation and leaving group stabilization. Both P_{OH^-} and P_{WAT} show a phosphorane-like TS in which the associative character of the TS is supported by the two ions, which get closer to each other in the TS geometry. More recently, Rosta et al. (2009, 2011) have studied the reaction mechanism of RNase H with a different QM/MM approach, which uses the DFT/B3LYP level of theory for the QM part. Overall, the TS geometry seems fairly consistent with that found in previous studies, although different proton transfer pathways are reported.

2. *QM/MM Studies of Phosphatase Activity in Soluble Epoxide Hydrolase*

A novel metal (Mg^{2+})-dependent phosphatase activity was discovered recently in the N-terminal domain of the dual-domain protein sEH. This N-terminal domain adopts an α/β-fold homologous to that of the haloacid dehalogenase superfamily, the majority of which comprises

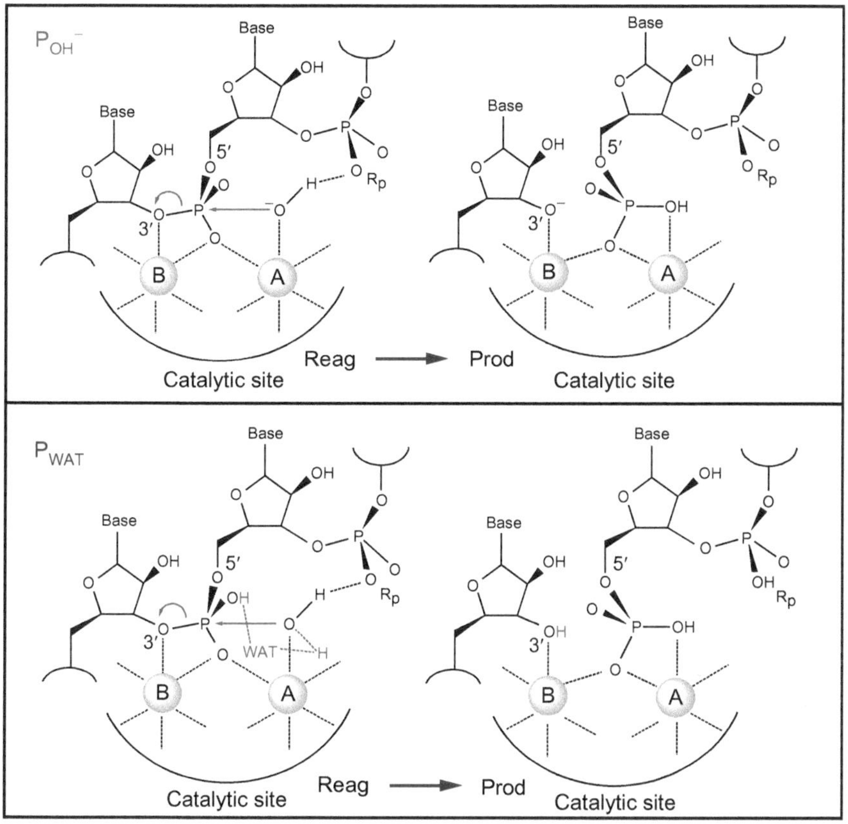

FIG. 3. The two schemes represent two possible pathways with different nucleophilic agents. Above: A mechanism where one hydroxide ion (P_{OH-} pathway) attacks the phosphorus atom. Below: A water molecule as a nucleophilic agent (P_{WAT} pathway). In both cases, an in-line SN_2-like nucleophilic attack by the nucleophilic group on the scissile phosphorus atom leads to the inversion of the phosphate stereo configuration and formation of the 5′-phosphate and 3′-hydroxy function of the RNA strand. (For color version of this figure, the reader is referred to the online version of this chapter.)

phosphotransferases. Interestingly, this finding could open up a new branch of fatty acid metabolism and provide new promising targets for drug discovery (Moghaddam et al., 1997; Node et al., 1999; Schmelzer et al., 2005).

Two QM/MM CPMD computational studies (De Vivo et al., 2005, 2007b), based on the crystallographic evidence (Cronin et al., 2003;

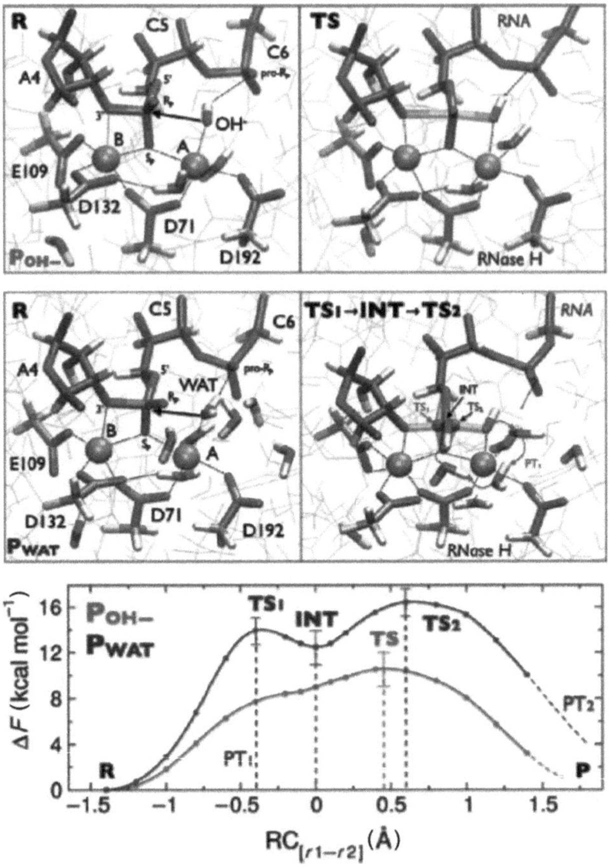

FIG. 4. Structural evolution and energetic profile of the two reaction pathways P_{OH-} and P_{WAT} for RNase H catalysis (only QM atoms are shown explicitly, the rest of the system is shown in thinner lines). In the first row, the nucleophilic group is one hydroxide ion (P_{OH-} path): **R** is the reagent, while **TS** shows the phosphorane-like transition state. In the second row, a water molecule is the nucleophilic agent (P_{WAT} path): **R** is the reagent, while the nucleophilic attack leads to **TS1**, where a proton shuttle (**PT1**) involves three water molecules that bridge the scissile phosphate and WAT. The protonation of the scissile phosphate stabilizes the phosphorane group, leading to the formation of the metastable intermediate **INT**. Then, **TS2** leads to the final products, in which the cleavage of the RNA strand is definitely completed, and the protonation of the 3′-hydroxy function of the RNA strand takes place (**PT2**, not shown here). In the third row, the two competitive free energy profiles of the investigated pathways for RNase H catalysis are shown.

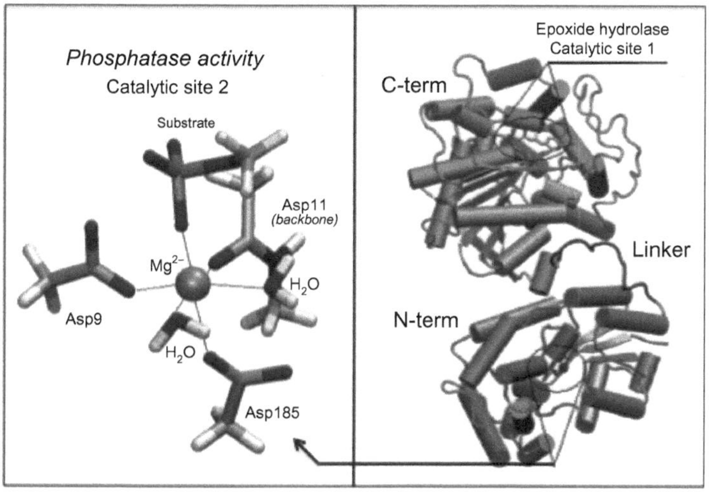

FIG. 5. Right: The crystallographic structure of the soluble epoxide hydrolase (sEH). Left: A close view of the catalytic site in the N-terminal domain of the sEH. The Mg^{2+} ion forms an octahedral with the ligands. The scissile phosphate group of the substrate is located on top of the Mg^{2+} ion.

Gomez et al., 2004; Newman et al., 2005), have characterized the two-step reaction mechanism for the two phosphoryl transfers catalyzed by the sEH phosphatase (Fig. 5). The first step is a nucleophilic attack on the phosphate group of the phosphoester substrate by Asp9, and protonation of the leaving group by either an intervening water molecule or an Asp11. The second step is the hydrolysis of the phosphoenzyme intermediate via a nucleophilic attack at the scissile phosphorus atom by a water molecule.

Both steps show an in-line nucleophilic substitution, which presents a rather dissociative character in the TS regions, with a planar metaphosphate-like TS that does not structurally resemble a phosphorane species. The free energy barriers are in good agreement with experimental data, suggesting step 1 ($\sim$19 kcal mol^{-1}) as the rate-determining step of the catalytic cycle (Fig. 6).

Interestingly, based on these QM/MM studies (De Vivo et al., 2005, 2007a,b, 2008), the phosphoryl transfer mechanism seems to be induced according to the geometry and stoichiometry of the metal(s) during catalysis. sEH stabilizes the transfer of the metaphosphate group via its apical coordination to the only Mg^{2+} ion present in the catalytic site. RNase H, on the other hand, uses the two Mg^{2+} cations to stabilize the

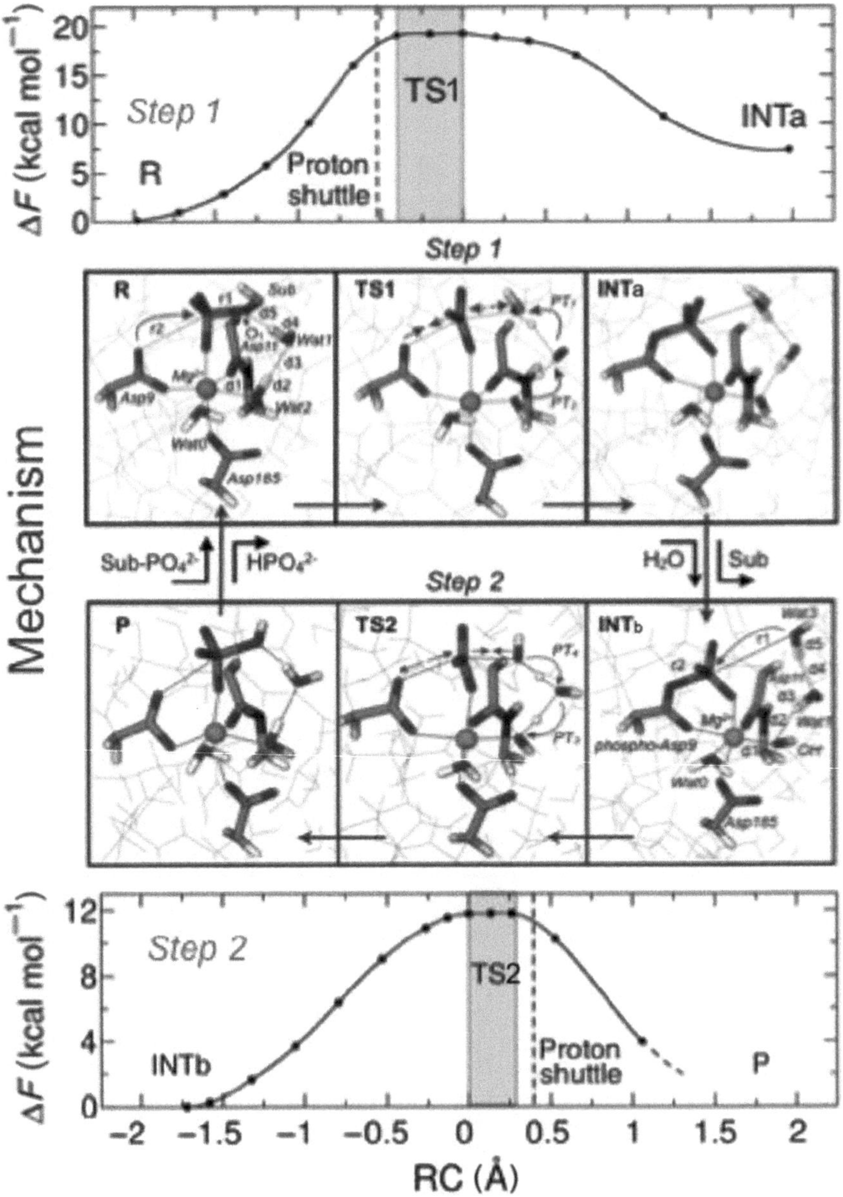

FIG. 6. The two central rows report the enzymatic mechanism. Only QM atoms are highlighted. First central row: Nucleophilic attack of Asp9 at the Mg^{2+}-coordinated

attacking and leaving groups, while the metaphosphate group is located between the two metals, showing a phosphorane-like TS geometry. This hypothesis (De Vivo et al., 2008), however, needs to be validated by further investigations.

Both RNase H and sEH computational studies of the catalytic mechanism provide examples of the key role played by water molecules in solvating the metal center and facilitating the migration of protons involved in the phosphoryl transfer reaction. This aspect of the mechanism is very important in understanding the catalytic strategy. Basically, the ordered H-bond network formed by water molecules in the catalytic site facilitates the creation of better attacking and leaving groups in enzymatic phosphoryl transfers. In particular, these examples show how metal–substrate connecting WBs, mostly induced by the electrostatic effects of the metal ion, facilitate efficient transfer of the protons necessary for nucleophile formation (water deprotonation) and leaving group stabilization during the phosphoryl transfers.

B. *Recent Modeling Studies by Adiabatic Mapping*

Despite the known limitations of the adiabatic mapping approach, it has been successfully used to elucidate enzyme mechanism by means of QM/MM (Lodola et al., 2008b; Mulholland, 2008; Senn and Thiel, 2009; Lonsdale et al., 2010). Adiabatic mapping is simple to apply, and does not require demanding calculations. Further, it can be useful for extensively exploring PESs, and for generating approximate models of TSs and intermediates, in which some allowance is made for structural relaxation to chemical changes at the active site (Ranaghan and Mulholland, 2010).

Recently, the adiabatic mapping approach has been used to successfully characterize the reaction mechanism of enzymes that are considered "hot" drug discovery targets (Mulholland, 2005).

For instance, QM/MM methods have been used to model the reaction mechanism of FAAH for endogenous substrates and inhibitors (Palermo et al., 2011). FAAH is a key enzyme involved in the endocannabinoid metabolism (Labar and Michaux, 2007), and a promising target for the

---

phosphoryl group, with formation of the phosphoenzyme intermediate formation (INTa). Second central row: Formation of the products via phosphoryl transfer from the phospho-Asp9 to one attacking solvent water. The top graph reports the free energy profile for step 1. The bottom graph reports the free energy profile for step 2.

treatment of central and peripheral nervous system disorders (Petrosino and Di Marzo, 2010), including anxiety, pain, and depression (Kathuria et al., 2003; Mor and Lodola, 2009). The crystal structure of FAAH revealed that three conserved residues comprise a novel serine–serine–lysine catalytic triad (Ser_{241}-Ser_{217}-Lys_{142}; Bracey et al., 2002). Recently, an SCC-DFTB/CHARMM27 QM/MM potential was used to model the mechanism of an acylation reaction catalyzed by FAAH, in the presence of the substrates oleamide (OA) and oleoylmethyl ester (OME; Capoferri et al., 2011). This extensive mechanistic study expanded the knowledge of FAAH catalysis, previously gained with simulations at PM3/CHARMM22 (Lodola et al., 2009), B3LYP/6-31+G(d)//PM3/CHARMM22 (Lodola et al., 2005), and PM3-PDDG/OPLS levels (Tubert-Brohman et al., 2006).

Researchers now widely accept the mechanism of hydrolysis catalyzed by FAAH, with Lys_{142} serving as a key acid and base in distinct steps of the catalytic cycle (McKinney and Cravatt, 2005). As a base, Lys_{142} activates the Ser_{241} nucleophile for attack on the substrate amide carbonyl, with formation of a classical tetrahedral intermediate. As an acid, Lys_{142} readily protonates the leaving group (in the TI configuration), leading to its expulsion. The impact of Lys_{142} on Ser_{241} nucleophile strength and leaving group protonation occurs indirectly, via the bridging Ser_{217} of the triad, which acts as a ‘‘proton shuttle’’ (Fig. 7).

The exploration of the PESs for FAAH acylation indicates that the collapse of the tetrahedral intermediate is the rate-limiting step of the reaction for both OA and OME substrates, with calculated barriers of 18.8 and 20.8 kcal mol^{-1} at SCC-DFTB/CHARMM27 level. These are in good agreement with the experimentally deduced ones of 16.0 and 17.0 kcal mol^{-1} as derived from K_{cat} (McKinney and Cravatt, 2003)

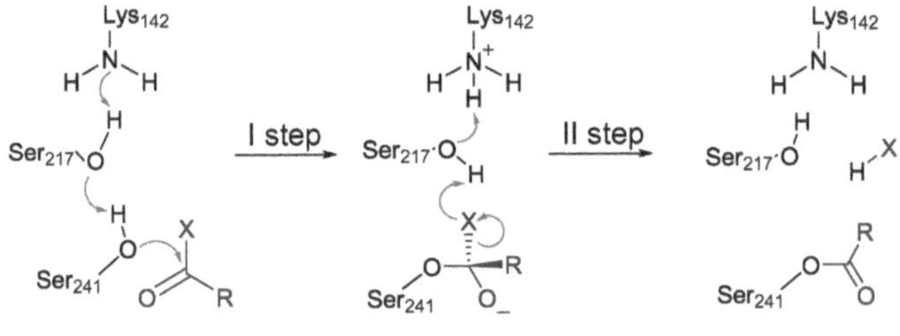

FIG. 7. Acylation mechanism of FAAH in the presence of oleamide (OA) and for oleoylmethyl ester (OME). X is $-NH_2$ for OA and $-OCH_3$ for OME. (For color version of this figure, the reader is referred to the online version of this chapter.)

by applying TS theory. Further, if the SCC-DFTB/CHARMM27 energy values are averaged over multiple reaction paths (obtained using six different starting structures, Fig. 8), the resulting potential energy barriers (21.1 ± 0.8 kcal mol^{-1} for OA and 24.7 ± 0.8 kcal mol^{-1} for OME) remain in reasonable agreement with experimental observations.

These calculations show that the collapse of the TI follows a concerted reaction mechanism, where Lys_{142} and Ser_{217} cooperate in the protonation of the leaving group heteroatom (nitrogen for OA, oxygen for OME). Protonation of the leaving group was identified as the key event in acylation for both OA and OME. This finding provides a theoretical explanation for the remarkable ability of FAAH to hydrolyze amides faster than esters.

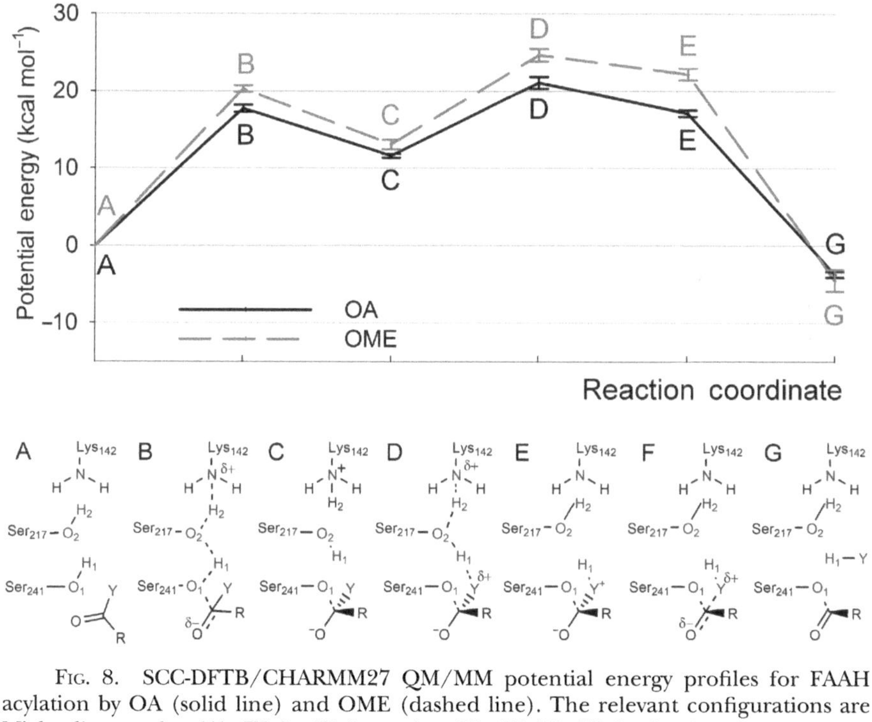

FIG. 8. SCC-DFTB/CHARMM27 QM/MM potential energy profiles for FAAH acylation by OA (solid line) and OME (dashed line). The relevant configurations are Michaelis complex (A), TS for TI formation (B), TI (C), TS for leaving group protonation (D), protonated TI (E), and acylenzyme (G). Potential energy values are reported as the mean value ± standard error of the mean (kcal mol^{-1}) of six independent simulations of the acylation reactions. (For color version of this figure, the reader is referred to the online version of this chapter.)

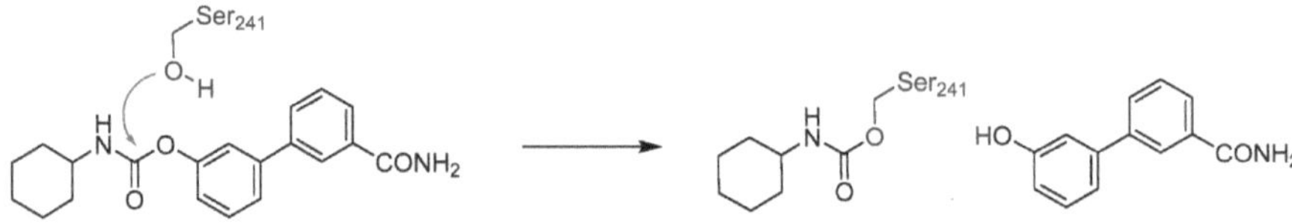

FIG. 9. Carbamoylation of FAAH by URB597 compound. (For color version of this figure, the reader is referred to the online version of this chapter.)

A QM/MM approach has also been used to describe, at the atomic level, the reaction between FAAH and carbamic acid aryl ester inhibitors (Lodola et al., 2008a). This class of compounds, including the reference compound URB597 (the 3′-$CONH_2$ derivative of URB524), were shown to inactivate FAAH by carbamoylation of the active nucleophile Ser_{241} (Fig. 9; Alexander and Cravatt, 2005).

In general, SBDD depends on the accuracy of ligand docking, and the ability to identify relevant binding modes. URB524 and its derivatives can be docked within the FAAH catalytic site in two possible orientations (called orientations *I* and *II*), both placing the carbamic group close to Ser_{241} (Basso et al., 2004; Mor et al., 2004). Traditional computational tools employed in drug discovery, such as docking and scoring (e.g., by classical interaction energies), failed to clearly discriminate between these two binding orientations.

The QM/MM approach was used to model the inhibitor binding process, using the PM3–CHARMM22 potential in combination with B3LYP/6-31G+(d) energy correction for a reliable description of reaction energetic. PESs were built for each binding orientation. TSs and intermediates were identified along the reaction profiles. The calculations clearly showed that the carbamoylation in orientation *II* was energetically preferred, thus identifying *II* as the productive binding mode.

These calculations suggest that carbamoylation of Ser_{241} likely starts from binding orientation *II*. In orientation *I*, the reaction has a significantly higher barrier and leads to an unstable product. This prediction was recently confirmed by the crystallographic resolution of the FAAH–URB597 carbamoylated structure (Mileni et al., 2010; Fig. 10). This suggests that QM/MM-based mechanistic modeling can contribute practically to inhibitor design.

The same QM/MM protocol has also been used to rationalize *in vivo* biological data for a focused series of FAAH inhibitors, based on a

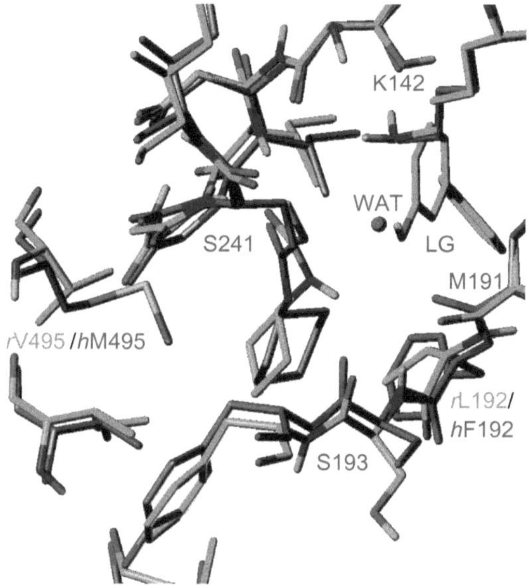

FIG. 10. Superposition of the rat FAAH–URB524 carbamoylated adduct, as obtained from early QM/MM simulation (green carbon) on the X-ray structure of the humanized rat variant of FAAH in its covalent adduct with URB597 (gray carbon). In evidence, the biphenol leaving group (LG) obtained from simulation, with its oxygen in proximity to the deacylating water molecule (WAT) identified by Mileni et al. (2010). (For interpretation of the references to color in this figure legend, the reader is referred to the online version of this chapter.)

carbamic template structure. In particular, *N*-alkylcarbamic acid biphenyl-3-yl esters are a highly versatile class of FAAH inhibitors. Their intrinsic reactivity can easily be tuned by chemical manipulation (Vacondio et al., 2009). It is possible to enhance their chemical (and metabolic) stability by introducing electron–donor substituents in conjugated positions of the proximal phenyl ring. This increases the electron density around the carbonyl carbon, limiting its reactivity toward nucleophiles. The introduction of electron–donor groups (e.g., *p*-OH or *p*-NH_2) significantly improves the stability of these carbamates versus nucleophiles, including those present in liver and plasmatic carboxylesterases (Clapper et al., 2009). However, the same substitution does not affect the interaction with FAAH.

This unexpected lack of correlation between reactivity and FAAH inhibitory potency might be due to the "unique" catalytic mechanism of FAAH. QM/MM was used to mechanistically model FAAH carbamoylation in the presence of the cyclohexylcarbamic acid biphenyl-3-yl ester URB524 and its *p*-OH (URB694) and *p*-NH_2 (URB618) analogs. This showed that FAAH is insensitive to the intrinsic reactivity of the carbamate group. This is because the crucial TS of the reaction is dominated by a proton transfer and not by a nucleophilic attack (Lodola et al., 2011a).

This finding could help in developing a new generation of "stabilized" carbamate inhibitors that, while retaining good *in vitro* potency for FAAH, would display longer half-life in plasma. This would make them significantly more potent *in vivo* than current inhibitors, and more selective than off-target carboxylesterases.

IV. Conclusions

The use of QM/MM methods in drug discovery is still in its infancy. Few research groups are seeking to apply mechanistic insights, arising from QM/MM calculations, to the design of new inhibitors. In recent years, however, there have been significant advances in the use of QM/MM methods for computational enzymology. Nowadays, QM/MM computations of pharmaceutically relevant enzyme systems are run routinely, providing results that can be directly compared with experiments. We predict that QM/MM will have a practical impact on rational drug design. In many cases, an atomistic understanding of the enzymatic catalytic mechanism can indeed suggest new scaffolds or modifications of a compound. This can lead to the design of covalent and noncovalent inhibitors, which can then be used as starting point for drug discovery programs.

Here, we have reported some representative examples of QM/MM studies of pharmaceutically relevant targets, based on our research. We have shown how QM/MM can predict the binding orientation of a reference inhibitor (as with the interaction between FAAH and URB524). We have also shown how QM/MM can rationalize the role played by the intrinsic of carbamated-based inhibitors in its interaction with the FAAH. These findings have significantly contributed to the design of a second generation of FAAH inhibitors. They thus offer promising evidence of the increasing role of QM/MM in the field of drug design.

Acknowledgments

A. L. is grateful to all the people involved in the QM/MM work on FAAH, particularly Prof. Marco Mor, Prof. Silvia Rivara, Luigi Capoferri (Università degli Studi di Parma) and Prof. Adrian Mulholland, Dr. Jitnap Sirirak, Ewa Chudyk (University of Bristol). M. D. V. thanks all the colleagues involved in the computational studies of RNase H and sEH catalysis, in particular, Prof. M.L. Klein (Temple University, Philadelphia), Prof. D.W. Christianson (University of Pennsylvania, Philadelphia), Prof. Matteo Dal Peraro (EPFL, Lausanne), and Dr. Bernd Ensing (University of Amsterdam). The authors thank Grace Fox for proofreading the manuscript.

References

Acevedo, O., Jorgensen, W. L. (2010). Advances in quantum and molecular mechanical (QM/MM) simulations for organic and enzymatic reactions. *Acc. Chem. Res.* **43**, 142–151.

Alexander, J. P., Cravatt, B. F. (2005). Mechanism of carbamate inactivation of FAAH: implications for the design of covalent inhibitors and in vivo functional probes for enzymes. *Chem. Biol.* **12**, 1179–1187.

Bachovchin, D. A., Cravatt, B. F. (2011). The pharmacological landscape and therapeutic potential of serine hydrolases. *Nat. Rev. Drug Discov.* **11**, 52–68.

Basso, E., Duranti, A., Mor, M., Piomelli, D., Tontini, A., Tarzia, G., et al. (2004). Tandem mass spectrometric data-FAAH inhibitory activity relationships of some carbamic acid O-aryl esters. *J. Mass Spectrom.* **39**, 1450–1455.

Bracey, M. H., Hanson, M. A., Masuda, K. R., Stevens, R. C., Cravatt, B. F. (2002). Structural adaptations in a membrane enzyme that terminates endocannabinoid signaling. *Science* **298**, 1793–1796.

Branduardi, D., De Vivo, M., Rega, N., Barone, V., Cavalli, A. (2011). Methylphosphate dianion hydrolysis in solution characterized by path collective variables coupled with DFT-based enhanced sampling simulations. *J. Chem. Theory Comput.* **7**, 539–543.

Broccoli, S., Rallu, F., Sanscartier, P., Cerritelli, S. M., Crouch, R. J., Drolet, M. (2004). Effects of RNA polymerase modifications on transcription-induced negative supercoiling and associated R-loop formation. *Mol. Microbiol.* **52**, 1769–1779.

Brooks, B. R., Brooks, C. L., 3rd, Mackerell, A. D., Jr., Nilsson, L., Petrella, R. J., Roux, B., et al. (2009). CHARMM: the biomolecular simulation program. *J. Comput. Chem.* **30**, 1545–1614.

Capoferri, L., Mor, M., Sirirak, J., Chudyk, E., Mulholland, A. J., Lodola, A. (2011). Application of a SCC-DFTB QM/MM approach to the investigation of the catalytic mechanism of fatty acid amide hydrolase. *J. Mol. Model.* **17**, 2375–2383.

Car, R. (2002). Introduction to density-functional theory and ab-initio molecular dynamics. *Quant. Struct. Act. Rel.* **21**, 97–104.

Car, R., Parrinello, M. (1985). Unified approach for molecular—dynamics and density-functional theory. *Phys. Rev. Lett.* **55**, 2471–2474.

Carloni, P., Rothlisberger, U., Parrinello, M. (2002). The role and perspective of a initio molecular dynamics in the study of biological systems. *Acc. Chem. Res.* **35**, 455–464.

Cavalli, A., Carloni, P., Recanatini, M. (2006). Target-related applications of first principles quantum chemical methods in drug design. *Chem. Rev.* **106**, 3497–3519.

Claeyssens, F., Harvey, J. N., Manby, F. R., Mata, R. A., Mulholland, A. J., Ranaghan, K. E., et al. (2006). High-accuracy computation of reaction barriers in enzymes. *Angew. Chem. Int. Ed. Engl.* **45**, 6856–6859.

Clapper, J. R., Vacondio, F., King, A. R., Duranti, A., Tontini, A., Silva, C., et al. (2009). A second generation of carbamate-based fatty acid amide hydrolase inhibitors with improved activity in vivo. *ChemMedChem* **4**, 1505–1513.

Cramer, C. J. (2004). Essentials of Computational Chemistry: Theories and Models. John Wiley and Sons, New York.

Cronin, A., Mowbray, S., Durk, H., Homburg, S., Fleming, I., Fisslthaler, B., et al. (2003). The N-terminal domain of mammalian soluble epoxide hydrolase is a phosphatase. *Proc. Natl. Acad. Sci. USA* **100**, 1552–1557.

Dal Peraro, M., Ruggerone, P., Raugei, S., Gervasio, F. L., Carloni, P. (2007). Investigating biological systems using first principles Car-Parrinello molecular dynamics simulations. *Curr. Opin. Struct. Biol.* **17**, 149–156.

De Vivo, M. (2011). Bridging quantum mechanics and structure-based drug design. *Front. Biosci.* **16**, 1619–1633.

De Vivo, M., Ensing, B., Klein, M. L. (2005). Computational study of phosphatase activity in soluble epoxide hydrolase: high efficiency through a water bridge mediated proton shuttle. *J. Am. Chem. Soc.* **127**, 11226–11227.

De Vivo, M., Cavalli, A., Carloni, P., Recanatini, M. (2007a). Computational study of the phosphoryl transfer catalyzed by a cyclin-dependent kinase. *Chemistry* **13**, 8437–8444.

De Vivo, M., Ensing, B., Dal Peraro, M., Gomez, G. A., Christianson, D. W., Klein, M. L. (2007b). Proton shuttles and phosphatase activity in soluble epoxide hydrolase. *J. Am. Chem. Soc.* **129**, 387–394.

De Vivo, M., Dal Peraro, M., Klein, M. L. (2008). Phosphodiester cleavage in ribonuclease H occurs via an associative two-metal-aided catalytic mechanism. *J. Am. Chem. Soc.* **130**, 10955–10962.

de, M. S. G., Walker, R. C., Elstner, M., Case, D. A., Roitberg, A. E. (2007). Implementation of the SCC-DFTB method for hybrid QM/MM simulations within the amber molecular dynamics package. *J. Phys. Chem. A* **111**, 5655–5664.

Dupureur, C. M. (2008). Roles of metal ions in nucleases. *Curr. Opin. Chem. Biol.* **12**, 250–255.

Dzeja, P. P., Terzic, A. (2003). Phosphotransfer networks and cellular energetics. *J. Exp. Biol.* **206**, 2039–2047.

Ferrer, S., Ruiz-Pernia, J., Marti, S., Moliner, V., Tunon, I., Bertran, J., et al. (2011). Hybrid schemes based on quantum mechanics/molecular mechanics simulations goals to success, problems, and perspectives. *Adv. Protein Chem. Struct. Biol.* **85**, 81–142.

Field, M. J., Pa, Bash, Karplus, M. (1990). A combined quantum-mechanical and molecular mechanical potential for molecular-dynamics simulations. *J. Comput. Chem.* **11**, 700–733.

Friesner, R. A., Guallar, V. (2005). Ab initio quantum chemical and mixed quantum mechanics/molecular mechanics (QM/MM) methods for studying enzymatic catalysis. *Annu. Rev. Phys. Chem.* **56**, 389–427.

Gao, J., Ma, S., Major, D. T., Nam, K., Pu, J., Truhlar, D. G. (2006). Mechanisms and free energies of enzymatic reactions. *Chem. Rev.* **106**, 3188–3209.

Garcia-Viloca, M., Gao, J., Karplus, M., Truhlar, D. G. (2004). How enzymes work: analysis by modern rate theory and computer simulations. *Science* **303**, 186–195.

Gomez, G. A., Morisseau, C., Hammock, B. D., Christianson, D. W. (2004). Structure of human epoxide hydrolase reveals mechanistic inferences on bifunctional catalysis in epoxide and phosphate ester hydrolysis. *Biochemistry* **43**, 4716–4723.

Ho, M. H., De Vivo, M., Dal Peraro, M., Klein, M. L. (2010). Understanding the effect of magnesium ion concentration on the catalytic activity of ribonuclease H through computation: does a third metal binding site modulate endonuclease catalysis? *J. Am. Chem. Soc.* **132**, 13702–13712.

Hohenberg, P., Kohn, W. (1964). Inhomogeneous electron gas. *Phys. Rev. B* **136**, B864–B871.

Hornak, V., Abel, R., Okur, A., Strockbine, B., Roitberg, A., Simmerling, C. (2006). Comparison of multiple Amber force fields and development of improved protein backbone parameters. *Proteins* **65**, 712–725.

Hu, H., Yang, W. (2008). Free energies of chemical reactions in solution and in enzymes with ab initio quantum mechanics/molecular mechanics methods. *Annu. Rev. Phys. Chem.* **59**, 573–601.

Jackson, M. D., Denu, J. M. (2001). Molecular reactions of protein phosphatases—insights from structure and chemistry. *Chem. Rev.* **101**, 2313–2340.

Jensen, F. (1999). Introduction to Computational Chemistry. John Wiley and Sons, New York.

Johnson, L. N., Lewis, R. J. (2001). Structural basis for control by phosphorylation. *Chem. Rev.* **101**, 2209–2242.

Karplus, M., McCammon, J. A. (2002). Molecular dynamics simulations of biomolecules. *Nat. Struct. Biol.* **9**, 646–652.

Kathuria, S., Gaetani, S., Fegley, D., Valino, F., Duranti, A., Tontini, A., et al. (2003). Modulation of anxiety through blockade of anandamide hydrolysis. *Nat. Med.* **9**, 76–81.

Kennelly, P. J. (2001). Protein phosphatases—a phylogenetic perspective. *Chem. Rev.* **101**, 2291–2312.

Klahn, M., Braun-Sand, S., Rosta, E., Warshel, A. (2005). On possible pitfalls in ab initio quantum mechanics/molecular mechanics minimization approaches for studies of enzymatic reactions. *J. Phys. Chem. B* **109**, 15645–15650.

Klumpp, K., Hang, J. Q., Rajendran, S., Yang, Y., Derosier, A., Wong Kai In, P., et al. (2003). Two-metal ion mechanism of RNA cleavage by HIV RNase H and mechanism-based design of selective HIV RNase H inhibitors. *Nucleic Acids Res.* **31**, 6852–6859.

Knowles, J. R. (1980). Enzyme-catalyzed phosphoryl transfer-reactions. *Annu. Rev. Biochem.* **49**, 877–919.

Kohn, W., Sham, L. J. (1965). Self-consistent equations including exchange and correlation effects. *Phys. Rev.* **140**, 1133–1138.

Labar, G., Michaux, C. (2007). Fatty acid amide hydrolase: from characterization to therapeutics. *Chem. Biodivers.* **4**, 1882–1902.

Lahiri, S. D., Zhang, G. F., Dunaway-Mariano, D., Allen, K. N. (2003). The pentacovalent phosphorus intermediate of a phosphoryl transfer reaction. *Science* **299**, 2067–2071.

Laio, A., VandeVondele, J., Rothlisberger, U. (2002). A Hamiltonian electrostatic coupling scheme for hybrid Car-Parrinello molecular dynamics simulations. *J. Chem. Phys.* **116**, 6941–6947.

Leach, A. R. (1996). Molecular Modeling. Principles and Applications. Addison-Wesley, Reading, MA.

Lodola, A., Mor, M., Hermann, J. C., Tarzia, G., Piomelli, D., Mulholland, A. J. (2005). QM/MM modelling of oleamide hydrolysis in fatty acid amide hydrolase (FAAH) reveals a new mechanism of nucleophile activation. *Chem. Commun. (Camb.)* **35**, 4399–4401.

Lodola, A., Mor, M., Rivara, S., Christov, C., Tarzia, G., Piomelli, D., et al. (2008a). Identification of productive inhibitor binding orientation in fatty acid amide hydrolase (FAAH) by QM/MM mechanistic modelling. *Chem. Commun. (Camb.)* **2**, 214–216.

Lodola, A., Woods, C., Mulholland, A. J. (2008b). Advances of QM/MM methods in computational enzymology. *Annu. Rep. Comput. Chem.* **4**, 155–169.

Lodola, A., Mor, M., Sirirak, J., Mulholland, A. J. (2009). Insights into the mechanism and inhibition of fatty acid amide hydrolase from quantum mechanics/molecular mechanics (QM/MM) modelling. *Biochem. Soc. Trans.* **37**, 363–367.

Lodola, A., Sirirak, J., Fey, N., Rivara, S., Mor, M., Mulholland, A. J. (2010). Structural fluctuations in enzyme-catalysed reactions: determinants of reactivity in fatty acid amide hydrolase from multivariate statistical analysis of quantum mechanics/molecular mechanics paths. *J. Chem. Theory Comput.* **6**, 2948–2960.

Lodola, A., Capoferri, L., Rivara, S., Chudyk, E., Sirirak, J., Dyguda-Kazimierowicz, E., et al. (2011a). Understanding the role of carbamate reactivity in fatty acid amide hydrolase inhibition by QM/MM mechanistic modelling. *Chem. Commun. (Camb.)* **47**, 2517–2519.

Lodola, A., Rivara, S., Mor, M. (2011b). Application of computational methods to the design of fatty acid amide hydrolase (FAAH) inhibitors based on a carbamic template structure. *Adv. Protein Chem. Struct. Biol.* **85**, 1–26.

Lonsdale, R., Ranaghan, K. E., Mulholland, A. J. (2010). Computational enzymology. *Chem. Commun. (Camb.)* **46**, 2354–2372.

Marti, S., Roca, M., Andres, J., Moliner, V., Silla, E., Tunon, I., et al. (2004). Theoretical insights in enzyme catalysis. *Chem. Soc. Rev.* **33**, 98–107.

Masgrau, L., Roujeinikova, A., Johannissen, L. O., Hothi, P., Basran, J., Ranaghan, K. E., et al. (2006). Atomic description of an enzyme reaction dominated by proton tunneling. *Science* **312**, 237–241.

McKinney, M. K., Cravatt, B. F. (2003). Evidence for distinct roles in catalysis for residues of the serine-serine-lysine catalytic triad of fatty acid amide hydrolase. *J. Biol. Chem.* **278**, 37393–37399.

McKinney, M. K., Cravatt, B. F. (2005). Structure and function of fatty acid amide hydrolase. *Annu. Rev. Biochem.* **74**, 411–432.

Menikarachchi, L. C., Gascon, J. A. (2010). QM/MM approaches in medicinal chemistry research. *Curr. Top. Med. Chem.* **10**, 46–54.
Mileni, M., Kamtekar, S., Wood, D. C., Benson, T. E., Cravatt, B. F., Stevens, R. C. (2010). Crystal structure of fatty acid amide hydrolase bound to the carbamate inhibitor URB597: discovery of a deacylating water molecule and insight into enzyme inactivation. *J. Mol. Biol.* **400**, 743–754.
Moghaddam, M. F., Grant, D. F., Cheek, J. M., Greene, J. F., Williamson, K. C., Hammock, B. D. (1997). Bioactivation of leukotoxins to their toxic diols by epoxide hydrolase. *Nat. Med.* **3**, 562–566.
Mor, M., Lodola, A. (2009). Pharmacological tools in endocannabinoid neurobiology. *Curr. Top. Behav. Neurosci.* **1**, 87–110.
Mor, M., Rivara, S., Lodola, A., Plazzi, P. V., Tarzia, G., Duranti, A., et al. (2004). Cyclohexylcarbamic acid 3′- or 4′-substituted biphenyl-3-yl esters as fatty acid amide hydrolase inhibitors: synthesis, quantitative structure-activity relationships, and molecular modeling studies. *J. Med. Chem.* **47**, 4998–5008.
Morrison, J. F., Heyde, E. (1972). Enzymic phosphoryl group transfer. *Annu. Rev. Biochem.* **41**, 29–54.
Mulholland, A. J. (2005). Modelling enzyme reaction mechanisms, specificity and catalysis. *Drug Discov. Today* **10**, 1393–1402.
Mulholland, A. J. (2008). Computational enzymology: modelling the mechanisms of biological catalysts. *Biochem. Soc. Trans.* **36**, 22–26.
Newman, J. W., Morisseau, C., Hammock, B. D. (2005). Epoxide hydrolases: their roles and interactions with lipid metabolism. *Prog. Lipid Res.* **44**, 1–51.
Node, K., Huo, Y. Q., Ruan, X. L., Yang, B. C., Spiecker, M., Ley, K., et al. (1999). Anti-inflammatory properties of cytochrome P450 epoxygenase-derived eicosanoids. *Science* **285**, 1276–1279.
Nowotny, M., Yang, W. (2006). Stepwise analyses of metal ions in RNase H catalysis from substrate destabilization to product release. *EMBO J.* **25**, 1924–1933.
Nowotny, M., Gaidamakov, S. A., Crouch, R. J., Yang, W. (2005). Crystal structures of RNase H bound to an RNA/DNA hybrid: substrate specificity and metal-dependent catalysis. *Cell* **121**, 1005–1016.
Nowotny, M., Gaidamakov, S. A., Ghirlando, R., Cerritelli, S. M., Crouch, R. J., Yang, W. (2007). Structure of human RNase H1 complexed with an RNA/DNA hybrid: insight into HIV reverse transcription. *Mol. Cell* **28**, 264–276.
Nowotny, M., Cerritelli, S. M., Ghirlando, R., Gaidamakov, S. A., Crouch, R. J., Yang, W. (2008). Specific recognition of RNA/DNA hybrid and enhancement of human RNase H1 activity by HBD. *EMBO J.* **27**, 1172–1181.
Palermo, G., Branduardi, D., Masetti, M., Lodola, A., Mor, M., Piomelli, D., et al. (2011). Covalent inhibitors of fatty acid amide hydrolase: a rationale for the activity of piperidine and piperazine aryl ureas. *J. Med. Chem.* **54**, 6612–6623.
Peters, M. B., Raha, K., Merz, K. M. (2006). Quantum mechanics in structure-based drug design. *Curr. Opin. Drug Discov. Devel.* **9**, 370–379.
Petrosino, S., Di Marzo, V. (2010). FAAH and MAGL inhibitors: therapeutic opportunities from regulating endocannabinoid levels. *Curr. Opin. Investig. Drugs* **11**, 51–62.

Raha, K., Peters, M. B., Wang, B., Yu, N., WollaCott, A. M., Westerhoff, L. M., et al. (2007). The role of quantum mechanics in structure-based drug design. *Drug Discov. Today* **12**, 725–731.

Ranaghan, K. E., Mulholland, A. J. (2010). Investigation of enzyme-catalysed reactions with combined quantum mechanics/molecular mechanics (QM/MM) methods. *Int. Rev. Phys. Chem.* **29**, 63–133.

Rosta, E., Woodcock, H. L., Brooks, B. R., Hummer, G. (2009). Artificial reaction coordinate "tunneling" in free-energy calculations: the catalytic reaction of RNase H. *J. Comput. Chem.* **30**, 1634–1641.

Rosta, E., Nowotny, M., Yang, W., Hummer, G. (2011). Catalytic mechanism of RNA backbone cleavage by ribonuclease H from quantum mechanics/molecular mechanics simulations. *J. Am. Chem. Soc.* **133**, 8934–8941.

Schmelzer, K. R., Kubala, L., Newman, J. W., Kim, I. H., Eiserich, J. P., Hammock, B. D. (2005). Soluble epoxide hydrolase is a therapeutic target for acute inflammation. *Proc. Natl. Acad. Sci. USA* **102**, 9772–9777.

Schramm, V. L. (2005). Enzymatic transition states and transition state analogues. *Curr. Opin. Struct. Biol.* **15**, 604–613.

Senn, H. M., Thiel, W. (2009). QM/MM methods for biomolecular systems. *Angew. Chem. Int. Ed.* **48**, 1198–1229.

Shaw-Reid, C. A., Feuston, B., Munshi, V., Getty, K., Krueger, J., Hazuda, D. J., et al. (2005). Dissecting the effects of DNA polymerase and ribonuclease H inhibitor combinations on HIV-1 reverse-transcriptase activities. *Biochemistry* **44**, 1595–1606.

Singh, J., Petter, R. C., Baillie, T. A., Whitty, A. (2011). The resurgence of covalent drugs. *Nat. Rev. Drug Discov.* **10**, 307–317.

Tubert-Brohman, I., Acevedo, O., Jorgensen, W. L. (2006). Elucidation of hydrolysis mechanisms for fatty acid amide hydrolase and its Lys142Ala variant via QM/MM simulations. *J. Am. Chem. Soc.* **128**, 16904–16913.

Vacondio, F., Silva, C., Lodola, A., Fioni, A., Rivara, S., Duranti, A., et al. (2009). Structure-property relationships of a class of carbamate-based fatty acid amide hydrolase (FAAH) inhibitors: chemical and biological stability. *ChemMedChem* **4**, 1495–1504.

Warshel, A. (2003). Computer simulations of enzyme catalysis: methods, progress, and insights. *Annu. Rev. Biophys. Biomol. Struct.* **32**, 425–443.

Warshel, A., Levitt, M. (1976). Theoretical studies of enzymic reactions—dielectric, electrostatic and steric stabilization of carbonium-ion in reaction of lysozyme. *J. Mol. Biol.* **103**, 227–249.

Warshel, A., Sharma, P. K., Kato, M., Xiang, Y., Liu, H., Olsson, M. H. (2006). Electrostatic basis for enzyme catalysis. *Chem. Rev.* **106**, 3210–3235.

Yang, W., Lee, J. Y., Nowotny, M. (2006). Making and breaking nucleic acids: two-Mg^{2+}-ion catalysis and substrate specificity. *Mol. Cell* **22**, 5–13.

Zhou, T., Huang, D., Caflisch, A. (2010). Quantum mechanical methods for drug design. *Curr. Top. Med. Chem.* **10**, 33–45.

ALLOSTERY AND BINDING COOPERATIVITY OF THE CATALYTIC SUBUNIT OF PROTEIN KINASE A BY NMR SPECTROSCOPY AND MOLECULAR DYNAMICS SIMULATIONS

By LARRY R. MASTERSON,[*,†] ALESSANDRO CEMBRAN,[†] LEI SHI,[†] AND GIANLUIGI VEGLIA[*,†]

*Department of Biochemistry, Molecular Biology, and Biophysics, University of Minnesota, Minneapolis, Minnesota, USA
†Department of Chemistry, University of Minnesota, Minneapolis, Minnesota, USA

Abstract

The catalytic subunit of cAMP-dependent protein kinase A (PKA-C) is an exquisite example of a single molecule allosteric enzyme, where classical and modern views of allosteric signaling merge. In this chapter, we describe the mapping of PKA-C conformational dynamics and allosteric signaling in the free and bound states using a combination of NMR spectroscopy and molecular dynamics simulations. We show that ligand binding affects the enzyme's conformational dynamics, shaping the free-energy landscape toward the next stage of the catalytic cycle. While nucleotide and substrate binding enhance the enzyme's conformational entropy and define dynamically committed states, inhibitor binding attenuates the internal dynamics in favor of enthalpic interactions and delineates dynamically quenched states. These studies support a central role of

DOI: 10.1016/B978-0-12-398312-1.00012-3

Copyright 2012, Elsevier Inc.
All rights reserved.

conformational dynamics in many aspects of enzymatic turnover and suggest future avenues for controlling enzymatic function.

I. Introduction

Protein phosphorylation is a central signaling event in eukaryotes and is orchestrated by protein kinases that catalyze the transfer of the γ-phosphoryl group of a nucleoside triposphate (usually adenosine triphosphate, ATP) to an amino acid hydroxyl group (commonly serine, threonine, or tyrosine) of a protein substrate. The protein kinase family makes up $\sim$2000 enzymes in humans, and $\sim$30% of all proteins encoded by the human genome undergo reversible phosphorylation (Cohen, 2001). Since aberrant phosphorylation plays a role in many diseases, from cancer to cardiovascular disease, a complete elucidation of the molecular mechanisms that regulate reversible phosphorylation would have a major impact on designing innovative treatments (Cohen, 1999, 2002).

Protein kinase A (PKA, EC: 2.7.1.37) was the first protein kinase ever to be studied (Walsh, et al., 1968). PKA exists as a holoenzyme of two regulatory subunits and two catalytic subunits. An allosteric conformational change triggered by 3′,5′-cyclic adenosine monophosphate causes the regulatory subunits to release active catalytic subunits and phosphorylate protein substrates. The catalytic core of this enzyme (PKA-C, 350 residues, Fig. 1A) is highly conserved throughout the protein kinase superfamily and has thus been used as a model enzyme for the other members (Johnson, et al., 2001). PKA-C recognizes the amino acid sequence R-R-X-S/T-Y (where X is any residue and Y is a hydrophobic residue) present in hundreds of cellular substrates (Boeshans et al., 1999; Shabb, 2001). PKA-C undergoes several co- and posttranslational modifications such as deamidation and phosphorylation at the N-terminus (Johnson, et al., 2001), and myristoylation at the N-terminal glycine (Fig. 1A; Taylor et al., 2004). In addition to catalyzing phosphoryl transfer, PKA-C also serves as a hub for several interacting partners that constitute complex signaling machinery (Taylor et al., 2004). It has been hypothesized that co- and posttranslational modifications of PKA-C may constitute chemical signals for substrate recognition and localization within the cell, allowing this ubiquitous enzyme to select specific substrates (Taylor et al., 2004).

In spite of the plethora of structural and functional studies on this kinase, several key mechanistic questions on phosphorylation remain

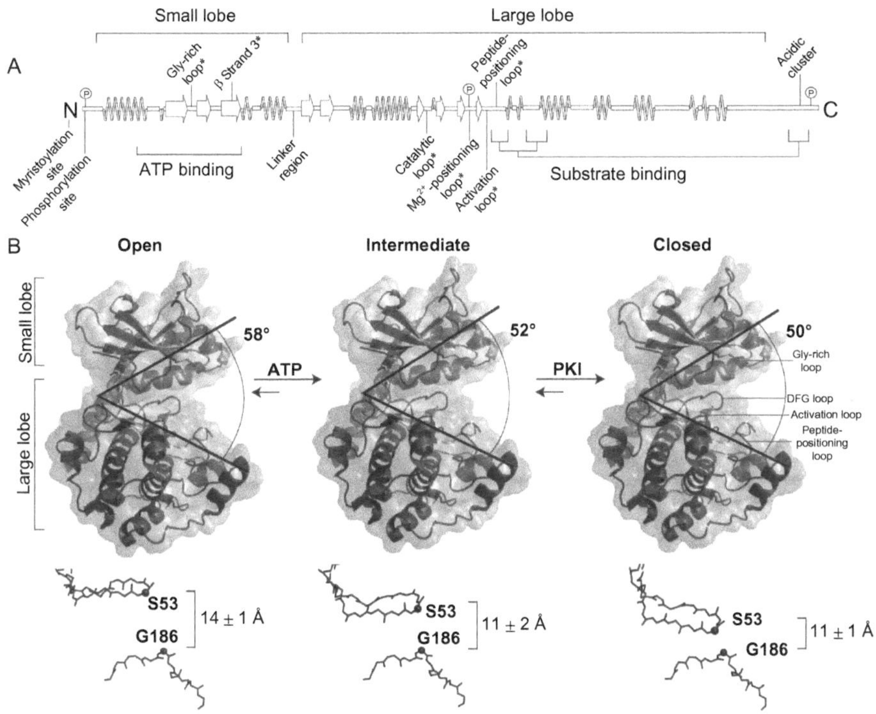

FIG. 1. Architecture of PKA-C. (A) Secondary structure, domains, conserved motifs (see asterisks), and ligand-binding regions of PKA-C. (B) Conformational changes observed in PKA-C by X-ray crystallography. The major conformational states are defined by the angle between the large and small lobes and by the S53/G186Cα distance (open, intermediate, and closed conformational states). The average angles and distances are calculated from open (PDB: 3O7L, 1CMK, 1CTP, 1J3H, and 2QVS), intermediate (PDB: 1BKX, 1BX6, 1STC, 1JLU, 1RE8, 1REK, 3DND, 3DNE, 3IDB, and 3IDC), and closed (PDB: 1JBP, 1ATP, 1APM, 1YDS, 1YDR, and 1YDT) X-ray structures. (For color version of this figure, the reader is referred to the online version of this chapter.)

unanswered: What governs substrate recognition? What determines substrate specificity? What are the driving forces for substrate release? How can PKA-C recognize soluble and membrane bound substrates? Answering these central questions will help our understanding of the etiology of diseases such as cancer, diabetes, rheumatoid arthritis, and various cardiomyopathies (Cohen, 2001).

X-ray crystallography has offered an essential contribution to understanding the structural architecture of protein kinases. The pioneering work of Taylor and coworkers resulted in the first atomic resolution structure of PKA-C in the inhibited state (Zheng et al., 1993). Following this study, several crystal structures were obtained, which mimicked major conformational states of the enzyme along the reaction coordinates, by using inhibitors, pseudosubstrates, and nucleotide analogs (Johnson, et al., 2001; Taylor et al., 2004). Based on these structures and kinetic data on mutant forms of PKA-C, Taylor and coworkers hypothesized the existence of a sophisticated intramolecular allosteric network that controls substrate recognition, chemical step, and product release (Kornev et al., 2008). Although supported by spectroscopic and computational studies, the allosteric network of communication had never been probed experimentally until recently (Yang et al., 2005, 2009; Kornev et al., 2006; Masterson et al., 2008, 2010; Hyeon et al., 2009; Kornev and Taylor, 2009, 2010; Karginov et al., 2010; Gaffarogullari et al., 2011).

Our group has studied the structural dynamics of PKA-C using liquid-state NMR spectroscopy with the objective of mapping the allosteric regulation of the kinase along the catalytic cycle (Masterson et al., 2008, 2010, 2011a,b; Gaffarogullari et al., 2011; Ha et al., 2011). This chapter summarizes these studies, highlighting our most recent progress in connecting conformational dynamics (molecular motion) of the kinase with catalysis, defined here as the sum of three main steps: substrate recognition, chemical step, and product release. Using a combination of NMR studies and molecular dynamics (MD) simulations, we attempted to trace the conformational and dynamic states of the enzyme. We found that ligand binding affects the conformational equilibrium of the enzyme and defines the different dynamical states that regulate kinase activity: committed, uncommitted, and quenched states (Masterson et al., 2010, 2011a,b). Our studies emphasize the importance of both conformational and dynamic changes for allostery to take place.

II. Major Conformational States Identified by X-Ray Crystallography

A wealth of crystallographic data is available on PKA-C, which is a direct result of the ability to express this enzyme as a recombinant protein in *Escherichia coli* (Slice, L.W. and Taylor S.S., 1989). Well over a dozen structures have been deposited for PKA-C in the protein data bank (PDB) (Johnson, et al., 2001). The fold of PKA-C is shared throughout the protein kinase family (Johnson, et al., 2001; Fig. 1A,B). The enzyme is bean shaped, comprising a small lobe, containing mainly β-sheets, and a mostly helical large lobe. The small lobe includes the nucleotide binding site, glycine-rich loop, β-strand 3, and the C-helix (Johnson, et al., 2001; Taylor et al., 2004). The large lobe contains docking sites for substrates, regulatory subunits, and colocalization elements such as A-kinase anchoring proteins (Wong and Scott, 2004). Key motifs such as the catalytic, DFG, and peptide-positioning loops are situated at an active site cleft formed between the two lobes, where phosphoryl transfer takes place (Fig. 1B; Johnson, et al., 2001; Taylor et al., 2004). The crystallographic data also provided snapshots for the arrangement of the functional groups involved in the phosphorylation process, identifying the position of the γ-phosphate of ATP and the acceptor hydroxyl group of the phosphorylatable serine in the substrate (Taylor et al., 2004). These static representations of the enzyme, along with fluorescence anisotropy data, suggested that the enzyme toggles between three major conformational states: an *open state* (typically observed with no ligand bound), an *intermediate state* (typically observed with a singly bound ligand), and a *closed state* (observed in the ternary complexes) (Taylor et al., 2004; Fig. 1B). Because significant conformational changes are observed upon nucleotide binding, it has been hypothesized that the allosteric signal initiated upon nucleotide binding travels *via* the hydrophobic core from the small to the large lobe (Kornev et al., 2008). Structural changes are also observed for regions peripheral to the active site cleft. Based on these observations and computational studies, it has been proposed that two hydrophobic spines buried in the catalytic core may work as a transmission line for intramolecular allosteric signaling, thus modulating binding interactions and catalysis (Kornev et al., 2006, 2008; Kornev and Taylor, 2009, 2010). Analyzing how the signal is transmitted within the catalytic core is central to understanding enzyme function and regulation. To this extent, modern NMR spectroscopy methods provide atomic views of short- and long-range conformational and dynamic changes, offering a unique perspective on allostery (Kern and Zuiderweg, 2003; Boehr et al., 2006; Tzeng and Kalodimos, 2009; Kalodimos, 2011).

III. NMR Analysis of Ligand Binding Shows Positive Allosteric Cooperativity in PKA-C

A stepping stone for carrying out NMR studies is the optimization of functional samples that also behave well spectroscopically (i.e., provide well-resolved fingerprint resonances). Under our experimental conditions, PKA-C is functional as established by ^{31}P NMR, with kinetics similar to those measured by coupled enzyme assays (Masterson et al., 2008). The assignment of the backbone resonances, however, was far from trivial. The line shapes of the amide fingerprint in the $[^{1}H,^{15}N]$-TROSY-HSQC spectra are inhomogeneous, due to intermediate exchange broadening in the chemical shift time scale. In addition to the classical triple resonance experiments, a combination of selectively labeled samples and new spectral-editing techniques (i.e., CCLS-HSQC (Tonelli et al., 2007)) was necessary to attain ~83% of the total backbone resonance assignments (Masterson et al., 2009). All of the key regions were identified in the enzyme fingerprint, which enabled us to trace the changes in the backbone amide resonances upon ligand binding. The NMR titrations carried out with 5′-adenylylimido-diphosphate (AMP-PNP) and a standard kinase substrate, Kemptide (LRRASLG), allowed us to map the effects of the ligands on both wild-type PKA-C (PKA-C^{WT}) and PKA-C^{Y204A}, a mutant with a severely impaired catalytic efficiency (Fig. 2; Moore et al., 2003; Yang et al., 2005; Masterson et al., 2008). Since the formation of the ternary complex occurs via random sequential order (Kong and Cook, 1988; Adams, 2001), we performed the NMR titrations by adding AMP-PNP first followed by Kemptide and then reversed the order of ligands. The *on*-rate for nucleotide binding was approximately two orders of magnitude slower than the diffusion rate, suggesting that binding is likely to occur via structural rearrangements. These conformational changes are necessary for proper alignment and positioning of the nucleotide and substrate, along with shielding the active site from water. The PKA-C^{WT} ligand-binding experiments revealed a positive *K*-type cooperativity (Table I), with dissociation constants similar to those reported previously (Taylor et al., 2004). Although weak (~fourfold), the *K*-type cooperativity is mutual between substrate and nucleotide: binding of the nucleotide enhances the substrate affinity, and substrate binding enhances the affinity for the nucleotide. Interestingly, titrations with the catalytically impaired mutant PKA-C^{Y204A} showed that the ligand-binding events are decoupled, that is, there is no binding cooperativity. The chemical

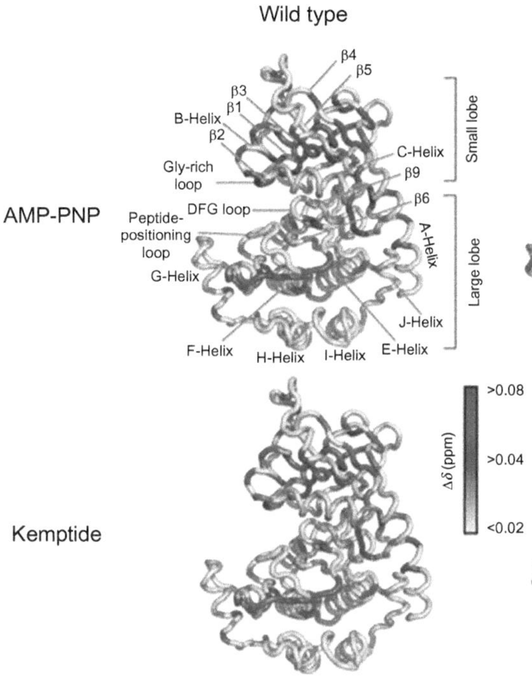

FIG. 2. Chemical perturbations upon ligand binding for PKA-C^{WT} and PKA-C^{Y204A}. The binding of nucleotide and substrate indicates attenuated local and long-range (allosteric) perturbations for PKA-C^{Y204A}. (For color version of this figure, the reader is referred to the online version of this chapter.)

TABLE I
Binding Constants (K_d, in units of micromolar) Measured from the NMR Titrations of PKA-C

	AMP-PNP first		Kemptide first	
	$K_d^{AMP\text{-}PNP}$	$K_d^{Kemptide}$	$K_d^{Kemptide}$	$K_d^{AMP\text{-}PNP}$
PKA-C^{WT}	39 ± 6	292 ± 61	980 ± 163	12 ± 3
PKA-C^{Y204A}	78 ± 20	834 ± 92	700 ± 105	68 ± 4

shift perturbations observed in the enzyme fingerprint spectra revealed the presence of local and long-range (allosteric) changes in PKA-C^{WT}. Based on the X-ray structures, these changes occur as far as ~20 Å from the ATP-binding site, including the N- and C-terminal regions of the small lobe and residues near the peptide-positioning loop of the large lobe (Fig. 2). Interestingly, the chemical shift changes for PKA-C^{Y204A} were substantially smaller than for the PKA-C^{WT}. A few long-range perturbations were detected in the large lobe, although with a different pattern (Fig. 2), and substrate binding resulted in very small changes with no detectable allosteric effects. Overall, AMP-PNP binding did not produce the allosteric changes observed for PKA-C^{WT}. Structurally, PKA-C^{Y204A} is almost superimposable to PKA-C^{WT} as assessed by X-ray crystallography (Yang et al., 2004). However, the deletion of a key hydrogen bond between the hydroxyl group of Y204 and E230 of the F-helix in the mutant disrupts the interaction network at the interface of the N- and C-lobes, as well as the interaction with the P-2 site of the substrate (Moore et al., 2003). The NMR fingerprints of the wild-type and the mutant were nearly superimposable. However, the elimination of these key interactions dismantles the allosteric communication (i.e., no long-range effects were observed) and disrupts the *K*-type cooperativity between nucleotide and substrate observed for PKA-C^{WT}.

From these measurements, two spectroscopic features were prominent. First, the extent of chemical shift changes ($\Delta\delta$) upon ligand binding was exiguous ($\sim$0.03 ppm), which was unexpected based on the structural differences observed in the X-ray structures (Johnson, et al., 2001; Taylor et al., 2004). The latter may be due to dynamic averaging of the backbone chemical shifts caused by the conformational equilibrium of the kinase in solution. Second, a number of resonances from highly conserved regions (glycine-rich, activation, and peptide-positioning loops) became exchanged broadened upon nucleotide binding, indicating that the enzyme is interconverting between at least two different conformational states in the microsecond to millisecond time scale. Although our initial NMR analysis was limited to chemical shift changes and warranted further investigation, we could infer that conformational dynamics played a key role in PKA-C function.

IV. Ligand Binding and Enzyme Motion

A. *Thermodynamics and Kinetics of Ligand Binding*

To understand the recognition process of the kinase, we compared the binding of the endogenous inhibitor of the kinase, protein kinase inhibitor (PKI), with that of a physiological substrate of PKA-C, phospholamban (PLN). PKI controls PKA-C function and cellular localization (Walsh and

Van Patten, 1994; Dalton and Dewey, 2006) acting as a physiological break to enzyme activity. PKI's recognition sequence occupies the active site and prevents phosphorylation of cellular targets. In contrast, PLN is a substrate of PKA-C and is reversibly phosphorylated during the heart muscle contraction-relaxation cycle (Traaseth et al., 2008). In both studies, we utilized either synthetic or recombinant peptides, each corresponding to either the 5–24 fragment of PKI (PKI_{5-24}) or the 1–20 fragment of PLN (PLN_{1-20}). Steady-state kinetic data showed that the PKA-C catalytic efficiency for the PLN_{1-20} peptide was comparable to that of full-length PLN (Masterson et al., 2010, 2011a,b) and had kinetics similar to Kemptide (Masterson et al., 2010, 2011a,b). The *K*-type cooperativity of PLN_{1-20} was slightly higher than that of Kemptide, with a fivefold increase in binding affinity in the presence of nucleotide. In contrast, PKI_{5-24} binding affinity was significantly higher ($K_d \sim 2$ μM for the apo enzyme). The latter was accompanied by ~40-fold increase in binding affinity when the nucleotide was present (Fig. 3; Masterson et al., 2011a,b). Thermocalorimetric measurements showed that while the binding of the PLN_{1-20} peptide substrate is *entropically driven*, the binding of the PKI_{5-24} peptide inhibitor binding is *enthalpically driven*. A substantial increase in the overall enzyme thermostability was observed upon binding PKI_{5-24}, whereas the nucleotide-bound form, or nucleotide and substrate bound form, caused only a negligible increase. Stabilization effects were also present upon addition of excess amounts of Mg^{2+} ions. Under these conditions, Mg^{2+} behaves as a noncompetitive inhibitor (Cook et al., 1982; Adams and Taylor, 1993; Herberg et al., 1999; Zimmermann et al., 2008) and two Mg^{2+} ions bind the nucleotide binding cleft. The kinetic effects of the competitive (PKI_{5-24}) and noncompetitive (secondary Mg^{2+} ion) inhibitors were synergistic, with an overall decrease of K_d to ~12 nM. The latter is in agreement with the additive changes observed for the two inhibitors by thermal melting curves (Zimmermann et al., 2008; Masterson et al., 2011a,b). From these data, it is possible to extrapolate that *as the enzyme stability increases, reactivity decreases.*

B. *X-Ray Structures of PKA-C in Complex with Substrate and Inhibitors*

The crystal structures of the ternary complexes of PKA-C with AMP-PNP and PLN_{1-19} or PKI_{5-24} displayed striking differences between the substrate- and inhibitor-bound complexes (Zheng et al., 1993; Masterson et al., 2010). In spite of excess amounts of substrate and nucleotide, the

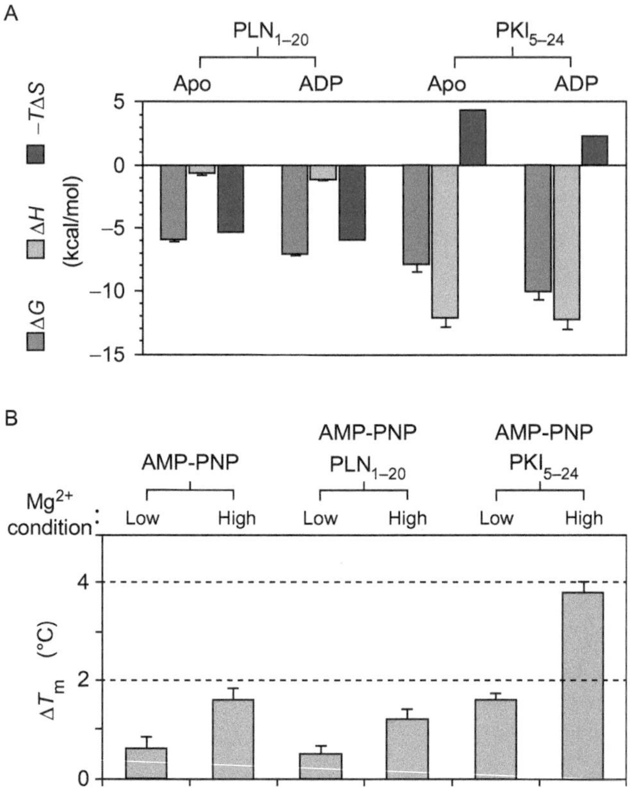

FIG. 3. Thermocalorimetric parameters for the binding of the substrate PLN_{1-20} and inhibitor PKI_{5-24} to PKA-C. (A) Binding energy for substrate and inhibitor to the apo or nucleotide-bound PKA-C. PLN_{1-20} is entropy driven, while PKI_{5-24} binding is enthalpy driven. (B) Thermal melting of PKA-C complexes. A small increase in thermal stability (relative to the apo enzyme melting temperature, ΔT) is present upon nucleotide or substrate binding under low and high Mg^{2+} concentration. A substantial increase in stability is measured in the presence of PKI_{5-24} and high Mg^{2+}. (For color version of this figure, the reader is referred to the online version of this chapter.)

PKA-C/AMP-PNP/PLN_{1-19} complex crystallized with two enzyme molecules in the asymmetric unit: an enzyme in a closed conformation bound to both ligands and an *apo* form enzyme in an open conformation. Interestingly, the *apo* form had lower B-factors than those in the ternary complex and, with the exception of the first few residues at the N-terminus, showed a well-defined electron density for all regions of the enzyme. On the other hand, the electron density of the complex with the substrate was undefined for the highly conserved glycine-rich loop and part of the C-terminal tail, which is believed to stabilize interactions with the substrate. These regions were observed in the crystal structures formed with PKI_{5-24} (Johnson, et al., 2001; Taylor et al., 2004), in which the inhibitor likely traps these regions into a stable conformational minimum. Qualitatively, there is an agreement between the thermodynamic data and the structural features observed in the crystals of the substrate versus the inhibitor; when the enthalpic contribution is dominant, the structure is more resolved (i.e., lower B-factors), whereas when the entropic contribution is dominant, static and dynamic disorder decrease the resolution (i.e., higher B-factors).

C. *Dynamic States of PKA-C Define Catalysis and Inhibition*

To follow the conformational changes along the enzymatic coordinates, we analyzed five different forms of PKA-C: (1) apo enzyme, (2) binary form (AMP-PNP bound), (3) ternary complex with low Mg^{2+} bound to PLN_{1-20} (which mimics the *productive* complex), (4) a ternary complex bound to PKI_{5-24} with low Mg^{2+} (*nonproductive* complex), and (5) a ternary complex bound to PKI_{5-24} with high Mg^{2+} (super-inhibited, *non-productive* complex) (Masterson et al., 2011a,b). The overall trend in chemical shift changes in the enzyme fingerprint correlates qualitatively with the displacement of C^{α} atoms observed by X-ray crystallography for the apo, binary and ternary forms. Therefore, in agreement with the crystal structures (Fig. 1B), only minimal changes occur in the overall fold for the transitions from nucleotide-bound binary form to the ternary complexes, and the average conformation of the ternary complex with peptide inhibitor closely resembles the ternary complex with substrate.

Although small, the chemical shift changes affect the catalytically important regions of the enzyme that are conserved throughout all protein kinases, including the glycine-rich, activation, DFG, and peptide-positioning loops (Fig. 4). Many of the changes in the amide fingerprint followed linear progressions from the *apo*, to the *intermediate*, to the *closed* super-inhibited ternary complex (PKI_{5-24} plus two Mg^{2+}, Fig. 4A). The linearity of chemical shifts revealed that the *apo* form exists as an ensemble of interchanging conformers (conformational equilibrium) that includes all of the different

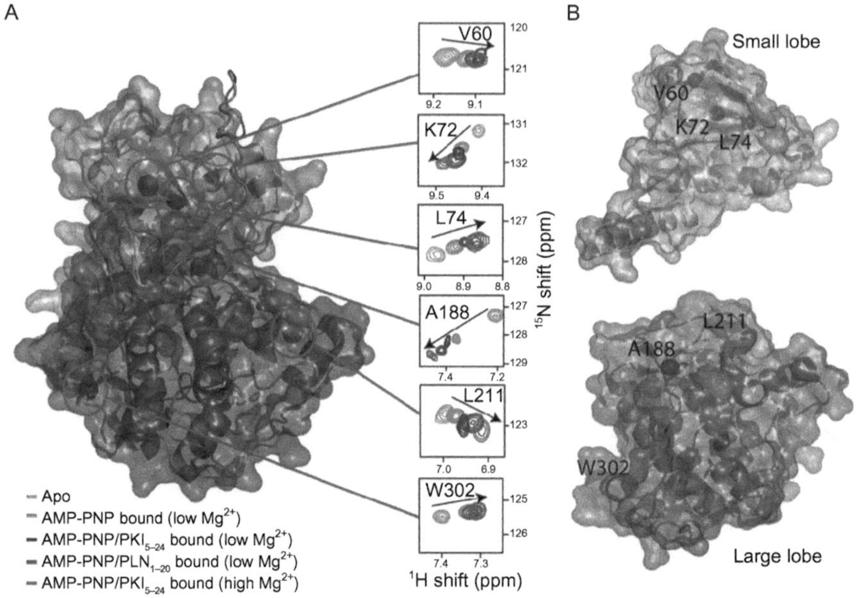

FIG. 4. Linearity of chemical changes among different forms of PKA-C. (A) Linearly correlated chemical shift changes in different regions of the small and large lobes. The two extremes of the correlations are the apo and the super-inhibited ternary complex of PKA-C, suggesting a fast conformational exchange between closed and open conformational states. (B) The location of these representative residues included regions near the active site and allosteric sites as far as 20 Å (W302) away. (For color version of this figure, the reader is referred to the online version of this chapter.)

states of the kinase. Ligand binding acts on the enzyme equilibrium, shifting the populations of the conformational ensemble. Nucleotide binding shifts the conformational ensemble of open toward the intermediate state, and substrate or inhibitor peptide binding skews the population toward the closed state. To corroborate these data, we carried out NMR relaxation measurements, which are able to image ground and excited states of proteins (Palmer, 2001; Boehr et al., 2006, 2009; Popovych et al., 2006; Berlow et al., 2007; Henzler-Wildman et al., 2007; Loria et al., 2008; Baldwin and Kay, 2009; Korzhnev et al., 2010; Mittag et al., 2010; Villali and Kern, 2010; Kalodimos, 2011). In agreement with previous spectroscopic data (Taylor et al., 2004), our results show that the enzyme becomes more compact upon ligand binding. Small but significant differences in stability are observed between the ternary complexes, with the most stable complex (inhibited enzyme under high Mg^{2+}) showing that the most compact structure has the shortest overall correlation time. These results follow the same trend as the thermocalorimetric data. The apo form of the enzyme displayed the expected flexibility in the nanosecond to picosecond (fast motion in the NMR time scale), with high mobility of residues in regions between structural elements such as the acidic cluster (residues 328–334) and the glycine-rich, catalytic, DFG, activation, and peptide-positioning loops. Surprisingly, using the conventional experiments for spin relaxation (Farrow et al., 1994), we could not detect microsecond to millisecond motions (slow motion in the NMR time scale) in the apo form. It should be noted, however, that the current NMR techniques are sensitive to a specific window of conformational dynamics within the microsecond to millisecond time scale and we cannot rule out any motions outside this range (Jarymowycz and Stone, 2006). Remarkably, upon nucleotide binding, we detected a noticeable increase in conformational dynamics, on both fast and slow time scales, for several key regions of the enzyme (Fig. 5). This was most apparent for the glycine-rich and the DFG loops, which are directly involved in nucleotide binding, as well as the activation loop, the peptide-positioning loop, and the acidic cluster (Fig. 6). Overall, the conformational dynamics of residues surrounding the active site were found to increase upon nucleotide binding. A diagnostic parameter of the conformational exchange is the relaxation rate of exchange (R_{ex}), which quantifies motion in the microsecond to millisecond time scale (Palmer, 2001). We found R_{ex} values greater than 25 Hz for several residues located in regions involved in catalysis. Although attenuated and interspersed, these conformational dynamics persist when PLN_{1-20} is bound. The latter indicates that conformational dynamics persists in the ternary complex and might be required for opening the active site and product release (Fig. 6). In fact, the linear correlation observed between R_{ex} and $\Delta\omega^2$, where $\Delta\omega$ is the difference in chemical shifts between free and bound enzyme, confirms that several residues experiencing slow motions undergo synchronous conformational dynamics (i.e., opening and closing) (Massi et al., 2006). From these data, we estimated an average rate constant for opening the active site cleft of $\sim$20–30 s^{-1}, which agrees remarkably well with the catalytic constant ($k_{cat} \sim 23\ s^{-1}$) (Adams and Taylor, 1992; Lew et al., 1997). Taken together, these data support that a global conformational change is the rate-determining step for enzyme turnover.

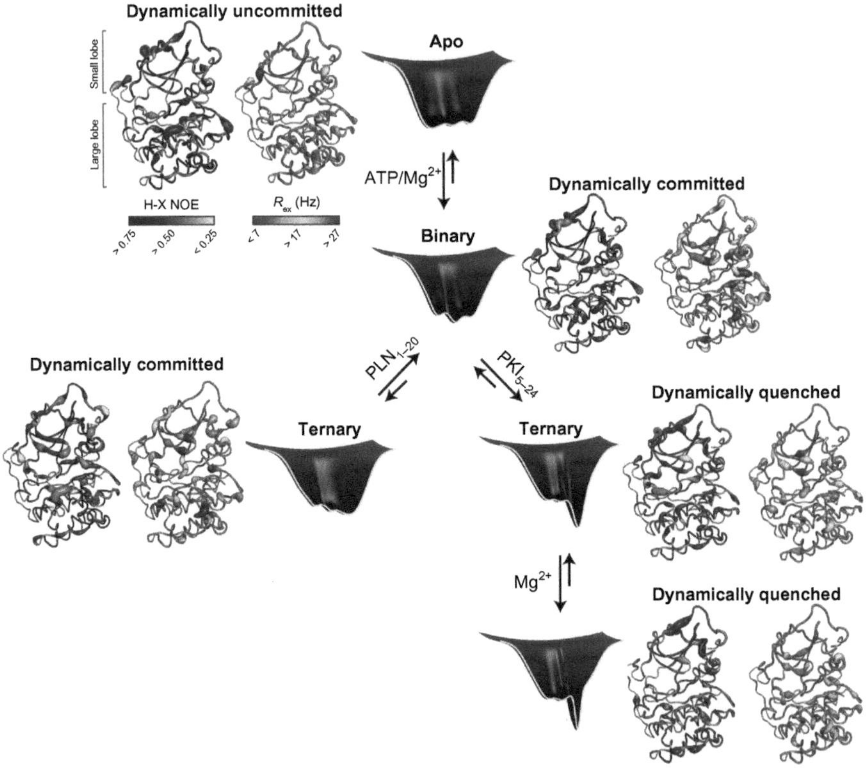

FIG. 5. Free-energy landscape of PKA-C and ligand binding. Ligand binding defines the conformational equilibrium of the kinase, skewing the populations in a dynamically uncommitted (apo form) basin with motions not synchronous to turnover, dynamically committed basin (nucleotide-bound and ternary complex with substrate) with motions synchronous with turnover, and dynamically quenched basin (inhibited complexes) with attenuated motions and low free energy. (For color version of this figure, the reader is referred to the online version of this chapter.)

From the above data, it is possible to speculate that the *apo* form of PKA-C experiences conformational dynamics outside the microsecond to millisecond time scale. This motion is not productive, that is, not synchronous with the opening and closing frequency necessary for catalysis. Nucleotide binding dynamically activates the enzyme, completing the C-spine architecture and engaging the small and large lobes (Kornev et al., 2006, 2008; Kornev and Taylor, 2009, 2010). The adenine group acts as a pivot point

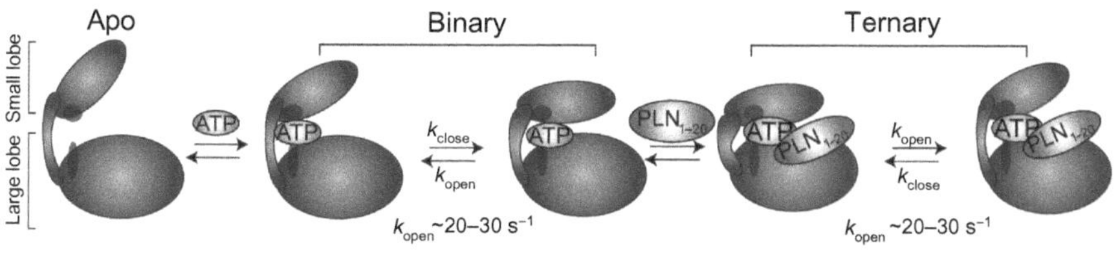

FIG. 6. Schematic for the formation of the catalytically competent ternary complex. The apo enzyme constitutes a dynamically uncommitted state, with its C-spine (red) disengaged between the two lobes. The nucleotide acts as an allosteric effector which is sensed throughout the enzyme. This binding event completes the C-spine architecture and induces a dynamically committed state. Fluctuations between open and closed conformations take place at a time scale which is synchronous with turnover. These fluctuations persist in the ternary complex with substrate, limiting the rate of catalytic turnover. (For interpretation of the references to color in this figure legend, the reader is referred to the online version of this chapter.)

between the two major enzyme domains (Kornev and Taylor, 2009) and activates the microsecond to millisecond conformational dynamics necessary for catalysis to occur (Fig. 6). Changes in conformation and dynamics upon nucleotide binding prime the active site for catalysis; that is, while the active site may be preorganized for the chemical step, these internal motions initiated by the nucleotide may contribute to surmounting the free-energy barrier for phosphoryl transfer to take place (Nashine et al., 2010). Although direct experimental proof of the role of dynamics in the chemical step is lacking, it has been shown that several mutations that disrupt the C-spine region lead to a significant decrease in catalytic rate (Kornev and Taylor, 2009). The latter strongly supports the hypothesis that the C-spine is a necessary element for allosteric dynamic transmission of the signals from the small to the large lobe. The opening and closing of the active cleft persists in the ternary complex with the substrate. These motions are probably responsible for product release, enabling opening of the active site cleft after phosphoryl transfer and allowing the exit of ADP and the phosphorylated product.

This process described by the NMR data epitomizes a conformational selection mechanism (Csermely et al., 2010). Nucleotide binding skews the preexisting equilibrium toward the next step of the enzymatic cycle, increasing the enzyme affinity for the substrate (positive allosteric cooperativity) and enhancing substrate recognition (Figs. 3 and 5).

We also mimicked the inhibited and super-inhibited forms of the enzyme with ternary complexes of PKA-C bound to PKI_{5-24} under low and high Mg^{2+} concentrations, respectively. PKI_{5-24} binding to the enzyme quenched both fast and slow motions throughout the entire enzyme backbone, trapping it in a dynamically inert or quenched state (Fig. 5). The attenuation of the conformational dynamics is even more apparent when two Mg^{2+} ions are bound in the nucleotide binding cleft. In the presence of substrate, the kinase preserved most of the flexibility in the proximity of the enzyme active site. In contrast, the presence of the two inhibitors reduced these pervasive dynamics to a handful of residues located in the conserved loops. The linear chemical shifts progressed toward the closed state, suggesting a shift of the enzyme conformational ensemble toward a well-defined basin with quenched dynamics and unable to carry out catalysis (Fig. 4A).

D. *Interpretation of Molecular Motions by MD Calculations*

Although NMR can detect the rates of conformational motions, it cannot interpret how these motions occur. Therefore, we carried out MD simulations in explicit solvent with the different forms of PKA-C: apo, nucleotide-bound (with one or two bound Mg^{2+} ions), and ternary complexes with either substrate (PLN_{1-20}) or inhibitor (PKI_{5-24} with one or two bound Mg^{2+} ions) (Masterson et al., 2011a,b). The analysis of the MD trajectories showed that the enzyme regions with the largest root mean square fluctuations (RMSF) coincide with the trends observed in the heteronuclear NOE measurements for both nucleotide-bound or ternary complex (Masterson et al., 2010, 2011a,b). Specifically, a good agreement between simulated and experimentally determined conformational dynamics was observed for the glycine-rich, activation, DFG, and peptide-positioning loops. Moreover, we observed marked differences between the motions experienced by the ternary complexes of PKA-C with nucleotide and substrate (PLN_{1-20}) and those corresponding to the complex with nucleotide and inhibitor (PKI_{5-24}). In particular, the RMSF for the complex with the substrate are on average higher than those with the inhibitors (Fig. 7A). Compared to the complex with PKI_{5-24}, the ternary complex with the substrate displays significantly higher RMSF values for the glycine-rich loop, β3, B helix, the loop connecting the F and G helices, as well as the C-terminal region. On average, the RMSF

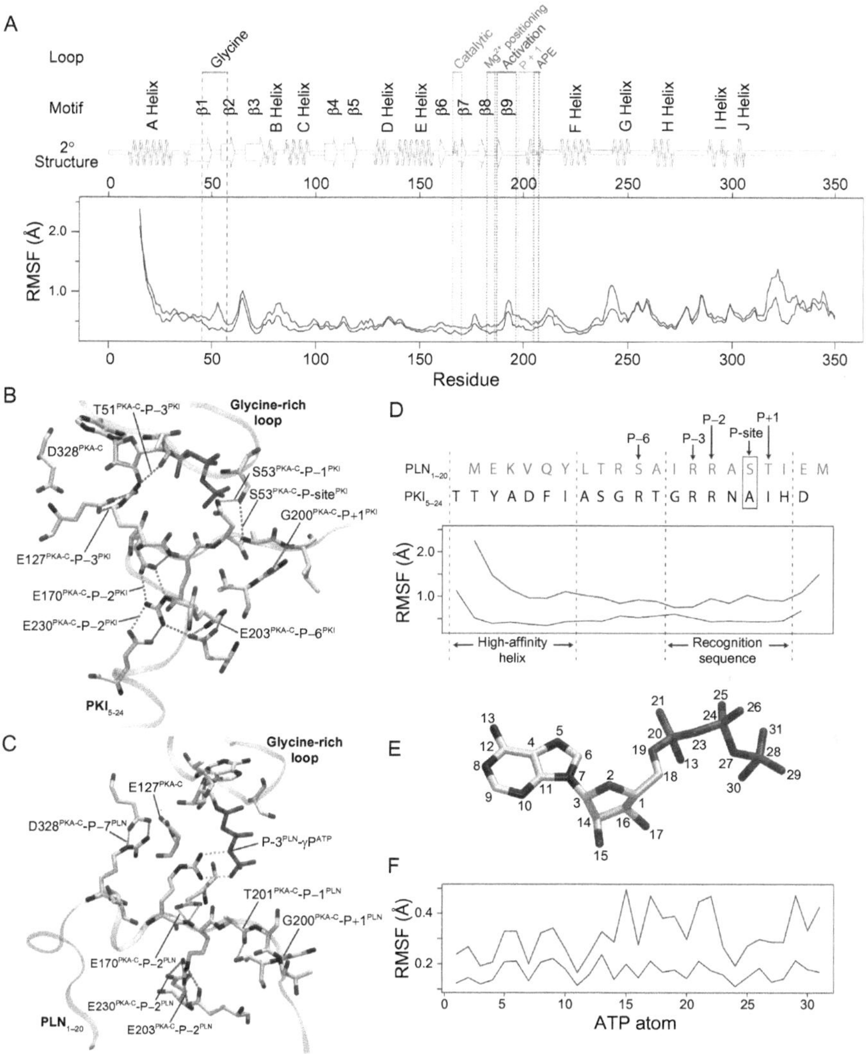

FIG. 7. MD simulations of PKA-C ternary complexes with PLN_{1-20} or PKI_{5-24}. All plots indicate ternary complexes with PLN_{1-20} (red) or PKI_{5-24} (black). (A) Backbone sub-nanosecond RMSF of PKA-C ternary complexes. Strong (red) and weak (green) hydrogen bonding networks which persisted during the simulations when PKI_{5-24} (B) or PLN_{1-20} (C) was present. (D) RMSF of the peptide backbone of PLN_{1-20} or PKI_{5-24}. (E) The structure and atom number of ATP, and (F) RMSF of ATP during simulations with the substrate or inhibitor. (For interpretation of the references to color in this figure legend, the reader is referred to the online version of this chapter.)

values are higher for all of the catalytically important loops. Also, substantially higher RMSF values are obtained for both the nucleotide and the substrate in their respective binding sites. In sharp contrast, the average fluctuations of the inhibited complex are rather attenuated (Fig. 7D–F). The latter is apparent from the perspective of the enzyme, the inhibitor, and the nucleotide. Qualitatively, these data follow the same trend observed by the thermocalorimetric measurements as well as the NMR relaxation data. The analysis of intermolecular hydrogen bond and electrostatic interactions between the enzyme and bound peptides revealed strong interactions between the PKI_{5-24} (from the $P+1$ through $P-6$ sites) and several residues constituting the binding groove. In contrast, a lower occurrence of these types of interactions is detected for the substrate during the MD simulations (Fig. 7B–C). These results are in agreement with the enthalpy- and entropy-driven binding modes for inhibitor and substrate, respectively, determined by isothermal titration calorimetry.

The large amplitude motions were projected using principal component analysis (PCA) (Amadei, A. et al., 1993; Miyashita et al., 2003; Maragakis and Karplus, 2005; Henzler-Wildman et al., 2007). We found that the first two principal components accounted for $\sim$60% of the variance in coordinates during the MD simulations (Fig. 8A). The first principal component (PC1) corresponded to the predicted opening and closing motion between the small and large lobes. The second component (PC2) corresponded to a shearing motion between these two lobes. This motion is favored by the greasy interface between the two major domains, conferring to the enzyme an "asymmetric bite" as monitored during MD simulations. A common parameter used in X-ray crystallography for the identification of the open and closed conformations is the distance between Ser53 and Gly186 ($d_{S53-G186}$, Fig. 8B) in PKA-C, which was used to probe active site accessibility during the MD simulations. We combined PC1 with $d_{S53-G186}$ to describe the relative motion of the two lobes. Based on 29 X-ray crystallographic structures, these two parameters are diagnostic of the kinase conformational state (Fig. 8C). The 2D plots reported in Fig. 8B display well-separated clusters, corresponding to the open, intermediate, and closed conformations. Among the different forms, the apo enzyme sampled a broad distribution of conformations that spanned all of the states: open, intermediate, and closed. The binary form with one Mg^{2+} ion bound had a distribution of conformations that sampled open and intermediates states, while addition of the substrate shifted the ensemble

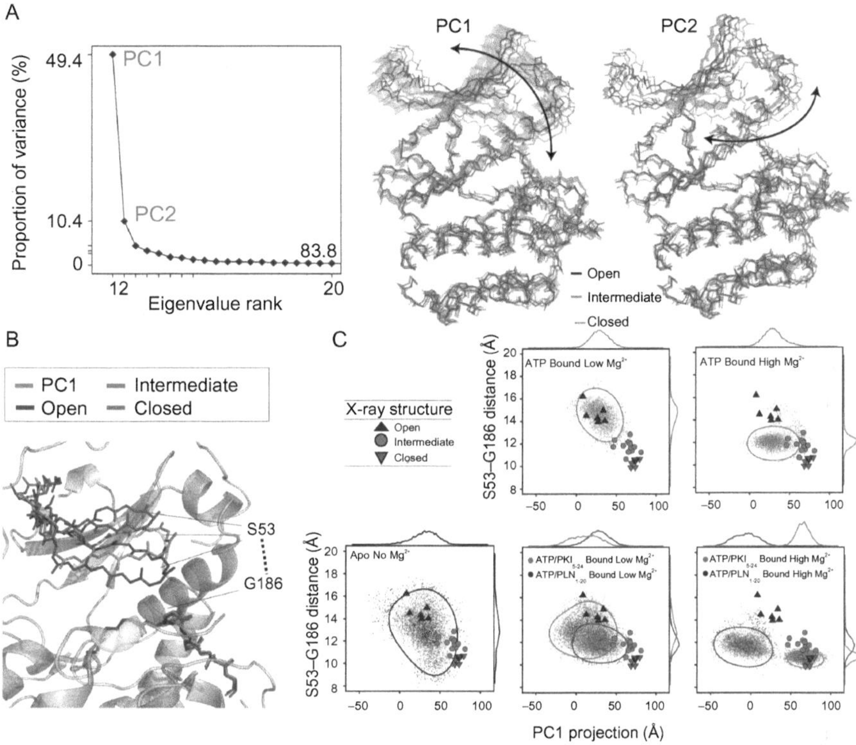

FIG. 8. MD simulations of PKA-C. (A) PCA analysis of the MD trajectories indicated two orthogonal motions between the small and large lobes which described ~50% of the variance: an opening/closing motion (PC1) and a shearing between the lobes (PC2). (B) The occlusion of the active site was probed using the distance between S53 and G186 C^{α} atoms and monitored with PC1 to assess active site accessibility. (C) 2D plots indicated that inhibitors restrict the conformational space accessible by PKA-C, while the other forms accessed these states more frequently. (For color version of this figure, the reader is referred to the online version of this chapter.)

toward the closed state. The second Mg^{2+} ion narrowed the minima of the conformational landscape explored by the enzyme for both binary and ternary complexes. Interestingly, the simulations with PKI_{5-24} and two Mg^{2+} ions (super-inhibited complex) showed the narrowest conformational distribution, which converged toward the closed conformations identified by X-ray structures. Collectively, the three major conformational states hypothesized by the X-ray data correspond to well-defined dynamic states of the enzyme. Together with the NMR data, we can give a qualitative description of the energy landscape of the enzyme (Fig. 6), assigning the apo enzyme to a dynamically uncommitted state located in a broad basin, where motions are present but not on a timescale relevant for turnover; and to the nucleotide-bound or Michaelis complex with the substrate, a dynamically committed state, with a more defined basin and where motions are synchronous with the rate-limiting step of enzyme turnover; and finally to the inhibited and super-inhibited complexes a dynamically quenched state with a basin where the enzyme dynamics are hindered (Fig. 8). The latter state is more populated when two Mg^{2+} are bound, and conformational dynamics are substantially attenuated. Previous H/D exchange (Yang et al., 2005) and fluorescence measurements (Li et al., 2002) are consistent with our definitions of the energy landscape.

E. Mutual Conformational Selection Mechanism for Enzyme and Substrate

Studies of full-length PLN revealed a preexisting equilibrium between four major conformational states (Gustavsson et al., 2011) driven by the reversible binding of cytoplasmic domain Ia to membranes (Fig. 9): a highly populated T state (L-shaped configuration), with domain Ia adsorbed on the membrane surface; a T′ state, with domain Ib partially unfolded; an R′ state that is partially unfolded at the N-terminus and in the C-terminal part of domain Ia; and an R state, which has domain Ia and Ib unfolded. The T′, R′, and R states are less populated and represent excited states, with the T and R states representing two extremes of conformational dynamics in PLN. Kinetic and structural measurements using a series of PLN analogues showed a clear correlation between phosphorylation efficiency with the extent of R state (Masterson et al., 2011a,b). We found that in the ground state (T state), PLN is less accessible to the protein kinase, and this decreases the catalytic efficiency of phosphorylation. By shifting the conformational equilibrium toward the R state, PKA-C catalytic efficiency increases. A plausible explanation is that the enzyme binds to the R state more readily and selects the extended conformation for phosphoryl transfer to occur. NMR and sequence homology-driven molecular docking provided a view of the structural ensemble of the complex between PKA-C and PLN_{1-20} (Masterson et al., 2011a,b). These results corroborate the kinetic preference of the membrane detached and extended conformation (R state) of PLN. The structural ensemble of

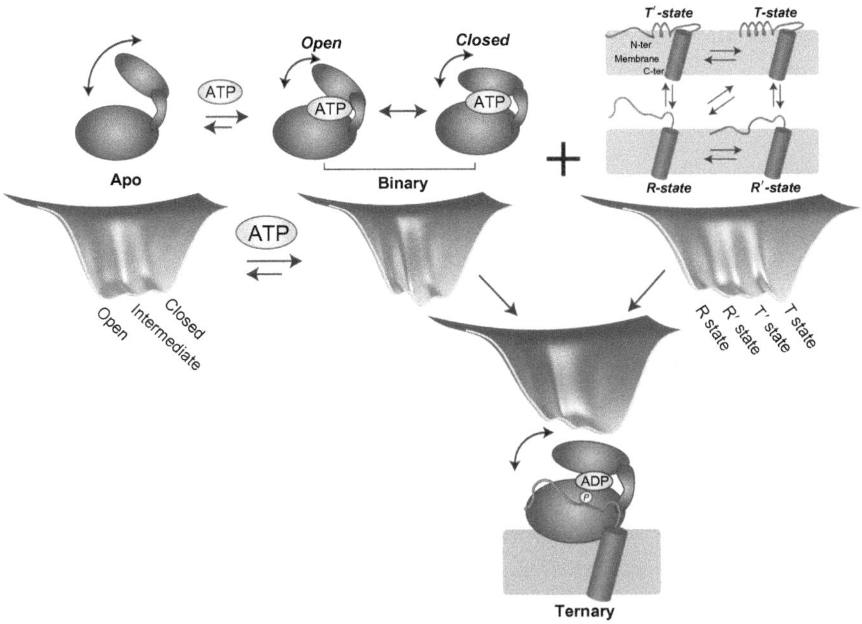

FIG. 9. Proposed *mutual conformational selection mechanism* between PKA-C and PLN. Both PKA-C and PLN undergo preexisting equilibria between major conformational states. PKA-C fluctuates between open and closed conformations, induced by nucleotide binding. PLN undergoes conformational fluctuations between various degrees of folded and unfolded conformations. These equilibria influence one another upon interaction, leading to a mutual adaptation and allowing the proteins to reach a minimum in the free-energy landscape of the complex. (For color version of this figure, the reader is referred to the online version of this chapter.)

the complex agrees with the nuclear spin relaxation data. The substrate remains quite dynamic when bound in the active site, with the exception of the recognition sequence (Masterson et al., 2010, 2011a,b). On the other hand, PLN binding shifts the PKA-C conformational equilibrium toward the closed state of the enzyme (Masterson et al., 2010, 2011a,b). The energy landscape for both PLN and PKA-C affects one another, leading to *mutual adaptation*. This recognition process exemplifies the recently theorized *mutual conformational selection model* reported by Nussinov and coworkers (Csermely et al., 2010; Fig. 9). During the encounter (i.e., recognition process), the two binding partners affect each other's energy landscape with a mutual adjustment in conformations. The binding process causes both proteins to reach a minimum in the free-energy landscape, allowing proper positioning of the chemical groups for catalysis to occur.

V. Conclusions and Future Perspectives

Our work highlights the prominent role of conformational dynamics in defining the energy landscape of PKA-C and its substrates. We found that local and long-range conformational dynamic changes contribute to the propagation of allosteric signals in this single protein enzyme. Both fast and slow dynamics (ps–ms) are modulated by ligand binding, emphasizing the necessity of high-frequency vibrations, rotations, and librations as well as low-frequency and large amplitude (i.e., breathing) motions for catalysis to occur. PKA-C internal motions increase upon nucleotide binding. Not only does ATP have an essential role for the chemistry occurring in the active site, but it also engages the two lobes of the enzyme structurally, completing the hydrophobic C-spine via its adenine ring. We propose that this event synchronizes the motions of small and large lobes, positioning the binding groove for substrate recognition, guiding it along the reaction coordinates. Taken with the kinetic measurements, the analysis of the enzyme and substrate structural dynamics within the Michaelis complex underscores the mutual dynamic adaptation of the two proteins upon binding. This mutual conformational selection mechanism is not surprising, given the large number of substrates targeted by PKA-C. Substrate and inhibitor binding have opposite effects. While substrate binding preserves the opening and closing of the active cleft, inhibitor binding traps the enzyme in a dynamically quenched state, where most of the motions are rather attenuated. This aspect opens up the possibility of tuning the structural dynamics of PKA-C via allosteric modulators, controlling enzyme function.

However, many aspects of allosteric propagation and signaling within PKA-C and other enzymes remain obscure. While most of the allosteric network can be clearly traced via contiguous residues (either in sequence or in space), there are many noncontiguous paths that are difficult to rationalize based on only physical contacts between atoms. Recent interpretations of allosteric signaling rely on statistical correlations of the chemical shifts to describe the propagation of allosteric signaling (Lipchock and Loria, 2010; Selvaratnam et al., 2011). Although state-of-the-art, these approaches still fall short for the complete description of allosteric phenomena. From a statistical thermodynamic point of view, allostery is a nonequilibrium relaxation process, where both entropy and enthalpy changes are concurrent and define the energetic states of a system (the protein and the environment) (Tsai et al., 2008; Vendruscolo, 2011). Indeed, the NMR analysis presented here is limited to the quantification of the conformational entropy of the protein backbone, while side chains dynamics and solvent effects remain uncharacterized. The combination of methyl group NMR spin relaxation methods with more sophisticated and longer time scale MD simulations promises to fill this apparent lacuna.

Acknowledgments

This work was supported by the NIH grants GM72701 and GM 100310 to G. V. and T32DE007288 to L. R. M. NMR data were collected at NMRFAM (NIH: P41RR02301, P41GM66326, RR02781, and RR08438; NSF: DMB-8415048, OIA-9977486, BIR-9214394) and the U. of Minnesota NMR Facility (NSF BIR-961477). Modeling and calculations were carried out at the Minnesota Supercomputing Institute.

References

Adams, J. A. (2001). Kinetic and catalytic mechanisms of protein kinases. *Chem. Rev.* **101**, 2271–2290.

Adams, J. A., Taylor, S. S. (1992). Energetic limits of phosphotransfer in the catalytic subunit of cAMP-dependent protein kinase as measured by viscosity experiments. *Biochemistry* **31**, 8516–8522.

Adams, J. A., Taylor, S. S. (1993). Divalent metal ions influence catalysis and active-site accessibility in the cAMP-dependent protein kinase. *Protein Sci.* **2**, 2177–2186.

Amadei, A., Linssen, A. B. M., Berendsen, H. J. C. (1993). Essential dynamics of proteins. *Proteins.* **17**, 412–425.

Baldwin, A. J., Kay, L. E. (2009). NMR spectroscopy brings invisible protein states into focus. *Nat. Chem. Biol.* **5**, 808–814.

Berlow, R. B., Igumenova, T. I., Loria, J. P. (2007). Value of a hydrogen bond in triosephosphate isomerase loop motion. *Biochemistry* **46**, 6001–6010.

Boehr, D. D., Dyson, H. J., Wright, P. E. (2006). An NMR perspective on enzyme dynamics. *Chem. Rev.* **106**, 3055–3079.

Boehr, D. D., Nussinov, R., Wright, P. E. (2009). The role of dynamic conformational ensembles in biomolecular recognition. *Nat. Chem. Biol.* **5**, 789–796.

Boeshans, K. M., Resing, K. A., Hunt, J. B., Ahn, N. G., Shabb, J. B. (1999). Structural characterization of the membrane-associated regulatory subunit of type I cAMP-dependent protein kinase by mass spectrometry: identification of Ser81 as the in vivo phosphorylation site of RI-alpha. *Protein Sci.* **8**, 1515–1522.

Cohen, P. (1999). The development and therapeutic potential of protein kinase inhibitors. *Curr. Opin. Chem. Biol.* **3**, 459–465.

Cohen, P. (2001). The role of protein phosphorylation in human health and disease. *Eur. J. Biochem.* **268**, 5001–5010.

Cohen, P. (2002). Protein kinases—the major drug targets of the twenty-first century? *Nat. Rev. Drug Discov.* **1**, 309–315.

Cook, P. F., Neville, M. E., Jr., Vrana, K. E., Hartl, F. T., Roskoski, R., Jr. (1982). Adenosine cyclic 3′,5′-monophosphate dependent protein kinase: kinetic mechanism for the bovine skeletal muscle catalytic subunit. *Biochemistry* **21**, 5794–5799.

Csermely, P., Palotai, R., Nussinov, R. (2010). Induced fit, conformational selection and independent dynamic segments: an extended view of binding events. *Trends Biochem. Sci.* **35**, 539–546.

Dalton, G. D., Dewey, W. L. (2006). Protein kinase inhibitor peptide (PKI): a family of endogenous neuropeptides that modulate neuronal cAMP-dependent protein kinase function. *Neuropeptides* **40**, 23–34.

Farrow, N. A., Muhandiram, R., Singer, A. U., Pascal, S. M., Kay, C. M., Gish, G., et al. (1994). Backbone dynamics of a free and phosphopeptide-complexed Src homology 2 domain studied by 15N NMR relaxation. *Biochemistry* **33**, 5984–6003.

Gaffarogullari, E. C., Masterson, L. R., Metcalfe, E. E., Traaseth, N. J., Balatri, E., Musa, M. M., et al. (2011). A myristoyl/phosphoserine switch controls cAMP-dependent protein kinase association to membranes. *J. Mol. Biol.* **411**, 823–836.

Gustavsson, M., Traaseth, N. J., Karim, C. B., Lockamy, E. L., Thomas, D. D., Veglia, G. (2011). Lipid-mediated folding/unfolding of phospholamban as a regulatory mechanism for the sarcoplasmic reticulum Ca(2+)-ATPase. *J. Mol. Biol.* **408**, 755–765.

Ha, K. N., Masterson, L. R., Hou, Z., Verardi, R., Walsh, N., Veglia, G., et al. (2011). Lethal Arg9Cys phospholamban mutation hinders Ca2+-ATPase regulation and phosphorylation by protein kinase A. *Proc. Natl. Acad. Sci. USA* **108**, 2735–2740.

Henzler-Wildman, K. A., Lei, M., Thai, V., Kerns, S. J., Karplus, M., Kern, D. (2007). A hierarchy of timescales in protein dynamics is linked to enzyme catalysis. *Nature* **450**, 913–916.

Herberg, F. W., Doyle, M. L., Cox, S., Taylor, S. S. (1999). Dissection of the nucleotide and metal-phosphate binding sites in cAMP-dependent protein kinase. *Biochemistry* **38**, 6352–6360.

Hyeon, C., Jennings, P. A., Adams, J. A., Onuchic, J. N. (2009). Ligand-induced global transitions in the catalytic domain of protein kinase A. *Proc. Natl. Acad. Sci. USA* **106**, 3023–3028.

Jarymowycz, V. A., Stone, M. J. (2006). Fast time scale dynamics of protein backbones: NMR relaxation methods, applications, and functional consequences. *Chem. Rev.* **106**, 1624–1671.

Johnson, D. A., Akamine, P., Radzio-Andzelm, E., Madhusudan, and Taylor S. S. (2001) Dynamics of cAMP-dependent protein kinase. *Chem. Rev.* 101, 2243-2270.

Kalodimos, C. G. (2011). NMR reveals novel mechanisms of protein activity regulation. *Protein Sci.* **20**, 773–782.

Karginov, A. V., Ding, F., Kota, P., Dokholyan, N. V., Hahn, K. M. (2010). Engineered allosteric activation of kinases in living cells. *Nat. Biotechnol.* **28**, 743–747.

Kern, D., Zuiderweg, E. R. (2003). The role of dynamics in allosteric regulation. *Curr. Opin. Struct. Biol.* **13**, 748–757.

Kong, C. T., Cook, P. F. (1988). Isotope partitioning in the adenosine 3′,5′-monophosphate dependent protein kinase reaction indicates a steady-state random kinetic mechanism. *Biochemistry* **27**, 4795–4799.

Kornev, A. P., Taylor, S. S. (2009). Defining the conserved internal architecture of a protein kinase. *Biochim. Biophys. Acta* **1804**, 440–444.

Kornev, A. P., Taylor, S. S. (2010). Protein Kinases: evolution of a dynamic regulatory protein. *Trends Biochem. Sci.* **36**, 65–77.

Kornev, A. P., Haste, N. M., Taylor, S. S., Ten Eyck, L. F. (2006). Surface comparison of active and inactive protein kinases identifies a conserved activation mechanism. *Proc. Natl. Acad. Sci. USA* **103**, 17783–17788.

Kornev, A. P., Taylor, S. S., Ten Eyck, L. F. (2008). A helix scaffold for the assembly of active protein kinases. *Proc. Natl. Acad. Sci. USA* **105**, 14377–14382.

Korzhnev, D. M., Religa, T. L., Banachewicz, W., Fersht, A. R., Kay, L. E. (2010). A transient and low-populated protein-folding intermediate at atomic resolution. *Science* **329**, 1312–1316.

Lew, J., Taylor, S. S., Adams, J. A. (1997). Identification of a partially rate-determining step in the catalytic mechanism of cAMP-dependent protein kinase: a transient kinetic study using stopped-flow fluorescence spectroscopy. *Biochemistry* **36**, 6717–6724.

Li, F., Juliano, C., Gorfain, E., Taylor, S. S., Johnson, D. A. (2002). Evidence for an internal entropy contribution to phosphoryl transfer: a study of domain closure, backbone flexibility, and the catalytic cycle of cAMP-dependent protein kinase. *J. Mol. Biol.* **315**, 459–469.

Lipchock, J. M., Loria, J. P. (2010). Nanometer propagation of millisecond motions in V-type allostery. *Structure* **18**, 1596–1607.

Loria, J. P., Berlow, R. B., Watt, E. D. (2008). Characterization of enzyme motions by solution NMR relaxation dispersion. *Acc. Chem. Res.* **41**, 214–221.

Maragakis, P., Karplus, M. (2005). Large amplitude conformational change in proteins explored with a plastic network model: adenylate kinase. *J. Mol. Biol.* **352**, 807–822.

Massi, F., Wang, C., Palmer, A. G., 3rd (2006). Solution NMR and computer simulation studies of active site loop motion in triosephosphate isomerase. *Biochemistry* **45**, 10787–10794.

Masterson, L. R., Mascioni, A., Traaseth, N. J., Taylor, S. S., Veglia, G. (2008). Allosteric cooperativity in protein kinase A. *Proc. Natl. Acad. Sci. USA* **105**, 506–511.

Masterson, L. R., Shi, L., Tonelli, M., Mascioni, A., Mueller, M. M., Veglia, G. (2009). Backbone NMR resonance assignment of the catalytic subunit of cAMP-dependent protein kinase A in complex with AMP-PNP. *Biomol. NMR Assign.* **3**, 115–117.

Masterson, L. R., Cheng, C., Yu, T., Tonelli, M., Kornev, A. P., Taylor, S. S., et al. (2010). Dynamics connect substrate recognition to catalysis in protein kinase A. *Nat. Chem. Biol.* **6**, 821–828.

Masterson, L. R., Shi, L., Metcalfe, E., Gao, J., Taylor, S. S., Veglia, G. (2011a). Dynamically committed, uncommitted, and quenched states encoded in protein kinase A revealed by NMR spectroscopy. *Proc. Natl. Acad. Sci. USA* **108**, 6969–6974.

Masterson, L. R., Yu, T., Shi, L., Wang, Y., Gustavsson, M., Mueller, M. M., et al. (2011b). cAMP-dependent protein kinase A selects the excited state of the membrane substrate phospholamban. *J. Mol. Biol.* **412**, 155–164.

Mittag, T., Kay, L. E., Forman-Kay, J. D. (2010). Protein dynamics and conformational disorder in molecular recognition. *J. Mol. Recognit.* **23**, 105–116.

Miyashita, O., Onuchic, J. N., Wolynes, P. G. (2003). Nonlinear elasticity, proteinquakes, and the energy landscapes of functional transitions in proteins. *Proc. Natl. Acad. Sci. USA* **100**, 12570–12575.

Moore, M. J., Adams, J. A., Taylor, S. S. (2003). Structural basis for peptide binding in protein kinase A. Role of glutamic acid 203 and tyrosine 204 in the peptide-positioning loop. *J. Biol. Chem.* **278**, 10613–10618.

Nashine, V. C., Hammes-Schiffer, S., Benkovic, S. J. (2010). Coupled motions in enzyme catalysis. *Curr. Opin. Chem. Biol.* **14**, 644–651.

Palmer, A., 3rd (2001). Nmr probes of molecular dynamics: overview and comparison with other techniques. *Annu. Rev. Biophys. Biomol. Struct.* **30**, 129–155.

Popovych, N., Sun, S., Ebright, R. H., Kalodimos, C. G. (2006). Dynamically driven protein allostery. *Nat. Struct. Mol. Biol.* **13**, 831–838.

Selvaratnam, R., Chowdhury, S., VanSchouwen, B., Melacini, G. (2011). Mapping allostery through the covariance analysis of NMR chemical shifts. *Proc. Natl. Acad. Sci. USA* 108, 6133–6138

Shabb, J. B. (2001). Physiological substrates of cAMP-dependent protein kinase. *Chem. Rev.* **101**, 2381–2411.

Slice, L. W., Taylor, S. S. (1989). Expression of the catalytic subunit of cAMP-dependent protein kinase in Escherichia coli. *J. Biol. Chem.* **264**, 20940–20946.

Taylor, S. S., Yang, J., Wu, J., Haste, N. M., Radzio-Andzelm, E., Anand, G. (2004). PKA: a portrait of protein kinase dynamics. *Biochim. Biophys. Acta* **1697**, 259–269.

Tonelli, M., Masterson, L. R., Hallenga, K., Veglia, G., Markley, J. L. (2007). Carbonyl carbon label selective (CCLS) (1)H-(15)N HSQC experiment for improved detection of backbone (13)C-(15)N cross peaks in larger proteins. *J. Biomol. NMR* **39**, 177–185.

Traaseth, N. J., Ha, K. N., Verardi, R., Shi, L., Buffy, J. J., Masterson, L. R., et al. (2008). Structural and dynamic basis of phospholamban and sarcolipin inhibition of Ca (2+)-ATPase. *Biochemistry* **47**, 3–13.

Tsai, C. J., del Sol, A., Nussinov, R. (2008). Allostery: absence of a change in shape does not imply that allostery is not at play. *J. Mol. Biol.* **378**, 1–11.

Tzeng, S. R., Kalodimos, C. G. (2009). Dynamic activation of an allosteric regulatory protein. *Nature* **462**, 368–372.

Vendruscolo, M. (2011). Protein regulation: the statistical theory of allostery. *Nat. Chem. Biol.* **7**, 411–412.

Villali, J., Kern, D. (2010). Choreographing an enzyme's dance. *Curr. Opin. Chem. Biol.* **14**, 636–643.

Walsh, D. A., Perkins, J. P., Krebs, E. G. (1968). An adenonsine 3′,5′-monophosphate-dependant protein kinase from rabbit skeletal muscle. *J. Biol. Chem.* **243**, 3763–3765.

Walsh, D. A., Van Patten, S. M. (1994). Multiple pathway signal transduction by the cAMP-dependent protein kinase. *FASEB J.* **8**, 1227–1236.

Wong, W., Scott, J. D. (2004). AKAP signalling complexes: focal points in space and time. *Nat. Rev. Mol. Cell Biol.* **5**, 959–970.

Yang, J., Ten Eyck, L. F., Xuong, N. H., Taylor, S. S. (2004). Crystal structure of a cAMP-dependent protein kinase mutant at 1.26A: new insights into the catalytic mechanism. *J. Mol. Biol.* **336**, 473–487.

Yang, J., Garrod, S. M., Deal, M. S., Anand, G. S., Woods, J., Virgil, L., et al. (2005). Allosteric network of cAMP-dependent protein kinase revealed by mutation of Tyr204 in the P+1 loop. *J. Mol. Biol.* **346**, 191–201.

Yang, J., Kennedy, E. J., Wu, J., Deal, M. S., Pennypacker, J., Ghosh, G., et al. (2009). Contribution of non-catalytic core residues to activity and regulation in protein kinase A. *J. Biol. Chem.* **284**, 6241–6248.

Zheng, J., Knighton, D. R., ten Eyck, L. F., Karlsson, R., Xuong, N., Taylor, S. S., et al. (1993). Crystal structure of the catalytic subunit of cAMP-dependent protein kinase complexed with MgATP and peptide inhibitor. *Biochemistry* **32**, 2154–2161.

Zimmermann, B., Schweinsberg, S., Drewianka, S., Herberg, F. W. (2008). Effect of metal ions on high-affinity binding of pseudosubstrate inhibitors to PKA. *Biochem. J.* **413**, 93–101.

USING A COMBINATION OF COMPUTATIONAL AND EXPERIMENTAL TECHNIQUES TO UNDERSTAND THE MOLECULAR BASIS FOR PROTEIN ALLOSTERY

By WANTING JIAO AND EMILY J. PARKER

Biomolecular Interaction Centre, Department of Chemistry, University of Canterbury, Christchurch, New Zealand

Abstract

Allostery is the process by which remote sites of a system are energetically coupled to elicit a functional response. The early models of allostery such as the Monod–Wyman–Changeux model and the Koshland–Némethy–Filmer model explain the allosteric behavior of multimeric proteins. However, these models do not explain how allostery arises from atomic level in detail. Recent developments in computational methods and experimental techniques have led the beginning of a new age in studying allostery. The combination of computational methods and experiments is a powerful research approach to help answering questions regarding allosteric mechanism at atomic resolution. In this review, three case studies are discussed to illustrate how this combined approach helps to increase our understanding of protein allostery.

I. Introduction

Allostery is the process by which remote sites of a system are energetically coupled to elicit a functional response. This mechanism of regulation is widely observed in biological processes such as signal transduction, metabolism, catalysis, and gene regulation (Changeux and Edelstein, 2005; Kuriyan

ADVANCES IN PROTEIN CHEMISTRY AND STRUCTURAL BIOLOGY, Vol. 87
DOI: 10.1016/B978-0-12-398312-1.00013-5

Copyright 2012, Elsevier Inc.
All rights reserved.

and Eisenberg, 2007; Goodey and Benkovic, 2008; Boehr et al., 2009; Smock and Gierasch, 2009). Allostery is associated with cooperativity of substrate binding and can be either homotropic (when the substrate acts as allosteric effector) or heterotropic (when the allosteric effector is not the substrate of the enzyme).

The early models of allostery include the Monod–Wyman–Changeux (MWC) model (also known as the concerted model) (Monod et al., 1965) and the Koshland–Némethy–Filmer (KNF) model (also known as the sequential model) (Koshland et al., 1966). Both models account for allosteric behavior of oligomeric proteins. In the MWC model, two interconvertible states of the oligomeric protein (referred to as the T and R states) exist in thermal equilibrium with all the subunits in one oligomeric protein adopting in the same state. Allostery is the result of allosteric effectors binding which causes a shift of this thermal equilibrium. The KNF model differs from the MWC model in that subunits within one oligomeric protein can adopt different states. In this sequential model, the binding of the first allosteric effector causes local conformational changes in one subunit, which in turn induces conformational changes of the nearby subunits. This results in changed affinity for the allosteric effector in those other binding sites, and cooperativity of binding can be observed. In the past, the investigation of allosteric regulation in proteins was mainly driven by experimental techniques, including X-ray crystallography to observe structural changes upon binding of allosteric effectors and kinetic measurements of substrate binding to observe cooperativity. Accompanied with experimental observations, the MWC model and KNF model have been applied to many biological systems and processes to explain the allosteric behavior of multimeric proteins, for example, hemoglobin, aspartate transcarbamoylase, membrane receptors, and signal transduction (Monod et al., 1965; Changeux and Edelstein, 1998, 2005; Eaton et al., 1999; Bai et al., 2010). However, both the MWC model and the KNF model are phenomenological (Cui and Karplus, 2008), although these models can be used to explain experimental observations, they do not explain how the signal of allosteric effector binding can affect the function of the enzyme at atomic level.

Recent developments in computational methods and experimental techniques have led the beginning of a new age in studying allostery. Monomeric proteins have been shown to be able to exhibit allosteric regulation (Ascenzi et al., 2005; Ascenzi and Fasano, 2010), and allosteric mechanisms associated with far more subtle changes in protein conformational states have been described (Popovych et al., 2006). The role of protein dynamics in determining function is becoming more recognized, and there are several reviews in the literature that discuss the importance of dynamics (Kern and Zuiderweg, 2003; Gunasekaran et al., 2004; Swain and Gierasch, 2006; Bahar et al., 2007; Cui and Karplus, 2008; Tsai et al., 2008, 2009; Peracchi and Mozzarelli, 2011; Tzeng and Kalodimos, 2011).

The combination of computational methods and experiments is a powerful research approach with elucidating the precise mechanisms allosteric control at the atomic resolution. To illustrate this, we will consider three case studies where computational and experimental techniques used together help increase the understanding of protein allostery.

II. Case Studies

A. *Allosteric Inhibition of Glutamate Racemase*

The enzyme glutamate racemase (GR) catalyzes the interconversion of L-glutamate (L-glu) and D-glutamate (D-glu); the nonproteogenic amino acid D-glu is a key building block for peptidoglycan which is an essential component for the bacterial cell wall (Doublet et al., 1993; Gallo and Knowles, 1993; Gallo et al., 1993; Tanner et al., 1993). GR therefore has been considered as a target for antibacterial drug discovery and has been the subject of a number of studies (Glavas and Tanner, 1997; de Dios et al., 2002; Lundqvist et al., 2007; May et al., 2007; Spies et al., 2009; Whalen et al., 2009). While most of these studies focused on developing competitive substrate-based inhibitors for GR, Lundqvist et al. successfully discovered an uncompetitive allosteric inhibitor specifically targeting GR from *Helicobactor pylori* (Lundqvist et al., 2007); however, the natural inhibitor molecule that binds in the same allosteric pocket is unknown. The case study discussed below concerns the allosteric inhibition of GR from *Bacillus anthracis*, which is the causative agent of inhalational anthrax. *B. anthracis* is unique in possessing two functional GR isozymes (RacE1 and RacE2) (Dodd et al., 2007). Both isozymes have been characterized, and both can catalyze the reaction that converts L-glu to D-glu (Dodd et al., 2007; May et al., 2007). RacE1 is monomeric in the absence of the substrate glutamate but dimerizes in its presence; however, RacE2 is a dimer in its native state with or without the presence of glutamate (May et al., 2007). *B. anthracis* genetic knockout studies showed that the absence of RacE2 resulted in the more severe defect in bacterial growth than the RacE1 isozyme, indicating that RacE2 isozyme is a more promising drug target than RacE1 (Shatalin and Neyfakh, 2005).

Dipicolinic acid (DPA) is a major component of *Bacillus* spores and was first found to exhibit low millimolar inhibition for GR from *Bacillus subtilis* (Spies et al., 2009). Recently, Whalen et al. tested DPA on the two GR isozymes from *B. anthracis* and showed that it has low micromolar inhibition for GR activity in the D-glu→L-glu direction, with clear uncompetitive behavior, suggesting DPA may be acting as an allosteric inhibitor of GR (Whalen et al., 2011). In order to fully understand the binding mode and mechanism of inhibition by DPA of GR from *B. anthracis*, Whalen et al. employed an approach that combines experiments (site-directed mutagenesis, circular dichroism, and electrophoresis) and computational calculations (ligand locking, molecular dynamics, and free-energy and pK_a calculations).

Since no crystal structure has been obtained for GR in complex with the inhibitor DPA, the first step in understanding the binding mode of DPA was to identify its binding site in GR. To this end, a series of blind-docking calculations targeting the RacE2 dimer were carried out using three different docking programs (GOLD v4.1, FRED v2.2.5, and AutoDock 4) with different types of docking and scoring methods (FRED, 2010; GOLD, 2010; YASARA, 2010). All three docking programs predicted the same binding site for DPA, located at the dimer interface of the RacE2 dimer, away from the enzyme active site (Fig. 1A). The docking poses predict that DPA makes key hydrogen-bonding contacts with Lys106 and the backbone of Ser207 from one monomer and the Lys106 from the second monomer (Fig. 1B). Energy minimization on the docking pose of DPA showed that the interactions with Lys106 remain intact, but the hydrogen bond with Ser207 backbone is switched to a hydrogen bond with Ser207 β-hydroxyl group (Fig. 1C). The role of the residue Ser207 was further probed by a molecular dynamics simulation using the Ser207Ala mutant of RacE2 dimer with DPA bound. As a result of the mutation, the MD simulation predicted DPA in a manner that is highly solvent exposed, due to a complete loss of contact with the whole region surrounding residue Ser207 (Fig. 1D). This indicates that Ser207 is critical for the formation of the DPA-binding pocket.

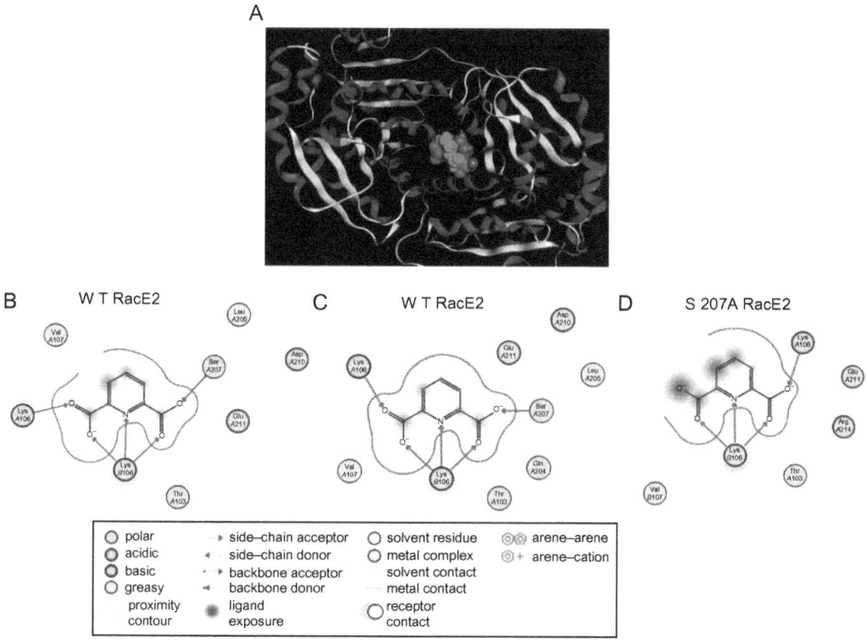

FIG. 1. (A) Superpose of top-docked positions of DPA (space-filling) to the RacE2 dimer (ribbon, 2GZM) as determined by GOLD v4.1 (magenta), Autodock v4 (blue), and FRED v2.2.5 (green). (B) The binding pocket is located at the dimer interface and is composed of residues from both monomers, as detailed by the interaction map. (C) After minimization, the backbone contact of Ser207 is swapped for a contact with the β-hydroxyl group. (D) After MD simulation of the top docked complex with Ser207 replaced by Ala, the binding site lacks any contact with the region previously containing Ser207. Letters immediately preceding the residue numbers indicate the monomer, A or B. Ligand interaction maps were constructed using the LigX function of MOE v2009.10. Figure taken from original paper (Whalen et al., 2011). (For interpretation of the references to color in this figure legend, the reader is referred to the online version of this chapter.)

The allosteric binding site of DPA predicted by these docking calculations was then validated by site-directed mutagenesis and inhibition assay experiments. Two mutants were made to test the role of the residues that were predicted to be important in the docking calculations, Lys106Ala and Ser207Ala, and the activity and inhibition of these two mutants were then assayed. Experimental results showed that the mutations had no effect on enzyme activity, but the K_i values observed for DPA binding to the mutants were shown to increase by 24- and 13-fold, respectively, thus confirming the allosteric binding site for DPA in GR.

The second step in understanding the binding mode of DPA in GR was to investigate the mechanism of allosteric inhibition. Since DPA binds in the dimer interface of the RacE2 dimer, it is possible that the DPA-binding event may shift the oligomeric equilibrium by disrupting the dimerization of RacE2 and thereby exerting an inhibitory effect. Whalen et al. ruled out this possibility by analyzing the oligomeric equilibrium of RacE2 in the presence of DPA, using blue native polyacrylamide gel electrophoresis supplemented with DPA. Results showed that DPA had no effect on the monomer–dimer equilibrium of RacE2. The effect of DPA binding at the allosteric site on the binding of D-glu in the active site was investigated by performing two MD simulations (20 ns each) for RacE2 dimer with D-glu bound to both active sites in the presence (abbreviated as E_2·D-glu$_2$·DPA) and absence of DPA (abbreviated as E_2·D-glu$_2$) in the allosteric site. The trajectories of the MD simulations were analyzed by QR factorization (O'Donoghue and Luthey-Schulten, 2005; Amaro et al., 2008), and the unique structures explored during the simulation (those structures with structural homology value of below 0.90) were selected from both systems. The QR factorization resulted in 14 unique structures from E_2·D-glu$_2$ and only two unique structures from E_2·D-glu$_2$·DPA, suggesting that the binding of DPA resulted in a limitation of the conformational diversity for RacE2. The binding free energy was estimated for D-glu in both monomers of RacE2 in both systems from MD simulations using an "end point" approach (Steinbrecher and Labahn, 2010) with the method MM-BEMSA (Warwicker and Watson, 1982; Shaw, 1985). The results indicated that DPA did not cause a significant difference in the binding energy of D-glu in monomer B (free-energy difference of -5.0 ± 7.0 kcal/mol), but a less negative binding energy for D-glu in monomer A is observed (free-energy difference of -6.0 ± 3.0 kcal/mol). Analysis of the final structures from MD simulations of the two systems showed different interactions between the enzyme active site and the bound D-glu, and different active site conformations in the absence and presence of DPA. However, the differences in binding energy could not fully explain the loss of catalytic activity by the enzyme. Since the basicity of residue Cys74 in the active site is quite important for the reaction converting D-glu to L-glu (Glavas and Tanner, 2001; Puig et al., 2007), Whalen et al. then sought to utilize this fact to explain the change in catalytic activity of RacE2 upon DPA binding by examining the changes in pK_a of the key catalytic residue, Cys74, the basicity of which is considered important for catalyzing the reaction. The MEAD algorithm was used to calculate pK_a values of Cys74 in each of the selected unique structures obtained from QR factorization (Gordon et al., 2005; Anandakrishnan and Onufriev, 2008). The comparison of the pK_a values of the two systems (E_2·D-glu_2·DPA and E_2·D-glu_2) illustrated that DPA binding was associated with a decrease in pK_a value of Cys74, and the limitation in the range of pK_a values possessed by Cys74, due to rearrangements in the active site conformation. This suggests that the loss of high basicity for the catalytic Cys74 caused by allosteric DPA-induced conformational change at the active site may be the mechanism of allosteric inhibition of RacE2.

B. *Allsoteric Signaling in Protein Kinase A*

Protein kinase is a family of enzymes that catalyzes the transfer of a phosphate group from ATP to a protein or peptide substrate. This phosphorylation usually results in a functional change of the substrate and is among the most common mechanisms of cell signaling in both eukaryotes and prokaryotes (Walsh and Van Patten, 1994; Smith et al., 1999; Shabb, 2001). Protein kinase A (PKA), also known as cAMP-dependent protein kinase, was one of the first protein kinases to be discovered (Walsh et al., 1968). PKA is a heterotetramer containing two catalytic subunits (PKA-C) and a dimer of regulatory subunits (R). The tetrameric PKA is inactive, but upon β-adrenergic stimulation, the heterotetrameric assembly is disrupted and the catalytic subunits PKA-C are released (Walsh and Van Patten, 1994). The protein kinase inhibitor (PKI) controls PKA activity by binding competitively to the substrate binding site (Dalton and Dewey, 2006). Magnesium ions also regulate the function of PKA; at low Mg^{2+} concentrations, PKA-C binds one primary Mg^{2+} ion which is required for activity, but at high Mg^{2+} concentration, PKA-C binds a further secondary Mg^{2+} ion which inhibits the enzyme noncompetitively (Cook et al., 1982; Zimmermann et al., 2008). PKA-C, the catalytic subunit of PKA, adopts a bean-shaped fold which is highly conserved throughout the protein kinase family (Taylor et al., 2004). Two lobes (small and large) flank the active site of PKA-C (Fig. 2A) and interconvert between different conformational states along various stages of the catalytic cycle (Kornev and Taylor, 2010). Three major conformational states have been identified by X-ray

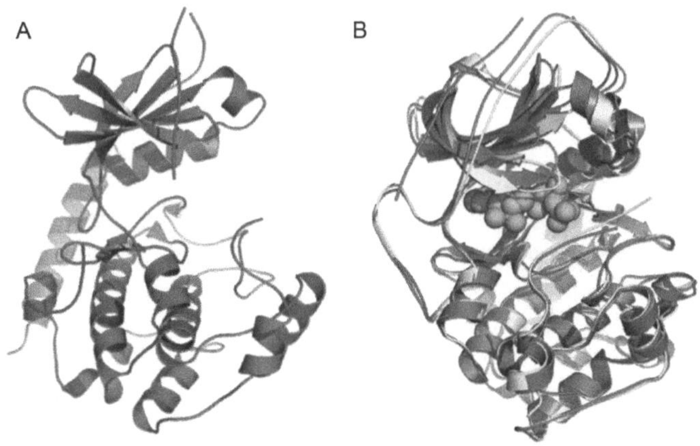

Fig. 2. (A) Crystal structure of PKA-C in the open state (PDB code 1CTP). (B) Superposition of crystal structures of PKA-C in its open (red, PDB code 1CTP), intermediate (blue, PDB code 1BKX), and closed (yellow, PDB code 1ATP) states. The ATP molecule that binds in the active site of PKA-C in the closed state is displayed as spheres in green. (For interpretation of the references to color in this figure legend, the reader is referred to the online version of this chapter.)

crystallography (Fig. 2B): an open (apo form), an intermediate (binary form, nucleotide bound), and a closed (ternary form, nucleotide/inhibitor bound) (Johnson et al., 2001). NMR dynamic measurements, X-ray crystallography, and thermodynamic measurements have been used to study these major conformational changes along the reaction coordinates, and it was found that both nucleotide and substrate bind with positive cooperativity, and that the binding of the nucleotide activates the enzyme dynamically and causes an allosteric effect on the distant regions by coupling the two lobes of apo PKA-C, inducing microsecond–millisecond conformational fluctuations and priming the enzyme for catalysis. These fluctuations are synchronous with the enzyme rate-limiting step (product release step), indicating the prominent role of conformational dynamics in substrate recognition and catalysis (Masterson et al., 2008, 2010).

In order to investigate the allosteric signaling of nucleotide and inhibitor binding on the conformational dynamics of PKA-C, Masterson et al. utilized the power of MD simulations in combination with the dynamic measurements obtained from NMR spectroscopy to define the changes in energy landscape of PKA-C upon ligand binding (Masterson et al., 2011).

First, the thermodynamics of binding and enzyme stability upon substrate and inhibitor binding was investigated using isothermal titration calorimetry (ITC) and thermostability experiments. The substrate peptide used in these experiments corresponds to the cytoplasmic domain of phospholamban (PLN_{1-20}), which is the native substrate of PKA-C in cardiac muscle (Traaseth et al., 2008). The inhibitor peptide used corresponds to the competitive inhibitor PKI_{5-24}. Results from these experiments indicated that substrate (PLN_{1-20}) binding is entropically driven and does not affect the enzyme stability, whereas inhibitor (PKI_{5-24}) binding is enthalpically driven and significantly enhances the thermostability of the enzyme. Chemical shift perturbations from NMR spectroscopy (TROSY) were used to map the conformational transition of PKA-C from open to closed states upon first binding of nucleotide, followed by either substrate or inhibitor binding. The results suggested that the initial nucleotide binding shifts the conformational ensemble from the open to the intermediate state, and subsequent binding of the substrate or inhibitor can both shift the population toward the closed states. The differences in chemical shifts between the two ternary complexes (PKA-C with nucleotide and either the inhibitor or the substrate bound) were found to be minimal. This observation combined with those obtained from thermostability measurements suggested that the effects of inhibitor and substrate binding differ in the internal dynamics of the enzyme (as reflected by thermostability measurements), but not significantly in conformation.

To complement the experimental observations from NMR spectroscopy, MD simulations were set up for PKA-C in water starting from the apo form, binary form (with ATP bound), and ternary form (ATP bound and with either substrate or inhibitor bound), in each form, the enzyme was modeled with either one or two Mg^{2+} ions bound to represent low and high Mg^{2+} ion concentrations, respectively. A total of seven MD simulations (75 ns each) were conducted in this study. MD simulations can produce a series of snapshots of the system along the simulation time, and from these snapshots, information regarding the internal dynamics of the enzyme can be extracted by applying different analytical methods. Principal component analysis (PCA) is one such method (Karplus and Kushick, 1981). It can be used to analyze motions of the enzyme during MD simulations, to produce a number of principal components of enzyme motion with different directions and amplitudes. Usually, the first few principal components can account for most of the motions explored by the enzyme system (Miyashita et al., 2003; Maragakis and Karplus, 2005; Cheng et al., 2007; Henzler-Wildman et al., 2007). In order to monitor the conformational transition from open to closed states of PKA-C upon ligand binding, Masterson et al. calculated the PCA for PKA-C in complex with the substrate PLN_{1-20}, and identified the top component of the observed enzyme motion (PC1) which represents opening and closing of the two lobes of PKA-C.

This mode of motion in PC1 (Fig. 3A) was further probed by monitoring the atomic distances between two residues, Ser53 and Gly186, which provided a measure of the openness of the active site. The distance was plotted against PC1, and the resultant scatter plot represented the distribution of conformational ensemble (in terms of opening and closing of the active site) explored by the enzyme during MD simulation (Fig. 3B). The distances obtained from crystal structures of open, intermediate, and closed forms of the enzyme were also plotted as a point of reference.

Combining the observations from previous experimental results (Masterson et al., 2010) and from the distribution of conformational ensembles sampled by the enzyme in different forms from the MD calculations, Masterson et al. attempted to explain the effect of substrate and inhibitor binding in terms of the energy landscape. The scatter plot indicates that when PKA-C is in its apo form, it can sample a broad range of conformations, including all three major conformational states identified from X-ray crystallography. This suggests that the apoenzyme is characterized by a dynamically uncommitted state, although it can explore a range of conformations on the energy landscape, these conformations do not interconvert on a timescale relevant for catalysis (Masterson et al., 2010). In the binary form with one Mg^{2+} ion bound, the enzyme is in dynamically committed states. The sampled conformations show similar distributions to open and intermediate conformations, and these conformations interconvert on a timescale optimal for catalysis (Masterson et al., 2010). In the ternary form with substrate (PLN_{1-20}) bound and one Mg^{2+} ion, the enzyme shows similar distributions along PC1 to the binary form; however, the active site conformation is shifted toward the closed state (as shown by distributions of Ser53-Gly186 distances). In this form, the enzyme retained sufficient conformational motions, and the enzyme was maintained at dynamically committed state. In contrast to substrate binding, the ternary form of the enzyme with inhibitor bound (PKI_{5-24}) showed that the distribution along PC1 is shifted toward open and

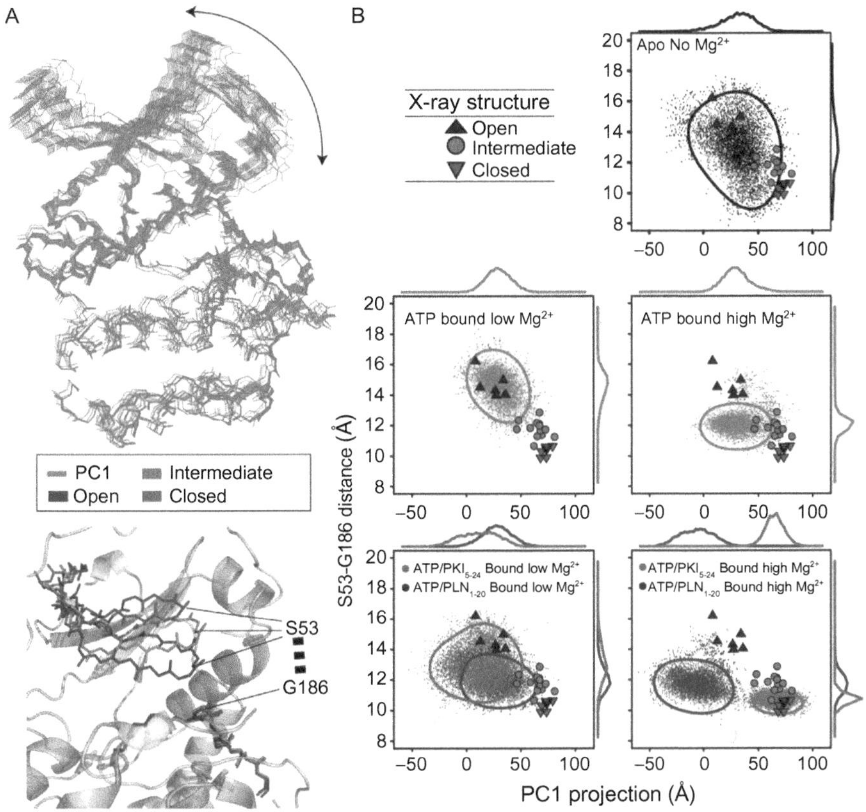

Fig. 3. Comparison of MD simulations for PKA-C. (A) Global motions suggested by PCA analysis of MD trajectories correspond to opening and closing of the active site (PC1), which compared well with the distances between residues S53 and G186 in crystal structures of open (1CMK), intermediate (1BX6), and closed (1ATP) conformations. (B) A map of the interatomic distances versus the PC1 from MD simulations indicates that PKA-C accessed the major crystallographic conformations frequently, except in the presence of inhibitors. Figure taken from original paper (Masterson et al., 2011). (For color version of this figure, the reader is referred to the online version of this chapter.)

intermediate states, but the distribution of the distance between Ser53 and Gly186 is not completely open (Fig. 3). Under high Mg^{2+} concentrations, the distribution of conformations sampled by the enzyme becomes more discrete, suggesting more localized minima exist on the energy surface. Interestingly, the conformational ensemble sampled for the ternary complex with inhibitor bound under high Mg^{2+} concentrations overlaps with the closed states observed from crystal structures. Upon binding of two Mg^{2+} ions and the inhibitor PKI_{5-24}, the enzyme is shifted to dynamically quenched states, in which the energy barriers between different conformations are raised resulting in deep, narrow energy minima, which hinder the interconversion between different conformational states.

This case study illustrates a successful example in examining protein allostery by the combination of molecular dynamics simulations and the experimental technique of NMR spectroscopy. The overall analysis of the behavior of this protein is strengthened by the ability of both these techniques to access the dynamic information of the protein system.

C. *Conformational Transition in the Allosteric Process of Human Glucokinase*

Glucokinase (GK) belongs to the hexokinase family (Grossbard and Schimke, 1966). It catalyzes the phosphorylation of a glucose molecule to produce glucose 6-phosphate and plays an important role in glucose utilization and metabolism in the liver and pancreas (Al-Hasani et al., 2003). GK expressed in the liver and pancreas is an important glucose sensor and helps in reducing plasma glucose levels in two ways: glucose-mediated activation of the enzyme in pancreatic β cells ultimately stimulates insulin secretion, while that in the liver enhances hepatic glucose uptake and glycogen synthesis (Van Schaftingen, 1994; Matschinsky et al., 1998). Both increased and decreased activity of GK variants are associated with diseases such as diabetes of the young type 2 (MODY-2) and persistent hyperinsulinemic hypoglycemia of infancy (PHHI), making this enzyme an important target for drug design (Vionnet et al., 1992; Froguel et al., 1993; Glaser et al., 1998; Christesen et al., 2002). Unlike other members of the hexokinase family, GK binds glucose with low affinity and exhibits positive cooperativity with regard to glucose binding and these two properties were considered to be critical for its role as glucose sensor in pancreas and liver (Storer and Cornish-Bowden, 1976; Gloyn et al., 2005). GK is a monomer with two domains (small and large) separated by a deep cleft which constitutes the active site. Crystal structures of human GK in both the active (closed) and inactive (superopen) states have been obtained which reveal large conformational changes of GK from the active to inactive states (Kamata et al., 2004; Fig. 4).

A kinetic model for GK activity based on existing experimental results and structural information was proposed by Kamata et al. (2004) (Fig. 5).

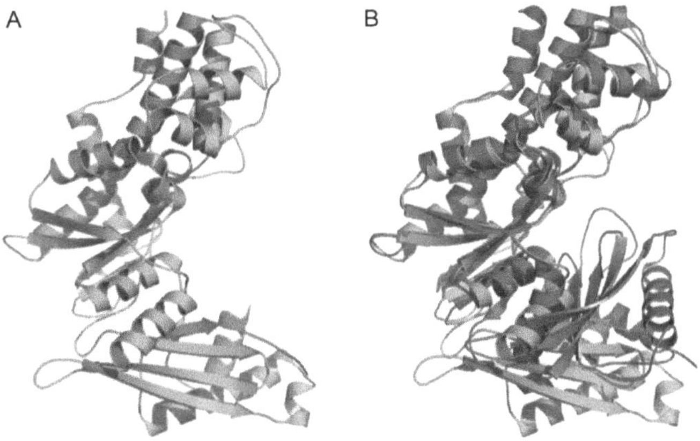

FIG. 4. (A) Crystal structure of GK in its superopen state (PDB code 1V4T). (B) Superposition of crystal structures of GK in the closed (red, PDB code 1V4S) and superopen (green, PDB code 1V4T) states. (For interpretation of the references to color in this figure legend, the reader is referred to the online version of this chapter.)

The proposed kinetic model is consistent with the mnemonical mechanism for positive cooperativity in a monomeric enzyme (Storer and Cornish-Bowden, 1977). In this model, GK has two catalytic cycles: a "slow cycle" which involves conformational transition between closed and superopen states and a "fast cycle" which involves conformational interconversion between the closed and open states. The ratio of the two cycles is determined by the concentration of glucose. When glucose concentrations are low, the slow cycle is dominant because the superopen form is more stable than other forms in the absence of glucose. When glucose concentrations are sufficiently high, GK uses the fast cycle with its high-affinity structure (Kamata et al., 2004).

There are three conformational states involved in the proposed kinetic model for GK activity, and structural information has been determined for two states (closed and superopen). Due to difficulty in capturing intermediate states during the conformational transitions of GK, the structural information of GK in the intermediate "open" conformation has not been obtained by experimental techniques to date. The crystal structure of GK in complex with an activator molecule, named "compound A" was also obtained, and the structure showed that compound A binds in an allosteric site on the back of the active site cleft which is 20 Å away from

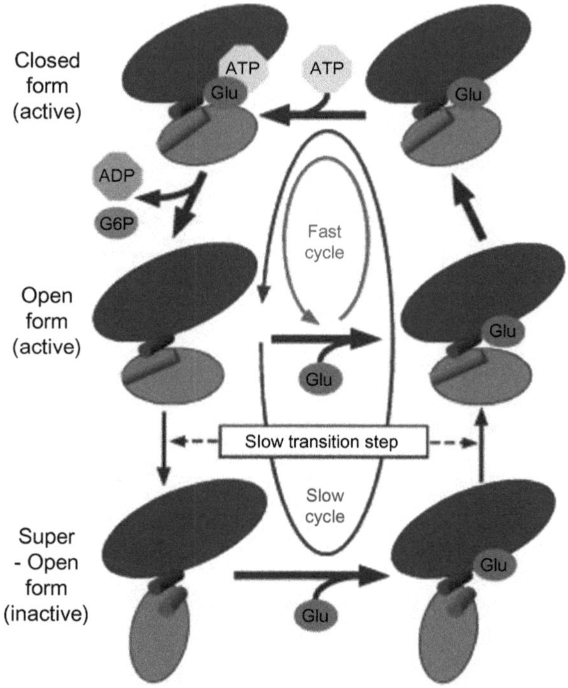

FIG. 5. Kinetic model of glucokinase (GK). GK appears to exist in three conformations and to have two catalytic cycles. The ratio of two catalytic cycles is responsible for the sigmoidal response to glucose. Although ordered process in the slow cycle of GK was indicated by isotope-exchange experiment, the addition of substrates in the fast cycle may be random. Figure taken from original paper (Kamata et al., 2004). (For color version of this figure, the reader is referred to the online version of this chapter.)

the active site (Kamata et al., 2004); however, the underlying molecular basis for the enhanced GK activity in the presence of compound A is unclear based purely on crystal structure.

To address questions regarding the pathway involved in conformational transitions of human GK, Zhang et al. employed a series of conventional MD (CMD) simulations and targeted molecular dynamics (TMD) on both wild type and mutants of human GK (Zhang et al., 2006).

In order to understand the mechanism of activation of GK by binding of the allosteric activator compound A, two CMD simulations were conducted (10 ns each) for GK at the closed state with and without compound

A bound. During the simulations, the opening and closing motions of the enzyme were monitored by measuring the angles between the two domains that make up the active site cleft. Results showed that in the activator-free closed state, GK is intrinsically dynamic with the open and close motions between the two domains and the active site cleft can open up to 10°. However, this motion was not observed in the activator-bound GK complex. PCA on the trajectory of this simulation showed that the opening and closing motions of the active site were quenched upon binding of the activator, suggesting that GK was locked into the catalytically active closed conformation. These two MD simulations indicate that the activator binding in the distant allosteric site can activate GK via conformational restriction.

Zhang et al. utilized TMD simulations to investigate the conformational transition between closed and superopen states of GK. In TMD simulations, the structures of the beginning (closed state) and end (superopen state) points of the simulation are known, and the process of transition is driven by the application of an external force. In order to find a reliable transition pathway between the two conformational states, Zhang et al. conducted a total of nine TMD simulations, each starting with different initial velocities and force constants. It was found that the different initial velocities and force constants used in each TMD simulations did not result in significantly different transition pathways, indicating the pathway identified was reliable (Fig. 6). Zhang et al. divided the pathway into three stages (0–360, 360–700, and 700–1000 ps). The transition pathway starts with a slow opening of the active site up to 9° during the first 160 ps of the simulation. This slow opening of the active site was thought to due to the energy barrier caused by strong interactions between loop I in the small domain and β1 and β2 in the large domain (Fig. 6). When the continued application of external force in the simulation finally overcame the energy barrier by breaking these strong interactions, the opening process of the active site began to speed up.

During the second stage of the pathway, the active site cleft undergoes rapid opening, while strands β6 and β7 gradually become more solvent exposed. In the third stage, the active site opens up further to resemble the superopen conformation, and this is accompanied by the complete conversion of strands β6 and β7 to coils and the withdrawal of helix α13 from the small domain.

In the kinetic model proposed by Kamata et al., the presence of an intermediate (open) conformation between the closed and superopen

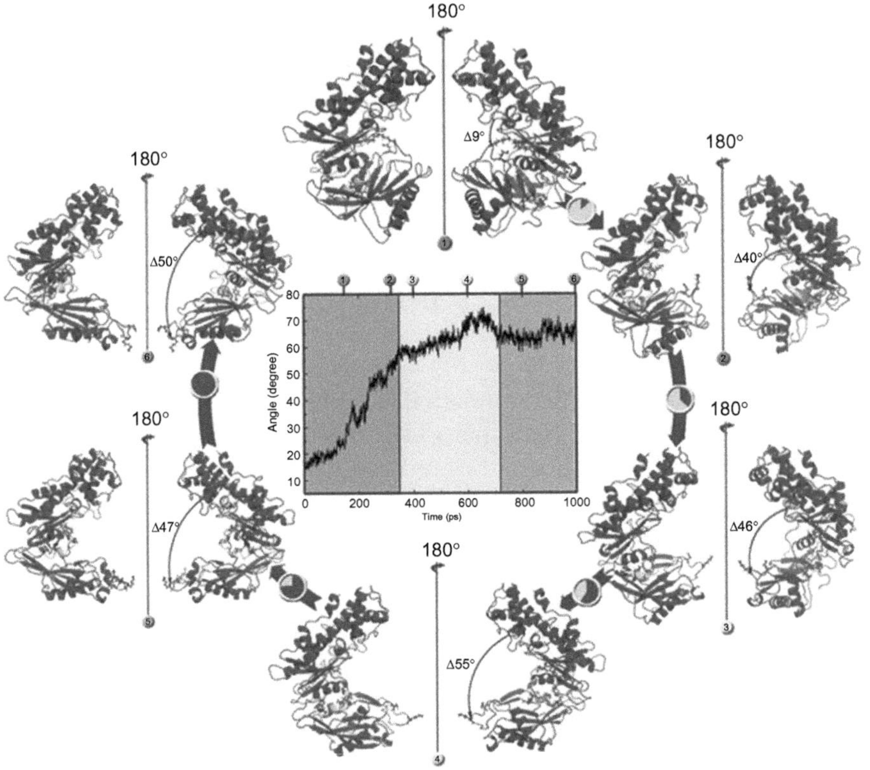

FIG. 6. The conformational transition pathway of GK from the closed to the superopen state. The center image is the time dependence of the cleft angle between the two domains. Around the center image, six snapshots are extracted from the trajectory at the times of 150 (1), 320 (2), 400 (3), 600 (4), and 800 (5) ps, and superopen state (6), respectively. The loop I between α4 and β7 segments is in red; the β6 and β7 strands are in cyan; helix α13 is in yellow. Figure taken from original paper (Zhang et al., 2006). (For interpretation of the references to color in this figure legend, the reader is referred to the online version of this chapter.)

conformations is critical to explain the positive cooperativity of GK with respect to glucose binding (Kamata et al., 2004). Kamata et al. proposed that this "open" conformation of GK may resemble the open conformation of another member of the hexokinase family, namely hexokinase I (Fig. 7). However, there was no structural information for such intermediate state in GK obtained by experiments. Zhang et al. therefore sought

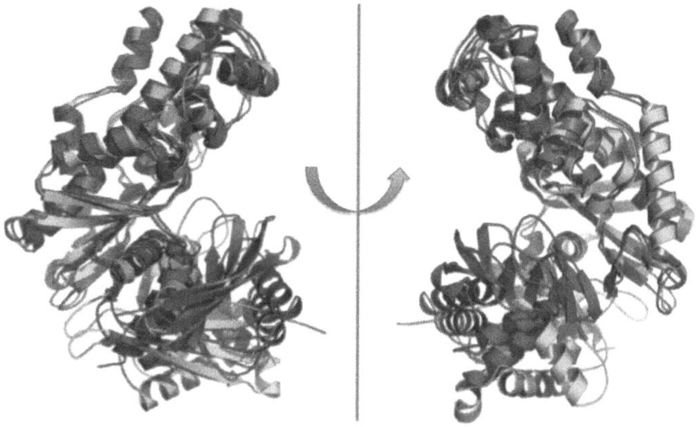

FIG. 7. Superposition of crystal structures of GK in the superopen (green, PDB code 1V4T) and closed (red, PDB code 1V4S) states with the open form of hexokinase I (blue, PDB code 1HKC). The open form of hexokinase I adopts an intermediate position compare to the superopen and closed states of GK. (For interpretation of the references to color in this figure legend, the reader is referred to the online version of this chapter.)

to identify possible intermediate conformational states along the conformational transition pathway by using free-energy calculations along the transition pathway with MM-PBSA-Nmode method from AMBER (Case et al., 2005). The results reveal three energy wells in the energy landscape occurring at time periods 130–160, 370–550, and 695–705 ps, and from these energy wells, three possible intermediate conformational states (named P1, P2, and P3) between the closed and superopen conformations of GK were identified.

Conformations from the energy well of the first intermediate state P1 were superimposed with the crystal structure of hexokinase I with the open conformation. Results showed that the P1 intermediate state of GK identified from this computational study highly resembles the open conformation of hexokinase I, indicating that the intermediate P1 corresponds to the "open" conformation as predicted by Kamata et al. in their kinetic model.

Zhang et al. then identified important interactions such as hydrogen bonds and hydrophobic contacts during each stage of the conformational transition pathway. Mutations were designed to either break or strengthen those interactions. The effects of these mutations were tested both by TMD simulations for the mutant systems and mutagenesis experiments, and simulation results in general correlate very well with the outcome of mutagenesis experiments.

Although structural information can be obtained for two forms of the same enzyme, the information regarding the transition between different forms of the enzyme and possible intermediate states involved during the transition can be hard to observe using only experimental techniques. This case study illustrates that the conformational transition pathway between the two forms can be reliably simulated by the use of computational simulation, TMD, and from which the key interactions were identified and validated with experiments such as site-directed mutagenesis. Results from this combined approach provided more detailed insight into the molecular basis of conformation transition in the allosteric process of human glucose.

III. Conclusions

In this review, we have outlined three examples to illustrate how an understanding of the principles and mechanisms of allostery can be advanced by combining the power of computational techniques with experimental findings. By combining computational methods such as docking, molecular dynamics, and pK_a calculations with experimental techniques including mutagenesis, circular dichroism, and gel electrophoresis, Whalen et al. not only successfully identified the binding site and binding mode of a natural allosteric inhibitor (DPA) of GR from *B. anthracis* but also proposed the likely mechanism of allosteric inhibition that may be general to all GR enzymes. In the study of PKA, Masterson et al. combined the power of MD simulations with NMR spectroscopy. They identified dynamically committed, uncommitted, and quenched states in PKA and outlined a possible mechanism for the activation and inhibition of PKA caused by ligand binding. Zhang et al. utilized the combination of CMD simulations and TMD simulations, with mutagenesis experiments to understand the allosteric response in GK. They investigated the conformational transition pathway in the allosteric process of human GK and identified three possible intermediate conformational states from closed to superopen states.

Proteins systems are dynamic in nature. Proteins exist as ensembles of many different conformational states with slightly different energies, which constantly interconvert through dynamic motions. Dynamics plays a role in many protein functions and regulation mechanisms, such as catalysis and allostery (Goodey and Benkovic, 2008). Most of the existing experimental techniques can only access information on the whole protein ensemble, and the observations from experiments are usually averaged over the protein ensemble. For example, the coordinates from an X-ray crystal structure of a protein molecule only represents its average conformation. Computation is a powerful tool as it allows understanding and interpretation of properties of single molecules, allowing intermediate states and transitional conformations to be visualized. The combination of computational techniques and experiments therefore becomes a robust approach that can explain observations from experiments at atomic resolution, predict outcomes from designed experiments, and provide guidance to future research directions.

References

Al-Hasani, H., Tschop, M. H., Cushman, S. W. (2003). Two birds with one stone: novel glucokinase activator stimulates glucose-induced pancreatic insulin secretion and augments hepatic glucose metabolism. *Mol. Interv.* **3**, 367–370.

Amaro, R., Baron, R., McCammon, J. (2008). An improved relaxed complex scheme for receptor flexibility in computer-aided drug design. *J. Comput. Aided Mol. Des.* **22**, 693–705.

Anandakrishnan, R., Onufriev, A. (2008). Analysis of basic clustering algorithms for numerical estimation of statistical averages in biomolecules. *J. Comput. Biol.* **15**, 165–184.

Ascenzi, P., Fasano, M. (2010). Allostery in a monomeric protein: the case of human serum albumin. *Biophys. Chem.* **148**, 16–22.

Ascenzi, P., Bocedi, A., Bolli, A., Fasano, M., Notari, S., Polticelli, F. (2005). Allosteric modulation of monomeric proteins. *Biochem. Mol. Biol. Educ.* **33**, 169–176.

Bahar, I., Chennubhotla, C., Tobi, D. (2007). Intrinsic dynamics of enzymes in the unbound state and relation to allosteric regulation. *Curr. Opin. Struct. Biol.* **17**, 633–640.

Bai, F., Branch, R. W., Nicolau, D. V., Jr., Pilizota, T., Steel, B. C., Maini, P. K., et al. (2010). Conformational spread as a mechanism for cooperativity in the bacterial flagellar switch. *Science* **327**, 685–689.

Boehr, D. D., Nussinov, R., Wright, P. E. (2009). The role of dynamic conformational ensembles in biomolecular recognition. *Nat. Chem. Biol.* **5**, 789–796.

Case, D. A., Cheatham, T. E., 3rd, Darden, T., Gohlke, H., Luo, R., Merz, K. M., Jr., et al. (2005). The Amber biomolecular simulation programs. *J. Comput. Chem.* **26**, 1668–1688.

Changeux, J. P., Edelstein, S. J. (1998). Allosteric receptors after 30 years. *Neuron* **21**, 959–980.
Changeux, J. P., Edelstein, S. J. (2005). Allosteric mechanisms of signal transduction. *Science* **308**, 1424–1428.
Cheng, X., Ivanov, I., Wang, H., Sine, S. M., McCammon, J. A. (2007). Nanosecond-timescale conformational dynamics of the human alpha7 nicotinic acetylcholine receptor. *Biophys. J.* **93**, 2622–2634.
Christesen, H. B., Jacobsen, B. B., Odili, S., Buettger, C., Cuesta-Munoz, A., Hansen, T., et al. (2002). The second activating glucokinase mutation (A456V): implications for glucose homeostasis and diabetes therapy. *Diabetes* **51**, 1240–1246.
Cook, P. F., Neville, M. E., Jr., Vrana, K. E., Hartl, F. T., Roskoski, R., Jr. (1982). Adenosine cyclic 3′,5′-monophosphate dependent protein kinase: kinetic mechanism for the bovine skeletal muscle catalytic subunit. *Biochemistry* **21**, 5794–5799.
Cui, Q., Karplus, M. (2008). Allostery and cooperativity revisited. *Protein Sci.* **17**, 1295–1307.
Dalton, G. D., Dewey, W. L. (2006). Protein kinase inhibitor peptide (PKI): a family of endogenous neuropeptides that modulate neuronal cAMP-dependent protein kinase function. *Neuropeptides* **40**, 23–34.
de Dios, A., Prieto, L., Martin, J. A., Rubio, A., Ezquerra, J., Tebbe, M., et al. (2002). 4-Substituted d-glutamic acid analogues: the first potent inhibitors of glutamate racemase (MurI) enzyme with antibacterial activity. *J. Med. Chem.* **45**, 4559–4570.
Dodd, D., Reese, J. G., Louer, C. R., Ballard, J. D., Spies, M. A., Blanke, S. R. (2007). Functional comparison of the two *Bacillus anthracis* glutamate racemases. *J. Bacteriol.* **189**, 5265–5275.
Doublet, P., van Heijenoort, J., Bohin, J. P., Mengin-Lecreulx, D. (1993). The *murI* gene of *Escherichia coli* is an essential gene that encodes a glutamate racemase activity. *J. Bacteriol.* **175**, 2970–2979.
Eaton, W. A., Henry, E. R., Hofrichter, J., Mozzarelli, A. (1999). Is cooperative oxygen binding by hemoglobin really understood? *Nat. Struct. Biol.* **6**, 351–358.
FRED (2010). 2.2.5 ed. OpenEye Scienctific Software, Sante Fe, NM, USA.
Froguel, P., Zouali, H., Vionnet, N., Velho, G., Vaxillaire, M., Sun, F., et al. (1993). Familial hyperglycemia due to mutations in glucokinase. Definition of a subtype of diabetes mellitus. *N. Engl. J. Med.* **328**, 697–702.
Gallo, K. A., Knowles, J. R. (1993). Purification, cloning, and cofactor independence of glutamate racemase from *Lactobacillus*. *Biochemistry* **32**, 3981–3990.
Gallo, K. A., Tanner, M. E., Knowles, J. R. (1993). Mechanism of the reaction catalyzed by glutamate racemase. *Biochemistry* **32**, 3991–3997.
Glaser, B., Kesavan, P., Heyman, M., Davis, E., Cuesta, A., Buchs, A., et al. (1998). Familial hyperinsulinism caused by an activating glucokinase mutation. *N. Engl. J. Med.* **338**, 226–230.
Glavas, S., Tanner, M. E. (1997). The inhibition of glutamate racemase by D-*N*-hydroxyglutamate. *Bioorg. Med. Chem. Lett.* **7**, 2265–2270.
Glavas, S., Tanner, M. E. (2001). Active site residues of glutamate racemase. *Biochemistry* **40**, 6199–6204.

Gloyn, A. L., Odili, S., Zelent, D., Buettger, C., Castleden, H. A., Steele, A. M., et al. (2005). Insights into the structure and regulation of glucokinase from a novel mutation (V62M), which causes maturity-onset diabetes of the young. *J. Biol. Chem.* **280**, 14105–14113.

GOLD (2010). 4.1 ed. Cambridge Crystallographic Data Centre, Cambridge, UK.

Goodey, N. M., Benkovic, S. J. (2008). Allosteric regulation and catalysis emerge via a common route. *Nat. Chem. Biol.* **4**, 474–482.

Gordon, J. C., Myers, J. B., Folta, T., Shoja, V., Heath, L. S., Onufriev, A. (2005). H++: a server for estimating pK_as and adding missing hydrogens to macromolecules. *Nucleic Acids Res.* **33**, W368–W371.

Grossbard, L., Schimke, R. T. (1966). Multiple hexokinases of rat tissues. Purification and comparison of soluble forms. *J. Biol. Chem.* **241**, 3546–3560.

Gunasekaran, K., Ma, B., Nussinov, R. (2004). Is allostery an intrinsic property of all dynamic proteins? *Proteins* **57**, 433–443.

Henzler-Wildman, K. A., Lei, M., Thai, V., Kerns, S. J., Karplus, M., Kern, D. (2007). A hierarchy of timescales in protein dynamics is linked to enzyme catalysis. *Nature* **450**, 913–916.

Johnson, D. A., Akamine, P., Radzio-Andzelm, E., Madhusudan, M., Taylor, S. S. (2001). Dynamics of cAMP-dependent protein kinase. *Chem. Rev.* **101**, 2243–2270.

Kamata, K., Mitsuya, M., Nishimura, T., Eiki, J., Nagata, Y. (2004). Structural basis for allosteric regulation of the monomeric allosteric enzyme human glucokinase. *Structure* **12**, 429–438.

Karplus, M., Kushick, J. N. (1981). Method for estimating the configurational entropy of macromolecules. *Macromolecules* **14**, 325–332.

Kern, D., Zuiderweg, E. R. (2003). The role of dynamics in allosteric regulation. *Curr. Opin. Struct. Biol.* **13**, 748–757.

Kornev, A. P., Taylor, S. S. (2010). Defining the conserved internal architecture of a protein kinase. *Biochim. Biophys. Acta* **1804**, 440–444.

Koshland, D. E., Nemethy, G., Filmer, D. (1966). Comparison of experimental binding data and theoretical models in proteins containing subunits. *Biochemistry* **5**, 365–385.

Kuriyan, J., Eisenberg, D. (2007). The origin of protein interactions and allostery in colocalization. *Nature* **450**, 983–990.

Lundqvist, T., Fisher, S. L., Kern, G., Folmer, R. H. A., Xue, Y., Newton, D. T., et al. (2007). Exploitation of structural and regulatory diversity in glutamate racemases. *Nature* **447**, 817–822.

Maragakis, P., Karplus, M. (2005). Large amplitude conformational change in proteins explored with a plastic network model: adenylate kinase. *J. Mol. Biol.* **352**, 807–822.

Masterson, L. R., Mascioni, A., Traaseth, N. J., Taylor, S. S., Veglia, G. (2008). Allosteric cooperativity in protein kinase A. *Proc. Natl. Acad. Sci. USA* **105**, 506–511.

Masterson, L. R., Cheng, C., Yu, T., Tonelli, M., Kornev, A., Taylor, S. S., et al. (2010). Dynamics connect substrate recognition to catalysis in protein kinase A. *Nat. Chem. Biol.* **6**, 821–828.

Masterson, L. R., Shi, L., Metcalfe, E., Gao, J., Taylor, S. S., Veglia, G. (2011). Dynamically committed, uncommitted, and quenched states encoded in protein kinase A revealed by NMR spectroscopy. *Proc. Natl. Acad. Sci. USA* **108**, 6969–6974.

Matschinsky, F. M., Glaser, B., Magnuson, M. A. (1998). Pancreatic beta-cell glucokinase: closing the gap between theoretical concepts and experimental realities. *Diabetes* **47**, 307–315.

May, M., Mehboob, S., Mulhearn, D. C., Wang, Z., Yu, H., Thatcher, G. R., et al. (2007). Structural and functional analysis of two glutamate racemase isozymes from *Bacillus anthracis* and implications for inhibitor design. *J. Mol. Biol.* **371**, 1219–1237.

Miyashita, O., Onuchic, J. N., Wolynes, P. G. (2003). Nonlinear elasticity, proteinquakes, and the energy landscapes of functional transitions in proteins. *Proc. Natl. Acad. Sci. USA* **100**, 12570–12575.

Monod, J., Wyman, J., Changeux, J. P. (1965). On the nature of allosteric transitions: a plausible model. *J. Mol. Biol.* **12**, 88–118.

O'Donoghue, P., Luthey-Schulten, Z. (2005). Evolutionary profiles derived from the QR factorization of multiple structural alignments gives an economy of information. *J. Mol. Biol.* **346**, 875–894.

Peracchi, A., Mozzarelli, A. (2011). Exploring and exploiting allostery: models, evolution, and drug targeting. *Biochim. Biophys. Acta* **1814**, 922–933.

Popovych, N., Sun, S., Ebright, R. H., Kalodimos, C. G. (2006). Dynamically driven protein allostery. *Nat. Struct. Mol. Biol.* **13**, 831–838.

Puig, E., Garcia-Viloca, M., Gonzalez-Lafont, A., Lluch, J. M., Field, M. J. (2007). New insights into the reaction mechanism catalyzed by the glutamate racemase enzyme: pH titration curves and classical molecular dynamics simulations. *J. Phys. Chem. B* **111**, 2385–2397.

Shabb, J. B. (2001). Physiological substrates of cAMP-dependent protein kinase. *Chem. Rev.* **101**, 2381–2412.

Shatalin, K. Y., Neyfakh, A. A. (2005). Efficient gene inactivation in *Bacillus anthracis*. *FEMS Microbiol. Lett.* **245**, 315–319.

Shaw, P. B. (1985). Theory of the Poisson Green's function for discontinuous dielectric media with an application to protein biophysics. *Phys. Rev. A* **32**, 2476–2487.

Smith, C. M., Radzio-Andzelm, E., Madhusudan, M., Akamine, P., Taylor, S. S. (1999). The catalytic subunit of cAMP-dependent protein kinase: prototype for an extended network of communication. *Prog. Biophys. Mol. Biol.* **71**, 313–341.

Smock, R. G., Gierasch, L. M. (2009). Sending signals dynamically. *Science* **324**, 198–203.

Spies, M. A., Reese, J. G., Dodd, D., Pankow, K. L., Blanke, S. R., Baudry, J. (2009). Determinants of catalytic power and ligand binding in glutamate racemase. *J. Am. Chem. Soc.* **131**, 5274–5284.

Steinbrecher, T., Labahn, A. (2010). Towards accurate free energy calculations in ligand protein-binding studies. *Curr. Med. Chem.* **17**, 767–785.

Storer, A. C., Cornish-Bowden, A. (1976). Kinetics of rat liver glucokinase. Co-operative interactions with glucose at physiologically significant concentrations. *Biochem. J.* **159**, 7–14.

Storer, A. C., Cornish-Bowden, A. (1977). Kinetic evidence for a "mnemonical" mechanism for rat liver glucokinase. *Biochem. J.* **165**, 61–69.

Swain, J. F., Gierasch, L. M. (2006). The changing landscape of protein allostery. *Curr. Opin. Struct. Biol.* **16**, 102–108.

Tanner, M. E., Gallo, K. A., Knowles, J. R. (1993). Isotope effects and the identification of catalytic residues in the reaction catalyzed by glutamate racemase. *Biochemistry* **32**, 3998–4006.

Taylor, S. S., Yang, J., Wu, J., Haste, N. M., Radzio-Andzelm, E., Anand, G. (2004). PKA: a portrait of protein kinase dynamics. *Biochim. Biophys. Acta* **1697**, 259–269.

Traaseth, N. J., Ha, K. N., Verardi, R., Shi, L., Buffy, J. J., Masterson, L. R., et al. (2008). Structural and dynamic basis of phospholamban and sarcolipin inhibition of Ca^{2+}-ATPase. *Biochemistry* **47**, 3–13.

Tsai, C. J., del Sol, A., Nussinov, R. (2008). Allostery: absence of a change in shape does not imply that allostery is not at play. *J. Mol. Biol.* **378**, 1–11.

Tsai, C. J., Del Sol, A., Nussinov, R. (2009). Protein allostery, signal transmission and dynamics: a classification scheme of allosteric mechanisms. *Mol. Biosyst.* **5**, 207–216.

Tzeng, S. R., Kalodimos, C. G. (2011). Protein dynamics and allostery: an NMR view. *Curr. Opin. Struct. Biol.* **21**, 62–67.

Van Schaftingen, E. (1994). Short-term regulation of glucokinase. *Diabetologia* **37**(Suppl 2), S43–S47.

Vionnet, N., Stoffel, M., Takeda, J., Yasuda, K., Bell, G. I., Zouali, H., et al. (1992). Nonsense mutation in the glucokinase gene causes early-onset non-insulin-dependent diabetes mellitus. *Nature* **356**, 721–722.

Walsh, D. A., Van Patten, S. M. (1994). Multiple pathway signal transduction by the cAMP-dependent protein kinase. *FASEB J.* **8**, 1227–1236.

Walsh, D. A., Perkins, J. P., Krebs, E. G. (1968). An adenosine 3′,5′-monophosphate-dependant protein kinase from rabbit skeletal muscle. *J. Biol. Chem.* **243**, 3763–3765.

Warwicker, J., Watson, H. C. (1982). Calculation of the electric potential in the active site cleft due to α-helix dipoles. *J. Mol. Biol.* **157**, 671–679.

Whalen, K. L., Pankow, K. L., Blanke, S. R., Spies, M. A. (2009). Exploiting enzyme plasticity in virtual screening: high efficiency inhibitors of glutamate racemase. *ACS Med. Chem. Lett.* **1**, 9–13.

Whalen, K. L., Tussey, K. B., Blanke, S. R., Spies, M. A. (2011). Nature of allosteric inhibition in glutamate racemase: discovery and characterization of a cryptic inhibitory pocket using atomistic MD simulations and pK_a calculations. *J. Phys. Chem. B* **115**, 3416–3424.

YASARA (2010). 9.11.9 ed. YASARA Biosciences GmbH, Vienna, Austria.

Zhang, J., Li, C., Chen, K., Zhu, W., Shen, X., Jiang, H. (2006). Conformational transition pathway in the allosteric process of human glucokinase. *Proc. Natl. Acad. Sci. USA* **103**, 13368–13373.

Zimmermann, B., Schweinsberg, S., Drewianka, S., Herberg, F. W. (2008). Effect of metal ions on high-affinity binding of pseudosubstrate inhibitors to PKA. *Biochem. J.* **413**, 93–101.

AUTHOR INDEX

Note: Page numbers followed by "*f*" indicate figures, and "*t*" indicate tables.

A

B

C

D

E

G

H

I

J

K

L

M

P

Q

S

T

W

X

Y

Z

SUBJECT INDEX

Note: Page numbers followed by "*f*" indicate figures, and "*t*" indicate tables.

A

E

F

I

K

L

M

Q

R

S

T

U